Handbook of Research on Emerging Technologies for Effective Project Management

George Leal Jamil
Informações em Rede, Brazil

Fernanda Ribeiro
University of Porto, Portugal

Armando Malheiro da Silva
University of Porto, Portugal

Sérgio Maravilhas Lopes
SENAI/CIMATEC, Brazil

A volume in the Advances in Logistics, Operations, and Management Science (ALOMS) Book Series

Published in the United States of America by
IGI Global
Business Science Reference (an imprint of IGI Global)
701 E. Chocolate Avenue
Hershey PA, USA 17033
Tel: 717-533-8845
Fax: 717-533-8661
E-mail: cust@igi-global.com
Web site: http://www.igi-global.com

Copyright © 2020 by IGI Global. All rights reserved. No part of this publication may be reproduced, stored or distributed in any form or by any means, electronic or mechanical, including photocopying, without written permission from the publisher. Product or company names used in this set are for identification purposes only. Inclusion of the names of the products or companies does not indicate a claim of ownership by IGI Global of the trademark or registered trademark.

Library of Congress Cataloging-in-Publication Data

Names: Jamil, George Leal, 1959- editor.
Title: Handbook of research on emerging technologies for effective project management / George Leal Jamil [and three others], editors.
Description: Hershey, PA : Business Science Reference, [2020]
Identifiers: LCCN 2019014233| ISBN 9781522599937 (hardcover) | ISBN 9781522599944 (ebook)
Subjects: LCSH: Project management. | Technological innovations--Management.
Classification: LCC HD69.P75 H35464 2020 | DDC 658.4/04--dc23 LC record available at https://lccn.loc.gov/2019014233

This book is published in the IGI Global book series Advances in Logistics, Operations, and Management Science (ALOMS) (ISSN: 2327-350X; eISSN: 2327-3518)

British Cataloguing in Publication Data
A Cataloguing in Publication record for this book is available from the British Library.

The views expressed in this book are those of the authors, but not necessarily of the publisher.

For electronic access to this publication, please contact: eresources@igi-global.com.

Advances in Logistics, Operations, and Management Science (ALOMS) Book Series

John Wang
Montclair State University, USA

ISSN:2327-350X
EISSN:2327-3518

Mission

Operations research and management science continue to influence business processes, administration, and management information systems, particularly in covering the application methods for decision-making processes. New case studies and applications on management science, operations management, social sciences, and other behavioral sciences have been incorporated into business and organizations real-world objectives.

The **Advances in Logistics, Operations, and Management Science** (ALOMS) Book Series provides a collection of reference publications on the current trends, applications, theories, and practices in the management science field. Providing relevant and current research, this series and its individual publications would be useful for academics, researchers, scholars, and practitioners interested in improving decision making models and business functions.

Coverage

- Decision analysis and decision support
- Computing and information technologies
- Risk Management
- Networks
- Production Management
- Finance
- Political Science
- Services management
- Information Management
- Marketing engineering

IGI Global is currently accepting manuscripts for publication within this series. To submit a proposal for a volume in this series, please contact our Acquisition Editors at Acquisitions@igi-global.com or visit: http://www.igi-global.com/publish/.

The Advances in Logistics, Operations, and Management Science (ALOMS) Book Series (ISSN 2327-350X) is published by IGI Global, 701 E. Chocolate Avenue, Hershey, PA 17033-1240, USA, www.igi-global.com. This series is composed of titles available for purchase individually; each title is edited to be contextually exclusive from any other title within the series. For pricing and ordering information please visit http://www.igi-global.com/book-series/advances-logistics-operations-management-science/37170. Postmaster: Send all address changes to above address. ©© 2020 IGI Global. All rights, including translation in other languages reserved by the publisher. No part of this series may be reproduced or used in any form or by any means – graphics, electronic, or mechanical, including photocopying, recording, taping, or information and retrieval systems – without written permission from the publisher, except for non commercial, educational use, including classroom teaching purposes. The views expressed in this series are those of the authors, but not necessarily of IGI Global.

Titles in this Series

For a list of additional titles in this series, please visit: www.igi-global.com/book-series

Sales and Distribution Management for Organizational Growth
Rahul Gupta Choudhury (International Management Institute, India)
Business Science Reference • ©2020 • 290pp • H/C (ISBN: 9781522599814) • US $195.00

Business Management and Communication Perspectives in Industry 4.0
Ayşegül Özbebek Tunç (Istanbul University, Turkey) and Pınar Aslan (Bursa Technical University, Turkey)
Business Science Reference • ©2020 • 337pp • H/C (ISBN: 9781522594161) • US $215.00

Handbook of Research on Urban and Humanitarian Logistics
Jesus Gonzalez-Feliu (Ecole Nationale Superieure des Mines de Saint-Étienne, France) Mario Chong (Universidad del Pacifico, Peru) Jorge Vargas Florez (Pontificia Universidad Católica del Perú, Peru) and Julio Padilla Solis (Universidad de Lima, Peru)
Information Science Reference • ©2019 • 497pp • H/C (ISBN: 9781522581604) • US $295.00

Industry 4.0 and Hyper-Customized Smart Manufacturing Supply Chains
S.G. Ponnambalam (University Malaysia Pahang, Malaysia) Nachiappan Subramanian (University of Sussex, UK) Manoj Kumar Tiwari (Indian Institute of Technology Kharagpur, India) and Wan Azhar Wan Yusoff (University Malaysia Pahang, Malaysia)
Business Science Reference • ©2019 • 347pp • H/C (ISBN: 9781522590781) • US $225.00

Additive Manufacturing Technologies From an Optimization Perspective
Kaushik Kumar (Birla Institute of Technology Mesra, India) Divya Zindani (National Institute of Technology Silchar, India) and J. Paulo Davim (University of Aveiro, Portugal)
Engineering Science Reference • ©2019 • 350pp • H/C (ISBN: 9781522591672) • US $205.00

Handbook of Research on Women in Management and the Global Labor Market
Elisabeth T. Pereira (University of Aveiro, Portugal) and Paola Paoloni (Sapienza University of Rome, Italy)
Business Science Reference • ©2019 • 423pp • H/C (ISBN: 9781522591719) • US $285.00

Imagination, Creativity, and Responsible Management in the Fourth Industrial Revolution
Ziska Fields (University of Johannesburg, South Africa) Julien Bucher (Chemnitz University of Technology, Germany) and Anja Weller (Saxon State Ministry of Education and Cultural Affairs, Germany)
Business Science Reference • ©2019 • 333pp • H/C (ISBN: 9781522591887) • US $225.00

701 East Chocolate Avenue, Hershey, PA 17033, USA
Tel: 717-533-8845 x100 • Fax: 717-533-8661
E-Mail: cust@igi-global.com • www.igi-global.com

Not only this work, but all my life, I dedicate to you, my beloved mother. To Marlene Jamil (In memoriam). "To God be all the honor and the glory".

Editorial Advisory Board

Maria Manuela Cruz-Cunha, *Instituto Politécnico do Cávado-Ave, Portugal*
Liliane Carvalho Jamil, *Independent Researcher, Brazil*
Antonio Briones Peñalver, *Universidad Politécnica de Cartagena, Spain*
Domingo Garcia Perez de Lema, *Universidad Politécnica de Cartagena, Spain*
Cláudio Roberto Magalhães Pessoa, *InfoAction, Brazil*
José de Poças Rascão, *Independent Researcher, Portugal*
Werner Silveira, *Instituto Filarmônica de Minas Gerais, Brazil*
Maria José Souza, *Gabinete de Estudos Estratégicos, Portugal*
João Eduardo Varajão, *Universidade do Minho, Portugal*
Augusto Alves Pinho Vieira, *Independent Researcher, Brazil*

List of Contributors

Almeida, Josenildo / *Universidade Salvador, Brazil* 255
Antunes, Adelaide Maria de Souza / *National Institute for Industrial Property & Chemical
  School, University of Rio de Janeiro, Brazil* 141
Anunciação, Pedro / *Research Center in Business Science, College of Business Administration,
  Polytechnic Institute of Setúbal, Portugal* 185
Barros, Manoel Joaquim / *Universidade Salvador, Brazil* 255, 271
Bastos, Luiz Eduardo Marques / *Braslift - Brasil Eletromecânica, Brazil* 83
Cairrão, Alvaro / *Polytechnic Institute of Viana do Castelo, Portugal* 158
Cardoso, António / *Universidade Fernando Pessoa, Portugal* 158
Castilla, Jet / *Universidad del Pacífico, Peru* 323
Chong, Mario / *Universidad del Pacífico, Peru* 323
Da Costa, Valéria Rocha / *Fundação Dom Cabral, Brazil* 124
Durrani, Rabia Imtiaz / *Institute of Management Sciences, Pakistan* 215
Durrani, Zainab / *Institute of Management Sciences, Pakistan* 215
Ferrari, Filippo / *Independent Researcher, Italy* 97
Figueiredo, Jorge / *Universidade Lusiada, Portugal* 158
Filho, José Márcio Diniz / *Fundação Dom Cabral, Brazil* 124
Fonseca, Platini / *Federal University of Bahia, Brazil* 200
Frederico D´Orey / *Universidade Portucalense, Portugal* 158
Geada, Nuno / *College of Business Administration, Polytechnic Institute of Setúbal, Portugal* 185
Hartz, Zulmira / *Institute of Hygiene and Tropical Medicine (IHMT). GHTM, NOVA University
  of Lisbon, Portugal* 141
Igarashi, Juliana Satie Oliveira / *Daudt Oliveira Pharmaceutical Laboratory, Brazil* 141
Jamil, George Leal / *Informações em Rede, Brazil* 1
Junior, Antonio Eduardo de Albuquerque / *Oswaldo Cruz Foundation, Gonçalo Moniz
  Institute, Brazil* 200
Macedo, Elizabeth Valverde / *Federal Fluminense University (UFF), Brazil* 141
Magalhães, Jorge Lima de / *Instituto de Tecnologia em Fármacos Farmanguinhos, Brazil* 141
Magalhães, Miguel / *Universidade Portucalense, Portugal* 158
Malheiro da Silva, Armando / *Faculty of Arts and Humanities, University of Porto, Portugal* 20
Maravilhas, Sérgio / *IES-ICS, Federal University of Bahia, Brazil* 200, 287
Maravilhas-Lopes, Sérgio / *IES-ICS, Federal University of Bahia, Brazil* 34, 169, 255, 271
Marques dos Santos, Ernani / *Federal University of Bahia, Brazil* 200
Medeiros, Cintia / *Universidade Salvador, Brazil* 271
Pennefather, Patrick / *University of British Columbia, Canada* 340

Pereira, Manuel / *Universidade Portucalense, Portugal* 158
Perez, Eduardo / *Universidad del Pacífico, Peru* 323
Pesqueira, António / *Takeda, Zurich, Switzerland* 237
R., Rajadurai / *Junior Research Fellow, Department of Civil Engineering, National Institute of*
    Technology, Warangal, India 63
Ralph, Rachel / *Centre for Digital Media, Canada* 340
Ribeiro, Fernanda / *Faculty of Arts and Humanities, University of Porto, Portugal* 20
Rodriguez, Vanessa Brasil Campos / *Universidade Salvador, Brazil* 271
Rosario, Hernan / *Universidad del Pacífico, Peru* 323
Santos de Miranda, Morjane Armstrong / *Federal University of Bahia, Brazil* 200
Silva, Daniella Barbosa / *Faculdade de Tecnologia e Ciências, Brazil* 200
Silva, Jersone Tasso Moreira / *Universidade Fumec, Brazil* 1
Sousa, Maria José / *ISCTE, Instituto Universitário de Lisboa, Portugal* 237
Tadeu, Hugo Ferreira Braga / *Fundação Dom Cabral, Brazil* 1
Thi, Kien Truong / *Academy of Journalism and Communication, Vietnam* 300
Vilventhan, Aneetha / *Assistant Professor, Department of Civil Engineering, National Institute*
    of Technology, Warangal, India 63

Table of Contents

Preface xxi

Section 1
A Word From the Editors

Chapter 1
Real Options Theory: An Alternative Methodology Applicable to Investment Analyses in R&D Projects 1
    Hugo Ferreira Braga Tadeu, Fundação Dom Cabral, Brazil
    Jersone Tasso Moreira Silva, Universidade Fumec, Brazil
    George Leal Jamil, Informações em Rede, Brazil

Chapter 2
Information Management in Project Management: Theoretical Guidelines for Practical Implementation 20
    Fernanda Ribeiro, Faculty of Arts and Humanities, University of Porto, Portugal
    Armando Malheiro da Silva, Faculty of Arts and Humanities, University of Porto, Portugal

Chapter 3
Strategic Information Management: Implementing and Managing a Digital Project 34
    Sérgio Maravilhas-Lopes, IES-ICS, Federal University of Bahia, Brazil

Section 2
Theoretical and Fundamental Contributions for a New Way to Understand Project Management

Chapter 4
4D BIM for the Management of Infrastructure Projects 63
    Aneetha Vilventhan, Assistant Professor, Department of Civil Engineering, National Institute of Technology, Warangal, India
    Rajadurai R., Junior Research Fellow, Department of Civil Engineering, National Institute of Technology, Warangal, India

Chapter 5
Industry 4.0 in Pumping Applications: Achievements and Trends 83
Luiz Eduardo Marques Bastos, Braslift - Brasil Eletromecânica, Brazil

Chapter 6
Are Family Businesses a Good Environment for Project Management? Non-Technological Factors Affecting Project and Knowledge Management Practices Within Family Firms 97
Filippo Ferrari, Independent Researcher, Italy

Chapter 7
Restructuring the Production Process: Use of Technology and Value Creation for a Law Firm 124
Valéria Rocha Da Costa, Fundação Dom Cabral, Brazil
José Márcio Diniz Filho, Fundação Dom Cabral, Brazil

Chapter 8
Project Management in Risk Analysis for Validation of Computer Systems in the Warehouse System 141
Jorge Lima de Magalhães, Instituto de Tecnologia em Fármacos Farmanguinhos, Brazil
Juliana Satie Oliveira Igarashi, Daudt Oliveira Pharmaceutical Laboratory, Brazil
Zulmira Hartz, Institute of Hygiene and Tropical Medicine (IHMT). GHTM, NOVA University of Lisbon, Portugal
Adelaide Maria de Souza Antunes, National Institute for Industrial Property & Chemical School, University of Rio de Janeiro, Brazil
Elizabeth Valverde Macedo, Federal Fluminense University (UFF), Brazil

Chapter 9
Canvas Marketing Plan: How to Structure a Marketing Plan With Interactive Value? 158
Miguel Magalhães, Universidade Portucalense, Portugal
Frederico D´Orey, Universidade Portucalense, Portugal
Manuel Pereira, Universidade Portucalense, Portugal
António Cardoso, Universidade Fernando Pessoa, Portugal
Alvaro Cairrão, Polytechnic Institute of Viana do Castelo, Portugal
Jorge Figueiredo, Universidade Lusiada, Portugal

Chapter 10
Sustainable Innovation Projects From Patent Information to Leverage Economic Development 169
Sérgio Maravilhas-Lopes, IES-ICS, Federal University of Bahia, Brazil

Chapter 11
Change Management Projects in Information Systems: The Impact of the Methodology Information Technology Infrastructure Library (ITIL) 185
Nuno Geada, College of Business Administration, Polytechnic Institute of Setúbal, Portugal
Pedro Anunciação, Research Center in Business Science, College of Business Administration, Polytechnic Institute of Setúbal, Portugal

Chapter 12
Information Management for the University-Enterprise Interaction: Considerations From the Research Groups Directory of the CNPQ in Brazil 200
    Morjane Armstrong Santos de Miranda, Federal University of Bahia, Brazil
    Sérgio Maravilhas, IES-ICS, Federal University of Bahia, Brazil
    Ernani Marques dos Santos, Federal University of Bahia, Brazil
    Antonio Eduardo de Albuquerque Junior, Oswaldo Cruz Foundation, Gonçalo Moniz Institute, Brazil
    Daniella Barbosa Silva, Faculdade de Tecnologia e Ciências, Brazil
    Platini Fonseca, Federal University of Bahia, Brazil

Section 3
Case Studies About Emerging Technologies for Project Management

Chapter 13
Information Technology Study Cases 215
    Rabia Imtiaz Durrani, Institute of Management Sciences, Pakistan
    Zainab Durrani, Institute of Management Sciences, Pakistan

Chapter 14
Pharmaceuticals and Life Sciences: Role of Competitive Intelligence in Innovation 237
    António Pesqueira, Takeda, Zurich, Switzerland
    Maria José Sousa, ISCTE, Instituto Universitário de Lisboa, Portugal

Chapter 15
A Business Intelligence Maturity Evaluation Model for Management Information Systems Departments 255
    Josenildo Almeida, Universidade Salvador, Brazil
    Manoel Joaquim Barros, Universidade Salvador, Brazil
    Sérgio Maravilhas-Lopes, IES-ICS, Federal University of Bahia, Brazil

Chapter 16
From the Interview "Eye in the Eye" to the "Eye in the WhatsApp": The Impact of Social Media on the Praxis of the Press Office in Organizational Communication Projects 271
    Cintia Medeiros, Universidade Salvador, Brazil
    Vanessa Brasil Campos Rodriguez, Universidade Salvador, Brazil
    Manoel Joaquim Barros, Universidade Salvador, Brazil
    Sérgio Maravilhas-Lopes, IES-ICS, Federal University of Bahia, Brazil

Chapter 17
Patent Information Project to Leverage Innovation: The Use of Social Media for Its Selective Dissemination 287
    Sérgio Maravilhas, IES-ICS, Federal University of Bahia, Brazil

Chapter 18
Internet Technology Application in Production of Internet Radio Programs in Vietnam 300
    Kien Truong Thi, Academy of Journalism and Communication, Vietnam

Chapter 19
Blockchain Technology Applied to the Cocoa Export Supply Chain: A Latin America Case 323
    Mario Chong, Universidad del Pacífico, Peru
    Eduardo Perez, Universidad del Pacífico, Peru
    Jet Castilla, Universidad del Pacífico, Peru
    Hernan Rosario, Universidad del Pacífico, Peru

Chapter 20
Agile Teams in Digital Media: A 13 Week Retrospective 340
    Rachel Ralph, Centre for Digital Media, Canada
    Patrick Pennefather, University of British Columbia, Canada

Compilation of References 360

About the Contributors 401

Index 406

Detailed Table of Contents

Preface xxi

Section 1
A Word From the Editors

Chapter 1
Real Options Theory: An Alternative Methodology Applicable to Investment Analyses in R&D
Projects 1
    Hugo Ferreira Braga Tadeu, Fundação Dom Cabral, Brazil
    Jersone Tasso Moreira Silva, Universidade Fumec, Brazil
    George Leal Jamil, Informações em Rede, Brazil

The objective of this chapter is to present the real options theory (ROT) as an alternative methodology applicable to investment analyses in research and development projects (R&D). The authors intend to simulate the evaluation of an R&D project as a real option, compare real options theory outcomes to a conventional R&D project evaluation technique, and review real options theory as a trend in innovation project evaluation. The outcomes were compared to those obtained via the traditional net present value (NPV) method and a brief practical discussion regarding project management decision making is held. Finally, although ROT is still in a developmental and consolidation stage, the authors suggests that it can be used as a promising tool in the decision-making process concerning R&D projects. ROT is presented as a research field that would integrate a set of emerging management technologies, becoming a theoretical base for new tools and methods to support project management (PM) decision making.

Chapter 2
Information Management in Project Management: Theoretical Guidelines for Practical
Implementation 20
    Fernanda Ribeiro, Faculty of Arts and Humanities, University of Porto, Portugal
    Armando Malheiro da Silva, Faculty of Arts and Humanities, University of Porto, Portugal

The authors start by presenting a disciplinary positioning regarding the information science (IS) that is being developed, investigated, and taught at the University of Porto, Portugal and rigorously spelling out the cross-cutting approach to information management (IM), which is also shared with other social sciences and technology-oriented disciplines. From this point, it is presented a short overview and a literature review about the nature of project management (PM) to emphasize the need for an effective and fully assumed approach between IM and PM. It is observed that for PM specialists, info-communicational flows are important, and they seek to manage them intuitively, but without feeling compelled to draw on the know-how of the IS and IM specialists. However, the opposite is true, that is, there are in course some interesting adaptations of PM procedures, applied to informational projects, as it is shown in the last part of this chapter.

Chapter 3
Strategic Information Management: Implementing and Managing a Digital Project............................ 34
 Sérgio Maravilhas-Lopes, IES-ICS, Federal University of Bahia, Brazil

The implementation of a digital information strategy project in a real estate company is analyzed in this chapter. This involves its historical and socio-economic framing and a brief description of the activity sector in which the company operates. Several well-managed digital projects have been developed to improve the competitive position of the company, but without focus, all the activities lose strength because they might not reach their proposed targets. Some tools to identify the information needs of business activity developed are described as well as the role of information as a promoter of competitive advantages. Social media tools were utilized and proved to be a great strategic decision. To conclude, a few reminders of the factors to consider in developing the information strategy to implement and that information management without a strategy could result in several diversified decisions without any positive consequence for the organization.

Section 2
Theoretical and Fundamental Contributions for a New Way to Understand Project Management

Chapter 4
4D BIM for the Management of Infrastructure Projects.. 63
 Aneetha Vilventhan, Assistant Professor, Department of Civil Engineering, National Institute of Technology, Warangal, India
 Rajadurai R., Junior Research Fellow, Department of Civil Engineering, National Institute of Technology, Warangal, India

Building information modelling has become a core topic in the architectural engineering and construction (AEC) industry, and its benefits have been realised over different phases of project construction. Adoption of nD BIM in the domain of infrastructure projects has provided challenges and is yet evolving. This chapter reviews the adoption of Building information modelling in the management of infrastructure projects. The use of nD planning (4D, 5D, 6D, 7D, and 8D planning) in infrastructure planning and management is discussed through Mapping n-D BIM with different applications in infrastructure projects. 4D BIM models are developed integrating the 3D models with the schedule and they support multiple construction management tasks. The implementation of 4D planning and management in infrastructure projects is demonstrated with the help of two case studies.

Chapter 5
Industry 4.0 in Pumping Applications: Achievements and Trends.. 83
 Luiz Eduardo Marques Bastos, Braslift - Brasil Eletromecânica, Brazil

This chapter addresses the so-called Industry 4.0 and some of its applications in industrial pumps, seeking to emphasize its characteristics and benefits. The introduction of 4.0 industry technologies in this traditional industry can cause profound changes in existing business models, providing greater customer satisfaction, either improving the effectiveness of equipment operation, contributing to better adjustment to working conditions, and also prolonging their life cycle. We are still in the early stages of these technologies and a lot is yet to evolve; however, there are already interesting examples developed by some pump manufacturers around the world, some of which will be mentioned in this chapter. It is subdivided into three main parts, namely brief historical panorama from the first industrial revolution to Industry 4.0, current applications in the industrial pump industry, and finally, future research directions and conclusion.

Chapter 6
Are Family Businesses a Good Environment for Project Management? Non-Technological Factors Affecting Project and Knowledge Management Practices Within Family Firms 97
    Filippo Ferrari, Independent Researcher, Italy

Relationships between project management, operations management, and organizational strategy are well-known, as well as organizational influences on project. Family businesses work on projects, but their unique nature makes family firms a challenging context for Project Management. This chapter aims to present and discuss the specific dynamics of family business that can impact project management practices. By definition, a project is a complex system, consisting of a set of dozens of interrelated sub-processes. As is known, the percentage of projects that satisfy both technical requirements, budget compliance and which meet the deadlines, is extremely low. This fact forces the researchers to equip themselves with more sophisticated tools to face the complexity of a project, in order to increase its chances of success.

Chapter 7
Restructuring the Production Process: Use of Technology and Value Creation for a Law Firm 124
    Valéria Rocha Da Costa, Fundação Dom Cabral, Brazil
    José Márcio Diniz Filho, Fundação Dom Cabral, Brazil

Process management, innovation, technology, and knowledge management are tools to achieve better results and create value for an organization, specifically for the law firm. This is why organizational processes, or business processes, have become fundamental structures for the management of modern organizations and to maintain the competitiveness of organizations. As a result, it was possible to identify that the use of process management techniques and tools is decisive for rational use of processes, increased productivity, and better customer service, presenting an ideal conceptual model.

Chapter 8
Project Management in Risk Analysis for Validation of Computer Systems in the Warehouse System 141
    Jorge Lima de Magalhães, Instituto de Tecnologia em Fármacos Farmanguinhos, Brazil
    Juliana Satie Oliveira Igarashi, Daudt Oliveira Pharmaceutical Laboratory, Brazil
    Zulmira Hartz, Institute of Hygiene and Tropical Medicine (IHMT). GHTM, NOVA
        University of Lisbon, Portugal
    Adelaide Maria de Souza Antunes, National Institute for Industrial Property & Chemical
        School, University of Rio de Janeiro, Brazil
    Elizabeth Valverde Macedo, Federal Fluminense University (UFF), Brazil

The informational and digital era of Big Data presents a non-trivial and unprecedented way in history for data and information management in organizations. Thus, to manage, protect, and ensure the validation of this data, it is imperative to develop new technologies for project management and their respective implementation in organizations. This chapter shows a case study in a pharmaceutical industry with the proposition of a methodology for validation of emerging technologies in the computerized systems. Data validation and security for project management in the organization is increasingly in demand. So, this implies that time and human resources in organizations are not infinite. It is necessary to prioritize the activities and resources dedicated to maintaining the validated state of the system. Authors propose a risk analysis to help companies with validation. They also present a proposed methodology for risk analysis from the point of view of the validation of computerized systems in a Warehouse Management module in a validated SAP ERP.

Chapter 9
Canvas Marketing Plan: How to Structure a Marketing Plan With Interactive Value? 158
    Miguel Magalhães, Universidade Portucalense, Portugal
    Frederico D´Orey, Universidade Portucalense, Portugal
    Manuel Pereira, Universidade Portucalense, Portugal
    António Cardoso, Universidade Fernando Pessoa, Portugal
    Alvaro Cairrão, Polytechnic Institute of Viana do Castelo, Portugal
    Jorge Figueiredo, Universidade Lusiada, Portugal

Canvas Marketing Plan is a design thinking tool to help companies build a marketing plan that allows them to make better decisions. It provides a simple structure that allows the user to visualize the dynamics and interaction of the different stages of the marketing plan and adapt the products and services to the needs of their clients, thus, "finding" the best position in relation to their competitors. This chapter presents a methodology of marketing that aligns the marketing plan with a highly connected and constantly changing market, but also online interaction vs. offline interaction, thus facilitating marketer planning. The canvas marketing model is validated by 146 marketeers from 17 distinct sectors of activity, allowing authors to gauge the timeliness and usefulness of this Framework.

Chapter 10
Sustainable Innovation Projects From Patent Information to Leverage Economic Development 169
    Sérgio Maravilhas-Lopes, IES-ICS, Federal University of Bahia, Brazil

Patent information can provide a growing competitiveness through the technology transfer it fosters, and be economically important because of the innovation it leverages. Organizations are not monetizing their potential related to the use of patent information that could encourage more innovation and the largest number of patent applications, resulting in more businesses and greater economic growth. This chapter sustains that a coherent and effective use of patent information, containing information from research and development (R&D) activities with industrial application, can contribute to solving problems, fostering innovation through the resulting products and processes. Sustainable solutions can be realized, using unexploited inventions, as by the formulation of new products based on R&D that can be adapted to new global needs, creating jobs and protecting the environment and its resources.

Chapter 11
Change Management Projects in Information Systems: The Impact of the Methodology
Information Technology Infrastructure Library (ITIL) 185
    Nuno Geada, College of Business Administration, Polytechnic Institute of Setúbal, Portugal
    Pedro Anunciação, Research Center in Business Science, College of Business
        Administration, Polytechnic Institute of Setúbal, Portugal

In the current economic and social context, management of change should not be framed by managers on a passive perspective and only when there are clear signs of changes in organizational or market factors. The management of change must be framed in a perspective of continuous improvement, which justifies the development of capacities of economic and social vision associated with the sector in which they are positioned. The information society and the impact that new IT technology has on the functioning of economic organizations and the modus operandi of the market and the economy have been evident. The IT competitiveness potential of companies has attracted managers to the increasing inclusion of more technology in organizations, challenging them in managing the implicit changes.

Chapter 12
Information Management for the University-Enterprise Interaction: Considerations From the
Research Groups Directory of the CNPQ in Brazil 200
    Morjane Armstrong Santos de Miranda, Federal University of Bahia, Brazil
    Sérgio Maravilhas, IES-ICS, Federal University of Bahia, Brazil
    Ernani Marques dos Santos, Federal University of Bahia, Brazil
    Antonio Eduardo de Albuquerque Junior, Oswaldo Cruz Foundation, Gonçalo Moniz
        Institute, Brazil
    Daniella Barbosa Silva, Faculdade de Tecnologia e Ciências, Brazil
    Platini Fonseca, Federal University of Bahia, Brazil

This chapter analyzes the importance of Information Management for the phenomenon of University-Enterprise (U-E) interaction, based on the Directory of Research Groups (DGP) in Brazil, of the National Council for Scientific and Technological Development (CNPq). The methodology used consisted in analyzing, by the empirical-analytic research and descriptive-analytical approach, the data available on this database. The data is about the activities of the research groups of the Federal University of Bahia (UFBA), interacting with companies from 2002 to 2010. Results show information management is important for this occurrence because it contributes to the recognition of interest and the conditions of interaction of the actors, enhancing the transfer of knowledge and technologies.

Section 3
Case Studies About Emerging Technologies for Project Management

Chapter 13
Information Technology Study Cases 215
    Rabia Imtiaz Durrani, Institute of Management Sciences, Pakistan
    Zainab Durrani, Institute of Management Sciences, Pakistan

This chapter focuses on the use of project management tools, techniques, and software in projects. The chapter includes a detailed discussion on the use of information communication technology within projects and provides a tour of the software and project management methodologies used to deploy projects.

To contextualize the discussion, a case study of four startup projects hosted by two different incubation centers is presented. The case study discussion is structured around four themes: financial aspects, family support, legal perspective, and project success and failure. Findings from the cases are then compared against the literature reviewed; finally, the chapter concludes by providing recommendations. However, the result divulges there is no proper mechanism that encompasses the use of project management software.

Chapter 14
Pharmaceuticals and Life Sciences: Role of Competitive Intelligence in Innovation 237
    António Pesqueira, Takeda, Zurich, Switzerland
    Maria José Sousa, ISCTE, Instituto Universitário de Lisboa, Portugal

This chapter analyzes innovation, knowledge, and competitive intelligence (CI). Besides these concepts, the focus will be on the role of innovation profiles. The innovation profiles include the creation, capture, organization, and integration of knowledge into the innovation process. The CI variable will be analyzed, demonstrating the potential for creating a context of competition for companies. A case study is presented about the pharmaceutical (pharma) industry with the application of the concepts of competitive intelligence, knowledge, and innovation to a real context.

Chapter 15
A Business Intelligence Maturity Evaluation Model for Management Information Systems
Departments 255
    Josenildo Almeida, Universidade Salvador, Brazil
    Manoel Joaquim Barros, Universidade Salvador, Brazil
    Sérgio Maravilhas-Lopes, IES-ICS, Federal University of Bahia, Brazil

One of the biggest challenges for managers today is decision making. The adoption of technological solutions to obtain information more easily and intuitively is increasing, so decisions are taken with greater coherence. In this aspect, Business Intelligence (BI) appears as a tool that extracts, transforms, and enables data to be crossed to assist managers in making decisions. This chapter proposes a BI maturity assessment model to assess the level of this phenomenon in the management of the Information Technology (IT) area to verify the main reasons why the IT managers of a company from the private sector in the city of Salvador, Bahia, Brazil, do not use BI tools in their management practices whereas their clients implemented such processes in the last two years. As a result, the level of maturity reached was 01, denominated empirical management or without maturity.

Chapter 16
From the Interview "Eye in the Eye" to the "Eye in the WhatsApp": The Impact of Social Media
on the Praxis of the Press Office in Organizational Communication Projects 271
    Cintia Medeiros, Universidade Salvador, Brazil
    Vanessa Brasil Campos Rodriguez, Universidade Salvador, Brazil
    Manoel Joaquim Barros, Universidade Salvador, Brazil
    Sérgio Maravilhas-Lopes, IES-ICS, Federal University of Bahia, Brazil

This study analyzes how technology and social media have transformed the praxis of press advisory activity and projects within the scope of the Communication of Organizations. To this end, it finds impacts on the functions of the activity facing this new scenario caused by the emergence of social media, updating the required profile of the new press advisor. The study adopts the conceptualization of the functions of the press officer in the organizational communication made by Duarte (2009), analyzing 17 of these functions in this new context. Authors studied the praxis of each function, before and after the advent of social media. They chose these functions because they stand for the dynamics of the Press Office, from the strategic to the operational level. The study found which social media are most used by press officers to publicize actions of their organization.

Chapter 17
Patent Information Project to Leverage Innovation: The Use of Social Media for Its Selective Dissemination 287
    Sérgio Maravilhas, IES-ICS, Federal University of Bahia, Brazil

This chapter describes the project for the development and implementation of a theoretical support model for the creation of an information system that will allow the dissemination of scientific and technical information contained in patent documents using the web sites of industrial property official entities. The support of information resources, available through libraries and information services in universities, will be crucial for the project and the success of university research centres (URC) in Science, Technology, and Medicine (STM). To achieve a coherent program of dissemination and make possible the access to patent information by the URC, social media network (SMN) tools (like RSS, Blogs, Wikis, Newsletters) will be used. The tools will also effectively achieve control to constantly improve the system implemented.

Chapter 18
Internet Technology Application in Production of Internet Radio Programs in Vietnam 300
    Kien Truong Thi, Academy of Journalism and Communication, Vietnam

This chapter introduces the concept and features of Internet radio, and expresses the status of Internet technology application in the production of Internet radio programs in Vietnam. Some solutions are proposed to help Vietnamese radio managers to improve the efficiency of Internet technology application in producing the programs.

Chapter 19
Blockchain Technology Applied to the Cocoa Export Supply Chain: A Latin America Case 323
    Mario Chong, Universidad del Pacífico, Peru
    Eduardo Perez, Universidad del Pacífico, Peru
    Jet Castilla, Universidad del Pacífico, Peru
    Hernan Rosario, Universidad del Pacífico, Peru

This chapter recommends applying block chain technology to the cocoa supply chain. Using this technology, it will be possible to show and guarantee the traceability of the final product. Traceability in the cocoa chain begins in the production stages (harvest and post-harvest) to obtain relevant data related to cocoa beans and their producers, promptly, until finding the raw material origin and inputs used during the process. The material provider's name must be considered, as well as the manufacturer's expiration date, the batch number, and the production area's reception date. This is why authors recommend using Block chain, which is a data structure that stores information chronologically in interlinked blocks. It works as a digital master book and the participants reach an agreement to register any information in the blocks. Throughout the chapter, authors show how to apply this technology.

Chapter 20
Agile Teams in Digital Media: A 13 Week Retrospective 340
Rachel Ralph, Centre for Digital Media, Canada
Patrick Pennefather, University of British Columbia, Canada

As we move towards the third decade of the 21st century, the development of emerging technologies continues to grow alongside innovative practices in digital media environments. This chapter presents a comparative case study of two teams (Team A and Team B) in a professional master's program during a 13-week, project-based course. Based on the role of documentation and the reflective practitioner, team blogs representing learner experiences of Agile practices were analyzed. This case study chapter focused on one blog post of a mid-term release retrospective. The results of this case study are framed around Derby and Larson's (2006) Agile retrospectives framework, including: set the stage, gather data, generating insights, deciding what to do, and closing the retrospective. The case study results suggest the need for public documentation of retrospectives and how this can be challenging with non-disclosure agreements. Also, the authors identify the importance of being a reflective practitioner. Future research on educational and professional practices needs to be explored.

Compilation of References 360

About the Contributors 401

Index 406

Preface

IGI Global offer to update our former publication, "Handbook of Research on Effective Project Management through the integration of knowledge and innovation", issued on 2015, presented a good opportunity to advance comprehension on matters that were productively addressed at that time and are continuously updated nowadays. As technology and its associated methods tend to contribute and even to revolutionize project management formulation principles, strategic alignment, design and execution, we understand that this is a topic where we have to keep discussions opened, evaluating new trends, tools and emerging scenarios to promote discussions around managerial changes in every production system or market in the world.

With this focus, we approached the former publication, invited authors to review their productions and called for some additional chapters to update our understanding about this almost unlimited matter, project management. The result is the work you have now, which we consider a real advance on considering technologies, from the basic to complex, as one of the main evolutionary sources for project management development.

Our book structure was composed mainly in three different sections. The first bring the views and main messages from editors, the same editorial team who published the former edition on 2015. The second section encompass studies which approach theoretically-focused chapters that attempt to discuss fundamentals, conceptual definitions and relationships and set new levels on understanding how these elements could be addressed by researchers and practitioners to develop further comprehension over changes in the project management field can be analyzed in the future. Finally, the third section opens a reflection space where practical results from field implementations can be detailed, exposed and evaluated by authors, defining an actual point of technological application in real cases, completing a context where theory reaches practices, enabling a deeper understanding about this complex and fast-changing context.

Chapter One, authored by editor George Leal Jamil and researchers Hugo Ferreira Braga Tadeu and Jersone Tasso Moreira Silva, addresses a method – Real Options Theory – to analyze investments and their potential returns in projects, specially Research and development ones, comparing this paradigm to classical ways, inviting the reader to consider ROT as a technique to be applied for new projects estimates and negotiations.

Editors Armando Malheiro da Silva and Fernanda Ribeiro contributed with an updated view on information management for project management, advancing their original view, published in the former 2015 book, adding considerations regarding new technologies influences and implications which are also impacting Information Science fundamental propositions in chapter Two. This chapter development shows how new changes can, potentially, bring phenomena to the context where classical and new theoretical prisms can serve for continuous assessments on these matters.

Ending the first section, chapter Three brings the views and reflections by editor Sérgio Maravilhas Lopes, regarding strategic aspects in the context of projects based on digital technologies. Sérgio Lopes analyze a practical case of a real state company which reacted to the market, implementing without the strategic framework, digital projects. This picture resulted in a challenge on how to integrate information and project management efforts itself. Lessons from this chapter brings light to the lack of strategic context and corporative results cohesion.

Section II, which is dedicated to promoting studies which can help to review theoretical basis for project management under application of emerging technologies, starts with the valuable contribution from authors Aneetha Vilventhan and Rajadurai R., from India in chapter Four. They analyze the growing application of nD BIM (Building information modelling) for architecture engineering and construction projects. In their chapter, authors also present an opportune study of 4D application of these techniques, allowing an open discussion for new project management aspects applying these technology-based elements.

As one remarkable major trend regarding emerging technologies, Industry 4.0 was addressed by Luiz Eduardo Marques Bastos in chapter Five. In this chapter, with a useful fundamental approach, the author observes how industrial pumps, a classical industrial market, presents signals of changes when exposed to the consolidation of technologies, as it is proposed by Industry 4.0 phenomena. Interestingly, the chapter also addressed business models issues for these applications, covering structural aspects which can be translated to favorable innovation enablers.

Pressuring business models' conceptions and, this way, impacting the strategic context for project management, entrepreneurial familiar structures were studied by author Fillipo Ferrari in chapter Six. As a cultural, traditional force in various economies, family-based organizations offer good political, social and economic contexts for project initiatives, although imposing external relationships, behaviors and tacit methods which result in a balance that must be precisely understood for future projects.

Chapter Seven enlightens an interesting study regarding law organizations services and, essentially, projects. As an interesting, globalized context, law firms have to address peculiarities approached by authors José Márcio Diniz Filho and Valéria Rocha da Costa, attempting to implement innovative project management techniques in order to become recognized by customers as competitive and addressing also ethical regulatory aspects.

Authors Jorge Magalhães, Juliana Satie, Adelaide Antunes, Elizabeth Macedo and Zulmira Hartz evaluated, in chapter Eight, the critical project management aspect of risk analysis in this new era of emerging technologies. Observing in the sensible Pharmaceutical sector, authors developed a study case where technologies such as Big Data and Analytics were considered to process and produce from huge sets of data, generated with frequency by any customer – company interaction and their implications with project management tasks.

As an innovative and expanding approach to adopt canvas-based methodologies for project management, authors Manuel Pereira, Miguel Magalhães, Frederico D'Orey, Jorge Figueiredo, Álvaro Cairrão and Antonio Cardoso produced chapter Nine, aiming to apply those methods for a service project: Marketing plans. As a usual planning action for companies, strategic marketing and its associated unfolding are now considered one of the most challenging and pressured tasks, becoming a source of efficient market response, as analyzed in this teamwork chapter.

In an interesting approach, Sérgio Maravilhas Lopes, one of this book editors, present the context of patent and licensing processes information to leverage project management in chapter Ten. As a complex and globalized arena, patent and licensing are strategic for national systems, organizations and,

Preface

obviously, for customers. In this chapter, the author addresses perspectives which help to understand possibilities presented by information management in this context, allowing a perspective of sustainable innovation process.

In chapter Eleven, authors Pedro Anunciação and Nuno Geada present a reflection encompassing powerful conceptual contexts: Information Systems, Project management, Change management and the methodology of Information Technology Infrastructure Library (ITIL). An observation is conducted on how ITIL principles are applied for change management projects that will allow, at last, one organization to better manage its data and information. This chapter introduced, this way, a new perspective for better comprehension, over the relationship between information and project management, as projects demand information, but also contribute for huge information production.

In an additional view on how information management relates to environmental conditions which will favor project management, authors Morjane Miranda, Sérgio Maravilhas Lopes, Ernani dos Santos, Antonio Júnior, Daniela Silva and Platini Fonseca, analyzed the specific phenomena of University – Enterprise interaction, one of the most potential and not precisely addressed contexts for information management, which implicate in better conditions for knowledge production and strategic innovation management. This is for chapter Twelve, which closes the Second Section of our book.

Section III, which aims mainly study and practical cases regarding new technologies applied to project management in modern context, starts with the research produced, in chapter Thirteen, by authors Rabia Durrani and Zainab Durrani, evaluating how inter-project information can benefit the overall project management. Authors opportunely analyzed this fact in a context of startups incubation program, reaching specifically software project management. Information interchange among projects is an opportune context to apply technology, which can, at last, effectively implement propositions of project programs and portfolio management, presented by professional institutions and academic researchers of project management for several years.

The role of competitive intelligence process, as a data, information and knowledge sources which can build up a structure for innovative project management was assessed by authors Maria José Sousa and António Pesqueira in chapter Fourteen. As an applied study case, authors reached the pharmaceutical sector, where those strategic processes have been practiced for decades, showing aspects of complexity related to competitive systems, external factors which will impact project propositions.

As a powerful set of tools and associated methods, Business Intelligence resources emerged in markets some decades ago. But, as a shared concept, sometimes, it is precariously identified and adopted, leading to errors when consolidating information and knowledge for decision-making. As a valuable source for innovation, BI elements and its strategic alternatives for decision-making were addressed by Josenildo Almeida, Manuel Barros and Sérgio Maravilhas Lopes in chapter Fifteen, updating this essential discussion and exploring an event – one emerging technology unbalanced adoption – that can happen again nowadays.

Approaching technology tools, methods and trends, it is undeniable to understand the power of social media contexts for actual projects and overall strategic initiatives. This phenomenon was studied by authors Cíntia Medeiros, Vanessa Rodriguez, Manuel Barros and Sérgio Maravilhas Lopes in chapter Sixteen, when they observed a traditional process in organizational communication and how social media implementations, such as instant messaging, is being increasingly adopted, producing impacts on a well-established conceptual methodology.

In chapter Seventeen, Sérgio Maravilhas Lopes brings back the observation of huge datasets regarding patent information for external environment strategic analysis, based on information management principles, analyzing how a theoretical model, which encompasses also social media resources, was applied to design a project for an information system dedicated to allow information dissemination. Interestingly, for the book purposes, this chapter approached technological resources, information management, strategic perspectives and two sides of project management – as a guidance for technological (information system) project and when project management is based on innovative technology.

Chapter Eighteen presents a peculiar project regarding technology over a classical resource – radio – shared through Internet, as a new powerful, global communication alternative. Reflections on how content, schedule and infrastructure are renewed under emergence of new technologies were approached by Kien Truong Tri, about Vietnamese radio stations. This chapter also brings some recommendations for managers, as to guide their communication projects in this new technological paradigm.

Projects to implement strategic to tactical planning remain a demanding challenge for organizations. Considering the important and competitive sector of Logistics and its associated project management demands, authors Eduardo Perez, Jet Castilla and Hernán Rosario evaluated, in chapter Nineteen, cocoa logistics chain in Peru, a newborn economic trend, which requires updated fashion of project management principles to be applied to expand its services as demanded, including the emerging Blockchain-supported mechanisms. As another source of competitiveness experience, as an external factor for strategic planning, this chapter also illustrates how technology and related methods must be applied to define overall management conditions, becoming a source for potential results.

Finally, in chapter Twenty, Agile project management, as a continuously evolving scenario, a fundamental background to offer a competitive alternative for strategic project responses, is studied by authors Rachel Ralph and Patrick Pennefather. Agile trends and efforts are present in the market for several years, demanding studies for their application definition for nowadays complex projects. In this study, from a comparative paradigm, authors contribute on analyzing how agile prospects are being applied for modern competitiveness, using retrospective tools. As a challenge, this tactic must be analyzed comparatively to initial affirmations by agility professionals and institutions.

As a final word, we witnessed how project management evolved in the recent years. Maybe, after being a leading topic for organizations, now this discipline is facing a demand of updating and review fundamentals, due to technological phenomena, posed not only by information and communication applications frequently put available by commercial players. Not only these components, but also their application methods and defined processes must also be rethought, assuring a continuous evolution for this important field.

Our book project, a though one itself, aimed to contribute with this important discussion, publishing studies of academic, researching, consulting and commercial practitioners around project management actual challenges and propositions. From one country to another, one perspective to real implementations, one possibility to challenges, these is our effort to keep the discussion alive and produce new horizons for project management with the strong support of emerging technologies.

Good project initiatives! Good reading!

George Leal Jamil
Informações em Rede, Brazil

Section 1
A Word From the Editors

Chapter 1
Real Options Theory:
An Alternative Methodology Applicable to Investment Analyses in R&D Projects

Hugo Ferreira Braga Tadeu
Fundação Dom Cabral, Brazil

Jersone Tasso Moreira Silva
Universidade Fumec, Brazil

George Leal Jamil
https://orcid.org/0000-0003-0989-6600
Informações em Rede, Brazil

ABSTRACT

The objective of this chapter is to present the real options theory (ROT) as an alternative methodology applicable to investment analyses in research and development projects (R&D). The authors intend to simulate the evaluation of an R&D project as a real option, compare real options theory outcomes to a conventional R&D project evaluation technique, and review real options theory as a trend in innovation project evaluation. The outcomes were compared to those obtained via the traditional net present value (NPV) method and a brief practical discussion regarding project management decision making is held. Finally, although ROT is still in a developmental and consolidation stage, the authors suggests that it can be used as a promising tool in the decision-making process concerning R&D projects. ROT is presented as a research field that would integrate a set of emerging management technologies, becoming a theoretical base for new tools and methods to support project management (PM) decision making.

INTRODUCTION

Conventional research and development (R&D) long term investment evaluation methods, such as the Net Present Value (NPV) and the Return on Investment (ROI) methods sustain basic shortcomings. These methods ignore outcome uncertainty, the choice of investment timing and the irreversibility of resource commitment, although being regarded and used by project managers for some time.

R&D project evaluation is often complex, due to substantial uncertainty found in different project phases, including the research, marketing and strategic planning and alignment phases. The stages can be sequentially evaluated through the differentiation between the many phases of an R&D program.

Each stage provides a gateway into the next stage. In addition, the time spent in each R&D phase affords the collection of other, relevant information to program evaluation. Essentially, each stage offers the manager an option to invest or not to invest in the next – and usually more expensive – phases of the R&D program, becoming an intricated and fundamental decision in the project management scope.

The value of technological "options" has been repeatedly used as a qualitative argument by researchers by the private and public sectors both, in supporting long-term strategic investigation. The technological option is the value of the opportunity broached by an R&D project in its initial stage, to invest later in a new technological area. Unfortunately, traditional empirical methods basing on cash flow estimates totally ignored the value of such opportunities; thus, risk research projects entailing substantial expected long-term returns were unduly penalized. Long term research has been traditionally supported at more modest levels than those preferred by their supporters.

Conversely, scenario-building and decision tree analyses are also often-used methods for the evaluation of R&D projects, since both allow for risk estimation in evaluation via the simplification of the complex return on risky projects problem. However, these traditional models show flaws as concerns the investment's potential profitability.

For this purpose, the Real Options Theory (ROT) has been given growing attention in financial theory and innovation management. Through the lenses of the ROT, the interaction among irreversibility, flexibility and uncertainty entails considerable difference in the evaluation of an investment alternative and should be considered in the pricing process. Modeling managerial uncertainties and flexibilities available during a project's life cycle is of the essence to establish what an investment risk is, reaching to an upper-level, qualified strategic context, enabling robustness to decisions made on investment and other issues when approaching project management processes and associated planning tasks.

The ROT is used to evaluate real assets, that is, those not traded in the marketplace. Capital investment projects, intellectual property evaluation, sources of natural resources and research and development project evaluation are examples of real assets that can be evaluated using this theory. A real option is the flexibility a manager has to make decisions involving real assets. As new information is obtained and cash flow uncertainties are cleared, managers can make decisions that will positively influence the project's final value (Dixit & Pindyck, 1994).

The decisions managers often have to contend with are: What is right time to invest, abandon or temporarily stop a project? When should a project's operating characteristics be changed, and also when should an asset be replaced by another? Thus, a capital investment project can be regarded as an ensemble of real options on a real asset: the project. This supporting context for better level of project management scope decisions can be regarded as an update of PM fundamentals and methods, as presented by institutional patronage of PMI and associated practitioners´ applications (PMBoK, 2018). As new aspects and events from the market happen, changes and competitive situations are imposed, a demanded scenario for practical application of scientific knowledge, such as project management must be permanently reviewed, and ROT can be integrated as a perspective theoretical field in this case.

Throughout this paper, the authors will seek to produce theoretical and empirical evidence using Geske's (1979) model, as adapted for real situations by Kemma (1993) and designated by Perlitz, Peske and Schrank (1999) as an alternative methodology to evaluate compounded options.

Authors such as Perlitz, Peske and Schrank (1999), Amram and Kulatilaka (2000), Boer (2000) and Silva et. al. (2012) have cast more attention to the application of real options compared to other traditional investment evaluation methods. Real option-based models provide a first step towards the integration between finance and strategy, to the extent that its results coincide with pre-judgment from a senior manager's experience. There is a need to understand how corporate strategy and execution interact with each other and how this affects business opportunities.

An emerging trend in research and development project evaluation contemplates the use of the options approach, affording a more flexible understanding of future growth opportunities throughout the process, presenting exactly aspects illustrated by PMBoK (2018) as projects continuity, maintenance and user-related services, reaching factors such as profitability and long-term returns. The options approach is the most appropriate in a world of uncertainties, since it sees the project as an initial investment that creates future commercial opportunities.

As postulated by Herath and Park (1999), it is usually difficult to justify R&D projects by means of the simple use of traditional methods; consequently, companies tend to under-invest in R&D. the real options approach bridges the shortcomings of the net present value (NPV) criterion when applied to projects under high uncertainty levels.

This paper is divided into five parts: an introduction, followed by the theoretical framework on the financial options theory, materials and methods, results and conclusion.

THEORETICAL BACKGROUND

This section will discuss and level the conceptual base intended for this study. First, project management will be approached, aiming to connect this base to organizational strategic decisions. Then, traditional financial decision-making theories for projects analysis are approached, followed by presentation of ROT – Real Options Theory – as a new fundamental context where PM analysis can be conducted. Supporting parameters and methods are also clarified in order to allow following comparative observations held in this chapter.

Project Management

According to PMI (2019), project management is the application of tools, techniques, methods and skills towards a complete comprehension and connection of all project-related activities. Interestingly, as remarked by Reich and Wee (2006) and Reich, Gemino and Sauer (2012), PM is a set of processes which demands and produce knowledge continuously, since the formulation, motivational phase of the project, up to its conclusion, when the project goal – a product, an assembly or a service – is put to operate to serve the desired purposes.

Observing from the knowledge generation, aiming to comprehend situations of decision-making in projects, as a main target for the further discussion about evaluation methodologies, Paramkusham and Gordon (2013) studied how knowledge transfer faces difficulties in (information) technology projects. As those authors analyzed, some factors are identifiable, as project dynamics, project environment to flexibility and fluid building blocks of these projects. It is opportune to state that some factors, classified by authors, are internal, others external to the organizational continent, also some are related to the project itself. These important signals show how decision-making aspects, which will conform the motivational criteria to choose for an innovative analysis method, are complex and project dependent.

As a specific area of technology-based projects, information technology projects are critical in terms of evaluation and investment analysis. For Jamil and Carvalho (2019), these projects represent the consolidation of several organizational functional, applicative and business needs and requirements. It is possible to understand this wideness of IT projects as a potential scenario for application of methods such as ROT, a factor which will be considered in the following discussion. Those projects also show an integrative perspective, understood as:

- IT projects, usually, interact with other organizational projects. This way, it is possible to estimate that return analysis from projects can also be applied to project programs and portfolios.
- IT projects also help generate other projects, or sub-projects, associated to the main definition, as one "umbrella" context. Analysis can be done in this scenario, applying to the main project itself and surrogates.
- Finally, IT projects can be derived, updated or repeated, as they are built over formalities and definitions which predict this opportune "template-like" characteristic, serving as a model to be re-implemented in several organizational processes and final, customer-oriented roles.

These definitions are worked by Turban, Rainer and Potter (2007) and Stair and Reynolds (2009), when observing how information systems and information technologies-related projects can be formulated, planned, implemented and monitored.

With observations like these in mind, it is possible to define a wider context for this chapter present analysis, regarding ROT-based methodology to evaluate potential returns from technological projects, evaluating also perspectives for these projects to serve as models, templates, origin of other projects or, finally, as an integrative focus for programs and portfolios.

Project management, as it is seen in this brief review, is a topic in a constant evolution and development, especially when considering R&D projects, a field that includes new economy efforts, innovations based both on technology and business models and updated economical forecasts. In this fast-changing arena, topics such as project return evaluation must be carefully focused, as risk management tools become an essential item for the modern project management itself.

One last word about project management is the potential relationship setting with organizational strategy and its associated planning capabilities and processes, such as strategic planning. As recommended by PMI (2019), projects can be seen as connected, coherent and integrated efforts towards strategic goals execution, becoming tools and methods for actions planning on promoting the strategic plan (Kaplan & Norton, 2007; Müller et al., 2013). Approaching project management by this view, it is possible to understand its attributes on supporting critical decision-making processes and options towards strategic problems solutions through well-dimensioned, formulated, coordinated and executed projects.

Financial Options Theory

According to Hull (1997), "options are asset purchase and sale agreements, whose prices depend on the value of the asset object of the agreement", that is to say, a purchase option (call) is a claim that the contract bearer has to buy the asset object of the agreement at a pre-established exercise price, on an established future date, on which this type of option presents a payment function as following:

$$C(T) = Max\ (S(T) - K, zero) \tag{1}$$

Where:
C(T) is the buy option value on date T;
T is the expiration date;
S(T) is the asset price at date T;
K is the actual price;
"Max" means "the greater of".

A sell option (put) gives its bearer the right to sell the asset object of the agreement at an exercise price on a future date. The function (put) given by the equation:

$$P(T) = Max\ (K - S(T), zero) \tag{2}$$

Where:
P(T) is the sell value option at date T;
T is the expiration date;
S(T) is the asset price at date T;
K is the actual price;
"Max" mean "the greater of".

The milestone in the ROT theoretical development is the effort produced by Black and Sholes (1973), who developed an analytical formulation to evaluate European call options. European options are those whose exercise can only happen as the security matures.

American options are those that can be exercised at any time until maturity. This characteristic confers upon American options a value at least equal to the value of similar European options. Option evaluation requires the establishment of an optimal investment policy, that is, the contract asset value from which the option will be exercised should be established, such as to maximize the present value of its remuneration. For the real options theory, the establishment of this policy is a core factor, since the best selling time in an investment project could be determined.

The main contribution from the Black and Sholes (1973) paper perhaps was not the equation in and by itself, but, rather, the proposed methodology. Building a dynamic asset portfolio independent from its owner's risk preferences allowed it to be used. The major difference resides in the analyses of discounted interest rates applicable to the stock's future remuneration.

Classically, options evaluation is made following two approaches: the Black and Scholes (1973) continuous time approach and the discrete-time model proposed by Cox, Ross and Rubinstein (1979) and the multi-stage binomial tree developed by Rendleman and Bartter (1979). The work of Cox, Ross and Rubinstein shows how to use a binomial tree to treat the early stock option exercise following a lognormal process. Its algorithm is a special lattice model used to solve control optimization problems. The trinomial tree was introduced by Clewlow and Strickland (1998).

An important parameter for investments and option evaluation models is volatility. An asset's volatility is the measure that seeks to identify the uncertainty of its future price movements, that is, the greater the volatility, the greater the risk to the investor; however, the chance to do good business will also be greater.

According to Hull (1997), this is due to two main causes:

- The random arrival at the marketplace of new information concerning the stock (or company) behavior.
- Trading involving the security causes the volatility, but empirical tests did not prove or disprove any of these theories.

Also, according to Hull (1997), volatility can be divided into:

- Historical volatility;
- Future volatility;
- Implicit volatility, this being the comparison between current and future prices.

Volatility estimations vary with the time horizon in consideration and, for risk management purposes, short term estimations are highly relevant.

It was, however, the pioneer work developed by Black and Sholes (1973) and Merton (1973) to evaluate financial options that provided the subsidies to the idea of incorporating pricing methods to the problem of evaluating real investments under uncertainty, which is presented in the following session.

The Real Options Theory (ROT)

Along the past decade, however, the effectiveness of the methodologies presented in the previous session was intensely challenged. Dixit and Pindyck (1994), for example, show that the applications of these methodologies may induce mistaken investment decisions, the reason is that they ignore two important characteristics of these decisions: irreversibility, that is, the fact that the investment is a sunk cost such that the investor will be unable to salvage his funds in full in case of regret; and the possibility of postponing the investment decision. These characteristics, together with future uncertainty, prompt the investment opportunity to be analogous to a financial option.

In the presence of uncertainty, a company having an irreversible investment opportunity carries one option: the company has the right – but not the obligation – to purchase an asset (the project) in the future, at an exercise price (the investment). When the company invests, it exercises or kills this investment option. The problem is that the investment option carries a value to be entered as an opportunity cost at the time the company invests. This value can be quite high and investment rules ignoring it – typically, NPV and IRR rules – may entail significant errors.

The ROT appears as a methodology to evaluate real assets as, for example, investment projects, which take into account the operating and managerial flexibilities along the project's working life. Differently than traditional techniques such as the NPV, its dynamic characteristic is conducive to more realistic outcomes.

Application of the ROT as an investment evaluation methodology is a novel practice. Its main concept is founded on the financial options theory, establishing an analogy between options and managerial decisions along the working life of an investment project.

The expression "Real Options" was initially used by Myers (1977), highlighting that a company's new expansion investments can be interpreted as being analogous to call options.

Managerial flexibility is a possibility of, but not an obligation to, changing a project at different stages of its operating working life. Myers (1987) proposed the Options Theory as the best approach to evaluate projects containing significant operating and strategic options, suggesting that the Theory can integrate strategy and finance.

The importance of investing in R&D was seen by Porter (1992) as being the most important factor of competitive advantage, due to the changes in the nature of competition and the increased pressure exerted by globalization upon organizations. The author observes that the investment in intangible assets (human resources, technology and corporate image) and in capabilities required for competitiveness, such as R&D, capacity-building of human resources, information technology, organizational development and customer and supplier relations to be a competitive differential.

What is seen from all this is that, without reinvestments, both the company's tangible assets (physical and financial assets) and intangible assets will depreciate. Much more than this: notably, investments are fundamental to maintain competitive advantage both in cost leadership and in differentiation.

A company will make decisions vis-à-vis a project throughout its entire life. Upon evaluating a project today, it is assumed that future decisions will be optimal; however, what these decisions will be is unknown, since most of the information remains to be found. Therefore, the possibility of postponing the investment represents an important option that should not be discarded upon evaluating an investment project. It was seen that the opportunity to invest in real assets presents characteristics of investments on options on financial assets, for which reason these investment opportunities are called "real options".

Net Present Value

Companies make capital investments to create and explore profit opportunities. Generally, a decision-making process to invest in a new project involves building a discounted cash flow through a very simple procedure. Initially, the present value of the expected sequence of cash flows that the project will yield is computed. Then the present value of the flow of expenses required to pursue the project is computed. Finally, the difference between the two values is computed, that is, NPV.

The NPV concept is considered the most consistent method with the company's objective of maximizing stockholders' wealth. Other alternative methods (such as the payback period and the internal rate of return), despite being considerably used in the corporate universe, have been deemed inferior to the NPV in existing literature.

According to Ross *et al.* (1998), the NPV looms as one of the most important concepts of financial management. At the time of an investment, only the amount to be invested in the project is known (the outlay) while the project's returns (inflows) are only estimated. Thus, we should know the relationship between $1 today and a possible uncertain $1 in the future prior to deciding about a specific project.

The NPV method is the present value of all cash flows discounted at the cost of capital, minus the cost of the investment also discounted at the cost of capital. Its greatest competitor is the Internal Rate of Return (IRR) method, which represents the discount rate that brings all cash flows to zero.

The main difference between the two methods is that the NPV method assumes reinvestments at the rate equal to the cost of capital, which the IRR method assumes reinvestment at its own rate. The first method has the advantage of showing how much value is added to the company for each investment contemplated.

In principle, each project has its own cost of capital. In practice, companies cluster similar projects in risk classes and apply the same cost of capital to projects of a same class. The existence of a positive NPV is defined as the basic criterion to accept or reject a given project and the order of NPVs is the choice criterion among different investment alternatives.

The critical variable in determining NPV and IRR are the cash flow and the cost of capital. When the investment evaluation is done at company level, all investment projects considered should be included, analysis, in this manger, uses the company's financial statements, such as the income statement and the balance sheet. Both can be used to explain the elements contained in the cash flow, as described annually to be capitalized. The sum of each annual cash flow minus the required investments, duly discounted, will compute the value of the company.

Trigeorgis (1996) argues in favor of an expanded or strategic criterion, reflecting two value components: the traditional NPV (static or passive) of discounted direct cash flows, and the value of the flexibility and strategic interactions option. Real options, in this manner, complement the net present value theory, adding an important dimension of flexibility to it.

The critical point of the NPV approach resides in the decision of what discount rate to use. Discount rates are influenced by the project's risk level and duration, and tend to increase on a par with interest and inflation rates.

THEORETICAL RELATIONSHIP WORK: R&D MODEL AS A REAL OPTION

Irreversibility, uncertainty and the possibility of postponement are three important characteristics of investment decision-making, which impact project management (Paramkusham & Gordon, 2013). In practice, investors' decisions take each one and its interactions into account, a significant PM communication factor. Since the options approach is an attempt at theoretically modeling the investors' decisions, understanding them requires, above all, a more careful analysis of these characteristics (Dixit & Pindyck,1994).

A basic question regarded to project management cost or financial basics, with immediate impact on its feasibility analysis is: "Why an investment decision can potentialize an irreversible cost?" First, because specific investments made by a company or industry are mostly sunk costs. Advertising investments, for example, are specific to each company and therefore unsalvageable. For example, we can consider an automaker which is specific of this industry. An ill-succeeded investment in this case would only contain changes of salvaging by the same of the plant to another company in the same industry, probably at a substantial discount.

Second, even non-company-or-industry-specific investments are partially irreversible. Computers, trucks and office equipment, for example, can be resold to companies from different industries, but at prices lower than their replacement cost.

Third, irreversibility may be produced by regulation or by institutional arrangements. Part of the investments in public service concessions reverts to the government at the end of the concession or in the case of breach of contract. Controls imposed on capitals may limit the sale of assets by foreign investors, while investments in human capital are also irreversible, due to high contracting, training and severance costs.

The uncertainty about the future is the second important characteristic of the investment decision-making on projects. Project and investment option values and the investment decision itself are affected by the uncertainty associated to relevant variables, such as product price, input costs, interest rates, exchange rates, credit supply and regulation. The importance of uncertainty to investment decision-making will be a recurrent theme throughout this paper.

The third characteristic is the possibility of postponing the investment. Evidently, companies not always have this possibility. Strategic considerations may prompt the company to antedate investments to inhibit effective competitor growth or the entry of potential competitors in the industry. It is important to remark that strategic risk management always operate at the organization´s most critical level, implicating, when resulting in scenarios of project postponement, for example, to likely harmful mitigation techniques, undesirable from several market aspects, such as financial results, corporate image and customer relationships.

In some cases, however, project postponement is doable. The company should always compare the cost of postponing – the risk of new companies entering the industry or cash flow losses – with the benefits of waiting for new information to subsidize the investment decision. These may be significant enough to justify postponements, implicating in a situation of decision-making where all variables must be considered, a case where efficient methodologies are welcomed by project managers.

ROT evaluates a research and development project as an option, to be or not to be exercised in the future, depending on the presence of favorable or unfavorable conditions. The R&D project can be seen as an option, for which a certain premium was paid (the investment in research) during an initial phase, should the project seem promising at the end of such phase (maturity), it will be exercised and the value of the investment in production and marketing will be paid.

According to Herath and Park (1999), an investment in R&D can be seen as a cost (of a real option in which the commercial project will proceed only if the R&D phase is successful). More specifically, the new project's marketing investment cost can be seen as the exercise price and the present value of the future cash flow ensuing from the sales, to be seen as the value of the underlying asset. The date of introducing the new product or resulting service in the marketplace can be seen as the exercise date. While it is assumed that the marketing decision will probably occur on a future date T_1, the decision maker could consider the option of postponing the marketing decision, according to Figure 1. The option to wait in this case has a value whilst the marketing option is not exercised.

A call option creates future opportunities (such as the development of new product lines or efficiency improvements) without compromising the company under its total investment burden. This is the issue the breeds the differences between traditional approaches and the ROT, since an eventual loss will be limited to the amount invested, the gain potential is limited and the greater the commercial uncertainty, the greater will the project value be.

R&D options have an important advantage over asset options, that is, the purchase of an asset option has no direct effect upon the exercise price or the asset price, while "the greater purpose of the R&D option is to influence future investments favorably by cost-curbing or by improving returns". Therefore, an R&D option is, arguably, more valuable than an asset option, since it is possible to act upon its future value.

Figure 1. Typical decision tree for investment processes

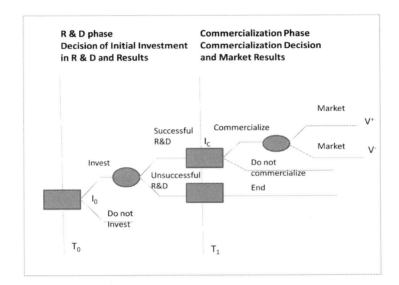

Consider the effect of the R&D option upon an asset value. Mitchell and Hamilton (1988) suggest an R&D option structure similar to that used for the asset option (Figure 2). It is assumed that the company expects to make future investments at a cost C (analogous to the exercise price), which will yield a return R (analogous to the asset value), when acquired.

The investment will be feasible for R > C, and the value of the investment is shown as "B". However, successful R&D programs may entail a cost reduction of the potential investment from C to C1. R&D programs can also improve return, from R to R1. The expected result is that R&D programs carry the potential to yield benefit "A", which increases the total value of the investment.

Figure 2. R & D impact option on future investments

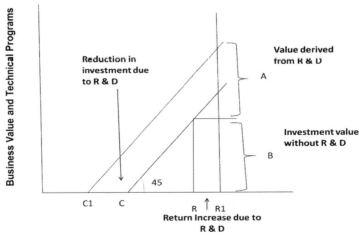

Such analysis corroborates what ROT literature has indicated, that is, investment analyses carried via the traditional manner has ignored the flexibilities present in the projects. This fact occurs because traditional analyses are done as if all decisions had to be made at the beginning of the project, which, as previously indicated, is naturally a false hypothesis.

R&D projects associated to new product launches have cast substantial challenges before the companies as concerns financial analysis. Literature criticizing traditional metrics abounds. Examples are Mitchell and Hamilton (1988), Faulkner (1996), Lint and Pennigs (1998), among others. According to these authors, metrics such as the IRR, ROI and NPV tend to underestimate the research's current value, since they do not consider the value of flexibility associated to this type of project.

Differently from traditional metrics that only contemplate the decision of investing or not investing in the project, placing investors in an inactive scenario, the real options method allows the consideration of other options along the project such as abandoning, proceeding or, should the case be, postponing the project depending on the scenario analyzed.

Another important issue addressed by ROT is uncertainty. According to Dixit and Pyndik (1994), the managerial flexibility incorporated to the uncertainty context that new product launch projects contain allows new information to be factored in throughout the project, and thence the course of action can be changed aiming to maximize project outcomes.

Market uncertainty is related to the future value of innovation, which in turn is strongly correlated to market demand. Subjectivity at the time of kicking off projects of this type brings about an intense difficulty in choosing a method. According to Luehraman (1998), although the authors have effectively demonstrated the advantages of the ROT vis-à-vis traditional metrics, they have not been equally successful in providing practical methodologies to approach this tool.

The Geske model presents an option on option pricing theory, or compound option model, which can be generalized for corporate liability values. The equations of a compound call option contemplate a purchase option on stocks, which in itself is an option of the company's assets.

Such perspective incorporates the effects of leverage upon option pricing and, consequently, the variance of the stock rate of return, which is not constant as assumed by Black-Scholes, but, rather, is a function of the stock price level. The Black-Scholes model is a special case in the compound option equation. This new model for calls and puts corrects some important biases in the Black-Scholes model.

Studies indicated that the value obtained through the Geske (1979) model, except for approximations, is 92% greater than that obtained by the traditional model. Traditional models compared by the study were: Net Present Value – Investment in Production and Marketing; Net Present Value – Sale of Rights; Decision Tree Analysis and Net Present Value by the Kallberg and Laurin (1997) method.

Such analysis corroborates what has been mentioned in ROT literature, that is, investment analysis as traditionally done has ignored the flexibilities contained in the projects. This fact occurs because traditional analyses are done as if all decisions had to be made at the beginning of the project, which, as previously indicated, is naturally a false hypothesis.

Figure 3. Generic model of the R & D process

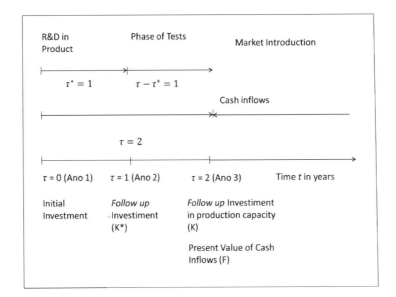

MATERIALS AND METHODS

The Geske Model

The solution to evaluate compound options was initially suggested by Geske (1979), who presents a situation containing the data from an R&D project with two growth opportunities with the initial investment made in year 1, followed by follow-up investments (K^* investment in the test phase) in year 2 and finally follow up investments in production capacity in year 3.

Perlitz, Peske e Schrank (1999) suggest that Geske's (1979) model be used in the computation of options containing the previously described project characteristics. According to the authors, assuming that the project value will follow a geometric Brownian motion, this compound option can be analytically evaluated in terms of the bivariate normal distribution. A compound option can be analytically evaluated by Geske's (1979) approach, basing on the Black-Scholes model, and adjusted for Kemna's (1993) real options evaluation, as follows:

$$C = Fe^{-r\ddot{A}}M\left(k,h;\hat{A}\right) - ke^{-r\ddot{A}}M\left(k-\tilde{A}\sqrt{\ddot{A}^*},k-\tilde{A}\sqrt{\ddot{A}};\hat{A}\right) - k^*e^{-r\ddot{A}}N(k-\tilde{A}\sqrt{\ddot{A}^*}) . \qquad (3)$$

Where, for the calculation of ρ . h and k have the following equations:

$$\rho = \left(\frac{\tau^*}{\tau}\right)^{\frac{1}{2}} \qquad (4)$$

Real Options Theory

$$h = \frac{ln\left(\frac{F}{K}\right) + \frac{1}{2}\sigma^2\tau}{\sigma\sqrt{\tau}} \qquad (5)$$

$$k = \frac{ln\left(\frac{F}{F_C}\right) + \frac{1}{2}\sigma^2\tau^*}{\sigma\sqrt{\tau^*}} \qquad (6)$$

$$N\left(k - \sigma\sqrt{\tau^*}\right) = \text{univariate cumulative normal distribution function} \qquad (7)$$

σ = .olatility of the rate of change of the second follow up investment;
K = present value of the capital expenditure of the second investment for time $\tau - \tau^*$.
K* = present value of the first year capital expenditure of the first investment for time τ^*.
r = riskless interest rate;
τ = time to maturity of the first option within the compound option;
τ^* = time to maturity of the simple option for the second venture;
F = present value of the cash inflow discounted to the beginning of the second investment;
F_c = critical value of the project above which the first call option will be exercised;
$M(k, h; \rho)$ = bivariate cumulative normal distribution function with k and h as the upper and lower integral limits, and correlation coefficient ρ.

CASE STUDIES

In this section, considerations regarding ROT analysis application are determined, a case study is conducted, and other decision-making scenarios are described, allowing better comprehension on how ROT can support planning sessions, regarding R&D projects.

The simulation method employed adheres to the following assumptions, implying to become pre requisite conditions to apply the real options theory. They are:

- **Irreversibility**: Once a certain amount has been invested in research – that is, a sunk cost – it is not possible to salvage such amount should the project not proceed.
- **Uncertainty**: A research and development project is carried out under uncertainties, be they technical (whether the product will work or not work is unknown) or economic (market conditions, for example). These uncertainties will only be dissipated through investing in research and proceeding with the project.
- **Timing**: Once the project has been kicked off, the company has the possibility of choosing the time to launch it in the marketplace, after having evaluated its feasibility. Having began the re-

search effort several options are found, such as: abandonment of the project should it look unpromising; temporarily mothballing the project waiting for the resolution of uncertainties and for a better moment to launch the product in the marketplace; development and marketing by the company owning the project; sale of the marketing rights to a third party.
- **Volatility**: The volatility of the net present value of future capital inflows is an important parameter to be considered, albeit difficult to estimate. Past data can be used to determine the expected volatility, assuming that past volatility will be replicated in the future. Depending on the project being analyzed, this information cannot be obtained because no prior projects exist; the data is then gathered without the company. Nicholas (1994) suggests that companies in the pharmaceutical industry generally assume a volatility rate between 40% and 60%, due to the high risks involved. Low volatility will be used for the example that follows.
- **Capital Inflows**: Begin after the second risk evaluation when the proposed product is approved and launched in the marketplace. However, to compute the option value, the capital inflow should be discounted back to the beginning of the second risk evaluation, thus using the company's cost of capital.

BETA Product Development Project

The example used here is from a company called ALPHA having an R & D project in which is divided into phases that are similar between the research departments of various companies. The company's name and product are not given because of confidentiality agreement.

It is assumed that the company ALPHA conducts research for the creation of a new product called BETA. The initial investment is (all values are expressed in Brazilian Reals) R$ 25,000,000.00. After completing the verification process, the board of directors decided to continue the promising composed option which is the continuity for the next phases or stages through an additional investment of R$ 90,000,000.00 (test phase).

The company will invest over R$ 500,000.00 (production and marketing). After 10 years, at the beginning of the 11th year, it is expected to enter the company's product in the market or sell the patent, resulting in a capital entry with present value for the 10th year of R$ 800,000,000.00 (see Lint and Pennings, (1998) ; Luehman (1998) to estimate future cash flows). The company assumes a volatility rate of 25% as being appropriate to the project. The interest rate for a risk-free application will be 4% and the company's cost of capital will be 14%. Figure 4 shows the R & D process model for BETA product, which contains information on the project.

Specifically, the information on Figure 4 can be expressed as follows:

Initial investment = R$ 25 millions
F = R$ 800 millions
$\sigma = 0,25$ .K = R$ 500 millions
K* = R$ 90 millions
F_c = R$ 613,978 millions
r = 4%
τ^* .= 3 years
τ .= 10 years
Capital cost = 14%

Figure 4. R & D process model for the ALPHA company's BETA product

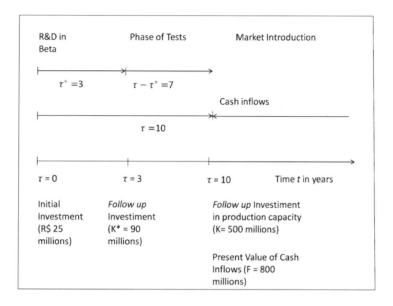

To solve the problem we first calculate the $\rho, k, h, N e M$ as indicated in the equations presented earlier.

$$\rho = \left(\frac{\tau^*}{\tau}\right)^{\frac{1}{2}} = \left(\frac{3}{10}\right)^{\frac{1}{2}} = 0,547.$$

$$h = \frac{\ln\left(\frac{F}{K}\right) + \frac{1}{2}\sigma^2\tau}{\sigma\sqrt{\tau}} = \frac{\ln\left(\frac{800}{500}\right) + \frac{1}{2}(0,25)^2 10}{0,25\sqrt{10}} = 0,99.$$

$$k = \frac{\ln\left(\frac{F}{F_C}\right) + \frac{1}{2}\sigma^2\tau^*}{\sigma\sqrt{\tau^*}} == \frac{\ln\left(\frac{800}{613,978}\right) + \frac{1}{2}(0,25)^2 3}{0,25\sqrt{3}} = 0,82.$$

$$N\left(k - \sigma\sqrt{\tau^*}\right) = N\left(0,82 - 0,25\sqrt{3}\right) = N\left(0,39\right) = .0,1517$$

$$M\left(k - \sigma\sqrt{\tau^*}, k - \sigma\sqrt{\tau}; \rho\right) = M\left(0,82 - 0,25\sqrt{3}, 0,82 - 0,25\sqrt{10}; 0,547\right) =$$

$$= M\left(0,39, 0,03; 0,547\right) \cong 0,42$$

$$M(k,h;\rho) = M(0,82, 0,99; 0,547) \cong 0,71$$

Then, using equation (3) option pricing composed of Kemna (1993), results in a value for the compound option:

$C = 800e^{-0,04(10)}0,71 - 500e^{-0,04(10)}0,42 - 90e^{-0,04(10)}0,1517$
$C = (800)(0,67)(0,71) - (500)(0,67)(0,42) - (90)(0,67)(0,1517)$
$C = 380,56 - 140,7 - 9,15$
$C = 230,71$

The value of this project consists of assets present value allocated (assets in place) and the present value of growth opportunities funded through the evaluation of option-based approach value.

Project value = sunk cost + growth opportunities

- The present value of the initial investment (sunk cost) is R$ 25 million
- The growth opportunities present value is equal to the value of the compound option G = R$ 230.71 million;

The value of investing in the BETA project, therefore, has a total value equal to:

Net Present Value (Geske) = R$ 230,71 – R$ 25 = R$ 205,71 milhões

Therefore,

The investment in the Beta project is R$ 205.71 million.

The net present value of the project calculated in the traditional way (Perlitz, Peske and Schrank, 1999), using the cost of capital of 14% would result in a value of:

$VPL_{Trad} = -25 - [90*(1,14^{-3})] + [(800-500)*(1,14^{-10})] = -4,82$ millions

Perlitz, Peske and Schrank (1999) present the following arguments for the fact that the traditional NPV is substantially smaller than the amounts obtained by the real options method. The authors feel there are some effects in favor of the investments in risky R&D projects when valuated by the real options pricing method, these being:

- NPV techniques are highly dependent on the discount rates applied. In the case of R&D projects these rates are often risk-adjusted, that is, they entail heavy discounts. The real options pricing method avoids the use of the risk-adjusted rate.
- In addition, the discount rate effect is strengthened by long time horizons applied to R&D investment decisions.
- Long time horizons allow more time to reach to changes in conditions. In the example, there is a possibility of stopping the investment or invest if the results from previous phases are known. This effect is taken into account in the evaluations of real options and not in the traditional NPV.
- The high volatility of R&D outlays positively influences the option value because great returns can be yielded, but small returns can also be avoided by a reaction to changes in conditions. In the NPV calculation, high volatilities entail a risk premium over the discount rate, and, consequently, a lower NPV.

As it is exposed above, ROT method, allowed its pre requisites to be applied, can serve better for R&D projects decisions, as it comprehensively solves market phenomena, changing scenarios and competitive issues, which take place all time in real world strategic product or service positioning. ROT development, with continuous application, can become a better alternative for projects like these, as it fits perfectly in those conditions, which apply to any market in the world.

Opportunely, these conditions to apply ROT also frequently are faced in new, innovative and disruptive strategic positioning, coming from technological and / or business model propositions, reaching innovation trends. This fact reinforces Real Options Theory as one up-to-date methodology, to be supported by continuous usage, adoption, refinement and, eventually, automated by software platforms, becoming an alternative to be considered by innovation sponsors and managers.

CONCLUSION

The static use of traditional investment evaluation techniques, mainly the NPV and the Discounted Cash Flow methods, has sustained harsh criticism, since they have not been able to capture the value of managerial flexibility contained in many projects. These models have been proposed for R&D application; however, they all have their advantages and disadvantages.

It is incumbent upon management to know which model should be chosen, seeking to apply the best fit to the peculiarities of a given project. This may be a barrier against the application of the theory, since there is no standard method to apply to each and every investment analysis. Since the decision-making process is not simple and, often times, involves thousands or even millions of dollars, management should be attentive to this new tool, paying some premium in the quest for the best solution.

ROT, as applied to a real research and development investment analysis proved satisfactory to bridge the gap between theory and practice, reaching innovative projects and entrepreneurship.

Finally, although the ROT is in a development and consolidation stage, the authors suggest that it should be used as a promising tool in the decision-making process.

REFERENCES

Amram, M., & Kulatilaka, N. (2000). Strategy and Shareholder Value Creation: The Real Options Frontier. *The Bank of America Journal of Applied Corporate Finance*, *13*(2), 8–21. doi:10.1111/j.1745-6622.2000.tb00051.x

Black, F., & Scholes, M. (1973). The Pricing of Options and Corporate Liabilities. *Journal of Political Economy*, *81*(3), 637–654. doi:10.1086/260062

Boer, F. P. (2000). Valuation of Technology Using "Real Options". *Research Technology Management*, *43*(July/August), 26–30. doi:10.1080/08956308.2000.11671365

Clewlow & Strickland. (1998). *Implementing Derivatives Models*. John Wiley & Sons.

Cox, J., Ross, S., & Rubinstein, M. (1979). Option Pricing: A Simplified Approach. *Journal of Financial Economics*, *7*(2), 229–264. doi:10.1016/0304-405X(79)90015-1

Dixit, A. K., & Pindyck, R. S. (1994). *Investment Under uncertainty*. Princeton, NJ: Princeton University Press.

Faulkner, T. W. (1996). Applying Options Thinking To R&D Valuation. *Research Technology Management*, *39*(3), 50–56. doi:10.1080/08956308.1996.11671064

Geske, R. (1979). The valuation of Compound Option. *Journal of Financial Economics*, *7*(1), 63–81. doi:10.1016/0304-405X(79)90022-9

Herath, H. S. B., & Park, C. S. (1999). Economic Analysis of R&D Projects: An Options Approach. *The Engineering Economist*, *44*(1), 1–35. doi:10.1080/00137919908967506

Hull, J. C. (1997). *Options, Futures and Other Derivatives Securities* (3rd ed.). New York: Prentice Hall.

Jamil, G. L., & Carvalho, L. F. M. (2019). Improving Project management decisions with Big data Analytics. In *Handbook of Research on Expanding Business opportunities with information systems and analytics*. Hershey, PA: IGI Global Publishers. doi:10.4018/978-1-5225-6225-2.ch003

Kaplan, R., & Norton, D. (2007). *The execution premium: linking strategies to operation for competitive advantage*. Harvard Business School Press.

Kemma, A. G. Z. (1993). Case Studies on Real Options, Financial Management: 259-270. Lint, O., E. Pennings. R&D as An Option on Market Introduction. *R & D Management*, *28*(4), 279–287.

Luehraman, T. A. (1998). Strategy as Portfolio of Real Options. *Harvard Business Review*, 89–99. PMID:10185434

Merton, R. C. (1973). Theory of Rational Option Pricing. *The Bell Journal of Economics and Management Science*, *4*(1), 141–183. doi:10.2307/3003143

Mitchell, G. R., & Hamilton, W. F. (1988). Managing R&D as a Strategic Option. *Research Technology Management*, *31*(3), 15–22. doi:10.1080/08956308.1988.11670521

Müller, R., Glücker, J., Aubry, M., & Shaun, J. (2013). Project management knowledge flows in network of project managers and project management offices: A case study in the pharmaceutical Industry. *Project Management Journal*, *44*(2), 4–19. doi:10.1002/pmj.21326

Myers, S. C. (1977). Determinants of Corporate Borrowing. *Journal of Financial Economics*, *5*(2), 147–175. doi:10.1016/0304-405X(77)90015-0

Paramkusham, R. B., & Gordon, J. (2013, Fall). Inhibiting factors for knowledge transfer in information technology projects. *Journal of Global Business and Technology*, *9*, 2.

Perlitz, M., Peske, T., & Schrank, R. (1999). Real Option Valuation: The New Frontier in R&D Project Evaluation? *R & D Management*, *29*(3), 255–269. doi:10.1111/1467-9310.00135

PMBoK. (2018). *Project Management Body of Knowledge* (6th ed.). Project Management Institute.

PMI – Project Management Institute. (n.d.). *What is project management*. Available at https://www.pmi.org/about/learn-about-pmi/what-is-project-management

Porter, M. E. (1992). Capital Disadvantage: America's failing capital investment system. *Harvard Business Review*, 65–82. PMID:10121317

Reich, B., Gemino, A., & Sauer, C. (2012). Knowledge management and Project-based knowledge in its projects: A model and preliminary results. *International Journal of Project Management, 30*(6), 663–674. doi:10.1016/j.ijproman.2011.12.003

Reich, B., & Wee, S. W. (2006). Searching for knowledge in the PMBoK Guide. *Project Management Journal, 37*(2), 11–27. doi:10.1177/875697280603700203

Silva, & Teixeira, & De Paula. (2012). Analysis of the Acquisition Process of a Autoparts Company Using Discounted Cash Flows and Real Options Models. *Perspectivas Contemporâneas, 7*, 11–43.

Stair, R., & Reynolds, G. (2009). *Principles of information systems*. Course Technology.

Trigeorgis, L. (1996). *Real Options: Managerial Flexibility and Strategy in Resource Allocation*. Cambridge, MA: The MIT Press.

Turban, E., Rainer, R. K. Jr, & Potter, R. E. (2007). *Introduction to information systems*. Hoboken, NJ: John Wiley and Sons.

Chapter 2
Information Management in Project Management:
Theoretical Guidelines for Practical Implementation

Fernanda Ribeiro
https://orcid.org/0000-0002-5641-9199
Faculty of Arts and Humanities, University of Porto, Portugal

Armando Malheiro da Silva
Faculty of Arts and Humanities, University of Porto, Portugal

ABSTRACT

The authors start by presenting a disciplinary positioning regarding the information science (IS) that is being developed, investigated, and taught at the University of Porto, Portugal and rigorously spelling out the cross-cutting approach to information management (IM), which is also shared with other social sciences and technology-oriented disciplines. From this point, it is presented a short overview and a literature review about the nature of project management (PM) to emphasize the need for an effective and fully assumed approach between IM and PM. It is observed that for PM specialists, info-communicational flows are important, and they seek to manage them intuitively, but without feeling compelled to draw on the know-how of the IS and IM specialists. However, the opposite is true, that is, there are in course some interesting adaptations of PM procedures, applied to informational projects, as it is shown in the last part of this chapter.

FROM INFORMATION SCIENCE TO INFORMATION MANAGEMENT: DISCIPLINARY POSITIONING

The practical epistemology, has been so designated by Jayme Paviani to mean the articulation of epistemological and methodological problems within a common horizon and, in this sense, its function would be "to make explicit the assumptions and purpose of science in an articulated manner with the rules, the

DOI: 10.4018/978-1-5225-9993-7.ch002

procedures and the research tools. Therefore, its contribution, as well as reflective, is programmatic" (Paviani, 2009:21), and has been behind the effort at the University of Porto, Portugal, of the precise and synthetic delimitation (joining theoretical concerns with the practical dimension of research) of the Information Science (IS) field. Hence, the proposed definition for it: "social science that investigates the problems, issues and cases related to info-communicational phenomenon, perceivable and knowable through confirmation or denial of the inherent properties of the informational flow genesis, organisation and behaviour" (Silva, 2006:141). It is also the science that, keeping alive the documentary tradition and practice inherited from previous disciplines such as librarianship, documentation and archivistics, studies the information cycle in its fullness and transversality: origin, collection, organisation, storage, retrieval, interpretation, transmission, processing and use of information.

Moreover, the question to ask is: How is the "information" defined? This is being defined in various ways and views. In the midst of such a variety of settings, we chose to take a definition that would allow to explore the fixed object: "Set of structured mental and emotional coded representations (signs and symbols) and modelled with/by social interaction, able to be recorded in any material medium and therefore communicated in an asynchronous and multi-directed way" (Silva, 2006:150).

To be communicated, information takes the form of a document, but it does not become identical to it; though our senses (visual and tactile) allow us to perceive the document as an artifact and inseparable (and symbiotic) from the mentefact (information). Perceiving the subtle but crucial difference between the content and the container, between the medium and the "substance of meaning" recorded in it, gives IS epistemic legitimacy.

The 60s pontificated the famous definition of IS exposed in the Georgia Institute of Technology conferences that took place in October 1961 and April 1962, and summarised in an article by Harold Borko (1968), in which the reference to the information properties emerged, listed firstly by Yves-François Le Coadic (1994). In the definition exposed and used in the University of Porto, they appear and mainly are specified in the book prepared to provide the theoretical and well-founded basis for the Bachelor in Information Science that began to be taught in the academic year 2001-2002 (Silva; Ribeiro, 2002). The properties of information, formalised as general axioms, are six and according to them, information: 1) is structured by an action (human and social) – the individual or societal act structurally establishes and models the information; 2) is integrated dynamically – the informational act is involved with, and results from, conditions and circumstances both internal and external to that action; 3) has potentiality – a statement (to a greater or lesser extent) of the act which founded and modelled the information is possible; 4) is quantifiable – the linguistic, numeric or graphic codification is capable of quantification; 5) is reproducible – the information can be reproduced without limit, therefore, its subsequent recording/memorising is made possible; 6) is transmissible – the informational (re)production is potentially transmissible or communicable. In addition, the properties are, somehow, intrinsic and "universal" characteristics of the info-communicational phenomenon. It is, moreover, in relation to this human and social phenomenon that the IS object (defined above) is (re)built.

The "steps" of the info-communicational cycle/process (object of study or *constructo*), listed above, form the IS object and can be distributed over three main specialised study areas or groups: the production of informational flow; the organisation and representation of information; and the informational behaviour (SILVA, 2006). This triple division of the IS object began with an ambiguity that was being addressed, and in principle solved, which was related to the inclusion of Information Management (IM). The connection of this topic – we will call it this from now on – with the trans and interdisciplinary IS, developed at the University of Porto, never offered any doubt. In fact, the training model that was shaped through Specialisation Programmes in Documentary Sciences, established in 1982 and closed in 2007[1], had to be deepened, radically rethought and reworked with the aim of creating a new hybrid professional, and in operating the symbiosis of the scientific-social and technological matrix. There was a clear inflection in affirming the information manager as the professional for a future that is increasingly present. The questions and difficulties were not, therefore, in the professional dimension, but in the epistemological framework. Our initial response was to dilute IM in the production of information: this orientation was to "present an alternative based on epistemological assumptions and conducive to place Information Management or explicit knowledge, as a segment of the IS object, in the way we conceive it from an epistemological and educational point of view at the University of Porto. Moreover, this perspective arises reflected, in condensed form, in Information Management and Knowledge Management DeltCI entries" (Silva, 2009:233)[2], and imposes itself. However, it is important to mention that, on Knowledge Management (KM), our position remains unchanged.

In later articles (Silva, 2005; Silva, 2009), the approach has become less simplistic and refuses the idea of identifying, without further thought, IM as one of those three areas of the IS object, which is associated with the production of informational flow and is clearly replaced by an "interdisciplinary topical of Information Management in the perspective of Information Science" (Silva, 2009: 246).

If we put disciplinary interrelations with maturity and complexity, as we currently understand them, it seems little doubt that, IM, as well as Information Knowledge (IK), and the more sectorial strategic ways of professional activity as Competitive Intelligence (CI) and Economic Intelligence (EI) (Paim, 2003; Tarapanoff, 2006; Santos; Leite; Ferraresi, 2007), do not constitute a scientific discipline in itself, but a "platform", essentially practical, for the application of ideas, theories, models and several solutions, which are condensed into multiple and different consulting "packages". This means, in plainer terms, that IM arises as a topical in pitch route from the traditional and instrumental view of Information and Communication Technologies (ICT). Indeed, much of the teaching of IM does not exist or has had a very strong emphasis on information encoded and stored in information technology (Rascão, 2008:14-15; RASCÃO, 2012). It is urgent, therefore, according to this author, for whom IM became, for better or worse, an academic discipline (this does not, of course, mean that it is a scientific discipline), that would break with the traditional view, in order to, alternatively, be constructed as a synthesis filled with contributions from Economics, Management, Strategic Management and Communication, which indicate other ways of looking at IM. These fields have a less technological vision of IM and turn the separation of the encoded, stored and accessed information through ICT, that is, in broad and comprehensive terms, into the process of decision making (Rascão, 2008:15).

Figure 1. Representation of the influence of IS in the practical dimension of IM

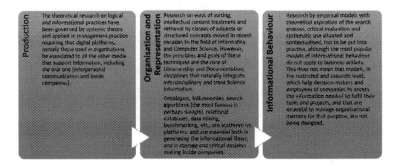

Being IM (and related variants) a topical moving towards the dynamic synthesis, it seems to settle down, more clearly, inside the inter-science Information Systems and in the interdisciplinary field of Communication and Information Sciences; thus summoning an increasingly strong approach from applied social sciences, and epistemic space where IS has to claim, with increasing force, its belonging and its own space (Silva, 2009:235-236).

Clarifying further, and to close this initiation, it can be said that IM corresponds entirely to the nature of applied social science with which the trans and interdisciplinary IS is defined and presented, *i.e.*, it is thus the application of IS across all areas of the field of study of this science and it is also composed by other different, but complementary and enriching scientific-technical, approaches.

It is evident that an information manager must have basic training in IS which is followed by the University of Porto. Their professional development and their expertise in the scientific and technological domains require additional skills through the deepening of other disciplines and knowledges such as Sociology of Organisations, Economy and Administration, Strategic Management, Business Development, Information Systems and Management Technology, for all of them, and others, find themselves intertwined in the complex and dynamic activity of IM.

Before moving on to the other sections of this chapter, it seems appropriate to concretely show how "pure" research on IS has a practical translation in the way(s) it produces, organises, mediates, searches, selects and uses information in many different situations, contexts and environments. This is intended following scheme:

PROJECT MANAGEMENT: ORIGINS AND AIMS

Management of projects has been a practical procedure for a long time, but this practice was widely recognised as a discipline only about two or three decades ago. Reference authors in the field, such as Jack Meredith and Samuel Mantel Jr. (2006; 2008), agree on this statement, considering that, in fact, project management (PM) developments are reinforced recently, "driven by quickly changing global markets, technology, and education. Computer and telecommunications technology, along with rapidly expanding higher education across the world allow the use of project management for types of projects and in regions where these sophisticated tools had never been considered before" (Mantel Junior, 2008:2-3).

Project management developed itself mainly in the scope of business organisations as a tool to achieve strategic goals and to accomplish routine tasks that could only be achieved with great difficulty if it is organised in traditional ways. The growth of PM can be well recognised not only because of its organisational implementation, but also because definition of methodologies appeared, professionals started to be hired and certified as project managers and numerous degree programmes were established in management schools and universities in several countries.

The recognition of PM was also reinforced by the creation of the Project Management Institute (PMI), the greatest and most respected association of project managers on a global level. PMI was founded in 1969 by five volunteers and its first workshop, held in Atlanta (USA), was attended by about eighty people. It grew rapidly and, at the end of 1970, nearly 2,000 members were part of the organisation. However, the first evaluation for certification as a project management professional (PMP) happened only in the 80s.

Furthermore, PMI implemented a code of ethics for the profession and, in the early 90s, the first edition of the *PMBOK Guide: a guide to the Project Management Body of Knowledge* (1st ed. 1996; 5th ed. 2013) was published. It soon became the basic pillar for the management and direction of projects in general. In 2000, PMI had more than 40,000 active members, had certified 10,000 PMPs and had sold about 300,000 copies of the *PMBOK Guide*.

Since then, PMI has grown to become the world's largest supporter of the profession of project manager. Currently, it has more than 650,000 members, in over 185 countries, and represents a lot of industry and business sectors, including information technology, aerospace and defence, financial services, telecommunications, engineering and construction, government agencies, insurance, health and many others.

In recent years, PM is no longer seen as an internal bureaucratic system of the organisations and it has come to be regarded as a strategic and competitive weapon that delivers levels of efficiency, quality and added value to business customers. On the other hand, PM started to be considered as an autonomous discipline and it is separated from general management. "Projects are more schedule-intensive than most of the activities that general managers handle" (Lewis, 2001:1).

Why then is PM considered as a discipline and why is it the object of teaching courses and programmes as a body of substantial knowledge? To answer these questions we must first discuss the significance of a project. There are several definitions, which are more or less identical, shorter or more extensive, and some of them can be chosen to illustrate the concept substance. For example, James Lewis considers that "a project is a multitask job that has performance, cost, time, and scope requirements and that is done only one time"; and adds that "if it is repetitive, it's not a project. A project *should* have definite starting and ending points (time), a budget (cost), a clearly defined scope – or magnitude – of work to be done, and specific performance requirements that must be met" (Lewis, 2001:2). This definition provides a quite objective delimitation of 'project' as a concept, and draws our awareness of the idea that a project is something dynamic, which requires procedures and processes to be developed.

According to PMI, a project is "a temporary endeavor undertaken to create a unique product or service" (Project Management Institute, 2004:5). This allows us to consider that a "temporary endeavor" can be a set of unique activities, simple or complex and related to each other, devoted to accomplish an objective or a purpose, in a defined period of time, with a particular budget and in accordance to certain requirements. Thus, a project can be anything/a process with the character of a temporary endeavour, if it fits the following issues: unique activities; sequence of activities; interrelated activities; a goal; a singular product or service; a defined period of time; a budget; is in accordance to a specification.

Complementing these characteristics, some authors consider that projects still have some attributes, such as rarity, restrictions, multidisciplinarity and complexity (Miguel, 2006.9). Furthermore, we can also add to all these components another structural and fundamental "ingredient" – **information** and its natural flow across and inside the project – which means that there is an effective need of IM directly related to PM. This is the simple and obvious element that is completely forgotten and ignored in PM literature. Therefore, the inclusion of IM in the standardised procedure of PM established by the *PMBOK Guide* would certainly make the difference regarding the success of a project.

According to Lewis, "there are many different models for the phases a project goes through during its life cycle", but he proposes an "appropriate life cycle" consisting of the following phases: definition phase, strategy, implementation planning, execution and control and closeout (Lewis, 2001:9-14). The management of a project also implies several steps to accomplish it in a successful way: definition of the problem; development of solution options; planning the project; executing the plan; monitoring and control progress; and closure of the project (Lewis, 2001:14-15).

Still according to the same author, "The Project Management Institute has attempted to determine a minimum body of knowledge that a project manager needs in order to be effective. At present, PMI identifies nine general areas of knowledge (…)" (Lewis, 2001:16-18) that are:

1. Project integration management
2. Project scope management
3. Project time management
4. Project cost management
5. Project quality management
6. Project human resource management
7. Project communications management
8. Project risk management
9. Project procurement management

An identical approach is taken by other experts in PM, such as Samuel Mantel Jr., Jack Meredith or Heloísa Lück, which develop in their books, in a detailed way, all the operative procedures that are implicit in the management of a project. Such operative knowledge and methodology started to be empirical but ended up being standardised and systematised in the *PMBOK Guide*, the *vade mecum* of PM.

On analysing the mentioned literature on PM, we realise that, in the same way as when a project is defined, the presence of IM is not referred to at any moment. IM does not figure among the "nine general areas of knowledge" considered as essential to PM. Nevertheless, a project generates information along its life cycle and such information needs to be managed. And if it is well managed (selected, recorded, organised, made available and preserved appropriately for future use), the project will be benefited from it and will attain better results.

An interesting contribution to this discussion appeared in the work of Barbara Allan (2004) who analyses the importance of PM tools and techniques in the information professionals' activities. Allan aims to "catch" knowledge from PM methodologies and expertise in order "to provide a guide and resource to project management within all types of information and library services (ILS)" (Allan, 2004:3). It is, obviously, an interesting perspective: to use PM knowledge to improve the success of ILS projects, that is, to use the project management literature, developed from work in industry and military projects, and add new elements to it in order to improve the results. "The author believes that project management is best achieved through co-operative and/or collaborative teamwork where the project manager's role is to facilitate and steer the project. (…) In addition, many project managers are not the line manager of people on their team and this means that they need to motivate and influence others through formal performance management processes. As a result, the 'soft' or human side of project management is considered an essential factor for success" (Allan, 2004:4).

The same author also shows which type of projects can be developed in the ILS field and gives a lot of illustrative examples, such as: "moving a library; developing a new information service; creating a new intranet site; digitizing a collection; merging two libraries; building and moving into a new learning resource centre; restructuring an information service; developing a new marketing campaign; re-cataloguing a collection; producing a common training programme for a number of ILS; developing a web-based information skills course; carrying out research in an innovative area" (Allan, 2004:5).

These examples show how information managers can be profited from PM to improve their work and establish a valuable relationship between both areas (IM and PM) in benefit of information professionals.

INFORMATION MANAGEMENT IN PROJECT MANAGEMENT: INFLUENTIAL OR ONLY IMPLICIT PRESENCE?

Given what was exposed in the previous two sections, it is inevitable to realise the embracing quality of PM, which does not constitute an axis of IM developed from the theoretical and methodological apparatus of IS. Nevertheless, the lack of centrality of this activity in the field of IS and, even in the broader and interdisciplinary field of Communication and Information Sciences, does not mean that the issue should not deserve all our attention. Our commitment is the most immediate contribution.

The transversal nature of PM coincides, moreover, with the identical transversality. It is not unique to IS and it has been appropriated by other approaches. However, in the operational context of IS, IM corresponds entirely to the applicational dimension, i.e., it corresponds to the search and implementation of theoretical and practical solutions that span through a broad spectrum of cases and contexts. Furthermore, we can, for a better explanation of our proposal, present two angles.

Firstly, we have the indisputable presence of IM in the dynamic process of any PM. This is clearly evidenced and implied in the main leading tools/models as, for example, PMBOK®. By simply glancing at its standards, we quickly understand the importance of recording, storing and retrieving information at each step, its evocation during the following steps and, especially, at the end of the whole management process of each project.

The complete cycle of IM, which has a clear beginning in the production or collection of information, is developed by a coordinated series of steps in which organisation – sorting, classifying and indexing –, as well as its interpretation for the dissemination, use and reproduction, gains special importance. Any entity, personal or institutional, individual or collective, that does not invest in clear and efficient procedures, at various points in time, loses information, and loses itself in the accumulated information, which by then becomes completely useless.

Consequently, the presence of good practices and, if possible, of suitable models of IM throughout the whole process of PM, should be noted. Speaking of models, it is also important to highlight the systemic model, which has several implications in any managed project. A key implication has to clarify the understanding of what constitutes information and if this makes up the triad "data-information-knowledge" or, on the contrary, absorbs the meaning of data and explicit knowledge. Operating with the definition presented at the beginning of the first section, it should be stated that it is not the medium or the technical form of registration that must be taken as a distinctive benchmark in relation to the information produced and accumulated within each project. The information is distinguished, therefore, by what it is or treats, which in documental terminology is called "typology". The information is distinguished therefore by its numerous typologies. Some examples of typologies that tend to form extensive series are: Emails (typological series merged to correspondence, previously only postage); the photographs, videos, reports, and personal notes of each member of the project team; the manuals or specific guidelines on PM; monographs published online on the particular subject or area that the project to be managed focuses on, and many others. Each PM process generates a plethora of informational typologies which, under the systemic viewpoint, are interdependent parts of a whole that is diverse, but one. Developing PM within this systemic perspective has visible and final consequences on the judgmental assessment of the work done.

Another implication, which is closely associated with the former, is to value the memory factor: the informational capital of a PM process has immediate utility, while running the case, but, after ended, this capital, if prolonged as permanent memory, has a potential utility of uncertain value. However, unlikely to occur, even if only occasionally and sometime far in the future, its potential utility is high, as shown by the personal and organisational experiences.

As demonstrated in the previous section, project managers do not need to have, and usually do not have, specific training in IS or in Communication and Information Sciences. However, it is interesting to note the emphasis that *Gerência de Projetos: Guia para o Exame Oficial do PMI Project Management Institute* (2nd edition) attaches to the skills, or communication skills, that a project manager must have:

One of the most important characteristics of a good project manager is to have good communication skills. Communicating by writing and by speech is the foundation of all successful projects. Throughout your project, you will use several modes of communication. As the person in charge of creating or managing most of the project's communication activities (documents, meeting updates, status reports, etc.), you must ensure that the information is complete, as well as fully and clearly expressed, so that your interlocutors have no trouble, whatsoever, understanding the purveyed message. After the information has been conveyed, it is the receiver's responsibility to make sure they have fully understood the information. (...)

Processes of Project Management

According to the PMBOK Guide, the processes of project management organise and describe the project's completion. They are carried out by people and, just like the project stages, they are interrelated and interdependent. For instance, it would be difficult to identify specific project activities without knowing the project's requirements beforehand.

PMI Process Groups

The PMBOK Guide states five process groups for project management: Initiating, Planning, Executing, Controlling and Closing. All of these process groups, except for Initiating, are comprised of individual processes which together constitute the group. For instance, the Closing Group comprises two processes: Contract Close-Out and Administrative Closure. Together, these process groups – including all its individual processes – shape the project's life cycle. (...)

Project management offices are a method of organising and designing patterns for project management techniques inside each organisation. They could also act as a project library by storing all documents for future reference. (...)

Project Communications Management

The processes of this knowledge area are related to general communication skills but go far beyond mere information exchange. Communication skills are the overall management skills that a project manager uses daily. The involved processes aim to ensure that all project information – including project plans, risk assessments, notes from meetings and such – is collected, documented, stored and disposed of when suitable. They also ensure information distribution and sharing with stakeholders, management and integral parts at appropriate times. At the end of the project, information is stored and used as reference for future projects – which is named as historical information on several project processes.

All project participants are somehow related to that knowledge area, as all of them send and/or receive communications throughout the project's life. It is of fundamental importance that all staff members and stakeholders understand how communication affects the project (Heldman, 2005; authors' translation).

In the same guide, the author includes a chapter entitled "Distribution of Project Information". This process

(...) aims to keep the stakeholders up to date regarding the project by using several methods: status reports, meetings about: information distribution, the plan for managing communications, based on the Communications Planning process, is implemented. Communication and Information Distribution act together to make all the information on the project staff's progress available (Heldman, 2005: 299; authors' translation).

Heldman recalls the three process inputs – results of the work, the communications management plan and project plan – addressed with advantages in the development of communication skills, in the exchange of information through the scheme "emitter–message–receiver", in the methods of information exchange, in the forms of communication, in the lines of communication, in the skills of effective listening, in conflict resolution, in information retrieval systems, in methods of distributing information and the outputs of the distribution of information (Heldman, 2005:300-306).

Under the term "communication", it is possible to develop a whole practice of IM supported by Information Technology (IT), which is a misconception to be undone. PM specialists need to be aware that operations such as recording, organisation and classification, storage and retrieval of information, which is produced in the scope of any project, imply skills and accumulated scientific basis for over a century that is condensed in IS. Strictly speaking, this is not merely a communication process, as it has true info-communicational requirements and specificities that are not fully absorbed and assimilated by IT technicians. During a project, the three areas – informational production, organisation and representation of information and information behaviour – in which it is possible to divide the IS object are interconnected and require special attention. The context of production, that is to say, the structural context of the project, has much to do with the informational typologies arising from the project, in addition to standardising and the crossing of the numerous different types of project information. Organising information into classes, descriptors and related terms is the basic assurance that information can be retrieved whenever there is such a need. This made possible to prove that these operations are not neutral and those who perform them are the producers themselves who should not, inadvertently, discard the role of mediator and conditioner of the subsequent search processes. Additionally, the present and future needs of those who use and search the information is a complex topic to consider.

If we seek examples of effective cooperation between IM specialists and PM specialists, in which the latter explicitly call upon the former to become fully effective with the projects' informational flow and its management, the frustration is great because there is a lack of cooperation between the two areas of activity. With our work, we intend to emphasise the urgency of an awareness on the part of project managers, towards the input that IM in particular, and IS in general, can bring to their work.

Secondly, we must face the presence of PM standards in projects related specifically to "serving information" – the construction of a public facility in a given municipality or community, as for example, a Municipal Library, is an eloquent example of how it is possible and inevitable to connect PM to informational aces/projects. First, because of and despite the immaterial nature of the information to be used, it tends to be materialised and this implies a set of concrete elements, which have to be designed, planned, budgeted and implemented with rigor and risk calculation – all are determined phases that enter the accumulated experience of PM.

If we put the emphasis on the informational projects developed on digital platforms, we can find, in the prescriptive background of PM, that the importing of all the specific know-how is done without difficulty. However, one must take into account that, for example, when building a portal on any one topic, whether or not the applicability of PM requirements is beyond dispute, and whether it is necessary to make important adjustments and to value the informational behaviour component, the research area of IS cannot obviously be dissociated from PM. What is more, the models arising in the course of research in this area, as well as the results obtained in academic studies, which have multiplied, show that there is no unique and universal standards on how users seek information according to needs and/or difficulty to be able to typify or fix it in a linear causality.

There is an important prevention point on how PM is linked to the management of projects that are primarily informational. The aim with which they are designed and implemented is to make access, search and the use of information by real people available, albeit diverse and inserted in more than one context, the context being an aggregating unit of elements that constrain or interfere with the dawning of an informational need. PM does not have to answer this complex specificity; IS comes in here with an increasing amount of exploratory studies that constitute a possible answer, which are always incomplete and always needing larger and successive developments.

Also, the function of PM is not to give answers to the problems of human mediation in the assembly of digital platforms while informational projects in themselves. It is a complex problem that involves two areas, as mentioned before, the field of IS – the study of information production logics and the organisation and representation of information. With regard, moreover, to this second area it is imperative to clarify that the Architecture and Design of Information, activities and disciplines that blend computer/technological training and artistic training (matrix of Fine Arts), flow into classical questions of organisation and representation of information to which the answers were found and developed from, of course, the core of IS. The issue of mediation therefore requires that, in the management of a specific informational project – e.g. a thematic site –, the accumulated contribution of IS, in both the productive aspect and in the representational aspect of contents that are intended to be accessed, searched and used, be openly and fully taken into account.

The search for examples illustrating the use of the precepts of PM in the construction of memory institutions, such as Archives, Public Libraries and Museums, in records management both in Records, as in "hybrid" services, and in the implementation of Information Systems in the broadest sense, i.e., not only the digital platforms, but services that concentrate all kind of information in the most various media, is more fruitful, since the popularity and the expansion of PM in multiple sectors and activities are increasing. It is easy to find courses that use PMBOK for the implementation of records management projects in archives. It is also natural that in the construction of buildings for Archives, Libraries and Museums, the same management "tool" be used – with indisputable advantages and benefits.

On the other hand, what was and still is less common is the adaptation of PMBOK to a specific informational project called SIMAI – *Sistema de Informação Municipal Ativa de Indaiatuba* [Active Municipal Information System of Indaiatuba], municipality of São Paulo, Brazil (Masson, Silva, 2000-2001:33-62). This is a project within another broader project that is entitled SIMAP – Active and Permanent Municipal Information System. They have a purpose, "under the systems approach, integrating an analysis of the functions of the organisational system of Municipal Administration with the information system thereunder, to objectively meet its established rules and its purpose/mission, thus ensuring, according to advanced methodology, effective internal and external informational communication" (Masson; Silva, 2000-2001:35). Later, in the article we are quoting, the author sums up in the following way:

It is suggested that SIMAP should be a process for assisting the Government Council Department or the Department itself, when it comes to the executive power, or for assisting the Municipal Presidency, when it comes to legislative power, due to the mandatory centralisation of the information that instruments, validates and subsidises the action of the municipal administration, despite its supra-structural function, which comprises all the organisational system (Masson; Silva, 2000-2001: 49; authors' translation).

SIMAI would be the application of the SIMAP model to the case of the Municipality of Indaiatuba. Moreover, as it has a concrete organic structure, inside the Municipal Executive or Legislative Power, it cannot be taken as a replica of an Archive Service or a Municipal Documentation Service. Strictly speaking, it is a system, or a whole, consisting of several parts that interdepend without losing their specificity or difference. An information system that is composed of multiple types of information in widely varying media, which is managed as a service that is installed in the "heart" of the municipal institution in order to make efficient its own management and public policies. In this sense, SIMAP is not an informational project, an archive or a library, which does not preclude that it can be implemented using the PMBOK.

This is precisely what Silvia Mendes Masson tried by producing documents and rehearsing an implementation strategy. She presented the first results in a paper at the *VI Congresso de Arquivologia do Mercosul*, sponsored by the *Associação de Arquivistas de São Paulo* and the *Centro de Documentação e Informação Científica da Pontifícia Universidade Católica de São Paulo*, and held at *Campos do Jordao*, State of São Paulo, Brazil, on 17-21 October 2005 (Masson, 2005). Her presentation focused on a work plan that followed the phases and procedures recommended by PM in general, and the PMBOK in particular. From a formal and theoretical standpoint, the results of this exercise, even today, after several years, are very exciting, but still await an effective application.

The evocation of this case confirms the curiosity and interest of experts and PM professionals and the use of PM in projects, even in the most innovative situations. However, the same interest does not appear explicit on the part of project managers in relation to the "core" of IS and IM. Hence, the timeless of this text, which constitutes a reminder and triggers the need for an open and deeper dialogue between the two specialties and disciplines.

REFERENCES

Allan, B. (2004). Project management: tools and techniques for today's ILS professional. London: Facet Publishing.

Borko, H. (1968). Information Science - what is it? *American Documentation*, *19*(1), 3-5.

da Silva, A. M., & Ribeiro, F. (2002). Das "Ciências" Documentais à Ciência da Informação: ensaio epistemológico para um novo modelo curricular. Porto: Edições Afrontamento.

da Silva, A. M. (2009). A Gestão da Informação na perspectiva da pesquisa em Ciência da Informação: retorno a um tema estratégico. In Coletânea Luso Brasileira: governança estratégica, redes de negócios e meio ambiente: fundamentos e aplicações. Anápolis: Universidade Estadual de Goiás.

Dicionário eletrônico de terminologia em Ciência da Informação. (n.d.). Retrieved from http://www.ccje.ufes.br/arquivologia/deltci/

Heldman, K. (2005). Gerência de projetos: guia para o exame oficial do PMI. Rio de Janeiro: Elsevier.

Le Coadic, Y-F. (1994). La Science de l'Information. Paris: PUF.

Lewis, J. P. (2001). Fundamentals of project management: developing core competencies to help outperform the competition (2nd ed.). New York: AMACOM - American Management Association.

Lück, H. (2013). Metodologia de projetos: uma ferramenta de planejamento e gestão (9th ed.). Petrópolis: Editora Vozes.

Mantel, S. J. Jr. (2008). *Project management in practice* (3rd ed.). Danvers: John Wiley & Sons.

Masson, S. M., & da Silva, A. M. (2001). Uma abordagem sistêmica da informação municipal: o projecto SIMAP e um caso de aplicação ainda incipiente: o SIMAI. *Cadernos de Estudos Municipais*, 33-62.

Masson, S. M. (2005). Projeto SIMAI-SIMAP: proposição de adoção da metodologia PMBOK para o gerenciamento de Sistema de Informação na Administração Municipal. In CONGRESSO DE ARQUIVOLOGIA DO MERCOSUL, 6º, Campos do Jordão, 2005 - Anais do VI CAM. São Paulo: CEDIC/PUC-SP.

Meredith, J. R.., & Mantel, S. J., Jr. (2006). Project management: a managerial approach (6th ed.). Danvers: John Wiley & Sons.

Miguel, A. (2006). Gestão moderna de projectos: melhores técnicas e práticas (2nd ed.). Lisboa: FCA – Editora de Informática.

Paim, I. (2003). A Gestão de informação e do conhecimento. Belo Horizonte: Escola de Ciência da Informação-UFMG.

Paviani, J. (2009). Epistemologia prática: ensino e conhecimento científico. Caxias do Sul: EDUCS.

Project Management Institute. (2004). *A Guide to the Project Management Body of Knowledge: PMBOK® Guide* (3rd ed.). PMI.

Project Management Institute. (2013). *A Guide to the Project Management Body of Knowledge: PMBOK® Guide* (5th ed.). PMI.

Rascão, J. (2008). Novos desafios da gestão da informação. Lisboa: Edições Sílabo.

Rascão, J. (2012). Novas realidades na gestão e na gestão da informação. Lisboa: Edições Sílabo.

dos Santos, S. A., Leite, N. P., & Ferraresi, A. A. (2007). Gestão do conhecimento: institucionalização e práticas nas empresas e instituições: pesquisas e estudos. Maringá, PR: Unicorpore.

Shera, J. H., & Cleveland, D. B. (1977). History and foundations of Information Science. *Annual Review of Information Science and Technology, 12*, 249-275.

da Silva, A. M. (2005). A Gestão da informação abordada no campo da Ciência da Informação. Páginas a&b: arquivos e bibliotecas, 16, 89-113.

da Silva, A. M. (2006). A Informação: da compreensão do fenômeno e construção do objecto científico. Porto: Edições Afrontamento; CETAC.COM.

Tarapanoff, K. (2006). Inteligência, informação e conhecimento em corporações. Brasília: IBICT; UNESCO.

KEY TERMS AND DEFINITIONS

Informational Behaviour: One of the areas of Information Science devoted to research on Informational needs and user's attitudes in technologically mediated contexts, adopting an interdisciplinary point of view that crosses information science, computer science, cognitive science and design.

Information Management: Cross activity based on the theoretical and methodological foundations of Information Science with a connection with other scientific disciplines, namely, the management and the sociology of organizations, which clearly lies in a dimension of practical and applied solutions and interventions.

Information Science: It is a social (applied) science that investigates the problems, issues and cases related info-communicational phenomenon perceptible and knowable through the confirmation, or not, of the properties inherent to the genesis, the flow, the organization and informational behaviour.

Organization and Representation of Information: One of the areas of information science devoted to research on technical and life cycle issues of information, from production to storage, focused on meta-information, controlled languages, information flow analysis and representation, access and search tools, classification, conceptual maps, information architectures and visualization.

Production of Information Flow: One of the areas of information science devoted to theoretical and practical research on logical and informational practices, governed by systemic theory and applied in management practice requiring that digital platforms, namely those used in organisations be associated to all the other media that support information, including the oral one (interpersonal communication and inside companies).

ENDNOTES

[1] These programmes lasted, with some technicist reinvigoration, the vocation of classical Librarian-Archivist High Programme, taught at the University of Coimbra (since 1935), to train librarians and archivists for the respective public cultural institutions.

[2] *Dicionário eletrônico de terminologia em Ciência da Informação* (url: http://www.ccje.ufes.br/arquivologia/deltci/ – accessed on 4-1-2013), the *online* entries which were published in Silva, 2006: 137-167). It is worth selecting from two of the mentioned entries – Information Management and Knowledge Management – the essential of the established and assumed definitions: "Management Information (...) In Information Science, the scientific component gains a considerable relief and Management Information becomes one of the three study areas extending in disciplinary branches of theoretical and practical application, such as archival science. By becoming the study area, its natural crossing or interaction with the other areas that act jointly in relation to treatment happens along with the Organisation and Representation of Information and regarding the use with Informational Behaviour" (Silva, 2006:149); and "Knowledge Management" (...) What is interesting in Information Science are the practices and techniques developed as knowledge management, which, after all, are simply practices and techniques of management, of organisation and use of information inside a more or less complex "entity" (Silva, 2006:149).

Chapter 3
Strategic Information Management:
Implementing and Managing a Digital Project

Sérgio Maravilhas-Lopes
https://orcid.org/0000-0002-3824-2828
IES-ICS, Federal University of Bahia, Brazil

ABSTRACT

The implementation of a digital information strategy project in a real estate company is analyzed in this chapter. This involves its historical and socio-economic framing and a brief description of the activity sector in which the company operates. Several well-managed digital projects have been developed to improve the competitive position of the company, but without focus, all the activities lose strength because they might not reach their proposed targets. Some tools to identify the information needs of business activity developed are described as well as the role of information as a promoter of competitive advantages. Social media tools were utilized and proved to be a great strategic decision. To conclude, a few reminders of the factors to consider in developing the information strategy to implement and that information management without a strategy could result in several diversified decisions without any positive consequence for the organization.

INTRODUCTION

The implementation of an information strategy is vital for any organization in the historical era in which we live, because we came from an industrial society to the information age (Gleick, 2011), where the efficient use of information as an economic resource and a production sector became a factor of strategic, economic, social and political (Best, 1996a) importance.

DOI: 10.4018/978-1-5225-9993-7.ch003

Information, along with natural and economic resources, proves to be an unprecedented social and strategic expedient (McGee & Prusak, 1995; Beuren, 1998). It has, therefore, a potential strategic value in identifying new market opportunities, and in identifying potential threats to the company (Gomes & Braga, 2001; Liautaud, & Hammond, 2001), two aspects to consider for any company that wants to achieve its goals.

Therefore, the importance of information for organizations is now universally accepted, being, if not the most important, at least one of the resources whose management influences the success of organizations (Ward & Griffiths, 1996). Information is also considered and used in many organizations as a structural factor and a tool for managing the organization (Zorrinho, 1991), as well as an essential strategic weapon for competitive advantage (Porter, 1985).

This information is called strategic information (Earl, 1998; Hinton, 2006) because it's indispensable for the full functioning of the organization and needed to sustain its competitive position. The systems that identify, store, organize and provide this information are called strategic information systems (Alturas, 2013; Amaral & Varajão, 2000; Davenport; Marchand & Dickson, 2004; Rascão, 2001, 2004; Varajão, 1998; Ward & Griffiths, 1996).

Information management relates to the organizational ability to make the right information available for use in decision making (Davenport, 1997; Rascão, 2008).

Hence, it is advantageous to any decision maker to hire professionals with these skills, since this function is not restricted to technological knowledge, but with the ability to properly organize and interpret the information obtained, allowing it to be used by those who need it, when they need it (Castells, 2001, 2004; Earl, 1998; Hinton, 2006; Ward & Griffiths, 1996).

Only someone with communication and language skills can more easily find and use the information available, making it profitable (Tapscott, 1999). Professionals in information management have a multidisciplinary training that involves information sciences, information technology and communication, management and several of its sub-disciplines (Hinton, 2006; Wilson, 1989a; Wilson, 2002).

This multidisciplinary approach allows a comprehensive and overall vision that enables not only the collection of relevant information, but also its analysis and, especially, their synthesis to support the decision making of managers and organizational leaders, transforming the informational chaos into useful and practical knowledge application, leading to benefits for the organizations.

The situation of an existing Real Estate company, BV[1], will be analyzed. Some advantages of bringing a properly defined information strategy, with the participation of all employees of the organization, will be discussed, leaving its coordination to the care of an information manager. The risks of leaving that task only to the care of the information technology (IT) department and the danger of information overload will be alerted. The conclusion will try to clarify some of these points stressing the benefits of conducting such activity.

All the information about the company and the activity described herein was obtained through participant observation and subsequent personal interviews with the three managing partners, five of the current ten branch directors and the head of the information systems department and his two collaborators.

Please remember that it's never easy to obtain confidential strategic information from an organization due to competitive reasons.

The information revealed here is just the one that had clearance and was authorized to be disclosed, given the need to keep certain confidential information outside the scope of competition, extremely aggressive in this sector in the current economic environment, and to maintain the image and reputation of the company among its customers, partners and collaborators.

BACKGROUND

For Edgar Morin (1996) "the whole universe is based on only three constituents and the relationships between them: matter, energy and information".

Generally, energy is defined, operationally, as the ability to produce work, not the work itself. Similarly, the information is defined, operationally, as the ability to store and transmit meaning or knowledge, not knowledge or meaning in themselves (Gatlin, 1972, p. 23). The ability to store and transmit knowledge should not be confused with the knowledge that it can produce (Choo, 2003; Gomes & Braga, 2001; McGee & Prusak, 1995).

Therefore, it is concluded that the very nature of information is mainly relational, organizational (Choo, 2003; Gleick, 2011).

That the human individual needs information is a common-sense observation and a factor generally accepted by experts, who attribute to cognitive aspects an important role in the genesis of behavior.

Gordon Pask (1969) supports the idea that the human being is characteristically an organization that processes information and learns. The amount of effort expended in the search for information differs, therefore, with its connection to diverse human needs and the intensity of the need in question.

Thus, it is understood that the energy expended in the search for information, such as a need to orient ourselves in the environment in which we operate, is higher in periods of growth of the individual, without direct connection with the satisfaction of other needs.

In a society in continuous technological and scientific development, it is also understood that the need to know is constantly activated and represents a predominant role in the life of the individual (Jorge, 1995).

Regarding information theory, what had fascinated me was to discover that the information could be defined physically. In reality it was a partial truth. To define the information is necessary to give a biophysical-anthropological definition: information, unquestionably, has something physical, but only appears with the living, and moreover, we are the ones who have named and discovered it (Morin, 1996, p. 107).

The importance of information stands out for its ability to allow management strategies that translate into competitive advantages of high importance (Porter & Millar, 1985), such as price leadership, differentiation and focus on a product, market area or specific consumer, which may involve a differentiation of costs, or the publication of differentiated services.

The Real Estate Brokerage Firm BV: Historical and Socio – Economic Background

BV is the second largest real estate company with Portuguese owners and capital in the Northern region of Portugal, and was one of the largest in the country, both in number of departments, stores and employees, as well as in sales and net income.

With over forty years of existence, its headquarters are based in Porto and the majority of its current ten stores are in the city or its surroundings. In 2001, it had fifty-one stores, which were distributed from Valença do Minho, in the North, to Leiria, in the lower center. The distribution of its stores was as can be seen in Table 1.

Strategic Information Management

Table 1. Number of BV stores in 2001 (fifty-one stores, from Valença do Minho to Leiria)

City/Location	Number of stores
Valença do Minho	1
Viana do Castelo	1
Ponte de Lima	1
Braga	1
Trofa	1
Vila do Conde	1
Póvoa do Varzim	1
Vila Nova de Famalicão	1
Guimarães	1
Stº Tirso	1
Paços de Ferreira	1
Maia	2
Porto	5 (one was also the Head office)
Matosinhos	1
Srª da Hora	1
Gondomar	2
Rio Tinto	2
Paredes	1
Penafiel	1
Felgueiras	1
Lousada	1
Valongo	2
Vila Nova de Gaia	3
Stª Maria da Feira	1
São João da Madeira	1
Estarreja	1
Espinho	1
Ovar	1
Aveiro	3
Ílhavo	1
Vagos	1
Águeda	1
Oliveira de Azeméis	1
Mealhada	1
Coimbra	2
Lousã	1
Figueira da Foz	1
Leiria	1

(Source: Author)

In 1999, the company recorded a turnover of around 100 million Euros[2]. This value has been declining every year, due to the socio-economic situation the country is facing, being currently barely sufficient to meet the expenses of maintaining the company. It consists primarily of sales and rent departments, supported by law and legislation departments, insurance, documentation, marketing and advertising, information technology, administrative support, accounting and also training (currently disabled due to the drop-in turnover and number of new employees to train).

The company can be very flexible in its decision making process because it's very flat and horizontal in its levels of decision and command, having only four levels: i) Three owners and directors who operate as CEO (chief executive officer), CFO (chief financial officer) and COO (chief operations officer); ii) Eight partners, who act as Heads of the commercial departments of each of the stores; iii) Supporting staff like accountants, lawyers, IT engineer and technicians, marketers, etc.; iv) Several sellers/brokers who work as independent professionals, allocated at one of the departments but selling where the customer wants to buy, even if it is in another department commercial area.

It had over six hundred employees, being the most brokers (realtors) that developed their activity in sales. This number currently is less than one hundred employees.

Nowadays, the distribution of its stores is as follows, in Table 2.

Since the company operates in an extremely aggressive and highly competitive industry, with a rotation of products for sale and human resources (HR) of the highest in the trade, services and related areas, it's essential for its survival to define the development and implementation of a strategy for organizational information in order to preserve and maintain its activity in an increasingly demanding market such as that of real estate brokerage.

INFORMATION

The efficient exploitation of information[3], as an economic resource and one of the production sectors, has become a factor of strategic, economic, social and political importance (Best, 1996b).

A new vision must be considered in relation to information and the role it plays in the organization.

Table 2. Number of BV stores in 2014 (ten stores, from Viana do Castelo to Coimbra).

City/Location	Number of stores
Viana do Castelo	1
Braga	1
Maia	1
Porto	2 (one is also the Head office)
Matosinhos	1
Rio Tinto	1
Vila Nova de Gaia	1
Aveiro	1
Coimbra	1

(Source: Author)

Strategic Information Management

If, on the one hand, an organization cannot function without information, on the other it is important to know how to use this resource to improve its functioning.

Thus, the faster the identification of the relevant information to the organizations and the quicker the access to this information more easily, their goals will be achieved.

Information is an intuitive, indefinable principle, such as energy, whose precise definition always seems to escape through the fingers like a shadow (Gleick, 2011; Jorge, 1995; Morin, 1996).

But most important is that information plays a key role in business since this affects the competition at three levels:

i) Modifies the industrial structure and, therefore, changes the rules of the competition;
ii) Creates competitive advantage for organizations offering new ways to overcome their rivals;
iii) Create new business opportunities, most often from the organization's internal processes (Porter, 1985; Porter & Millar, 1985).

It is important to remember that the primary function of information is to try to eliminate uncertainty (Wilson, 1985, 2001), a problem that plagues any area of activity.

Organizations should follow some recommendations to successfully achieve their goals, such as:

1. Value information as a resource so or more important than any other that it needs to function (Maravilhas, 2013b);
2. Give the employees the relevant and necessary information to the excellent performance of their function minimizing, where possible, the overhead with unnecessary information (information overload);
3. Pay attention not only to the internal information generated within the organization, necessary to carry out the organizational tasks it undertakes, but also external information, from various points of interest to the sector in order to maintain their activity profitable.

Effective managers and decision-makers should not ask about the cost of obtaining the information needed. They should ask instead how much the loss will be if they don't have it.

Information Management

Information management is usually defined as a comprehensive organizational capacity to create, maintain, retrieve and make available the right information, at the right place, at the right time and in the hands of the right person, at the lowest cost, in the best support for its use in decision making (Choo, 2003; Davenport, 1997; Earl, 1998; Hinton, 2006).

Information management must be based on developing a strategy that involves the entire organization, considering mainly the users, information resources and appropriate technology available (Wilson, 1985, p. 65).

Consequently, information management (IM) should be seen as the conscious process by which information is gathered and used to assist in decision making at all levels of an organization.

(...) A final point about the definition is that it makes no reference to computers or information technology. Information management is as much about paper-based systems, or even human voice-based systems, as it is about technology-based systems.

It is a popular misconception that information management is only concerned with information technology management (Hinton, 2006, pp. 2, 3).

It is also an economic coordination of the efficient and effective production, control, storage, retrieval and dissemination of information from internal and external sources, to improve the performance of a given organization (Best, 1996a; Beuren, 1998; Choo, 2003; Davenport, Marchand, & Dickson, 2004; Prusak & McGee, 1995).

The important role of information management and its integration into the organizational strategy must be emphasized, revealing itself as a key factor in creating added value for the company, as it helps detect new opportunities, create competitive advantages and enables to defend them from the competitors.

Information management is an activity that aims to regulate the information, the information and communication technologies (ICT) and their respective users through the application of management techniques to process and provide updated and relevant information, using electronic means or not, depending on the user's needs.

Information management is much more than the ability to obtain information from computer form (the computer will be just a tool here, as a catalog of a library can also be). Managing information means information processing so that, whoever needs it, can then get some help to achieve their goals.

Armed with relevant information, at the appropriate time, at the lowest possible cost, the organization will be better prepared to face the adversities of the industry.

The Information Manager Role: Importance and Attributions

Everybody associated with the business world knows that some function that saves time and allows to easily access information within the area of interest of the organization will be positively valued.

The functions assigned to an information manager are many and varied (Choo, 2003; Wilson, 1989a) and he may be asked to:

1. Develop and disseminate web sites that aim to optimize the service provided to customers (ex. bringing together all existing information published in any format) (Marchand & Horton Jr., 1986; McGee & Prusak, 1995);
2. Perform information audits, internal and external (Best, 1996b; Orna, 1999);
3. Use the internet as a base to search databases of companies and organizations, conducting competitive/market intelligence and technology watch (Gomes & Braga, 2001; Kahaner, 1997; Liautaud, & Hammond, 2001);
4. Develop skills that enable him to act as a consultant to his customers, deciding what is the best search engine for use in a research, what sites can be used to support specific information needs, or where are the best information sources for decision making support, etc. (Choo, 2003);
5. Implement systems where services could go from the traditional way of transmitting information to a form of information analysis, i.e., to filter, monitor and evaluate sources of information, and/

or prepare and gather summaries of executive decisions (Beuren, 1998; Brown & Duguid, 2000; Varian & Shapiro, 1999);
6. Train professionals that could build new paths in order to make available web site information to end-users (Adeoti-Adekeye, 1997).

In sum, the role of the information management professional should be to focus on optimization, as well as in service personalization, and not be just a mere unqualified worker, carrying only simple tasks, like the collection of e-mails and other types of unfiltered information.

For any organization, in any industry, will be vital to its success to have the capabilities of a information management professional that will give a quick and efficient access to information in a particular subject, saving time and resources, providing the company with innumerous benefits, including financial, increasing the competitiveness of its activity (Hildebrandt, 1997) with the quality of the information collected and filtered to allow managers a more coherent and efficient decision making (Wilson, 1989a).

With regard to support for the management and administration of organizations, and recruiting professionals who have this function, it's positive for any decision maker to hire professionals with training in information management, since this function does not only refer to technical knowledge and technology, but the ability to properly interpret the information obtained, as well as its organization, so it can be used by those who need it.

The information manager, by its extensive network of multidisciplinary knowledge, will be the key element in the collection, analysis and dissemination of relevant information to those responsible for decision making in organizations.

At this point, it should be noted that it is essential for the information manager, who should be responsible for the definition of the information strategy to implement, to know what are the aims of the organization, including its goals and strategic plan (Ward & Griffiths, 1996), or in other words, what is the business strategy of the organization, what are their goals for next year and what kind of problems the company is currently facing (Cleveland, 1985; Wilson, 2002).

If such issues are not available to the information manager, it will be extremely difficult to plan the obtainment and delivery of information services related to the strategic planning of the company or organization (Orna, 1999).

The information manager must ensure that this does not happen, if the organization gives him the necessary resources for the proper performance of its duties.

THE STRATEGIC MANAGEMENT OF INFORMATION

An information strategy defines the organization's information needs, ensuring that existing information services to meet those needs are properly organized and managed (Wilson 1987; 1989b, 1994b), and has access to ICT for storage, search, distribution, communication and information security.

Information must be managed like any other vital resource of the organization. In fact, sometimes, all other resources are dependent on this one and it is imperative for the organization to safeguard and take advantage of the information resources gathered internally with the same, or more relevance, of the ones gathered externally. Research and development (R&D) results, clients list and respective contacts, suppliers of sensitive or specific materials used in production, business models and marketing strategies, distribution and operations logistics, trade secrets, among other invaluable information's are vital for the success and survival of every organization (Liautaud, & Hammond, 2001).

The lack of information might induce mistakes or originate wrong and costly decisions. It is necessary to adapt and adequate the information content to the needs of the users. In every organization exists huge volumes of information but, frequently, the vital information arrives after it was needed.

Vital information exists but, occasionally, it is dispersed and it's very hard to find the answer for simple needs and questions. There is a lot of information without any value to the organization and the right one is not available in an accessible format. Sometimes, it is kept by someone in a department without being shared with all the organization and others who need it. So, it's necessary to develop three tasks to avoid or minimize those situations:

i) In all the organization, information sources available should be identified and centralized;
ii) These activities should be regarded as parts of a whole;
iii) These activities should be managed by someone capable of understanding the information needs of the company and each actor involved with its specific tasks to perform (Donnelly, Gibson, & Ivancevich, 2000).

The organization should start the strategic planning by preparing an industry profile (Markides, 2000, 2008). This is an essential management function, because it allows a more effective assessment of the situation of the company in the sector in which it operates or that the company depends on for survival, and involves:

i) Knowledge of the environment where the company operates (STEEP factors)[4];
ii) Statistics and forecasts, starting from the current state of the industry and related areas;
iii) Identification and analysis of the biggest companies in the industry, their main competitors and business practices developed by them.

This profile will allow building a coherent map of the position of the company in its industry sector and the situations that they will have to deal with and should be included in the strategy.

Instruments and Tools to be Used in the Information Strategy

Some instruments for detecting the information needs of the company are (Wilson, 1994a; Romagni, et al., 1999):

1. Comparative analysis of the sector to which the organization belongs (Benchmarking);
2. Value chain analysis and identification of critical areas of activity of the company (Porter, 1985);
3. Analysis of Porter five forces (1980), to understand from which direction comes the most important pressure for the organization;
4. SWOT[5] analysis, to understand the strengths, weaknesses, opportunities and threats that the company is subject to and should seek to prevent and improve;
5. STEEP analysis, to be aware of the risks, chances and benefits that can be exceeded if information is well managed.

Strategic Information Management

According to Porter (1980), it should be borne in mind the five forces that define, regulate and maintain the market in which the organization operates and acts, in order to avoid unforeseen situations that may endanger its economic and professional integrity.

The implications of the five forces to consider and its relation to the activity undertaken by the organization are:

1. Rival firms or direct competition within the sector;
2. Bargaining power of Suppliers;
3. Bargaining power of Customers;
4. Threat of Substitute products or services;
5. Threat of new entrants, or potential new competitors that may enter the organization's business sector.

This latest threat is, undoubtedly, one of the most pressing and cumbersome due to the lack of regulation and the effective legislation of the sector. At this conjuncture, information plays a key role, since this affects the competition at three levels:

1. Modifies the industrial structure and, therefore, changes the rules of the competition;
2. Creates competitive advantage by offering companies new ways to overcome their rivals;
3. Create new business opportunities, most often from the internal productive processes of the organization (Porter, 1985).

Analysis of Porter Five Forces Regarding the Organization Industrial Sector

The Real Estate Brokerage is, above all, an activity that provides services since whoever is selling the property is its owner and not the company itself. The company and its professionals operate as agents or brokers.

The core activity is to find the right property that the customer demands, according to the criteria desired for the house, apartment or land, and reach an agreement between the interested buyer and the owner. For that, there will be a monetary reward for the service provided by the company. That is the work done by the agents or brokers, also called sellers, which collaborate with the company.

Thus, for the analysis of Porter's value chain or rather the critical sectors of the same elaborated by Hunsicker (1989), BV situates his practice in sales and services (Porter, 1985; Grant, 2002; Picot, 1989).

For the five forces described by Porter (1980), this activity refers to:

1. Rival firms or direct competition within the sector:
    a. These are all the other real estate companies or other entities that sell or rent properties (ex. construction companies);
2. Bargaining power of Suppliers:
    a. The persons or entities who want to sell or rent their properties;
3. Bargaining power of Customers:
    a. The people or entities who intend to buy or rent a property;

4. Threat of substitute products or services:
   a. New ways of looking for a house through the Internet, bank websites or competition websites, fleeing from the traditional activity that the company is dedicated to;
5. Threat of potential new competitors that may enter the organization's business sector:
   a. Foreign companies that develop business, using Franchising models, in Portugal (ex. ERA, Remax, Century 21, etc.);
   b. Former employees of the biggest companies who engage on their own in the market;
   c. The banks trying to sell the properties with unpaid loan debts;
   d. Construction companies and builders themselves, who create their own real estate companies to operate in parallel with the construction;
   e. The owners trying to sell the property themselves using Social Media Networking (ex. Facebook, Hi5, Pinterest, Google+, Twitter, etc.), trying to avoid the payment of the service to the Real Estate companies.

As can be seen, there are several pitfalls to avoid and various situations that should be known so that the organization can continue to occupy the prominent place that has always occupied.

Following Brandenburger & Nalebuff (1996) advices, the company initiated several Co-opetition deals with construction companies and banks to obtain their products with exclusivity, being it houses or credit loans at the best price and interest rate to surpass their competitors in the market.

Grant (2002) called this the sixth force. It's not present on Porter five forces model but, contrary to the other five that withdraw value from the industry, this is the only one that aggregates value. It is the case of the software industry that aggregates value to the computer business, the ink cartridges that aggregate value to the printers, and the video game industry that aggregates value to certain hardware manufacturers, amongst others.

The company has a website that allows searching for properties available for sale or rent using various criteria, such as the type, location, price, purpose and the county where the property is located. Simultaneously, through a partnership with a bank, allows for a simulation of the amount of mortgage loans to be paid with online approval. Social Media has been adopted recently with hardware to support and facilitate its use.

This strategy allows the reduction of the costs in physical facilities and personnel. Being a first point of contact for potential customers, it provides that they can look through photos and short videos all the various options available with all the convenience and may then turn to a physical store for personalized support and make a visit to the properties of their interest and formalize the completion of the deal.

THE IMPORTANCE OF INFORMATION IN THE DIGITAL STRATEGY PROJECT

The importance of information[6] stands out for its ability to allow management strategies that translate into competitive advantages (Faulkner, & Bowman, 1995; Porter, 1985; Porter, Millar, 1985) of high importance, such as:

1. **Price Leadership**: Not the prices to be charged by the property, but possibly the low fees charged for services rendered;

Strategic Information Management

2. **Differentiation**: Which can provide a unique product or service, handling all legal documents and procedures required for the purchase or lease of a building or property;
3. **Focus**: In a particular product, in a particular area or specific consumer market, which may involve the differentiation of fees or the publication and advertisement of differentiated services.

Since October 2012, the Portuguese Government is conceding the so-called Gold Visa to attract foreign investors that can obtain citizenship by the investment of 1 million Euros in a business endeavor, creating more than 10 permanent jobs, or half a million in the purchase of a residence for the purpose of living in the Portuguese territory. Several citizens from China, Russia, Brazil and Angola are adhering to this proposal so they will need a professional service to advise them and take care of all the paperwork involved[7].

The Focus strategy can involve the creation of a special task force, highly educated and knowledgeable of the situation, to take proper care of these segments of customers and their needs. With an adequate price fee, dealing with all the bureaucracy to conclude the business satisfactorily, this can create a differentiation that make this type of customers to search for the company when they need to fulfill these acquisitions or others in the future, passing the good service by word-of-mouth to their associates, creating new customers and new businesses. Social Media will be an important part of this strategy.

The company strategists decided also to entail several negotiations with the Embassies and Consulates of the countries from the Portuguese official language region, like Brazil, Angola, Mozambique and Macau (now Chinese) that were former Portuguese colonies and have an excellent relationship with Portugal, in order to be certified as a trustable company to do business with in case one of their residents decides to buy a half million house or start a million Euros company. No decision has been made yet from the Embassies side.

But, for this, it is necessary to obtain relevant strategic information, such as:

1. Can the adaptation of the company business model to the consumers' needs be improved? (Debelak, 2006; Osterwalder, & Pigneur, 2010; Rappa, 2010; Watson, 2005).
2. Can the tenacity of the sales force and its location be improved? (Faulkner, & Bowman, 1995).
3. Can the consumer attention for products marketed by the company be improved? (Kumar, 2004).
4. Could a price change improve the market share of the company? (Cruz, 1990).
5. Could the use of new ICT (Smartphone, Tablets and Apps) and Social Media Networks (Facebook, Twitter, Instagram, Tumblr, Flickr, Google+, Hi5, Pinterest, etc.) result in increased consumer demand and loyalty? (Maravilhas, 2013a, 2013b; Ryan, & Jones, 2012, 2013).

These are the issues that the company should pay special attention to, and this is the type of situation that a well-defined information strategy will be able to respond, enabling the company to maintain its market share and, if possible, eventually improve it (Wilson 1987; 1989b, 1994a; 1994b).

Specifications of the Digital Information Strategy to Implement

Information management must be based on developing a strategy that involves the whole organization taking mainly into account:

1. The users;
2. The information resources;
3. The appropriate technology available (Wilson, 1985, p. 65).

According to Wilson (1985), a certain number of questions must be made when we start to develop a strategy, such as:

1. Which type of information is needed so people can be helped in the performance of their tasks?
2. Is that information available to them? If yes, at what cost? If not, at what prejudice?
3. Which priorities are there to direct who should get that information?
4. Which methods of acquisition, storage and distribution of information are used? And, at what cost?
5. Which alternatives exist for each of these processes? How much do they cost?
6. Which systems would be better? How much do they cost?
7. What is the proper balance between manual and computerized systems?

A strategy based on these issues means that a new vision, a new look, should be considered in relation to information and the role it plays in the organization.

The specifications that the information strategy must meet to be effectively implemented and successful are (Back, & Moreau, 2001):

1. The strategy to be implemented should be formally documented;
2. Must be initiated by the top management of the company, with a project sponsor and a project champion[8];
3. Should be monitored by planned revisions;
4. It should be based on the provision of information about key indicators, as well as detailed analysis of the information needs of top management.

It should be stressed that to be successful, the information strategy to develop and implement must include strategies of:

1. Information management;
2. Information systems;
3. Information technology (Wilson 1987, 1989b, 1994b).

The whole process described above should be formalized, since many studies (Kalseth, 1991; O'Connor, 1993; Earl, 1998) show a higher percentage of success if the Management Board is involved and committed, motivating all other actors in the process[9].

Also, if formal documentation does not exist, no one will know what the information strategy defined is, to be able to carry it out effectively and efficiently. Regarding monitoring, if it is not formally performed it's not possible to consider the defined strategy offered by environmental changes and controlling its effective implementation. Finally, without an effective needs analysis, there is no apparent reason why the strategy exists.

Strategic Information Management

Mintzberg (1987) suggests that outlining a strategy is more of an art than a skill or science, since this involves negotiating various barriers. These barriers may be contrary to the attempt to implement any innovation, such as hostile attitudes by the middle management level, as may be related to the problem of hiring the appropriate employees or even difficulties in quantifying the benefits of such implementation.

It should be noted that it is essential for the Information Manager, the person who will be in charge of the definition of the information strategy to be implemented, to know what are the objectives of the organization, including its goals and strategic plan (Ward & Griffiths, 1996; Rascão, 2001, 2008), being aware of what is the business strategy of the company, including the company's goals for the next five years, and what kind of problems the organization is currently facing (Cleveland, 1985; Wilson, 2002).

If such issues are not available to the Information Manager, it will be extremely difficult for him to plan to obtain and deliver the information services related to the strategic planning of the company or organization (Orna, 1999). However, he can only be aware of what is strategic information for the company if faced with the strategic issues for the efficient performance of the organization (Wilson, 1994b, p. 266).

Mistakes to Avoid in Implementing the Digital Information Strategy

For an organization to implement an effective information strategy, two errors must be avoided. The first concerns the use of IT and the people in charge for this sector being responsible for coordinating the strategy (Rogers, 2017). The second relates to the collection and provision of information in excess, or the so-called information overload (Dearlove, 1998). Both are likely to be related, which will be described next. Both assumptions will be analyzed in the light of an overall information strategy, engaging and comprehensive, not compartmentalized to the IT department of the company and the IT strategy.

Thus, it is usual to assign responsibility for implementing an information strategy to the department responsible for computers in the company, constantly confusing information management[10], with information systems management and information technologies (Hinton, 2006).

Even if everybody should cooperate and work together for the benefits it brings to the company, it is not a single issue to be taken into consideration (McGee & Prusak, 1995; Ward & Griffiths, 1996), and this confusion often originates that the investment made in IT does not become effectively profitable (King; Grover & Hufnagel, 1989; Premkumar & King, 1994; Oliveira, 2004), precisely due to the lack of coordination between the various sectors involved.

Although there is an IT department in the company and the person responsible for it must have an extensive knowledge of its industry, typically that person is unaware of all the other factors that could have implications on the real business benefit of these technologies for the company (Cattela, 1981; Hinton, 2006; Wilson, 2002).

Factors such as the degree of knowledge of employees regarding the use of IT, which is usually little or null (Rogers, 2017), activities involving the real estate business such as documentation required, taxes payable and other amounts involved in the process of buying and selling real estate properties, management expertise to respond adequately to the challenges posed by an activity in constant change and constant attacks from inside and foreign competition are essential for the implementation of the information strategy for the proper performance of the organization.

Therefore, unaware of the activity from the 'inside', and thinking that the interests of the company are being satisfied, disseminating and providing information electronically to all the users will occupy many of the employees with issues that are not relevant to their work and will make them lose time for the effective performance of the activity (Edmunds & Morris, 2000). For the same reason stated, it might happen that the information provided is not the one that really matters, and it's needed, disseminating unfiltered information from someone not knowledgeable of the intricacies of the profession (Choo, 2003; Gomes & Braga, 2001).

It can also be concluded that the mere purchase of technological equipment's alone does not lead to any competitive advantage (McGee & Prusak 1995; Earl, 1998; Davenport; Marchand & Dickson, 2004), unless they are appropriately framed in a wider plan, allowing their potential to be monetized (Oliveira, 2004).

This leads to another problem that should be seriously considered and regarding the issue of excess of information, or information overload. If, on the one hand, a company cannot function without information, on the other, it is important to know how to use this resource to improve its functioning. Since we live in an era of abundance of information, it is necessary to distinguish the useful, relevant and necessary information from the surplus (Dearlove, 1998; Wilson, 2001). Thus, the faster the identification of the relevant information to the enterprise, and the faster the access to this information, more easily, their goals will be achieved.

Since people and organizations, similarly, have a peak in the information processing capacity, there is a saturation point from which an increase of the available information does not correspond to an increase in its use, but rather corresponds to its contraction.

It is considered information overload, in personal terms, when someone has the perception (either the user or an observer) that the information associated with the tasks to perform in the workplace is in greater quantity than can be managed efficiently and effectively (Dearlove, 1998; Edmunds & Morris, 2000; Wilson, 2001). Also involves the perception that this excess creates a degree of stress[11] in the attempt to adapt to the strategies defined that proves itself ineffective.

Excess of information is considered, then, in organizational terms, when the situation in which the extent of perceived individual information is sufficiently dispersed in the organization, thus reducing the effectiveness of all operations management. Although not a new or recent phenomenon, it tends to worsen with the advent of IT, Social Media and the Internet being a risk factor in terms of organizational control and the health of the employees.

The impact of information overload causes bad time management, delaying decision-making, allowing distractions of the main tasks, causing stress and consequent lack of job satisfaction, reduced social activity, fatigue and, in more severe cases, effective disease (Wilson, 2001).

Finally, it should be kept in mind that the company is not an island (Davenport, 1997), doesn't live closed in on itself, being imperative to monitor everything that goes on the outside and that relates to the sector of activity of the company whether economic, political and social (Gomes & Braga, 2001), both nationally and internationally, without incurring in the error described in the previous point of an eventual excess of irrelevant information that could interfere with the performance of the company.

So, the company must invest heavily in an information strategy to allow the knowledge and analysis of all stakeholders' issues in its activity, both internally and externally.

DECISIONS ADOPTED AND THEIR RESULTS

According to Rogers (2017, pp. 15-35), there are five domains of the digital transformation, being:

1. Clients;
2. Competition;
3. Data;
4. Innovation;
5. Value.

All those issues were considered when developing the digitization of the processes, conducting to the information strategy developed.

After defining the information strategy, in conjunction with the overall strategy of the company, some decisions to adopt essential measures were taken. The essential decisions consisted in:

1. Increasing the loyalty of the clients and prospects, in a long-life personal and digital relationship, making them advertise the company to their friends and connections through word-of-mouth and word-of-web (Michael, & Salter, 2006; Ryan, & Jones, 2012, 2013);
2. Improve the work of the sellers to increase the number of deals concluded positively, reducing stress and maximizing profits for them and the company.

To achieve these goals, the company realized that a change in their traditional business model must be produced (Debelak, 2006; Markides, 2008; Osterwalder, & Pigneur, 2010; Rappa, 2010; Watson, 2005).

A project champion and a project sponsor (Boonstra, 2013; Earl, 1998) were chosen to warrant the implementation of the measures defined. The IT manager became the project sponsor for his technology expertise, and one Head of commercial department became the project champion for his leadership qualities and respect gained by his knowledge of the profession.

The Marketing manager applied his knowledge in the digital environment but, in the future, an experimented digital Marketing manager should be integrated to bring more value to the team.

The company Board members realized that they must adapt and change "from selling products to providing solutions" (Kumar, 2004).

Some measures adopted, and their results were:

1. Creation of company accounts on major Social Media Networking sites to communicate with prospects and clients, together with the existing Website and e-mail addresses;
2. Training the sales people on the use of the IT resources available on the company, to direct the prospects and customers to the choices available on the company Website and Social Media Networks;
3. The use of smart phones and/or tablets is now mandatory, so the company can keep the communication channels with the sellers and clients always open;
    a. Smart phones and/or tablets allow sellers to take pictures and send them to the company and/or customers to check their interest and make them dream about their new house or answer any question or doubt that they might have;
    b. Smart phones and/or tablets allow communicating via web, using e-mail and/or free Apps, reducing substantially the costs with phone calls and communication;

c. Smart phones and/or tablets allow the access to Social Media Networks, increasing the number of contacts through their personal connections and the endurance of relationships more permanent and durable with customers and prospects;
d. When accompanying clients on a visit to a property, the seller can show them in the smart phone and/or tablet other options, documents available on the company website, costs, taxes, and so on, to support the sale, making the time shorter for the client's decision;
e. Smart phones and/or tablets allow the seller to check competitors Websites, confirming prospects information about the products and prices, analyzing their market strategy, knowing their employees and avoiding unpleasant surprises;
f. The GPS available through the smart phones and/or tablets, operating with Google Maps and other geo referencing tools, allow arriving on time to meetings and property visits, avoiding the seller to get lost, as sometimes happened in the past (or giving that excuse);
g. Geo referencing tools, operating with Google Maps, allow the seller and the clients to check the infrastructures available nearby like: schools, hospitals, restaurants, gyms, ballet classes, green parks, museums, libraries, highways, supermarkets, and other equipment's required by the clients as a must to buy the property;
h. The seller can save time writing reports, checking e-mail from the company or clients, updating the products photos, price and characteristics, asking for documents, and so on, while waiting for a meeting or for a visit with prospects, reducing stress and workload;
4. Cold calls now are not so cold because the seller, almost every time, knows who he/she is talking to, making them more confident and motivated;
5. The company Intranet has a knowledge database made with a Wiki tool, allowing the seller to check for answers when confronted with a doubt. They can also contact the department responsible, by phone or e-mail, and solve the problem talking with the expert on the subject, being it a lawyer, the insurance expert, the credit/loan expert or another seller.
6. Market segmentation made possible a focus strategy with the constitution of a task force to deal with big accounts, with big numbers involved, and different needs to attend, based in the Porto Head office.

According to Markides (2008), a new company or spin-off could be created to address this last situation, but BV has a strong brand name and reputation and the Board decided to use all the synergies involved to give a quality assurance to the prospects with the same brand name they already know.

Until now the results are very positive, allowing the company to increase twofold the number of information requests by interested prospects and the number of connections/followers in their networks. Portugal is under International Monetary Fund (IMF) economical rescue so, in this environment, the results are considered good for the company.

Another situation that helped the company in financial terms was that each seller/agent/broker, being independent professionals, had to buy and pay for its smart phone and/or tablet, as a tool of the profession, making the company save several hundred Euros in the process.

For the customers it's also positive because, now they can track the seller and know if there are any complaints about his/her professionalism and ethical conduct.

Strategic Information Management

Some customers have written positive feedback about the company and the service received which can be confirmed by other prospects checking the commentators' profiles to confirm that they are real customers and not company employees faking good results. This improves the positive image of the company and its brand.

The company now fulfills a stronger service, even more professional, because the seller has everything needed at his/her fingertips, saving time and money, and increasing sales. With that information, the seller can better fulfill its tasks, feeling less stressed and with better results.

To increase the loyalty of the customers, the company helps them with all the legal documents, taxpaying, and even with the annual individual tax deduction, making the customers link and connect long after the sale is done.

The customers receive, by e-mail or a post in their preferred Social Media Network where they have connected with the company, or via really simple syndicate (RSS) tools, information that can be useful to them, like deadlines to pay the annual tax for house owners, how to minimize their costs, when to deliver certain documents, and so on and so forth.

Next time they need to sell their house and buy a new one they will, eventually, deal again with whom they know and trust, knowing that everything will be taken care with high professionalism.

FURTHER RESEARCH DIRECTIONS

It would be useful and important to continue to follow this case study on the advantages of implementing an information strategy project, the mistakes to avoid, and essential results that would allow other organizations in the same area of activity or complementary areas to benefit from this experience and can, themselves, implement their own information strategies faster and more consistently, avoiding mistakes and adapting the most effective solutions for their performance and continuity in the market.

The analysis of the future implementation in the same company of a Customer Relationship Management (CRM) solution, based in the Cloud, will also be very useful to allow the measuring of the results that such a solution will bring to this type of company and the benefits obtained with it.

CONCLUSION

For an organization to remain profitable in this new century, it is essential to be well prepared for the challenges that the information-based society entails (Shapiro & Varian, 1999; Brown & Duguid, 2000). To do this, companies should try to develop and implement an information strategy, which involves information management strategy, information systems strategy and information technology strategy, because only then can involve all the stakeholders needed to create a good, harmonious, understanding of the problem to overcome (Alturas, 2013; Amaral & Varajão, 2000; Rascão, 2004; Varajão, 1998).

For the strategy to prove itself effective, it must be properly documented and formalized as any other company initiative so that all employees may be aware of the measures to adopt, whatever they are after decided by everybody involved (Orna, 1999). It must also have a genuine commitment from the top management assuring its importance and allowing every collaborator to be motivated in their application (Earl, 1998).

Since the real estate sector has been deeply affected by successive external factors, such as rising interest rates, the entry of more experienced foreign companies in the business, excess of professionals operating in the sector, oversupply of products for consumption, acute crisis in the political, economic and social life of the country, international recession, among other factors that have serious repercussions on confidence and purchasing power of domestic consumers, it is of utmost importance for the company to develop protection mechanisms, enabling it to anticipate certain events that may influence the sector and make it less stable. This may be possible through the process described above to provide the organization with structures to channel relevant information to be used when needed.

Implementing such a strategy should consider the future business plans of the company, the activity that it develops and related areas, which will only become effective after a preliminary analysis of the sector and its implications for the company's plans (Marchand & Horton Jr., 1986; Wilson, 2002). Such a strategy should, under no circumstances, be left only to the IT sector of the organization, since it does not have knowledge of all the details involved, although it should also be consulted in relation to its industry and the changes to be made for effective performance of the organization (Ward & Griffiths, 1996; Hinton, 2006; Oliveira, 2004).

With relevant information, at the appropriate time, at the lowest possible cost, the organization will be better prepared to face the rigors of the industry. However, for this reason it should not overload the operation of the business with information in excess, superfluous and redundant, as this will only cause a delay in what could be a competitive advantage, transforming it in an organizational nightmare, leading to a situation more unfavorable than previously maintained (Dearlove, 1998; Edmunds & Morris, 2000; Wilson, 2001).

A well-designed information strategy is a step ahead of the closest competitors, which may well serve to avoid certain mistakes and make certain decisions that, otherwise, would be totally unknown to the decision makers (Best, 1996b). Always remember that the primary function of information is to try to eliminate uncertainty (Wilson, 1985).

Finally, some recommendations that the company must follow to successfully achieve its goals were described, such as: the company should consider the information as a resource equally or more important than any other that it needs to function; the company should think about investing in the training of its human resources so that they are in a position to be able to respond positively to the changes that the implementation of such a strategy may entail in terms of knowledge of new IT solutions and Social Networking; the management board of the company should set the value of information to the business strategy of the organization; the organization must provide a thorough knowledge of the company plans to its information manager so that he can intervene positively in obtaining valid resources; the mere acquisition of information technology is not sufficient to obtain the competitive advantages that it can provide so, there must be a more comprehensive planning, elaborated previously to the development or acquisition of technological equipment's; the organization must give its employees the relevant and necessary information to the excellent performance of their function minimizing, when possible, the unnecessary information to avoid overload; the company must be attentive not only to the internal information generated within the organization, the one it needs to carry out the business tasks it undertakes, but also external information, from various points of interest to the sector, in order to maintain its activity profitable; the company must provide their customers the access to its products and services through various technological solutions so as to meet various market segments that may have interest or need in the business area of the company.

Some measures were put in place, after defining the information strategy to adopt and implement, complementarily to the general strategy of the company (Back, & Moreau, 2001). The results improved twofold the number of prospects that contacted the company to obtain information about properties, augmenting the sales funnel that can increase the number of effective sales. The Social Media solutions also helped the customers become more loyal and improve the positive image of the company amongst other potential customers.

Decentralizing the access to the right information and disseminating useful information sources needed by their professionals, allowed the organization to avoid information overload in its sales force.

The limitations of this study are motivated to restrictions imposed by the organization regarding the confidentiality of their strategic and tactic/operational information, understandable situation given its sensitivity and value that it contains.

REFERENCES

Adeoti-Adekeye, W. (1997). The importance of management information systems. *Library Review*, *46*(5), 318–327. doi:10.1108/00242539710178452

Alturas, B. (2013). *Introdução aos sistemas de informação organizacionais* (1st ed.). Lisboa: Sílabo.

Amaral, L., & Varajão, J. (2000). *Planeamento de sistemas de informação* (2nd ed.). Lisboa: FCA.

Back, W., & Moreau, K. (2001). Information management strategies for project management. *Project Management Journal*, *32*(March), 10–20. doi:10.1177/875697280103200103

Best, D. (1996a). Business process and information management. In The fourth resource: Information and its management. Hampshire: Aslib/Gower.

Best, D. (1996b). *The fourth resource: information and its management*. Hampshire: Aslib/Gower.

Beuren, I. (1998). *Gerenciamento da informação: um recurso estratégico no processo de gestão empresarial*. São Paulo: Atlas.

Boonstra, A. (2013). How do top managers support strategic information system projects and why do they sometimes withhold this support? *International Journal of Project Management*, *31*(4), 498–512. doi:10.1016/j.ijproman.2012.09.013

Brandenburger, A., & Nalebuff, B. (1996). *Co-opetition: A revolution mindset that combines competition and cooperation: The game theory strategy that's changing the game of business*. New York: Bantam Doubleday.

Brown, J., & Duguid, P. (2000). *The social life of information* (1st ed.). Boston: Harvard Business School Press.

Castells, M. (2001). *A sociedade em rede: A era da informação: Economia, sociedade e cultura* (5th ed.). São Paulo: Paz e Terra.

Castells, M. (2004). *A galáxia internet: Reflexões sobre internet, negócios e sociedade*. Lisboa: Fundação Calouste Gulbenkian.

Cattela, R. (1981). Information as a corporate asset. *Information & Management*, 4(1), 29–37. doi:10.1016/0378-7206(81)90023-9

Choo, C. (2003). *Gestão de informação para a organização inteligente: A arte de explorar o meio ambiente*. Lisboa: Editorial Caminho.

Cleveland, H. (1985). *The knowledge executive: Leadership in an information society* (1st ed.). New York: Dutton.

Cruz, E. (1990). *Planeamento estratégico: Um guia para a PME* (3rd ed.). Lisboa: Texto Editora.

Davenport, T. (1997). *Information ecology: Mastering the information and knowledge environment*. New York: Oxford University Press.

Davenport, T., Marchand, D., & Dickson, T. (2004). *Dominando a gestão da informação*. Porto Alegre: Bookmann.

Dearlove, D. (1998). *Key management decisions: Tools and techniques of the executive decision-maker*. Wiltshire: Financial Times/Pitman.

Debelak, D. (2006). *Business models made easy*. Entrepreneur Press.

Donnelly, J. H., Gibson, J. L., & Ivancevich, J. M. (2000). *Administração: Princípios de gestão empresarial* (10th ed.). Lisboa: McGraw-Hill.

Earl, M. (1998). *Information management: The organizational dimension*. New York: Oxford University Press.

Edmunds, A., & Morris, A. (2000). The problem of information overload in business organizations: A review of the literature. *International Journal of Information Management*, 20(1), 17–28. doi:10.1016/S0268-4012(99)00051-1

Faulkner, D., & Bowman, C. (1995). *The essence of competitive strategy* (1st ed.). Exeter: Prentice Hall.

Gatlin, L. (1972). *Information theory and the living system*. Columbia University Press.

Gleick, J. (2011). The information: A history, a theory, a flood. St. Ives: 4th Estate.

Gomes, E., & Braga, F. (2001). *Inteligência competitiva: Como transformar informação em um negócio lucrativo*. Rio de Janeiro: Campus.

Grant, R. (2002). *Contemporary strategy analysis: concepts, techniques, applications*. Wiley-Blackwell.

Hildebrandt, D. (1997). *Internet tools of the profession: a guide for information professionals*. Chicago: Special Libraries Association.

Hinton, M. (2006). *Introducing information management: The business approach* (1st ed.). Burlington: Elsevier Butterworth-Heinemann.

Hunsicker, J. (1989). Strategies for European survival. *The McKinsey Quarterly*, (Summer): 37–47.

Jorge, M. (1995). *Biologia, informação e conhecimento*. Lisboa: F. C. Gulbenkian.

Kahaner, L. (1997). *Competitive intelligence: How to gather, analyze, and use information to move your business to the top* (1st ed.). New York: Simon & Schuster.

Kalseth, K. (1991). Business information strategy: The strategic use of information and knowledge. *Information Services & Use, 11*(3), 155–164. doi:10.3233/ISU-1991-11307

King, W., Grover, V., & Hufnagel, E. (1989). Using information and information technology for sustainable competitive advantage: Some empirical evidence. *Information & Management, 17*(2), 87–93. doi:10.1016/0378-7206(89)90010-4

Kumar, N. (2004). *Marketing as strategy: Understanding the CEO's agenda for driving growth and innovation*. Boston: Harvard Business School Press.

Liautaud, B., & Hammond, M. (2001). e-Business intelligence: Turning information into knowledge into profit. New York: McGraw-Hill.

Maravilhas, S. (2013a). A web 2.0 como ferramenta de análise de tendências e monitorização do ambiente externo e sua relação com a cultura de convergência dos media. *Perspectivas em Ciência da Informação, 18*(1), 126–137. doi:10.1590/S1413-99362013000100009

Maravilhas, S. (2013b). Social media tools for quality business information. In Information quality and governance for business intelligence. Hershey, PA: IGI Global.

Marchand, D. (1990). Infotrends: A 1990s outlook on strategic information management. *Industrial Management Review, 5*(4), 23–32.

Marchand, D., & Horton, F. Jr. (1986). *Infotrends: Profiting from your information resources*. Wiley.

Markides, C. (2000). *All the right moves: A guide to craft a breakthrough strategy*. Boston: Harvard Business School Press.

Markides, C. (2008). *Game-changing strategies: How to create new market space in established industries by breaking the rules* (1st ed.). San Francisco: Wiley.

McGee, J., & Prusak, L. (1995). *Gerenciamento estratégico da informação: Aumente a competitividade e a eficiência de sua empresa utilizando a informação como uma ferramenta estratégica*. Rio de Janeiro: Campus.

Michael, A., & Salter, B. (2006). *Mobile marketing: Achieving competitive advantage through wireless technology* (1st ed.). Oxford: Butterworth-Heinemann. doi:10.4324/9780080459653

Mintzberg, H. (1987, July). Crafting strategy. *Harvard Business Review*, 66–75.

Morin, E. (1996). *O Problema epistemológico da complexidade*. Publicações Europa-América.

O'Connor, A. (1993). Successful strategic information systems planning. *Journal of Information Systems, 3*(2), 71–83. doi:10.1111/j.1365-2575.1993.tb00116.x

Oliveira, A. (2004). *Análise do investimento em sistemas e tecnologias da informação e da comunicação*. Lisboa: Sílabo.

Orna, E. (1999). *Practical information policies* (2nd ed.). Cambridge: Gower.

Osterwalder, A., & Pigneur, Y. (2010). *Business model generation: A handbook for visionaries, game changers, and challengers*. Wiley.

Pask, G. (1969). *Learning strategies and teaching strategies*. Surrey: System Research Ltd.

Picot, A. (1989). Information management: The science of solving problems. *International Journal of Information Management*, 9(4), 237–243. doi:10.1016/0268-4012(89)90047-9

Porter, M. (1980). *Competitive strategy: Techniques for analyzing industries and competitors*. New York: Free Press.

Porter, M. (1985). *Competitive advantage: creating and sustaining superior performance*. New York: Free Press.

Porter, M., & Millar, V. (1985). How information gives you competitive advantage. *Harvard Business Review*, (July/August): 75–98.

Premkumar, G., & King, W. (1994). The evaluation of strategic information systems planning. *Information & Management*, 26(6), 327–340. doi:10.1016/0378-7206(94)90030-2

Rappa, M. (2010). Business models on the web. *Managing the Digital Enterprise*. Retrieved 25-05, 2012, from http://digitalenterprise.org/models/models.html

Rascão, J. (2001). *A análise estratégica e o sistema de informação para a tomada de decisão estratégica* (2nd ed.). Lisboa: Sílabo.

Rascão, J. (2004). *Sistemas de informação para as organizações: A informação chave para a tomada de decisão* (2nd ed.). Lisboa: Sílabo.

Rascão, J. (2008). *Novos desafios da gestão da informação* (1st ed.). Lisboa: Sílabo.

Rogers, D. (2017). Transformação Digital: Repensando o seu Negócio para a era Digital. S. Paulo: Autêntica Business.

Romagni, P. (1999). *10 instrumentos-chave da gestão* (1st ed.). Lisboa: D. Quixote.

Ryan, D., & Jones, C. (2012). *The best digital marketing campaigns in the world: Mastering the art of customer engagement* (1st ed.). Croydon: Kogan Page.

Ryan, D., & Jones, C. (2013). *Understanding digital marketing: Marketing strategies for engaging the digital generation* (2nd ed.). Croydon: Kogan Page.

Shapiro, C., & Varian, H. (1999). *Information rules: A strategic guide to the network economy* (1st ed.). Boston: Harvard Business School Press.

Tapscott, D. (1999). *Creating value in the network economy*. Boston: Harvard Business School Press.

Varajão, J. (1998). *A arquitectura da gestão de sistemas de informação* (2nd ed.). Lisboa: FCA.

Ward, J., & Griffiths, P. (1996). *Strategic planning for information systems* (2nd ed.). Wiley.

Watson, D. (2005). *Business models: Investing in companies and sectors with strong competitive advantage*. Harriman House Publishing.

Wilson, D. (2002). *Managing information: IT for business processes* (3rd ed.). Woburn: Butterworth-Heinemann.

Wilson, T. (1985). Information management. *The Electronic Library, 3*(1), 62–66. doi:10.1108/eb044644 PMID:2498741

Wilson, T. (1987). Information for business: The business of information. *Aslib Proceedings, 39*(10), 275–279. doi:10.1108/eb051066

Wilson, T. (1989a). Towards an information management curriculum. *Journal of Information Science, 15*(4/5), 203–209. doi:10.1177/016555158901500403

Wilson, T. (1989b). The implementation of information system strategies in UK companies: Aims and barriers to success. *International Journal of Information Management, 9*(4), 245–258. doi:10.1016/0268-4012(89)90048-0

Wilson, T. (1994a). Tools for the analysis of business information needs. *Aslib Proceedings, 46*(1), 19–23. doi:10.1108/eb051339

Wilson, T. (1994b). The nature of strategic information and its implications for information management. In New Worlds in Information and Documentation. Elsevier Science B. V.

Wilson, T. (2001). Information overload: Myth, reality and implications for health care. *ISHIMR - International Symposium on Health Information Management Research.*

Zorrinho, C. (1991). *Gestão da informação*. Lisboa: Editorial Presença.

ADDITIONAL READING

Baldam, R. (2005). *Que ferramenta devo usar? Ferramentas tecnológicas aplicáveis a: Gestão de empresas, racionalização do trabalho, gerenciamento do conhecimento*. Rio de Janeiro: Qualitymark.

Barabba, V., & Zaltman, G. (1992). *A voz do mercado: A vantagem competitiva através da utilização criativa das informações do mercado*. São Paulo: Makron Books.

Bedell, D. (2011). Business intelligence. *Global Finance,* (March), 46-47.

Cardoso, P. (2004). As informações em Portugal (1st ed.). Lisboa: Gradiva | Instituto da Defesa Nacional.

Christensen, C., Anthony, S., & Roth, E. (2004). *Seeing what's next: Using the theories of innovation to predict industry change*. Harvard Business School Press.

Cleveland, H. (1983). A informação como um recurso. Diálogo. Rio de Janeiro, 16, 7-11.

Davenport, T. (1997). *Information ecology: Mastering the information and knowledge environment*. New York: Oxford University Press.

Davenport, T. (2007). *Profissão: Trabalhador do conhecimento: Como ser mais produtivo e eficaz no desempenho das suas funções*. Amadora: Exame.

Davenport, T., & Prusak, L. (1998). *Working knowledge: How organizations manage what they know*. Boston: Harvard Business School Press.

Davenport, T., Prusak, L., & Wilson, J. (2003). *What's the big idea: Creating and capitalizing on the best management thinking*. Boston: Harvard Business School Press.

Dinis, J. (2005). *Guerra de informação: Perspectivas de segurança e competitividade* (1st ed.). Lisboa: Sílabo.

Drucker, P. (1987). *Inovação e gestão* (2nd ed.). Lisboa: Presença.

Drucker, P. (1988). *As fronteiras da gestão*. Lisboa: Editorial Presença.

Drucker, P. (1993). *Gerindo para o futuro*. Lisboa: Difusão Cultural.

Drucker, P. (1993). *Sociedade pós- capitalista*. Lisboa: Difusão Cultural.

Galliers, R. (1993). Towards a flexible information architecture: Integrating business strategies, information systems strategies and business process redesign. *Journal of Information Systems*, *3*(3), 199–213. doi:10.1111/j.1365-2575.1993.tb00125.x

Hampton, D. (1992). *Administração contemporânea* (3rd ed.). São Paulo: Makron Books.

Hess, P., & Siciliano, J. (1996). *Management: Responsibility for performance*. McGraw-Hill.

Instituto de Altos Estudos Militares. (2000). *A gestão da informação e a tomada de decisão*. Sintra: Atena.

Introna, L. (1997). *Management, information and power* (1st ed.). London: MacMillan. doi:10.1007/978-1-349-14549-2

Johannessen, J., & Olaisen, J. (1993). The information intensive organization. In J. Olaisen (Ed.), *Information management: A Scandinavian approach*. Oslo: Scandinavian University Press.

Lacalle, D. (2013). *Nós os mercados: O que são, como funcionam e porque são indispensáveis* (1ª ed.). Barcarena: Marcador.

Mintzberg, H. (1973). *The nature of managerial work*. New York: Harper and Row.

Mintzberg, H. (1994). *Rise and fall of strategic planning*. New York: Free Press.

Montgomery, C. A., & Porter, M. E. (1995). *Strategy: seeking and securing competitive advantage*. Harvard Business Review Book Series.

Olaisen, J., & Revang, O. (1991). Information management as the main component in the strategy for the 1990s in Scandinavian airline system (SAS). *International Journal of Information Management*, *11*(3), 185–202. doi:10.1016/0268-4012(91)90032-8

Penzias, A. (1989). *Ideas and information: Managing in a high-tech world*. New York: W. W. Norton.

Penzias, A. (1995). *Harmony: Business, technology & life after paperwork*. Harper Collins.

Petit, P. (2001). *Economics and information*. Dordrecht: Kluwer. doi:10.1007/978-1-4757-3367-9

Porter, M. (1986). *Competition in global industries*. Boston: Harvard Business School Press.

Porter, M. (1993). *A vantagem competitiva das nações*. Rio de Janeiro: Editora Campus.

Rascão, J. (2012). *Novas realidades na gestão e na gestão da informação* (1st ed.). Lisboa: Sílabo.

Reponen, T. (1993). Information management strategy: An evolutionary process. *Scandinavian Journal of Management*, 9(3), 189–209. doi:10.1016/0956-5221(93)90016-L

Roberts, N., & Clarke, D. (1989). Organizational information: Concepts and information management. *International Journal of Information Management*, 9(1), 25–34. doi:10.1016/0268-4012(89)90034-0

Rothschild, W. (1992). *Como ganhar (e manter) a vantagem competitiva nos negócios*. São Paulo: Makron Books.

Rue, L., & Holland, P. (1989). *Strategic management: Concepts and experiences* (2nd ed.). New York: McGraw-Hill.

Vickers, P. (1984). Information management: A practical view. *Aslib Proceedings*, 36(6), 245–252. doi:10.1108/eb050930

Walsham, G. (1993). *Interpreting information systems in organizations*. Chichester: Wiley.

Ward, J., & Peppard, J. (2002). *Strategic planning for information systems* (3rd ed.). Chichester: Wiley.

Webster, F. (2000). *Theories of the information society*. Cornwall: Routledge.

Wilmot, R. (1989). Change in management and the management of change. *Long Range Planning*, 20(6), 23–93. doi:10.1016/0024-6301(87)90128-2

KEY TERMS AND DEFINITIONS

Business Management: The activity of conducting all the necessary operations in order to make the organization grow, benefiting all the shareholders and stakeholders and the society in general. It involves several specialized functions like finance, marketing, operations, human resources, production, etc.

Competitiveness: Competence of an organization or country to produce and sell products/services that meet the quality of the markets at the same or lower prices and maximize returns on the resources consumed in producing them.

Competitors: A company in the same business or similar industry which offers a similar product or service. Competitors can lower the prices of products and services and gain a larger market share, reducing clients and lowering profits.

Data: The representation of facts, concepts or instructions. Formalized with a structure suitable for communication, interpretation or processing by humans or by automatic means. The raw material of information. The basic element for the production of new information.

Information: A set of data arranged in a certain order and form, useful to people to whom it is addressed. Reduces uncertainty and supports decision-making. Information is considered to support human knowledge and communication in the technical, economic and social domains. Results from the structuring of data in a given context and particular purpose.

Information and Communication Technologies: Electronic tools based on the Internet, that allow synchronous and asynchronous communication and exchange of information—using sound, text, images, and video—between several points in different locations. Improves and stimulates the economic and societal globalization.

Information Management: Usually defined as a comprehensive organizational capacity to create, maintain, retrieve and make available the right information, at the right place, at the right time and in the hands of the right person, at the lowest cost, in the best support for its use in decision making.

Knowledge: Is A fluid composed of experiences, values, context information and apprehension about their own field of action that provides a cognitive apparatus for evaluating and incorporating new experiences and information.

Market Intelligence: Data and information collected by commercial and industrial organizations about their competitive environment to support good decision making. Makes possible to compare our market share with our competitors and take actions to maintain or improve that share.

Strategy: The way an organization chooses to do his business in order to surpass competition and gain consumers preference. What is going to do, how, for whom, using what resources, and so forth. It's the plan to conduct business, maximize profits, allocate the needed resources, and stay in the market as long as possible. Includes foresee what can happen, using forecasting methodologies and tools, in order to avoid problems or, if they happen, be prepared to minimize their effects.

ENDNOTES

[1] For ethical reasons (by request of the management board of the company) the full name of the organization will not be mentioned, but only an acronym (BV) when referring to it during this work.

[2] We were not allowed to provide the latest financial information, being the current values a pale reminder of the values between 1994 and 1999. As mentioned above, currently the volume of turnover is practically to cover the fixed monthly expenses.

[3] "Information is a process aimed at knowledge, or, more simply, information is everything that reduces uncertainty... A tool for understanding the world and the action on it" (Zorrinho, 1991, p. 21).

[4] STEEP - Sociological, Technological, Environmental, Economic and Political factors.

[5] SWOT – Strengths, Weaknesses, Opportunities, Threats.

[6] "It is no longer appropriate to consider information as simply a fourth factor in production. Corporate management is nothing else but information work. This means that information has become the prime production factor, and the purpose of information management is to make sure that it is put to good use at both strategic and operational levels" (Picot, 1989, p. 238).

[7] Since October 2012 and March 2014, the Portuguese Government granted 772 Gold Visa, being 612 of them to Chinese citizens. That decision has brought to the State about 464 million of Euros. In January and February 2014, the purchase of properties of 500.000€ and above increased about 30% in relation to those months in the previous year. From those 772 Gold Visa granted, only two were for the creation of enterprises with more than 10 employees, one of the conditions for the grant (Cf. http://www.publico.pt/sociedade/noticia/sef-1629438).

> *"Em termos acumulados - desde que os vistos 'gold' começaram a ser atribuídos, em 08 de Outubro de 2012, até Junho último -, o investimento total captado ascende a 3.895.295.041,37 euros, dos quais 366.144.760,19 euros por transferência de capital e 3.529.150.281,18 euros pela compra de imóveis."* (https://www.sabado.pt/portugal/detalhe/vistos-gold-sef-diz-que-ha-1300-pedidos-de-investidores-pendentes-em-fase-de-instrucao).

[8] "Champions are more than leaders and they are different from sponsors. Sponsors have the funds and authority to accomplish their goals. Champions, on the other hand, bring about change in their organizations in spite of having less than the requisite authority or resources" (Earl, 1998, p. 347. Cf. also pp. 348-380).

[9] "If an ISS is to be properly related to the business strategy, it is important that the Board should be closely involved" (Wilson, 1989b, p. 248).

[10] "Information Management comprises activities including the acquisition, protection, utilization, accessibility and dissemination of information, and the promotion and management of thrusts to derive maximum benefit from the resource. It also incorporates the development, management and marketing of an enterprise-wide model, and application of the principles of data management" (Ward & Griffiths, 1996, p. 360).

[11] "The machines we have invented to produce, manipulate and disseminate information generate information much faster than we can process it. It is apparent that an abundance of information, instead of better enabling a person to do their job, threatens to engulf and diminish his or her control over the situation. It is now widely recognised that stress can be experienced from a feeling of lack of control" (Edmunds & Morris, 2000, p. 18).

Section 2
Theoretical and Fundamental Contributions for a New Way to Understand Project Management

Chapter 4
4D BIM for the Management of Infrastructure Projects

Aneetha Vilventhan
https://orcid.org/0000-0001-5614-3500
Assistant Professor, Department of Civil Engineering, National Institute of Technology, Warangal, India

Rajadurai R.
Junior Research Fellow, Department of Civil Engineering, National Institute of Technology, Warangal, India

ABSTRACT

Building information modelling has become a core topic in the architectural engineering and construction (AEC) industry, and its benefits have been realised over different phases of project construction. Adoption of nD BIM in the domain of infrastructure projects has provided challenges and is yet evolving. This chapter reviews the adoption of Building information modelling in the management of infrastructure projects. The use of nD planning (4D, 5D, 6D, 7D, and 8D planning) in infrastructure planning and management is discussed through Mapping n-D BIM with different applications in infrastructure projects. 4D BIM models are developed integrating the 3D models with the schedule and they support multiple construction management tasks. The implementation of 4D planning and management in infrastructure projects is demonstrated with the help of two case studies.

BACKGROUND

The construction of infrastructure project is complex to execute and involve participation of various tradesman to perform explicit activities on challenging design constraints and on-site conditions. Infrastructure projects involve massive sets of information and huge cost implications. They often require a lot of effort to understand and communicate various sets of information among the project team to obtain desired results. This necessitates collaboration, coordination of different participants and efficient exchange media for sharing and interpretation of information. Failing causes unnecessary delays and huge cost overruns. Hence, it is necessary to manage these projects to provide efficient and profitable results.

DOI: 10.4018/978-1-5225-9993-7.ch004

BIM (Building Information Modelling) is one such platform, which can be utilised to achieve better performances in the projects (Kymmell, 2008; Kamardeen, 2010). Building Information Modelling (BIM) is a set of interrelating policies, processes and technologies that generate a systematic approach for managing the critical information for building design and project data in digital format throughout the life cycle of a building (Penttila 2006). BIM not only stores different sets of data but also allows to perform various applications over the available data in its repository. BIM models are generally n-dimensional computer-generated models representing the physical and operational characteristics of a facility in digital environment (Kulasekara, Jayasena, & Ranadewa, 2013; Philipp, 2013). One of the main characteristics of BIM application is the transparency in the management process. This allows effective collaboration of multiple stakeholders to actively visualise, participate and manage during different stages of a project (Fanning, Clevenger, Ozbek, & Mahmoud, 2014).

One of the main challenges in construction is generation and communication of quality information (Crotty, 2012). The use of BIM overcomes this through 3D visualisation of the physical facility enabling to understand and share realistic information among the project team. The ability of BIM to visualise the facility in 3D geometry makes it easy to perform various operations. It enables to perform feasibility study over constructability, satisfying both functional requirements and budget of the project (Liu, Guo, Li, & Li, 2014). It allows collaboration of different set of functional models to perform clash checks (Lee, Lee, Shim, & Park, 2012; Liu et al., 2014; Chong, Lopez, Wang, Wang, & Zhao, 2016). This eliminates many errors during construction. Further, the visualisation using BIM is not only limited for 3D representation, rather extends to visualisation of projects risk (Chang Su Shim, Lee, Kang, Hwang, & Kim, 2012), traffic impacts (Zanen, Hartmann, Al-Jibouri, & Heijmans, 2013), alignment checks (H. Kim et al., 2014) and visualising of difficult construction process (Fanning et al., 2014).

It is often essential to control the schedule and cost overruns in a project. The BIM models allow users to perform earned value analysis to measure and control the performance using specific user defined algorithms (Marzouk & Hisham, 2014). BIM enables n-dimensional simulation of real time events to assess and formulate solution to current problems. The simulation during construction stage visualises on site construction progress, helps analyse and assess optimal improvement in schedule performance (Liu et al., 2014). It also helps visualize the erection and operational process of facilities over time (Chiu, Hsu, Wang, & Chiu, 2011). Application of BIM also allows to use modelled information to use for asset maintenance and management of infrastructure projects.

This chapter discuss on various uses of n-d BIM on infrastructure projects. Applications and insights of various research efforts present over the decade are discussed. This is followed by the discussion on 4D planning and management of infrastructure projects through case studies of two ongoing infrastructure projects namely a bridge project and a road project.

N-DIMENSIONAL BIM

BIM allows linking of n-dimensional information associated in a project. It starts with three-dimensional data and currently extends up to 8-dimensional data namely 3D, 4D, 5D, 6D, 7D and 8D BIM. Each dimensional value represents different sets of information added and specific task functions which they can be used over the project's life cycle. Thereby, BIM becomes a more centralized model, allowing to assess any type of project information over the project lifecycle. The multidimensional BIM is basically achieved through appending different sets of information with the 3D models.

3D BIM

Generally, a 3D BIM model represents the physical facility in three-dimensional data namely length, breath and height. They are mainly utilised to conduct visualisation, clash analysis and design coordination (Kulasekara et al., 2013). The conventional 2D drawing is difficult to understand and communicate as the scale of the project increases. The project team has to visualise mentally and communicate the information. This incurs problem in better communication as each person has own level of imagination and perception. 3D BIM overcomes these issues through its ability in visualising any complex facility in 3D virtual environment. This enables effective understanding and communication of information among the project team.

4D BIM

A 4D BIM represents active linking of 3D BIM with the schedule (time) information. The 4D BIM is an effective way to understand the construction schedule and identify logical errors in the construction sequence (J. H. Kang, Anderson, & Clayton, 2007). This aids to improve communication among the project team and facilitate to make proactive decision and provide multi-dimensional feedback during implementation (Russell, Staub-French, Tran, & Wong, 2009). The 4D BIM models are used to create a real time simulation of construction sequence in virtual environment. They aid as a tool to visualise, track, plan the work flow and coordinate different parties in a project (Kamardeen, 2010; Kulasekara et al., 2013; Philipp, 2013; Smith, 2014).

5D BIM

The 5D BIM represents linking of cost related information with the 3D models. It enables to obtain cost estimation reports and cash flow details over the sequence of time (Kamardeen, 2010; Kulasekara et al., 2013; Philipp, 2013; Smith, 2014). 5D BIM aids the construction of personnel to manage the project cost through planning of actual cost incurred over time with the planned budgeted cost. Further, 5D simulations can be used to optimise schedule with respect to the cost constraints of the project team. The use of 5D BIM is relatively little low when compared with 4D BIM for infrastructural projects. The 5D BIM models are mainly developed during the preconstruction stage to estimate the total project cost (Chang Su Shim et al., 2012) (Chong et al., 2016). During the course of construction, the 5D simulation of BIM models are used to identify cost incurred with respect to the progress over time (Cho et al., 2011).

6D BIM

6D BIM represents linking of 3D model with the project's life cycle information. It is used in performing operation and maintenance of the facility (Zhou, Ding, Luo, & Chen, 2010; Philipp, 2013). A 6D BIM database includes information such as geometry, details of its engineering services, safety, operational manuals, product specific information. The developed as built model geometry of a facility are helpful during inspection and evaluation procedures. The ability of BIM in visual presentation allows the facility mangers to visualise the damage in 3D and takes necessary maintenance process (McGuire, Atadero, Clevenger, & Ozbek, 2016). A detailed 6D models for roads are useful in facilitating management and maintenance of underground utility facilities (Chang & Lin, 2011).

7D BIM

A 7D BIM involves incorporation of sustainability standards to the 3D BIM models. The 7D BIM models enables to visualise a buildings performance with respect to its environment and allows to perform operations relating sustainable analysis (Abanda, Vidalakis, Oti, & Tah, 2015). BIM supporting sustainability models are utilised in identifying impacts of a facility over time and aid in decision making purposes (J. I. Kim, Kim, Fischer, & Orr, 2015). They generally include analysis of carbon emission, lifecycle cost, and ecological footprints of a facility. The BIM based sustainability analysis is useful in assessing the suitable structural design over prevailing alternatives (Oti, Tizani, Abanda, Jaly-Zada, & Tah, 2016). This enables to develop an energy efficient building through comparison of cost and energy impacts for different materials (Lewis et al., 2015).

8D BIM

Recent studies on appending 8 dimensions to BIM are proposed. 8D BIM represents 3D BIM incorporated with safety standards for design and construction (Kamardeen, 2010). Linking of safety management system with the 3D models enables project team to identify safety hazards visually involved in construction of activities. Thereby, the labour forces can be instructed and trained to work with necessary safety on site (Cho et al., 2011). Automatic hazard identification in structure and visual simulation over time enable safety engineers to plan and take preventive measures before execution of construction (Zhang, Teizer, Lee, Eastman, & Venugopal, 2013). However, they require user defined algorithms to operate.

N-D BIM APPLICATIONS IN INFRASTRUCTURE PROJECTS

Pre-Construction

Visualisation

The infrastructure projects use the principle of visualisation to understand the design idea of the project in a virtual environment. The developed models using BIM enable project participants to visualise the model in 3D and get a clear picture of the design drafted in 2D drawings. The use of 3DBIM in visualising has been applied to many infrastructure projects. In Bridge and road expansion projects in the countries of China and Australia, visualisation concepts are used in the pre-construction phases (Chong et al., 2016). BIM has also been used in virtual representation of 3D models to aid design check and collision tests (Liu et al. 2014; Teall 2014). On comparison with traditional blue print studies to identify clash or collision of various disciplines, BIM platforms enable to identify and visualise in 3D space in various colour representations. Digital mock-ups are another tool used in compliance with BIM to identify design conflicts through visualisation for bridge infrastructure projects. The digital mock-up enables to assemble the 3D modelled elements like reinforcements, tendons even concrete in virtual environment. This enable the project team to visually identify design errors, interfaces between parts and modify the design before the course of construction (Lee et al., 2012). 4D visualisation is done by linking the 3D model elements to a time schedule of these elements. Then, it is available to view and visualize what is planned to be constructed at any date by the project team to allow more control over the project (Marzouk, Hisham, Ismail, Youssef, & Seif, 2010). However, they are mainly used during the construction phase of the project to aid in monitoring and controlling the project.

Apart from the general discussion in use of BIM of nD models to visualize and optimize the problems (Kumar, Cai, & Hastak, 2017), BIM to visualise underground pipeline in road construction (Chang & Lin, 2011), visualise risk information for bridges (Chang Su Shim et al., 2012), visualise the traffic impacts (Zanen et al. 2013; Chong et al. 2016), visualizing the road facilities and earthwork in highway alignment selection project (Park, Kang, Lee, & Seo, 2014) are used. In road projects, application of BIM is also applied in modelling and visualising the underground utilities with exact topology, location, size of utilities present (Chang & Lin, 2011). BIM in visualising the associated risk information in a process of construction over time are also addressed using 4D BIM models for bridge infrastructures (Chang Su Shim et al., 2012). The digital representation enables to represent variations of colours depending on severity of associated risks. As like risk, BIM to identify and visualise the impacts over time period of construction over its surrounding are explored (Zanen et al., 2013). On comparison with building projects, infrastructure projects are horizontal projects and propose a greater difficulty in visualising and handling them in 3D space. BIM enables easy visualisation of such horizontal construction in virtual environments. BIM has been used in road projects to represent the entire alignment of road in 3D space. Integration of GIS enables BIM to visualise and mark the exact topographical surface and mark position of various infrastructures respectively (Park et al., 2014). Further, BIM allows to generate and visualise exact earthwork models in desired topography (Raza, Arshad, Soo, & Won, 2017).

Engineering and Environmental Impact Analysis

BIM provides a collaborative platform to perform engineering and impact analysis over infrastructural projects. The engineering analysis for infrastructure project involves structural analysis on shear, torsion, deflection (Fanning et al., 2014) and finite element analysis using 3D BIM models. Other impact analysis using BIM include impact of construction on environmental over time. The engineering analysis ensures the structural safety and serviceability on compliance with country defined standards and load criteria (Chong et al., 2016).

Analysis to determine the lifting points of a structure and tolerance level to reduce significant structural damage can be performed using 3D BIM (Fanning et al., 2014). BIM supports sensor network to analyse significant strain calculations on the members to ensure each component can meet its design level even after installation of the structures in bridge projects (Liu et al., 2014). Commercially available softwares, like Revit and Robot Structural analysis, collaborated to perform calculations and simulations on load analysis on bridge deck structures (Chong et al., 2016).

Construction activities often constitute to various types of impacts to their surrounding environment. They involve impact on greenhouse gas footprint, acidification potential, on human heath particulate, acidification potential, eutrophication potential, ozone depletion and impact on smog. The use of BIM models helps quantify these environmental impacts occurring over the life cycle of the project (Marzouk, El-zayat, & Aboushady, 2017).

Risk Assessment and Reduction

Compared to traditional methods of risk management, BIM provides better control over risk and improves risk management among different disciplines of a project (Omoregie & Turnbull, 2016). The collaboration of risk with BIM improves coordination, communication, visualisation and decision making with the project participants (Vossebeld & Hartmann, 2016; Zou, Kiviniemi, & Jones, 2016). The visualisation of risk is achieved by linking quantified risk information with the developed 3D model of the facilities. Further, BIM allows simulation of model with colour coded risk information to interpret risk involved over the span of construction (Chang Su Shim et al., 2012).

Feasibility Study

Feasibility study refers to check the viability of the project proposal and seek for better execution plan of the proposal. The BIM models enable feasibility study over design through the digital and parametric representation of structures and improves communication among participants in choosing design options with satisfying budget constraints. BIM has been used to analyse various design options and simulation techniques to figure suitable design and construction sequence of transportation projects (Liu et al., 2014). BIM allows integration of different platforms to perform different applications. The integration of BIM and GIS (Geographical Information System) is helpful in carrying feasibility study over design and alignment of roads, bridges and (Park et al., 2014).

Design Optimisation

BIM has been studied to perform design optimisation (Lee et al. 2012; Liu et al. 2014), easy identification of design errors (Marzouk et al., 2010) and enables to revise design corrections with less efforts (C. S. Shim, Yun, & Song, 2011). The design optimisation of the structure involves enhancing the design through climination of dcsign flaws.

The digital representation of 3D model facilitates easy check of design among design participants and enable to modify or remodel the modifications at ease in design before the initiation of the construction phase (Liu et al., 2014). Use of digital mock-up practices enable to identify any errors in design and constructability issues (C. S. Shim et al., 2011; Lee et al., 2012). This enables effective revisions in the design and optimize errors.

Cost/Quantity Estimation

The 3D BIM facilitates estimation of quantity (Kubota, 2011), in general the quantities are extracted directly from the designed BIM model (Kumar et al., 2017) and used for cost estimation. The cost estimation from 3D models are mostly performed through user defined customized algorithms or techniques.

4D BIM for the Management of Infrastructure Projects

However, in case of 5D models, separate cost database is integrated with the 3D model. Moreover, both modelling can be used for deriving quantities and estimating cost.

Use of 3D BIM models in bridge infrastructure projects utilizes separate user defined attributes to estimate the direct and indirect cost associated in the construction (Marzouk & Hisham, 2014). As like estimation for bridges, BIM supports estimation of quantity and cost for highway construction projects. 3D models are used in extracting the quantities of the earthwork from the BIM generated TIN (Triangulated Irregular Networks) surfaces (Park et al., 2014). Later, estimation of cost is carried through user defined programs supporting integration with BIM models.

Use of 5D BIM has been adopted in different transportation infrastructure projects (Chang Su Shim et al., 2012; Chong et al., 2016). A 5D BIM for bridge projects is used to estimate the cost of the project by extracting the quantities from the developed 3D model (Marzouk et al., 2010). Further, the cost database in the form of sheets is used in estimating the cost of the facility. BIM enables simulation of 5D models to calculate cost requirements for construction over time before the start of construction. A 5D BIM simulation for a bridge project uses simulation to predict the expenses of the project to complete and the cost to be incurred at different milestones of a bridge project (Marzouk et al., 2010).

Clash Detection

Clash detection is performed to overcome the interface problem between various different elements in the 2D design (Liau & Lin, 2017) and to validate the design to be error free (Chang Su Shim et al., 2012). The adoption of BIM in projects would enable easy clash detention (Kumar et al., 2017). Though clash detection identifies only the interface between various elements in a model, it indirectly helps in coordination and collaboration among different participants. The use of clash detection has seen to be used during the preconstruction process in bridge projects (Liu et al. 2014 and Chong et al. 2016). BIM supporting digital mock-ups of the structures is used to find the clash or the interface between the major parts like reinforcement, tendon ducts (Lee et al., 2012).

Construction Stage

Visualisation During Construction Phase

The visualisation during construction stage would help project participants overcome the difficulties in construction. The main difficulties for the project team during construction period are understanding the design drawings and the construction sequence for complex structures. Even in current practice, use of 2D drawings is mainly used in construction. It is shared to the project team to execute the work and disseminate information. This creates difficulties on execution and rework has been observed. The ability of BIM to visualise the structure in 3D would help the project team to get a clear picture on what they are about to execute on site. This also enables to share meaningful communication to the labour crews on site. The ability of the 4D BIM in simulating the construction progress over time helps the project team understand the sequence of construction. The use of 4D BIM has been used for complex erection of steel bridge project in mountains (Chiu et al., 2011). This enabled the project team to identify the spatial conflicts and review their existing erection process. BIM also supports modelling of the surrounding environment on site and equipment like cranes and other operating machinery can be visualised in simulation (Liu et al., 2014). Apart from simulation of construction sequence, the BIM is used to visualise the site setting out points in site for a road construction project. Adoption of theodolite and BIM is used to set out site. Further, visualisation of using laser scanning in BIM is used in detection of deviation of alignment in road infrastructure projects (Chong et al., 2016).

Progress Monitoring: Cost and Schedule Control

The control over the cost and schedule is done during the construction stage of the project. The 3D BIM models aids cost and schedule by allowing users to perform earned value concepts to find out the performance of the project (Marzouk & Hisham, 2014). Though earned value metrics cannot be calculated in standalone BIM software, the BIM platform enables users to integrate used defined platforms with BIM and allows to perform desired operations.

BIM also supports monitoring of onsite construction progress through linking of BIM systems with the onsite web cameras (Chong et al., 2016). Further, BIM supporting real time site monitoring and progress control can also be affected through integrating the onsite cameras with the simulation modules developed using BIM (L. S. Kang, Kim, Moon, & Kim, 2016). This enables real time tracking of the project with respect to the virtual simulation results of the BIM models. This can be used to estimate the amount of work completed on site and delays incurred in the project. They also can be used to take necessary actions to control the progress of the project.

Process Optimisation

Process optimisation deals with the works and measures taken to improve the existing process during the course of construction. The main objective of process optimisation is to maximize the output with minimal efforts. Use of simulation techniques is mainly adopted in process optimisation in infrastructure projects. The use of simulation of BIM has been adopted to analyse the process sequence and identify the construction problem. The simulation in this process mainly focuses on enhancing the process by creating better understanding of the process to its workers. BIM supports virtual environment to visualise the real time objects such as equipment used with the real time site conditions. BIM supporting sensor system is used in monitoring the stress value on steel bridge during erection and adopts suitable construction process (Liu et al., 2014). To optimize the process, resource allocation through nD CAD simulation were also discussed (L. S. Kang et al., 2016). The resource allocation was utilized to optimize the process in accordance to mobilization of resource levels at field. The study considers 5D simulation to optimize resource, further applied the schedule applied to perform 4D simulation to optimize the process during limited supply of resources.

Project Co-Ordination

A construction project involves participation of different parties, stake holders to execute the job. However, in many cases, coordination among different parties is not way effective. This causes significant delays and confusions in work execution. 4D BIM being a simulative tool creates a transparency in the operation principles and enable different project participants to share ideas and coordinate in the work processes. Use of BIM has been proven to improve collaboration and coordination among participants to solve problems in construction of steel arch bridge (Liu et al., 2014).

Conflict Management

BIM in infrastructure helps in identifying on site construction conflicts occurring during construction process (Marzouk et al., 2010). Workspace conflicts often reduce productivity and increases the chances of safety at site. 4D BIM enable project team to identify work space conflict which occurs over time through schedule information (Moon, Dawood, & Kang, 2014). This helps the project team identify overlapping activity and reschedule to minimize the conflicts at site.

Reduction in Change Orders and Rework

In construction projects, there is no direct metric relationship between BIM applications and its effect over change orders and rework. However, BIM supporting infrastructure projects in reducing change orders is explored in academic studies. A change order and rework generally occur due to unclear information and poor understanding of the work. Implementation of BIM overcomes the above difficulties and provide efficiencies. The study found the implementation of BIM over a bridge project resulted in lower changer orders and reworks when compared with non-BIM infrastructural projects (Fanning et al., 2014).

Post Construction Stage

Inspection and Evaluation

Inspection and evaluation operations are carried out to existing structure or facility to assess its damage severity. BIM supporting inspection and evaluation are mainly carried out in the maintenance phases. The purpose of inspection and evaluation is to identify the damage locations and to assess the structural conditions of the existing structural members. Use of BIM supports to model the existing facility in 3D environment and enables to visualize them with the structural deterioration of the facility (McGuire et al., 2016). In situations where 3D modelling or model data is not viable, use of laser scanning has been adopted in many places. Various BIM platforms in the industry enables to import point cloud data and process on them. Use of such techniques of BIM and laser scanning has been adopted in road inspection procedures (Heikkila, 2013). In case of evaluation on the data available, BIM supports various user defined custom plugins and perform structural analysis or various evaluation methods and techniques (Heikkila, 2013; McGuire et al., 2016).

Assets Management or Maintenance

Maintenance activities are targeted in achieving long-term benefit to the structure, maintaining service levels while minimising the life cycle costs (Dunn & Harwood, 2015). BIM supporting management and maintenance is observed in road projects where the underground utilities are mapped and modelled to assist during maintenance phases of projects (Chang & Lin, 2011). In a study, use of BIM integrated automated systems is used to perform road repair, rehabilitation and maintenance process are found (Heikkila, 2013).

The various BIM applications used in infrastructure projects are represented and mapped with different dimensions of BIM as shown in Table 1. This chapter focuses in detail only on the applications of 4D BIM in infrastructure projects and is demonstrated through two cases studies.

Table 1. Mapping N-D BIM with different application of BIM in Infrastructural projects

BIM Applications in Infrastructural projects	3D	4D	5D	6D	7D	8D
Pre-Construction phase						
Visualisation	●	●				
Engineering and Impact analysis	●	●			●	
Risk assessment and reduction		●				
Feasibility study	●					
Design optimisation	●					
Cost/quantity estimation	●		●			
Clash detection	●					
Construction phase						
Visualisation during construction phase	●	●				
Progress monitoring - cost and schedule control		●	●			
Process optimisation		●				
Project co-ordination		●				
Conflict management		●				
Reduction in change orders and rework	●					
Post-Construction phase						
Inspection and evaluation	●			●		
Assets management or maintenance	●			●		

APPLICATION OF 4D PLANNING

Though different levels of N-D BIM exist, in practice, use of 4D and 5D BIM models is established well as compared to other multi-dimensional BIM models for building and infrastructure projects, whereas 6D, 7D and 8D are still in state of infancy as shown in Table 1 and require detailed establishments (Charef, Alaka, & Emmitt, 2018).

The rapid developments in current project scenario requires efficient techniques to visualise project over time and manage them (Hallberg & Tarandi, 2011). Though 3D models provide sufficient visualisation of a facility, 4D planning is necessary to monitor, coordinate and manage projects to achieve desired deliverables. The 4D planning is mainly used in visualisation of construction sequence and monitoring of construction progress over time. However, they act as a tool to develop collaboration, coordination, helps enrich communication among team and manage overall progress of the project over time. This section gives a detailed outlay of different applications of 4D BIM in planning and management of infrastructure projects.

4D planning and visualization has been used at various instances in the projects. The 4D planning of bridge erection sequence is performed to visualise the complexity in construction of steel bridge in mountain regions (Chiu et al., 2011). The 4D simulation further is used to identify the conflicts, issues of crane operation and helps the project team develop rectification plans. The 4D planning also allows

4D BIM for the Management of Infrastructure Projects

its user to perform different levels of visualisation such as activity level, discrete operation and continuous operation levels for better field management (C. Kim, Kim, Park, & Kim, 2011). These levels are applied for visualisation of cable stayed bridge construction in identifying appropriate level so that effective planning and communication is possible.

Apart from visualisation, 4D planning also allows visualisation of work space conflict over time. Application of certain algorithms and bounding box techniques in 4D planning has been found to identify, analyse and reduce conflicts in work spaces for bridge projects (Moon et al., 2014). The 4D planning is also applied to visualise risk information associated with the construction of individual elements over time. Quantified risk data can be linked with the 4D BIM model for bridge structures and further simulated to represent risk possibilities in construction through visualisation of different colour codes based on the severity factors (Chang Su Shim et al., 2012). The use of 4D simulation over time is also used in visualisation of impacts associated with the road construction projects. In a study, the 4D system are used in creating a visual representation of different types of impacts namely traffic, noise, vibrations concerning to the public and surrounding (Zanen et al., 2013). 4D models of earthwork cutting and filling operations of highway projects can also be simulated (L. Kang, Pyeon, Moon, Kim, & Kang, 2013).

Application of 4D planning to monitor the progress can been applied in different scenarios. A 4D model can be simulated over time to identify the planned and actual construction of bridge projects (Marzouk et al., 2010). Further, 4D models are linked with the onsite webcams to develop real time site progress monitoring for bridge projects (L. S. Kang et al., 2016). This enables the project team to communicate and interact with real time progress through web-based telepresence systems.

CASE STUDIES

In the previous section, various applications of 4D planning and management for infrastructure projects were discussed. This section will illustrate the 4D planning and management with the help of two infrastructural projects in India. The 4D BIM was used in two real time projects in India and their benefits were measured.

Case 1: 4D BIM for Construction and Management of a Flyover Project

A case study of construction of road over bridge project in India is performed to apply and analyse the benefits of 4D planning in management of projects. The case study considered is an ongoing project, involving multiple stakeholders such as owner/client, design consultants, construction contractor and other sub-contractors. The cost of the project is estimated to be 80 crore Indian rupees. The adopted delivery model of the project is Built Operate Transfer (BOT). The bridge is a reinforced concrete structure, measuring a length of 711 metres. The construction is divided into a construction of main span of the bridge with a length of 270 metres and approach roads with length of 441 metres. The main span of the bridge is divided into 9 intermediate spans of length 30 metres each. The bridge site is located in the National Highway having major connectivity roads and aims to reduce the present traffic congestion. The considered bridge is shown in Figure 1.

In India, many infrastructure projects still use 2D CAD drawings for design and construction. Use of 3D models is rarely used in practice. The project under consideration uses 2D drawings from design to construction and do not use BIM in their project execution. Initially, site observations and data col-

Figure 1. Construction of flyover bridge

lection were executed. This involves observation of site environment, worker's attitude, site constraints, on site problems, collection of dimensional data, process or sequence of onsite construction, schedule information and other necessary details. The authors used softwares such as Autodesk Revit, Navisworks to develop 3D and 4D models and Microsoft Project to develop the schedule for the bridge project.

At first, a 3D realistic BIM model with sufficient level of detail (LOD 300) is generated. It is reviewed and modifications in the models are updated. Then, a detailed construction schedule is developed and submitted to the project team to use it in their day to day practice. Further, a detailed 4D BIM model for the bridge structure is developed and a 4D simulation is performed and demonstrated to the project team. The developed 3D and 4D models are shown in Figure 2 and Figure 3 respectively.

The project members were initially unaware of the use of 4D planning and have no experience in using BIM for their projects. A basic information regarding BIM and their use in management of projects is explained and the developed models were shown in their project meetings. The project team experienced the potential of BIM to visualise the bridge with detailed information on individual structural members. The developed model shows no ambiguity on comparison with the as built structure.

The 4D BIM models helped the project team visualise the sequence of construction activities over time. The use of 4D simulations allowed the contractor to visualise the difference in planned and actual progress of the project, making them realise how far the project is delayed. The models were used in the weekly meetings to communicate the progress to the shareholders. This made the project team to easily

Figure 2. Developed 3D model of the bridge

Figure 3. 4D simulation of the bridge

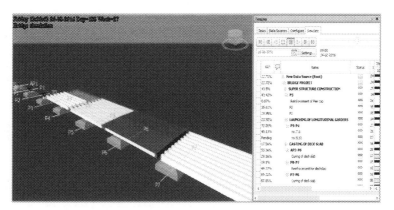

communicate the outcomes of the project and easily coordinate with different parties in improving the performance of the project.

Based on the review through simulations, catch-up plans to overcome the delays were deployed and simulated. An optimised best suitable plan of construction is selected from the simulations and used for further progress. During the simulation over time, logical errors were identified in the schedule and was updated. This reduced the RFI (Request for Information) by the contractors and improved the progress on site. Further, contractors used quantification tools for easy estimation of quantities for the future construction activities. The identified logical error in the project schedule is shown in the Figure 4.

On evaluation of the 4D models on site, the 4D planning was found to have potential in overall management of the project. They were very helpful to improve communication and coordination, enhance progress monitoring and control, optimize construction schedule and enable better documentation.

Figure 4. Identified logical errors in schedule

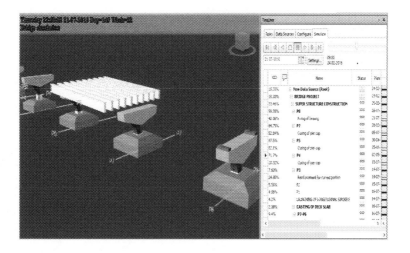

Case 2: 4D BIM for Management of Road Projects

In Indian context, application of BIM for road construction is scarcely used in practice. Hence, their benefits are not yet discovered. This case study presents the real time application of BIM and provides the potential benefits of 4D planning in effective management on road infrastructure projects.

An ongoing road construction project in India was selected for 4D BIM modelling. The potential of 4D BIM for management of road project was evaluated. The project is located in Jammu and Kashmir (J&K), India. The length of the proposed road is 7KM with a width supporting 2 lanes of 5.5 metres with 1.25 metre shoulders. The project is estimated to complete in a duration of 365 days. The purpose of the project is to provide connectivity for local regions in the state. The project is undertaken by a local contractor on BOT basis. The project team used 2D CAD drawings to execute the project and used schedule in excel sheet format.

Preliminary site observations were made and data was collected. The 2D drawings, survey point, reduced levels and schedule sequence were collected on site. The authors developed the BIM models using AutoCAD Civil 3D software. Further, a detailed schedule for the project was developed using Microsoft Project. Along with these softwares, applications such as Google earth, TCX converters were used to import and edit survey point data.

The data collected were analysed and imported to Civil 3D and a 3D road model was generated and provided to the project participants for implementation. The contractors were able to visualise the road structure to be constructed at ease. They used these models to communicate information to their labourers and coordinate other sub-contractors in work. The developed 3D geometry of the road is shown in Figure 5.

One of the main benefits for the project team is the ability of the developed models to perform cut and fill operations. This saved lot of time for the project team and helped to improve the speed of execution and reduced the work of calculation and documentation. The estimated quantity through cut and fill operations are shown in Figure 6.

Figure 5. 3D profile of the road structure

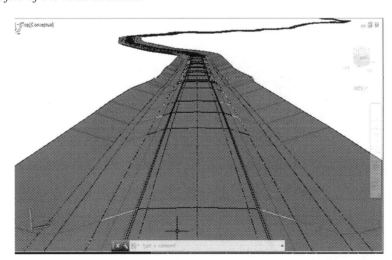

4D BIM for the Management of Infrastructure Projects

Figure 6. Estimation of cut and fill quantities

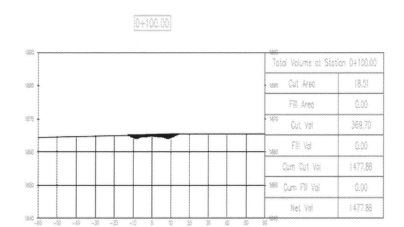

The developed detailed schedule was linked with the 3D model to create a 4D BIM model. The 4D model showcased the actual progress and helped the team list out works under delay. The team used the model to identify the quantum of delay, onsite constraints and opportunities to improve the performance of the project. These models were taken under serious considerations in the project meetings and measures to ensure on time completion were executed. There are 12 chainages with 625m length each and the simulation of layers of the chainages for the road construction is shown in Figure 7 and Figure 8.

To evaluate the real time benefits of 4D planning, a questionnaire session was finally conducted. Enquiry on the various problems faced and the solutions provided by the BIM were discussed. The project team agreed the ability of BIM to provide exact visualisation of the as built facility. The application of 4D planning helped identified the project delay and its associated issues. The results of the simulation were better utilised in meetings for communication to project participants, decision making and identifying missed opportunities on site. They provide a better solution to perform calculation and documentation.

The two case studies demonstrated in this chapter showed how 4D Modelling can be used in various applications such as visualisation of as-planned and as-built structure, schedule logical error identification, progress monitoring, improve coordination and project performance.

Figure 7. 4D simulation of the road structure

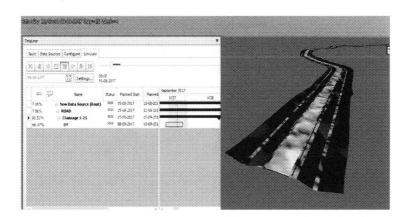

Figure 8. 4D simulation at the finish level of road

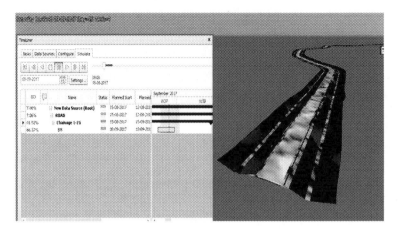

CONCLUSION

Mega complex infrastructure projects are constructed worldwide and they involve complex construction environment. Adoption of BIM in these complex infrastructure projects will increase efficiency, and enhance effective management of these projects. nD BIM namely 3D, 4D, 5D, 6D, 7D and 8D are the different dimensions of BIM in current practice of infrastructure projects. This chapter presents the various functional benefits of nD BIM through the project lifecycle. It was found that use of 4D and 5D BIM models are established well as compared to other multi-dimensional BIM models for building and infrastructure projects, whereas 6D, 7D and 8D are still in the state of infancy.

The next section discusses the various applications of 4D modelling and planning in infrastructure projects. Further, the application of 4D planning in the management of infrastructure projects is discussed with case studies of ongoing real time bridge and road projects. A full discussion on the development of 3D model, schedule, 4D model, softwares used, its applications, case study data and the evaluation of the 4D model are presented. The case studies demonstrate how 4D models can be used for visualisation during construction phase, progress monitoring, schedule control, and project co-ordination for effective management of projects.

BIM is used as a process and tool in the construction industry and forms a reliable basis for making decisions throughout the life cycle of the projects. 4D BIM Models are effective in evaluating the executability of the construction schedule and also the constructability of the designs.

ACKNOWLEDGMENT

The authors gratefully acknowledge the financial support by the Science and Engineering Research Board, Department of Science and Technology (DST- SERB), India for conducting the case studies presented in this chapter (Project File No: ECR/2017/000002).

REFERENCES

Abanda, F. H., Vidalakis, C., Oti, A. H., & Tah, J. H. M. (2015). A critical analysis of Building Information Modelling systems used in construction projects. *Advances in Engineering Software*, *90*, 183–201. doi:10.1016/j.advengsoft.2015.08.009

Chang, J.-R., & Lin, H.-S. (2011). Underground Pipeline Management Based on Road Information Modeling to Assist in Road Management. *Journal of Performance of Constructed Facilities*, *25*(August), 326–335.

Charef, R., Alaka, H., & Emmitt, S. (2018). Beyond the third dimension of BIM: A systematic review of literature and assessment of professional views. *Journal of Building Engineering*, *19*(April), 242–257. doi:10.1016/j.jobe.2018.04.028

Cho, H., Lee, K. H., Lee, S. H., Lee, T., Cho, H. J., Kim, S. H., & Nam, S. H. (2011). Introduction of construction management integrated system using BIM in the Honam high-speed railway lot no. 4-2. *Proceedings of the 28th International Symposium on Automation and Robotics in Construction, ISARC 2011*, (4), 1300–1305.

Chong, H. Y., Lopez, R., Wang, J., Wang, X., & Zhao, Z. (2016). Comparative Analysis on the Adoption and Use of BIM in Road Infrastructure Projects. *Journal of Management Engineering*, *32*(6), 0501602. doi:10.1061/(ASCE)ME.1943-5479.0000460

Crotty, R. A. Y. (2012). *The Impact of Building Information Modelling Transforming Construction*. New York: SPON Press.

Dunn, R., & Harwood, K. (2015). Bridge Asset Management in Hertfordshire - now and in the future. In *Proceedings of Asset Management Conference 2015*. London, UK: IET. 10.1049/cp.2015.1722

Fanning, B., Clevenger, C. M., Ozbek, M. E., & Mahmoud, H. (2014). Implementing BIM on Infrastructure : Comparison of Two Bridge Construction Projects. *Practice Periodical on Structural Design and Construction*, *20*(4), 04014044. doi:10.1061/(ASCE)SC.1943-5576.0000239

Hallberg, D., & Tarandi, V. (2011). on the Use of Open Bim and 4D Visualisation in a Predictive Life Cycle Management System for Construction Works. *Journal of Information Technology in Construction*, *16*, 445–466.

Heikkila, R. (2013). Development of BIM based rehabilitation and maintenance process for roads. In *Proceedigs fo 30th International Symposium on Automation and Robotics in Construction and Mining and Petroleum Industries* (pp. 1216–1222). Montreal, Canada: ISARC. 10.22260/ISARC2013/0136

Kamardeen, I. (2010). 8D BIM Modelling Tool for Accident Prevention Through Design. In *Proceedings of the 26th Annual ARCOM Conference*, (vol. 2, pp. 281–289). Leeds, UK: ARCOM.

Kang, J. H., Anderson, S. D., & Clayton, M. J. (2007). Empirical Study on the Merit of Web-Based 4D Visualization in Collaborative Construction Planning and Scheduling. *Journal of Construction Engineering and Management*, *133*(6), 447–461. doi:10.1061/(ASCE)0733-9364(2007)133:6(447)

Kang, L., Pyeon, J., Moon, H., Kim, C., & Kang, M. (2013). Development of Improved 4D CAD System for Horizontal Works in Civil Engineering Projects. *Journal of Computing in Civil Engineering*, *27*(3), 212–230. doi:10.1061/(ASCE)CP.1943-5487.0000216

Kang, L. S., Kim, H. S., Moon, H. S., & Kim, S. K. (2016). Managing construction schedule by telepresence: Integration of site video feed with an active nD CAD simulation. *Automation in Construction*, *68*, 32–43. doi:10.1016/j.autcon.2016.04.003

Kim, C., Kim, H., Park, T., & Kim, M. K. (2011). Applicability of 4D CAD in Civil Engineering Construction: Case Study of a Cable-Stayed Bridge Project. *Journal of Computing in Civil Engineering*, *25*(1), 98–107. doi:10.1061/(ASCE)CP.1943-5487.0000074

Kim, H., Orr, K., Shen, Z., Moon, H., Ju, K., & Choi, W. (2014). Highway Alignment Construction Comparison Using Object-Oriented 3D Visualization Modeling. *Journal of Construction Engineering and Management*, *140*(10), 05014008. doi:10.1061/(ASCE)CO.1943-7862.0000898

Kim, J. I., Kim, J., Fischer, M., & Orr, R. (2015). BIM-based decision-support method for master planning of sustainable large-scale developments. *Automation in Construction*, *58*, 95–108. doi:10.1016/j.autcon.2015.07.003

Kubota, S. (2011). Utilization of 3D Information on Road Construction Projects in Japan. In *Proceeding of 2nd International Conference on Construction and Project Management*, (vol. 15, pp. 21–25). Singapore: IACSIT Press.

Kulasekara, G., Jayasena, H. S., & Ranadewa, K. A. T. O. (2013). Comparative effectiveness of quantity surveying in a building information modelling implementation. In Proceedings of Socio-Economic Sustainability in Construction, (pp. 101–107). Colombo, Sri Lanka: Academic Press.

Kumar, B., Cai, H., & Hastak, M. (2017). An Assessment of Benefits of Using BIM on an Infrastructure Project. In *Proceedings of International Conference on Sustainable Infrastructure 2017* (pp. 88–95). New York: ASCE. 10.1061/9780784481219.008

Kymmell, W. (2008). *BIM - planning and manging construction projects with 4D CAD and simulations. Construction*. New York: The McGraw-Hill Companies, Inc.

Lee, K. M., Lee, Y. B., Shim, C. S., & Park, K. L. (2012). Bridge information models for construction of a concrete box-girder bridge. *Structure and Infrastructure Engineering*, *8*(7), 687–703. doi:10.1080/15732471003727977

Lewis, A. M., Valdes-Vasquez, R., Clevenger, C., & Shealy, T. (2014). BIM Energy Modeling: Case Study of a Teaching Module for Sustainable Design and Construction Courses. *Journal of Professional Issues in Engineering Education and Practice*, *141*(2), C5014005. doi:10.1061/(ASCE)EI.1943-5541.0000230

Liau, Y. H., & Lin, Y. C. (2017). Application of Civil Information Modeling for Constructability Review for Highway Projects. In *Proceedings of International Symposium on Automation and Robotics in Construction (ISARC)* (vol. 1, pp. 410-415). Taipei, Taiwan: IAARC. 10.22260/ISARC2017/0057

Liu, W., Guo, H., Li, H., & Li, Y. (2014). Using BIM to improve the design and construction of bridge projects: A case study of a long-span steel-box arch bridge project. *International Journal of Advanced Robotic Systems*, *11*(8), 1–11. doi:10.5772/58442

Marzouk, M., El-zayat, M., & Aboushady, A. (2017). Assessing Environmental Impact Indicators in Road Construction Projects in Developing Countries. *Sustainability*, *9*(5), 843. doi:10.3390u9050843

Marzouk, M., & Hisham, M. (2014). Implementing earned value management using bridge information modeling. *KSCE Journal of Civil Engineering*, *18*(5), 1302–1313. doi:10.100712205-014-0455-9

Marzouk, M., Hisham, M., Ismail, S., Youssef, M., & Seif, O. (2010). On the use of Building Information Modelling in infratructure bridges. In Proceedings of 27th International Conference-Applications of IT in the AEC Industry (CIB W78) (vol. 136, pp. 1-10). Cairo, Egypt: CIBW78.

McGuire, B., Atadero, R., Clevenger, C., & Ozbek, M. (2016). Bridge Information Modeling for Inspection and Evaluation. *Journal of Bridge Engineering*, *21*(4), 04015076. doi:10.1061/(ASCE)BE.1943-5592.0000850

Moon, H., Dawood, N., & Kang, L. (2014). Development of workspace conflict visualization system using 4D object of work schedule. *Advanced Engineering Informatics*, *28*(1), 50–65. doi:10.1016/j.aei.2013.12.001

Omoregie, A., & Turnbull, D. E. (2016). Highway infrastructure and Building Information Modelling in UK. *Proceedings of the Institution of Civil Engineers. Municipal Engineer*, *169*(4), 220–232. doi:10.1680/jmuen.15.00020

Oti, A. H., Tizani, W., Abanda, F. H., Jaly-Zada, A., & Tah, J. H. M. (2016). Structural sustainability appraisal in BIM. *Automation in Construction*, *69*, 44–58. doi:10.1016/j.autcon.2016.05.019

Park, T., Kang, T., Lee, Y., & Seo, K. (2014). Project Cost Estimation of National Road in Preliminary Feasibility Stage Using BIM/GIS Platform. *Computing in Civil and Building Engineering*, 423–430.

Penttilä, H. (2006). Describing the changes in architectural information technology to understand design complexity and free-form architectural expression. *ITcon.*, *11*, 395–408.

Philipp, N. H. (2013). Building information modeling and the consultant: Managing roles and risk in an evolving design and construction process. *Proceedings of Meetings on Acoustics Acoustical Society of America*, *133*(5), 3441–3441. doi:10.1121/1.4806085

Raza, H., Arshad, W., Soo, S., & Won, J. (2017). Flexible Earthwork BIM Module Framework for Road Project. In *Proceedings 34th International Symposium on Automation and Robotics in Construction (ISARC)* (vol. 1, pp. 410-415). Taipei, Taiwan: IAARC. 10.22260/ISARC2017/0056

Russell, A., Staub-French, S., Tran, N., & Wong, W. (2009). Visualizing high-rise building construction strategies using linear scheduling and 4D CAD. *Automation in Construction*, *18*(2), 219–236. doi:10.1016/j.autcon.2008.08.001

Shim, C. S., Lee, K. M., Kang, L. S., Hwang, J., & Kim, Y. (2012). Three-dimensional information model-based bridge engineering in Korea. *Structural Engineering International: Journal of the International Association for Bridge and Structural Engineering, 22*(1), 8–13. doi:10.2749/101686612X13216060212834

Shim, C. S., Yun, N. R., & Song, H. H. (2011). Application of 3D bridge information modeling to design and construction of bridges. *Procedia Engineering, 14*, 95–99. doi:10.1016/j.proeng.2011.07.010

Smith, P. (2014). BIM & the 5D Project Cost Manager. *Proceedia Social and Behavioral Sciences, 119*, 475–484.

Te Chiu, C., Hsu, T. H., Wang, M. T., & Chiu, H. Y. (2011). Simulation for steel bridge erection by using BIM tools. In *Proceedings of the 28th International Symposium on Automation and Robotics in Construction, ISARC 2011* (pp. 560–563). Seoul: I.A.A.R.C.

Teall, O. (2014). Building information modelling in the highways sector: major projects of the future. *Proceedings of the Institution of Civil Engineers - Management, Procurement and Law, 167*(3), 127–133. 10.1680/mpal.13.00018

Vossebeld, N., & Hartmann, T. (2016). Modeling Information for Maintenance and Safety along the Lifecycle of Road Tunnels. *Journal of Computing in Civil Engineering, 30*(5), C4016003. doi:10.1061/(ASCE)CP.1943-5487.0000593

Zanen, P. P. A., Hartmann, T., Al-Jibouri, S. H. S., & Heijmans, H. W. N. (2013). Using 4D CAD to visualize the impacts of highway construction on the public. *Automation in Construction, 32*, 136–144. doi:10.1016/j.autcon.2013.01.016

Zhang, S., Teizer, J., Lee, J. K., Eastman, C. M., & Venugopal, M. (2013). Building Information Modeling (BIM) and Safety: Automatic Safety Checking of Construction Models and Schedules. *Automation in Construction, 29*, 183–195. doi:10.1016/j.autcon.2012.05.006

Zhou, Y., Ding, L., Luo, H., & Chen, L. (2010). Research and Application on 6D Integrated System in Metro Construction Based on BIM. *Applied Mechanics and Materials, 26*(28), 241–245. doi:10.4028/www.scientific.net/AMM.26-28.241

Zou, Y., Kiviniemi, A., & Jones, S. W. (2016). Developing a tailored RBS linking to BIM for risk management of bridge projects. *Engineering, Construction, and Architectural Management, 23*(6), 727–750. doi:10.1108/ECAM-01-2016-0009

Chapter 5
Industry 4.0 in Pumping Applications:
Achievements and Trends

Luiz Eduardo Marques Bastos
Braslift - Brasil Eletromecânica, Brazil

ABSTRACT

This chapter addresses the so-called Industry 4.0 and some of its applications in industrial pumps, seeking to emphasize its characteristics and benefits. The introduction of 4.0 industry technologies in this traditional industry can cause profound changes in existing business models, providing greater customer satisfaction, either improving the effectiveness of equipment operation, contributing to better adjustment to working conditions, and also prolonging their life cycle. We are still in the early stages of these technologies and a lot is yet to evolve; however, there are already interesting examples developed by some pump manufacturers around the world, some of which will be mentioned in this chapter. It is subdivided into three main parts, namely brief historical panorama from the first industrial revolution to Industry 4.0, current applications in the industrial pump industry, and finally, future research directions and conclusion.

INTRODUCTION

The first Industrial Revolution arose about two hundred and fifty years ago and the industrial equipment production and application processes are in constant evolution: cost reduction, higher productivity rates, increased reliability and return on investment are constant drivers in the industry.

In this context Industry 4.0 comes up. It is characterized by the combination of various technologies that are gaining momentum today, such as the Internet of Things, Big Data, Cloud Computing, and the Integration of Physical and Cybernetic Systems, creating the so-called smart factories and businesses. Other technologies would be mentioned here such as artificial intelligence, advanced robotics, autono-

DOI: 10.4018/978-1-5225-9993-7.ch005

mous vehicles, 3D printing, nanotechnology, biotechnology, new forms of energy and energy storage, quantum computing, neurotechnologies, new materials, implantable and wearable technologies, connected houses, smart cities, bitcoins, blockchain and the shared economy (Schwab, 2015; Hermann, Pentek & Otto, 2015).

Two aspects distinguish Industry 4.0 from other previous revolutions: the speed, breadth, and depth of the transformations that will emerge from it; fusion of technologies and interaction between the physical, digital and biological domains (Schwab, 2015).

An important intention of such technologies is to decentralize production and allow decisions to be made in real time, based on accurate data or even autonomously by machines, according to the needs of each moment, bringing clear gains in productivity, material savings and reduction of maintenance costs, since the machines will have information to operate more efficiently and following demand fluctuations.

This is the main purpose of this chapter, i.e., to address the latest applications, advances and trends of Industry 4.0 in pumping fluids in various areas, such as process industry, water supply among others, exemplifying with case studies.

BACKGROUND

Industry 4.0: A Historical and Technological Perspective

Throughout history the technological advances have brought various benefits to mankind, among them, perhaps the most important is the labor transformation, which made it easier and more productive, through the use of devices, machines and equipments, from the simplest to the most complex, which made it possible to multiply human strength, perform tasks faster and more accurately, and harness energy sources to provide better living conditions.

Since prehistory, humans have developed artifacts to hunt better and, therefore, get more food which, together with the ability to generate fire and the use of the wheel, have provided relevant progress in that phase.

Therefore, in each historical moment, the human being develops technologies to solve the main problems of his time, but which also generate the effect of shaping social life and economic activities, thus presenting several developments that, over the centuries, have made possible the human evolution.

Regarding specifically the energy transitions, as an example, relating to what was mentioned, Smil (2010) shows that they do not occur through sudden revolutionary advances, which follow long periods of stagnation, but, on the contrary, they evolve continuously from processes that alter the composition of the sources used to generate movement, heat and light.

On the other hand, some technological advances have been striking enough to profoundly transform the world in the moments in which they've occurred. Steam machine, mass production, the parallel development of electricity and the use of petroleum as an energy source, the emergence of digital technologies and automation marked, respectively the first, second and third industrial revolutions.

Klaus Schwab, founder and president of the World Economic Forum, ratifying what was mentioned, says there are four distinct periods of industrial revolution throughout history, including the phase that is just beginning. Schwab (2015) describes an industrial revolution as the emergence of new technologies and new ways of perceiving the world that bring about profound changes in economic and social structures. For the author, the speed, breadth and depth of this revolution will force us to rethink how countries develop, how organizations create value, and what it means to be human being.

However, more recently, a major new technological transformation is taking place for the fourth time in history. It is the so-called Fourth Industrial Revolution, also known as Industry 4.0, boosted by impressive technological advances such as Cyber Physical System (CPS), Smart Factory, Smart Manufacturing, Machine to Machine (M2M), Advanced Manufacturing, Internet of Things (IoT), and Industrial Internet (Bahrin et al., 2016).

As mentioned before Industry 4.0 encompasses technologies such as artificial intelligence, advanced robotics, autonomous vehicles, 3D printing, nanotechnology, biotechnology, new forms of energy and its storage, quantum computing, neurotechnologies, new materials, implantable and wearable technologies, connected houses, smart cities, bitcoins, blockchain and the advent of shared economy (Schwab, 2015; Hermann, Pentek & Otto, 2015).

The first industrial revolution can be characterized as a profound shift from the agrarian and artisanal economy to a new phase dominated by industry and manufacturing processes. Historically, it emerged in the United Kingdom in the 18th century, and then spread throughout Europe and other parts of the world.

The main changes that took place after the so-called Industrial Revolution encompasses the technological, social, economic and cultural dimensions. Let us then succinctly describe each of these transformations. The changes in the technological field covered, among others, the following aspects: use of materials such as iron and steel; use of coal and steam as energy sources; the emergence of new machines that made it possible to increase labor productivity; new work organization that has absorbed principles such as division of labor and specialization; developments in transport and communications and, finally, the science and technology application in industrial production (Rafferty, 2018).

The same author has pointed out that there were also significant changes that encompassed other dimensions that surpassed the technological aspect, such as, among others: improvements in agriculture that made it possible to provide more food for a growing population; changes that resulted in a wider distribution of wealth and the decline of the land as the sole source of wealth, as a consequence of growing industrial production and international trade; political changes that reflected the modification in economic power, with the rise of an urban bourgeoisie; social changes, including the growth of cities and the emergence of a working class and the corresponding workers' movements, as well as the emergence of other broad cultural transformations comprising arts and behavioral aspects. Finally, there was also a significant change in the psychological nature, that is, the increase in confidence in the ability to use resources and in the nature control (Rafferty, 2018).

Although there is no consensus among authors regarding the coverage period of the second wave of transformations entitled Second Industrial Revolution and taking into account eventual overlaps with the previous phase, it is known that it began between the end of the 19th century and the beginning of the 20th century.

In the technological field some significant changes occurred compared to the previous period: the main energy sources became undoubtedly electricity, generated through various forms and petroleum, which contributed directly to the development of a new industry, the carmaker; new materials began to be used, such as metal alloys and rubber, among others; in the scope of production, besides the emergence of new tools and machines, perhaps the most relevant advent was the appearance of the assembly line, which among other aspects, enabled mass production. In the economic field there was the power shift from the United Kingdom to the United States, which consolidated its position as the great world power.

New economic theories emerged with a more interventionist bias than liberalism, especially after the Stock Market Crash in 1929 on the famous Black Tuesday when the New York Stock Exchange collapsed, triggering a deep crisis that lasted for more than one decade, entitled The Great Depression, which affected all Western countries. In the political field this period was also marked by the implantation of socialist regimes, particularly with the fall of the Russian empire in 1917, giving rise to the emergence of Soviet Union (Bryan, 2013).

In the period between the 1950s and 1970s, the Third Industrial Revolution, also called the digital revolution, began to be structured, characterized by the proliferation of semiconductors, computers, automation and robotization in production lines, with information stored and processed in digital form, as well as with the support of new communications technologies, such as mobile phones and the internet, more recently, especially since the 1990s. In parallel, changes in production methods, such as flexible manufacturing and quality control, as well as the emergence of new business strategies, with the intensification of mergers and acquisitions and the internationalization of companies, also are included those from emerging countries (Coelho,2016; Alem & Cavalcanti, 2005).

Coutinho (1992) presented an interesting synthesis of the main changes that occurred in the world since the second half of the 20th century and which intensified from the 1990s on the basis of the expansion of what he calls the increasing expansion of the electronic complex. They are: (1) the weight of the growing electronic complex; (2) a new paradigm of industrial production, with the adoption of flexible manufacturing; (3) revolution in work processes, observing greater empowerment and qualification of workers moving away from the Taylorist-Fordist paradigm; (4) transformation of corporate structures and strategies, intensifying the decentralization, outsourcing and formation of strategic alliances between companies; (5) new bases for competitiveness, concentrated in the synergistic association between research and development centers and private companies to generate innovation; (6) globalization as a deepening of internationalization, perceived, for example, through the intense interconnection between international capital and financial markets; and (7) technological alliances as a new form of competition, conducting cooperation agreements, conjoint projects, joint ventures and research consortia between two or more companies, often competing with each other.

The same author concludes his article by pointing out that the important trends of technological, business and financial reorganization intensified since the 1990s characterize the scenario of economic innovation, in the Schumpeterian terms, which was endogenously articulated (Coutinho,1992).

Finally, the Fourth Industrial Revolution arose. We will address at this point when and how the concept emerged, its main nine pillars and the ongoing technological changes and some of its economic and social impacts.

Initially, as a definition, one can understand that the combination of advanced technologies and the internet, enhancing connectivity are, once again and recently transforming the industrial landscape, what is being called the Fourth Industrial Revolution or also entitled Industry 4.0 (Perez, 2014).

Rüßmann et al. (2015) state that the Industry 4.0 concept actually represents the Fourth Industrial Revolution which is defined as a new level of organization and control over the entire value chain of the product life cycle and is geared to customer requirements increasingly individualized.

Shead (2013) notes that Industry 4.0 enables the connection between intelligent factories at all stages of the value chain and next generation automation, which began in the years 2010.

For Smit et al (2016), Industry 4.0 can be understood as the organization of production processes that are based on technology and devices that communicate autonomously in the various stages of the value chain. The so-called "smart" factories of the future will have computer-controlled systems monitoring physical processes, creating virtual copies of the physical world, and making decentralized decisions based on self-organizing mechanisms.

The term Industry 4.0 was initially created by the German government to describe the set of technological changes aimed at maintaining the global competitiveness of German industry. On the one hand, this initiative encompasses a conceptual aspect, insofar as it defines the contours of the ongoing transformations, but on the other one, it structures a set of policies to encourage research and development, with the support of government and companies (Smit et al, 2016; Santos et al, 2018).

It was in the 2011 edition of the famous technology fair which happens every year in Hannover, Germany, that the concept of Industry 4.0 was revealed to the public for the first time and, in 2013, was published the final report produced by the working group set up by the German government and composed by researchers and industry representatives (Santos et al, 2018).

According to Smit et al (2016) and Hermann et al. (2015), the main characteristics of Industry 4.0 are:

1. **Interoperability**: Cyber-physical systems (CPS) and humans in intelligent factories connect and communicate with each other.
2. **Virtualization**: It means that the CPS are able to monitor physical processes, that is, the data captured by sensors are linked to virtual plant models and simulation models, thus creating a virtual copy of the physical world.
3. **Decentralization**: It is related to the ability of CPS to make decisions on their own to produce locally through technologies such as 3D printing, not depending on centralized control systems, as the increasing demand for customized products makes it increasingly difficult to control systems centrally.
4. **Real-Time Capability**: The ability to collect and analyze data and provide data for immediate decision-making. For example, the plant may react to failure of one machine and route products to another machine.
5. **Service Orientation**: Means that companies, CPS and the humans are connected and can be triggered by other participants, which can occur internally and also across company boundaries. As a result, the operations of the product development process, including services, can be structured based on the customer's specific requirements.
6. **Modularity**: It is the ability to flexibly adapt to changing requirements by replacing or expanding individual modules that are part of intelligent plants.

Vaydia et al (2018), Coelho (2016) and Rüßmann et al. (2015) detail the nine pillars of Industry 4.0:

1. **Big Data and Analytics**: Comprehensive data collection and evaluation from a variety of sources, such as equipment and production systems, as well as enterprise and customer management systems, will become a standard to support real-time decision making.
2. **Autonomous Robots:** These machines are becoming more autonomous, flexible and cooperative, being able to interact safely with each other and with human beings, even learning from them. These are a new generation of robots with more capabilities and lower costs than those currently used by the industry.
3. **Simulation**: This concept is evolving to a level where real-time data from the physical world can be mirrored in the virtual world, enabling operators to test and fine-tune parameters in the virtual

world before being implemented in the physical environment, which will allow to increase productivity, through the reduction of setups and rework times, with a better quality.

4. **Systems Integrated Vertically and Horizontally**: In IT systems currently available, it is not always possible integration between companies, suppliers and customers, covering the whole value chain. However, in Industry 4.0 more and more effective integration will occur, providing the emergence of the true automatic value chain.
5. **The Industrial Internet of Things:** The term Internet of Things (IoT) refers to the connection of physical and virtual objects to the Internet. The expanded concept of Internet of Everything (IoE) encompasses the Internet of Service (IoS), Internet of Manufacturing Services (IoMs) and Internet of People (IoP), in other words, embedded systems that allow the integration of information technologies and communication. Characteristics such as context, which refers to the interactions that objects perform responses in case of changes in the existing environment, omnipresence, which provides information about the location, physical and atmospheric conditions of an object, as well as optimization, which corresponds to the improvements implemented in the integration between components of the system.

In summary, the value chain must be intelligent, agile and connected through the integration of objects, human beings, intelligent machines and sensors, manufacturing processes and production lines, with internal and external coverage to the company.

6. **Cybersecurity and Cyber Physical Systems (CPS):** With increasing connectivity in Industry 4.0, the need to protect critical systems and manufacturing lines has also increased significantly, resulting in more reliable and secure communications through access management systems and security. An important aspect of Industry 4.0 is the fusion between physical and virtual world, which is made possible by cyber physical systems (CPS), that is, they are computer integrations and physical processes. CPSs use sensors that connect to all the intelligence distributed in the environment in order to know it more deeply, thus enabling more precise answers.
7. **The Cloud:** Cloud computing refers to the use of memory, storage capacity, and computation of shared and interconnected computers and servers via the Internet. In summary, cloud computing enables the provision of computing services, such as servers, storage, databases, network, software, and other resources over the Internet called the cloud. Regardless of the type of business, cloud technology will act as a critical element in providing the means to innovate through new technologies. Companies that adopt cloud-based solutions will experience improvements related to response times of the order of a few milliseconds.
8. **Additive Manufacturing:** The concept of additive manufacturing refers to the generic term used to describe the manufacturing process by which various equipment known, as 3D printers operate. It is a mechanical process through which several layers of a given material are superimposed on one another in order to form an object, usually based on a digital model. Decentralized high performance additive manufacturing systems will contribute to reducing distances and eliminating the availability of materials in stock.
9. **Augmented Reality:** The concept of augmented reality is related to the technology that makes it possible to add elements of the virtual world to the physical world, that is, through a device such as a smartglasses, a tablet or even a smatphone, it becomes possible to visualize contents such as videos, images, holograms, animations and others, in a way that is superimposed on the physical

Industry 4.0 in Pumping Applications

reality, thus allowing access to tutorials related to a machine, performing inspections or maintenance assisted by specialists located remotely, as well as viewing manuals, avoiding mistakes or interventions that occur in a more effective way.

Rüßmann et al. (2015) pointed out that Industry 4.0 will surely produce impacts in some dimensions among others, based on the German experience, which can be extrapolated to the other countries' realities, with the necessary adaptations: (1) productivity; (2) revenue growth; (3) employment, and (4) investment.

The dimension of employment, given its importance, invites us to detail it. In other words, in the short term, increased automation should unleash low-skilled workers who perform simple and repetitive tasks. On the other hand, intensified use of software, increased connectivity and use of analytical tools will boost the demand for IT-qualified workers in areas such as mechatronics and robotics. In the specific case of Germany it is estimated a net increase of 6% in jobs over the next ten years (Rüßmann et al., 2015).

Cachon (2012) has developed three interesting hypothesis regarding the impact of Industry 4.0 encompassing structural, technological and organizational perspectives.

As regards the value chain, the same Rüßmann et al. (2015), in the report, point out that all of it will be affected, from the project to the post-sales service through the following changes:

1. Since manufacturing processes will be optimized through integrated IT systems, isolated manufacturing cells will be replaced by fully integrated and automated production lines.
2. Products and processes will be designed and implemented through integration and collaboration between producers and consumers.
3. Flexible integrated intelligent robots and machines can enable the economical production of small batches of customized products, which will provide a new perspective in meeting customers' needs.
4. Manufacturing processes can be modified through machines that learn and self-adjust, in an autonomous way.
5. Use of automated logistics, with the adoption of autonomous vehicles and robots, will be able to adjust that automatically to the production needs.

As a matter of synthesis, Industry 4.0 will enable faster responses to customer needs through faster, more flexible, more productive processes, increasing the quality and also creating new business models and manufacturing processes.

In synthesis, Industry 4.0 presents spectacular opportunities for innovative companies, as well as for countries and regions that are properly attuned to these changes, however, one can not ignore the enormous challenges that must be faced to implement that, as well as the existence of real threats that will be put to the laggards in such wave of transformation.

INDUSTRY 4.0 IN INDUSTRIAL PUMPS APPLICATIONS: SOME CASE STUDIES

Case studies are oftenly used in engineering because by their origin and characteristics, engineering analyzes and studies situations that already exist, i.e., real cases.

Jordan (1988) has described the basics of the problem-solving method known, which describes in a general way the stages that comprise a case study with application in the engineering as: situation — problem — solution — evaluation, which is detailed as follows:

1. Understanding the situation being studied;
2. Analyzing the specific problem to be tackled;
3. Creating, analyzing, and refining a solution;
4. And further evaluating, improving, and implementing.

This section presents some representative cases of Industry 4.0 in the application and use of industrial pumps, not related to design and manufacturing processes, which were developed by manufacturers in several parts of the world, some with global presence and others with a more concentrated presence locally or regionally. Bearing in mind that the examples and applications that will be described are not framed as trade secrets, instead, real cases are being reported and they are available in technical and commercial literature issued by manufacturers, their names will be mentioned providing additional information to interested readers.

Industrial hydraulic pumps are equipments that have wide use in the process industries, and in other applications. In fact, their use is extremely necessary in the various areas of engineering: oil and gas, water supply, sewage, drainage, irrigation, utilities and firefighting are just a few of the many possible applications for such equipments. Therefore, their use is at the heart of an industrial production process. Thus, it becomes evident that the use of the concepts of Industry 4.0 in these equipments will change intensely the way of dealing with them. It is possible to verify several cases where the application of these innovations already represents significant increases in the productivity, control and security, generating, mainly, benefits to the clients. That's exactly what this work is all about.

The market for industrial pumps has fluctuated a lot in recent years. Manufacturers around the world are looking to identify ways to recover profit margins and deliver more value to customers amid all these economic uncertainties. Issues such as Industry 4.0, Industrial IoT, Augmented Reality and others are steadily gaining momentum and pump makers are watching for opportunities which are emerging that can make better profits. In summary, this means that manufacturers and suppliers of industrial pumps are looking to improve their knowledge throughout the pumps value chain, aiming to leverage greater value delivery to customers (Goh, 2018).

Although adverse global economic conditions can lead to price wars in market disputes and thus cause sales losses, the need for reliable, economical and easy to install industrial centrifugal pumps will compensate for any decline in global demand (Hardcastle, 2015).

Case One: Smart Pumps and Services[1]

A large global manufacturer has developed a mobile app for smartphones and tablets that can determine the pump efficiency that operates with fixed rotation in just seconds. The application developed is based on an algorithm that has been integrated with a diagnostic platform offered by the manufacturer. This integrated system has already accounted for thousands of units sold worldwide. The APP does not require prior knowledge of the individual characteristic curve or any additional technical data of the motor-pump set to be measured and can also be used to evaluate pumps from other manufacturers. The following data must be entered for this purpose: electric motor rating, rated electric motor speed, head and pump flow rate.

The microphone on your smartphone or tablet is used for about twenty seconds to record the noise emitted by the electric motor fan. This noise is then analyzed by the APP to determine the exact speed of the motor-pump set and determine the torque. By correlating performance data entered by the user with the data available in a hydraulic pump systems database developed by the manufacturer, the application can conclude whether the pump is operating at partial load and informs the user if it is possible to save energy by optimizing the system.

The manufacturer has been developing a range of products and services to increase the productivity of pumping systems using Industry 4.0 principles. From the mobile application, as mentioned, which uses the electric motor noise to identify possible savings in operation, to the diagnostic platform - they all have one thing in common: they provide information and are based on functions in the form of a digital twin, which is the virtual image of a real object, such as a smart pump. Such features enable intelligent factory networking, thus maximizing productivity.

The products developed by this manufacturer seek to differ in their network capacity: QR codes made available on the pumps, which can be read by mobile devices, provide real-time process data through network connections, dynamically.

Identification occurs digitally using the serial number of the equipment, and, therefore, the amount of relevant information available about the product will obviously depend on the type of network and level of scanning.

Intelligent pumps contribute independently to optimize the operation of a manufacturing process through distributed intelligence. They do this by performing functions autonomously, such as operation control and then making these functions available to the digital twin on the network interfaces.

Smart services are based on valuable information from products that are made available on the network, such as electronic history files or monitoring operating conditions. This allows, for example, increasing efficiency during operation and maintenance.

Case Two: Predictive Condition Monitoring[2]

This is another application case developed by a large traditional manufacturer of global reach pumps. One of its energy industry customers has recently installed a multistage pump. In the first three months of operation, the equipment had two failures in the thrust bearing. In order to diagnose and solve the problem, the manufacturer's service team installed an operating variable monitoring system, which included sensors to measure bearing temperature, ambient temperature and pump shaft vibration. The data obtained were made available in real time on the site of the platform and had high alarm capacity, in case of detecting anomalies, from the analysis of the data.

Through this diagnostic system an abnormal elevation of the bearing temperature was identified when it was raised to room temperature. The alarm was given to factory operators who reported that, during the work period observed several hours earlier, the pump house where the equipment was located was closed. The collected data allowed to draw a trend graph, which can establish the direct correlation between the ambient temperature and the bearing temperatures. This information was provided to the client through a finding report, with suggestions for corrective actions. Improvements in pump house ventilation have been performed to reduce ambient temperature and the pump has been running smoothly ever since, providing satisfaction and value delivery to the end customer.

The use of cloud resources and the intelligent monitoring and diagnostics system developed by this manufacturer enabled the implementation of this solution.

Case Three: The Entire World of Pumps in One App[3]

This application was also developed by a traditional international manufacturer and aims to provide a kind of "assistant" to the client aiming to optimize design, operation and maintenance characteristics.

This tool allows offering the world of cutting-edge technology in high-efficiency pumps through smartphones and tablets sanitation, heating and air conditioning industries, as well as for pumping consultants.

It is an easy-to-use mobile application and supports the creation, customer consultation, installation and maintenance. It also offers a wide range of pump technology advantages such as energy efficiency, economy and environmentally friendly applications for the heating, cooling and air conditioning sectors.

One of the tools available in the application enables the user to calculate new projects, scale and select equipment, and finalize them later through a smartphone, tablet or desktop in the office or residence.

In addition, the solution developed by this manufacturer provides the user with direct access to the product information and technical pump data. Most content is installed on the smartphone and, therefore, is available to the user even when no Internet connection is available, such as in remote or hard-to-reach places.

There are also other features that are available through this tool:

- Identify potential savings in terms of energy costs and CO_2 emissions when using a high-efficiency, energy-saving pump.
- Pump installation, operation and manual instructions.
- Pump selection and sizing: when determining the desired operating point of the pump, the selected equipment is recommended in seconds.
- Tips and tricks to optimize certain applications.
- Unit converter for the most important physical units.
- Pipe calculator.
- Updated information and news for specialized technicians and consultants.
- QR scanner for obtaining information about the equipment.
- Augmented Reality - glasses with augmented reality technology that work as training devices. Glasses can replace paper documents and manuals, providing virtual work instructions that also display complex work steps.
- Contact the manufacturer through a quick connection.

Once again, this application uses the features of Industry 4.0 to increase perceived quality, through the delivery of added value to the customer.

Case Four: Integrated Electronic Pressurization System[4]

This application was developed by a traditional Brazilian pump manufacturer, which has a nationwide presence.

The product in question is an integrated system consisting of a self-priming multistage centrifugal motor pump, an electronic control circuit and an expansion vessel.

The main applications are related to water supply and pressurization for domestic and industrial use.

The control through the frequency inverter guarantees several functions, and perhaps most importantly, the possibility of keeping the pressure constant in the discharge, thus saving energy.

The frequency inverter allows the constant pressure of a hydraulic circuit to be kept constant by varying the frequency and, consequently, the rotation of the pump. If it would be operated without a frequency inverter, the pump could not change the rotation and when it increased the flow consumption would necessarily decrease the pressure, or vice versa, releasing high pressures with low flow rates or low pressures associated with the increase in flow.

With the rotation variation depending on the instant need of the point of consumption, the frequency inverter limits the power supplied to the equipment to the minimum condition to ensure that the consumption is satisfied. Without the inverter the equipment would always operate in the maximum power consumption.

The great advantage of this system is the possibility of remote control via wireless optimizing the operation, saving energy and protecting the equipment.

In future developments, similar systems will "learn" the operating conditions of a particular application, such as the variation of consumption during the day, or depending on the environmental conditions, thus optimizing the operation according to the variables of the applications.

FUTURE RESEARCH DIRECTIONS

This work explored the historical and technological development of Industry 4.0 and its applications in the use of industrial pumps.

As the technologies involved are quite broad and are still in their prelude, there are also numerous research possibilities to be developed within this theme.

Points that were not addressed in this work and that allow further analysis are, among others: (1) the use of Industry 4.0 resources in industrial pumps design and manufacturing processes, which is already occurring in several manufacturers; (2) the study about the possibilities offered by Industry 4.0 in relation to the integration of supply chain components, particularly in those with the potential to deliver more value to customers; (3) conducting research related to personnel training needs, particularly in undeveloped countries, which could be a huge challenge for the industrial pump business.

The suggestions registered here are intended to deepen the studies regarding Industry 4.0 and its innovative technologies, thus allowing taking advantages of its potentialities in the pump industry, benefiting the whole society.

CONCLUSION

As it was initially proposed, this work presented Industry 4.0 through a historical and technological perspective, passing through the steam engine, in the first industrial revolution until the Internet of Things and other technologies in the fourth industrial revolution, stage in which we are today, as also presented the applications of these new technologies in the pump industry, developed through four case studies.

The prevalence of pumps, their importance and their significant share in total energy consumption, as well as the degree of complexity of pump systems, require continuous optimization, leading to a reduction in the costs of pump use (Świtalski, 2014).

Modern smart pumps can make a considerable contribution towards increasing convenience and saving energy due to state-of-the-art technological advancements (Greitzke, 2007).

It can be seen, therefore, that the technological changes already reach the hydraulic pumps. As with the applications mentioned in this study, there are many others, both in industrial pumps and in other industrial equipment fields. Industry 4.0 is a present reality and an inevitable future. The speed with which innovations arise and with which changes occur requires professionals to be able to adapt themselves to new scenarios and to develop skills and competences that enable them to cope with these technologies and the challenges and opportunities they offer. Both professionals and companies that do not adapt will most likely be obsolete and will be exceeded in a short time.

Therefore, it is evident the need of study and development of devices and applications by companies, as well as the personal and intellectual training for professionals in all areas of engineering, manufacturing, operation, maintenance and commercialization of industrial pumps.

We conclude this study pointing out that we are facing a new world, practically still unexplored, that will present us with many challenges and opportunities.

REFERENCES

Alem, A. C., & Cavalcanti, C. E. (2005). O BNDES e o apoio à internacionalização das empresas brasileiras: Algumas reflexões. *Revista do BNDES, 12*(24), 46–76.

Bahrin, M.; Othman, F.; Azli, N.; Talib, M. (2016). Industry 4.0: A review on industrial automation and robotic. *Journal Teknologi, 78*(6-13),137–143.

Bryan, D. (2013). The Great (Farm) Depression of the 1920s. *American History USA*. Retrieved from https://www.americanhistoryusa.com/great-farm-depression-1920s/

Cachon, P. G. (2012). What Is Interesting in Operations Management? *Manufacturing & Service Operations Management: M & SOM, 14*(2), 166–169. doi:10.1287/msom.1110.0375

Coelho, P. M. N. (2016). *Rumo à Indústria 4.0. Tese de Mestrado*. Faculdade de Ciências e Tecnologia. Departamento de Engenharia Mecânica. Universidade de Coimbra.

Coutinho, L. (1992). A terceira revolução industrial e tecnológica: As grandes tendências de mudança. *Economia e sociedade. Revista do Instituto de Economia da UNICAMP., 1*(1), 69–87.

Goh, J. (2018). Internet of Things and the pump market: How the industry will evolve. *Market Insight. IHS Markit*. Retrieved from https://technology.ihs.com/599560/internet-of-things-and-the-pump-market-how-the-industry-will-evolve

Greitzke, S. (2007). Intelligent pumps for building automation systems. *World Pumps, 490*, 26–32.

Hardcastle, J. L. (2015). IoT Shakes Up Global Pumps Industry. *Environmental Leader*. Retrieved from https://www.environmentalleader.com/2015/07/iot-shakes-up-global-pumps-industry/

Hermann, M., Pentek, T., & Otto, B. (2015). Design principles for Industrie 4.0 scenarios: a literature review. *Working Paper nº 1. Technische Universität Dortmund*. Retrieved from https://www.researchgate.net/publication/307864150_Design_Principles_for_Industrie_40_Scenarios_A_Literature_Review

Jordan, M. P. (1998). How can problem-solution structures help writers plan and write technical documents? In L. Beene & P. White (Eds.), Solving Problems in Technical Writing. Academic Press.

Perez, C. (2010). Technological revolutions and techno-economic paradigms. *Cambridge Journal of Economics*, *34*(1), 185–202. doi:10.1093/cje/bep051

Rafferty, J. (2018). Industrial Revolution. *Encyclopaedia Britannica*. Retrieved from https://www.britannica.com/event/Industrial-Revolution

Santos, B. P., Lima, T.D.F.M., & Charrua-Santos, F.M.B. (2018). Indústria 4.0: Desafios e oportunidades. *Revista Produção e Desenvolvimento.*, *4*(1), 111–124.

Schwab, K. (2015). *The fouth industrial revolution*. New York: Penguin Random House.

Shead, S. (2013). Industry 4.0: the next industrial revolution. *Engineer (online edition)*. Retrieved from http://www.theengineer.co.uk

Smil, V. (2010). *Energy Trasitions: history, requirements, prospects*. London: Praeger Publishers.

Smit, J., Kreutzer, S., Moeller, C., & Carlberg, M. (2016). Industry 4.0. *Study issued by Policy Department A: Economic and Scientific Policy. European Parliament.* Retrieved from http://www.europarl.europa.eu/studies

Świtalski, P. (2014). Intelligent pumps: Do they exist? *World Pumps*, *2014*(3), 34–36. doi:10.1016/S0262-1762(14)70052-5

Vaidya, S., Ambad, P., & Bhosle, S. (2018). Industry 4.0 – A Glimpse. *Procedia Manufacturing 20. 2nd International Conference on Materials Manufacturing and Design Engineering*, *20*, 233-238.

KEY TERMS AND DEFINITIONS

App: A mobile application or mobile application is software designed to be installed on a mobile electronic device such as tablets and smartphones. Apps are commonly known as apps or mobile app.

Flow Rate: The capacity or flow rate is the fluid flow in volume per unit of time that the pump transfers from a point of lower energy condition to one of higher energy condition.

Frequency Inverter: Generally, a frequency inverter, also known as VFD, or Variable Frequency Drive is an electronic device capable of varying the speed of rotation of a electric motor.

Head: It can be defined as the energy per unit per unit of weight that the fluid will require from the pump to be pumped at a given flow rate.

Industrial Pump: It is an equipment which is used for a broad range of applications across many industries. An industrial pump is used to displace many different types of products, including water, chemicals, petroleum, wastewater, oil, sludge, slurry, food, or others. Usually the categories used in industrial applications include centrifugal pumps and positive displacement pumps.

Industry 4.0: Also called Fourth Industrial Revolution is an expression which encompasses some technologies for automation and robotics that uses concepts of cyber physical systems, internet of things, cloud computing, and other technologies.

Information Technology: This is abbreviated as IT. It is the broad theme that encompasses all aspects of generation, processing, distribution, and information management in an organization.

ENDNOTES

[1] This solution was developed and provided by KSB.
[2] This solution was developed and provided by ITT.
[3] This solution was developed and provided by Wilo.
[4] This solution was developed and provided by Famac.

Chapter 6
Are Family Businesses a Good Environment for Project Management?
Non-Technological Factors Affecting Project and Knowledge Management Practices Within Family Firms

Filippo Ferrari
https://orcid.org/0000-0002-7509-2320
Independent Researcher, Italy

ABSTRACT

Relationships between project management, operations management, and organizational strategy are well-known, as well as organizational influences on project. Family businesses work on projects, but their unique nature makes family firms a challenging context for Project Management. This chapter aims to present and discuss the specific dynamics of family business that can impact project management practices. By definition, a project is a complex system, consisting of a set of dozens of interrelated sub-processes. As is known, the percentage of projects that satisfy both technical requirements, budget compliance and which meet the deadlines, is extremely low. This fact forces the researchers to equip themselves with more sophisticated tools to face the complexity of a project, in order to increase its chances of success.

INTRODUCTION

Since family businesses have a high economic impact in most countries (Beckhard & Dyer Jr., 1983; Shanker & Astrachan, 1996; Feltham, Feltham, & Barnett, 2005; Astrachan & Carey Shanker, 2006; Donckels & Fröhlich, 1991; Corbetta & Montemerlo, 1999) such enterprises deserve scholars' attention not only from the economic and financial but also from the organisational and human resource management point of view. Nevertheless, a systemic understanding of the family firm's business model and

DOI: 10.4018/978-1-5225-9993-7.ch006

its relationship with performance has not yet been satisfactorily developed in the relevant literature. In particular, issues related to knowledge and project management within family firms are until now not conclusive (Sadkowska, 2017).

Relationships between project management, operations management, and organizational strategy are well known, as well as organizational and environmental influences on project management (PMBOK, 6th Edition, Chapter 2). The technologies and work processes proposed by the organization 4.0 are ineffective if the system of interpersonal relations hinders the free circulation and co-construction of knowledge.

Family businesses work on projects, but their unique nature makes family firms a challenging context for Project Management. For instance, family firms are peculiar regarding enterprise environmental factors as organizational culture, structure, and governance (vision, mission, values, beliefs, cultural norms, leadership style, hierarchy and authority relationships, organizational style, ethics, and code of conduct: Gomez-Mejia, Makri, and Larraza Kintana, 2010; Stewart, Hitt, 2012). Furthermore, they are peculiar about how they treats organizational capability (existing human resources expertise, skills, competencies, and specialized knowledge): They are unique also in managing organizational knowledge (Chirico, 2008; Chirico, Salvato, 2008) and in setting organizational objectives (Berrone et al., 2012; Chua, Chrisman, & Sharma, 1999). Moreover, family owners-entrepreneurs often show peculiar self-centred leadership style, potentially in contrast with project leader's management action (if the project leader doesn't belong to the family especially). In addition, family business literature clearly shows family firms are particular in managing relationships with most important stakeholders (Berrone et al., 2012; Oomen, Ooztaso, 2008; Sadkowska, 2017) and in managing procurement processes.

In a nutshell, family firms usually show a lower level of *professionalization* and *managerialization* (Stewart, Hitt, 2012) in comparison with non-family ones, and this fact can easily undermine project and knowledge management practices. From this point of view, family businesses represent an entirely unique arena. For instance, regarding the knowledge sharing practices, project team and, more in general, [small] family firms seem to be in a favourable position, due to their (small) size and strong everyday relationships among family members (Bjuggren et al. 2001; Kogut & Zander, 1992), therefore knowledge transfer is favoured in these firms (Sirmon & Hitt, 2003). Moreover, several factors such as commitment, confidence, trust, reputation, and strong sense of identity can play a role in making knowledge sharing easier during business succession between involved generations (Cabrera-Suñez et al., 2001; Sirmon & Hitt, 2003; Zahra et al., 2007).

However, in spite of this favourable situation, literature highlights that within family firms the knowledge sharing process often fails: an important reason for this is the lack of consideration for the organizational and interpersonal context influences (Carter & Scarbrough, 2001; Martinez-Barroso et al., 2013; Voelpel, Dous, & Davenport, 2005). Project management practices, by definition strongly based on quality, time and cost objectives, may be negatively affected by the overlapping enterprise-family, which emphasizes contextual, cultural non-economic aspects.

The overlapping between ownership and management is a pivotal contextual characteristic: since the seminal study of Levinson (1971), literature has suggested that there should be increased conflict in family businesses compared to non-family because of the underlying role conflict when family members work together. Moreover, the lack of managerial skills of family businesses (Cagliano, Spina, 2000) can hinder the adoption of advanced managerial practices such as those required by project management. For all these reasons, the family business becomes a particularly challenging context for project management. However, from a methodological point of view, the peculiar nature of family firms challenges scholars, and forces them to scratch beneath the surface of this organizational phenomena. The problem, for scholars and practitioners, is not simply to investigate managerial practices and technological-organizational innovations. The family business has specificities that make it unique, and that can hinder the implementation of a mature project management system. For instance, Berrone and colleagues (2010) call for major attention to be paid to some family firm specific features such as; feelings, emotions, and relationships within the family controlling a firm. These features may vary from one firm to another and, within the firm, from one point in time to another (Hoy & Sharma, 2010), even with the same level of ownership.

In summary, a family firm is a very complex system: multilevel crossroads involving business and family goals, different generations and often different blood lines and siblings. In a family firm, at organizational level, the analysis unit is the overlapping area between family and business. At this level, attempting to obtain an adequate balance between economic and non-economic goals plays a prominent role. For instance, recent literature has investigated the non-economic advantages that family member owners derive from corporate control (Berrone et al., 2010; Gomez-Mejia et al., 2007; Zellweger & Astrachan, 2008). However, literature suggests that there is also a *dark side* of trust and involvement that can lead to opportunism, complacency and blind faith, (Eddleston & Kidwell, 2012; Steier, 2001; Sundaramurthy, 2008), and can also negatively affect proactive stakeholder engagement (Kellermans et al., 2012). Since, as is known (PMBOK 6th Edition), contextual factors have gained increasing attention from PM scholars and practitioners, it becomes necessary to identify the specific features showed by a family firm oni adopting project management & knowledge sharing practices.

This chapter aims to present and discuss the specific features and dynamics of family business that can affect project and knowledge management practices.

By definition, a project is a complex system, consisting of a set of dozens of interrelated sub-processes (PMBOK, 6th Edition). As is known, the percentage of projects that satisfy both technical requirements, budget compliance and which meet the deadlines is extremely low (US Defense Department, 1989). This fact forces the researchers to equip themselves with more sophisticated tools to face the complexity of a project, in order to increase its chances of success.

The chapter is organized as follows. A first part presents and discusses the characteristics of the family business as a specific environment in which project management practices can be hindered or reinforced. In this paragraph, definitions of family business are given, and the strategic aspects and priorities of governance are highlighted.

The next paragraph presents the available evidence on project & knowledge management in family businesses, with particular attention to the socio-relational dynamics that can impact on project and knowledge management.

The third paragraph presents the differences between family and non-family businesses, identifying three non-economic objectives pursued by family businesses that can constitute an obstacle to project and knowledge management practices. Finally, in this section, we formulate three statements describing the specific relationship between family and project management.

The fourth paragraph highlights some attention points in project management body of knowledge related to the characteristics of family businesses, for each proposed theoretical proposal.

The fifth paragraph concludes.

FAMILY FIRMS AS A SPECIFIC ENVIRONMENT FOR PROJECT AND KNOWLEDGE MANAGEMENT

What a Family Firm is

Although family business is a well-developed field of research, scholars are not unanimous in defining a family firm. Current literature (Cortesi, Alberti, Salvato, 2004) however, suggests, at least three different definitions, from a broader one ("a company is defined as a family when it responds jointly to two criteria, a certain degree of control over the company's management by a family and that this control remains in the future") to a more restrictive one, which requires the possession of three other characteristics:

1. A single family must share at least the 50% of the ownership.
2. The strategic decisions must be managed by the family.
3. At least two generations belonging to the same family must be present/involved directly in the firm

The size of the company is not an important factor, even if in small companies a significant family control prevails. The family business is in fact characterized by the overlap of two different institutions (family and business) that have different objectives by definition. A company has economic objectives, which it must achieve in order not to be expelled from the market; the family, on the other hand, has objectives for the care and protection of its members. The consequences of this overlap have been widely debated in the literature, especially in the attempt to describe the differences between family and non-family businesses.

This paper follows Chua et al. (1999, p25), by defining a family business as: "a business governed and/or managed with the intention to shape and pursue the vision of the business held by a dominant coalition controlled by members of the same family or a small number of families in a manner that is potentially sustainable across generations of the family or families".

Obviously, in reality there are many family businesses that correspond to the aforementioned definition but which are different from each other, both on the business plan (total number of employees, level of structuring) and on the family plan (which and how many generations and siblings are present in the company). It is, therefore, very risky to put big companies on the same level, of which the family owns the shareholding, but the company is in fact managed by managers, and small businesses, in which the family owns, but is also involved in management and also carries an operating contribution.

Family businesses also differ significantly in terms of their cultural orientation, that is to say values that are given priority and importance. In this sense, there are business-oriented businesses, in which the family-relational element and related objectives (care and well-being of family members) is completely neglected. On the contrary, there may be companies strongly oriented towards the family, for which the highest priority is the conservation and protection of the well-being and subsistence of family members. Synthetically, it is possible to identify three different conceptions regarding the family-business relationship (Corbetta, 1995).

1. **The Family and the Business are Totally Overlapping:** This situation is typical, usually, in the very small enterprises, where the ownership coincides with the management and, often, with the direct supervision of production. The work and the patrimony of the members of the family are at the base of the existence of the company itself. The company is managed according to family

logic, which is questioned only on the occasion of the generational change, often very difficult, given the total identification between the founder and the company. In this situation it is likely that managerial practices are applied with greater difficulty. Furthermore, relations with managers outside the family can become problematic in the case of conflicting strategies.
2. **The Family Prevails Over the Company:** At the base there is a misunderstanding about the exact role of the company, considered submissive to the needs of the family. In this scenario, in the event of a conflict between business management and family dynamics, the latter will always be privileged, giving up opportunities for entrepreneurial development.
3. **The Company is Independent of the Family:** The company is seen as an independent entity, managed in the economic interest of those who invested time and money in it. The company must prosper regardless of the people represented in it and the family involved in it. In this third situation, in practice, there is no difference (with non-family companies) regarding knowledge and project management.

Many researches have considered this dualism between family and business to explain the functioning of family businesses (Dunn, 1995; Poza, 2007; Reid et al., 1999; Ward, 1987), emphasizing from time to time the advantages and disadvantages of the prevalence of an orientation to the family with respect to a business orientation. Other approaches have instead considered the two dimensions as systemically correlated (Dyer, 2006, Habbershon et alii, 2003, Lubatkin, Durand and Ling, 2007, Simon and Hitt, 2003, Stafford et al., 1999), underlining the mutual interactions between aspects of business and instances of care and protection of family members. A first question that the research has asked is the following: for a family business, is a business or family orientation better? Is it possible to obtain high performances both as a family (care and protection of members) and as a company (profits, market share, company longevity)? And, more specifically, how are the most advanced managerial practices (eg project and knowledge management) influenced by the specific context of the family business?

Governance and Managerial Practices in Family Firms

Indeed, the ability to work for projects and the attention paid to knowledge management practices represent an added value for companies. However, the literature suggests that the overall level of managerialisation of family businesses is very low (Cagliano, Spina, 1990). The research so far has started from the hypothesis (Ward, 1987, Poza, 2007) that family businesses that pay the same attention to both business and family relationships have better performances as families. Furthermore, it is generally believed that these family businesses have equal business performance compared to companies that focus solely on the business. The literature (Basco, Pèrez-Rodriguez, 2009), examined the management of both systems (family and company), in the four main areas of study in the field of family business: strategic processes, governance, human resources and business transfer. The choices made by family companies in these areas can be indicators of their overall orientation (to the family, to the business).

The *strategy* of family businesses (that is, everything concerning the company's medium to long-term fundamental goals) is perhaps the main topic of family business studies (Debicki et al., 2009). However, the literature has paid more attention to the formal aspects of the strategy (for example, the existence of a business planning or market positioning. (Daily, Thompson, 1994; Gudmunson et al., 1999) and much less to the management processes involving the whole family. Sharma, Chrisman and Chua (1997), in fact, suggest that these processes are strongly influenced by the family.

Regarding the structure and functioning of the management team, Cabrera Suàrez, Santana Martin (2000), as well as Schwartz, Barnes (1991) stress that the family-business relationship has a typical function of governance, in the aspects of control, service and resources. For example, the management team develops control functions related to the business (monitoring of results, assessment of managers) but also related to the family (controls the interests of the family in the company) as well as service functions (manages conflicts, plans the succession).

Research on the topic of *human resource management* has clearly highlighted the differences between family and non-family businesses, since the study of Lansberg (1983) and subsequent research (Astrachan, Kolenko, 1994; Cromie, Stephenson, Monteith, 1995; Lèon-Guerreno, McCann, Haley, 1998; Kok, Uhlaner, Thurik, 2006), who have shown that family businesses make less recourse to managerial practices in human resources management, probably due to the contradiction between merit principle and equality principle, with this second to characterize family businesses (Lansberg, 1983; Ferrari, 2014).

The business succession is extensively studied in literature (Ferrari, 2005), in particular, as regards the time of the successor's entry (Barach et al., 1988; Cabrera Suarez, Garcia-Falcòn, 2000), his experience (Fiegener et al., 1996; Kets de Vries, 1993; Ward, 1987), training and development (Boyd et alii, 1999; Fiegener et al., 1994; Goldberg, 1996; Lansberg, Astrachan, 1994; Seymour, 1993). Many authors have stressed that the succession process is determined, in form and content, by the decision to keep the company in control of the family, or to resort to an external manager: In the first case, the path will be focused on the needs of the future successor, in the second case, on the needs of the business.

In conclusion, Ward (1987) suggests that every company has three choices regarding the family-business relationship: the company first of all, the family first of all and the family business first of all. Poza (2007), in a very similar but perhaps more specific way, speaks of family priority, ownership priority (*ownership first*), management priority (*management first*). In agreement with Ward, families need to be aware of their type of approach when making decisions that affect the family business, as the type of approach influences or even determines decisions relating to issues such as ownership structure, strategy, staff policies (hiring, rewards).

Probably, as Dunn (1995) suggests, there is a theoretical continuum that goes from one extreme, dominating the priority to the business, to the other extreme, in which attention to the family dominates. The decisions made in the four areas mentioned above (strategic processes, governance, human resources and business transfer) reveal to what extent the continuum is a specific family business. In addition to being aware of their approach to the situation (family first of all, ownership first of all, management first of all) it is necessary that family businesses are aware of the possible conflicts of interest that arise from this approach. A favourable choice for the family can damage the company (the most typical example is the entry into the company of children with inadequate curriculum, thus generating a high and useless cost on the business plan) or a decision favourable to the management can be unfavourable to property (the most typical example is the decision to reinvest profits, a choice favourable to management, but not to property, which profits would like to enjoy). In the small enterprise, in which family, ownership, governance and operations overlapping, the conflict of interests is therefore, fourfold, and the criticalities are obviously greater.

It is therefore necessary to identify the impact that the specific dynamics of family businesses have on project & knowledge management practices. The next section deals with this.

PROJECT AND KNOWLEDGE MANAGEMENT IN FAMILY FIRMS

In order to apply the project management approach successfully, to create and share organizational knowledge repositories for storing and retrieving information is strategic. Examples are (PMBOK ® 6th Edition, p. 41) configuration management knowledge repositories containing the versions of software and hardware components and baselines of all performing organization standards, policies, procedures, and any project documents; financial data repositories containing information such as labor hours, incurred costs, budgets, and any project cost overruns; historical information and lessons learned knowledge repositories (e.g., project records and documents, all project closure information and documentation, information regarding both the results of previous project selection decisions and previous project performance information, and information from risk management activities); issue and defect management data repositories containing issue and defect status, control information, issue and defect resolution, and action item results; data repositories for metrics used to collect and make available measurement data on processes and products; and project files from previous projects (e.g., scope, cost, schedule, and performance measurement baselines, project calendars, project schedule network diagrams, risk registers, risk reports, and stakeholder registers).

Organizational Learning Theories applied to family business (Moores, 2009) assert that knowledge is a fundamental asset in ensuring a firm's development and survival (Penrose, 1959; Nonaka, 1991; Nonaka, Takeuchi, 1995; Grant, 1996; for a review see Wang & Noe, 2010). Therefore, transferring, sharing and even re-creating such knowledge between people who are involved in projects plays a fundamental role in ensuring reliability to organizational processes (Cabrera-Suarez, 2001; Koiranen, Chirico, 2006; 2008; Le Breton-Miller, 2004; Zahra et al., 2007). This knowledge transfer does not occur automatically within family firms: for instance, research shows that one of the main causes of firm failure, that occurs recursively, is the successor's business incompetence (Carter, Van Auken, 2006; Gibb, Webb, 1980; Kucharska & Kowalczyk, 2016), thus suggesting that knowledge is not always properly transferred between generations.

Literature suggests that family firms show several advantages in organizational learning processes. For instance, Cognitive Social Capital Theory highlights the central importance of networks of strong, cross-generational personal relationships developed over time that provide the basis for trust (Nahapiet & Ghoshal, 1998). Hoelscher (2002) described family capital as a special instance of social capital, more intense, enduring, and immediately available. Furthermore, transferring knowledge internally sets the basis for innovating and improving efficiency, thus realizing the potential, specific value of that knowledge (Davenport & Prusak, 1998). In summary, regarding the knowledge sharing process, small family firms seem to be in a favourable position, due to their (small) size and strong everyday relationships among family members (Bjuggren et al. 2001; Kogut & Zander, 1992), therefore, knowledge transfer is favoured in family firms (Sirmon & Hitt, 2003).

Organizational and Socio-Relational Dynamics Influencing Project and Knowledge Management

However, statistics show (Cabrera-Suarez, 2001; Koiranen, Chirico, 2006; Le Breton-Miller, 2004) that organizational failure can be explained by the scarce knowledge and skill levels of people involved in organizational processes. Often, family firm failures could be due also to the involved generations' lack of capacity and/or willingness to create, share, transfer and acquire the appropriate knowledge from generation to generation (Hatak, Roessl, 2015; Kellermanns & Eddleston., 2004; Szulanski, 1996). Ortenblad (2001) indicates that organizational learning refers to the activity of the learning process within organizations. Learning integrates new individual knowledge and/or combines existing knowledge in different ways, leading to innovation and fostering organizational long-term reliability. This leads to a double approach in the study of organizational learning: an *individual* one – focused on exploring organizational learning as individual learning in an organizational context - and an *organizational* approach – seen as more than the sum of individual learning experiences (Ortenblad, 2001; Antonacopoulou and Chiva, 2007). Not surprisingly, Wang and Chugh (2014) considered the relationship between individual and organizational learning a key aspect of overall entrepreneurial learning. From a *Cognitive Social Capital* point of view (Nahapiet and Goshal, 1998), weaknesses in knowledge management could be due to both personal characteristics of the involved people (e.g. see Barnes, Hershon, 1976) and contextual/environmental influence (e.g. see Overbecke et al., 2015). Family business literature has until now analysed many dynamics in relation to knowledge transfer (see Barroso- Martinez et al., 2013 for a review). In order to explore knowledge management process, literature analysis suggests following critical dynamics should be considered. In absence of planned formal/structural knowledge sharing processes (Cabrera-Suarez, 2001; De Massis et al., 2008), the knowledge management process does not activate. To facilitate effective social interactions among individuals, organizations need to have *effective systems for knowledge sharing* (Jones and Macpherson 2006). Furthermore, dynamics as *conflicts/rivalries/competition in parent-child relationships* (Sharma et al., 2001; Sveen & Lank, 1993; Venter et al., 2005) can hinder entrepreneurial knowledge sharing between family members. Empirical evidence also suggests *not clearing the role* played by involved people in a project may hamper the focus on the knowledge to be developed (Le Breton-Miller et al., 2004; Organ, 1990; Rosenberg, 1991).

Literature also considered further socio-relational dynamics undermining knowledge management and sharing within family firms. For instance, *lack of trust* in younger people by non-family members, or *conflicts* between family members and non-family members can create problems in the knowledge management process (Cespedes and Galford, 2004).

Regarding the specific process of entrepreneurial knowledge development, Politis (2005) emphasizes that entrepreneurial knowledge is based on the ability to transform experiences into knowledge. Therefore, a lack of next generation's *career planning*, or late or insufficient exposure of the potential successor(s) to the business (Cabrera-Suárez et al., 2001) may delay or prevent the knowledge management process. Moreover, *failure to train* potential successor(s), not giving the potential successor(s) sufficient feedback about the succession progress and *incorrectly evaluating the gaps* between the potential successor's needs and abilities, can undermine the knowledge management process (Le Breton-Miller et al., 2004).

An overall analysis of the SEW literature seems to suggest that a firm first orientation is appropriate for a family business. Socio-relational factors have a negative impact on managerial practices. In particular, the principle of belonging in contrast with the principle of merit conflicts with the PM as a system strongly oriented to responsibility and results. The family first logic could also reward, in the choice of the project manager, a family member rather than a non-member, although more prepared. Finally, clientelism could harm the relationship with stakeholders, especially in project procurement processes. In the light of these criticalities, which emerge from a comparison between family and non-family businesses, the next section then explores the differences between family and non-family firms.

WHY CAN'T A FAMILY BUSINESS BE MORE LIKE A NONFAMILY BUSINESS?

As explained in a previous paragraph, at the governance level, the overlap between family and business can give rise to three different governance structures: total overlapping, the family's prevails over the firm, and the firm is independent from the family. Historical, geographical and cultural factors could affect the resulted governance. In Latin culture, for instance, often the family prevails over the firm, showing a prevalence of relational aspects over professional, business-oriented ones (Pellegrino, Zingales, 2014).

Recently, some authors (Steward, Hitt, 2012) have investigated the difficulties for family businesses to develop an adequate level of managerialization. Many researchers believe that family businesses would perform better if they behaved like non-family businesses. In this regard, the focus is on managerialization, which family firms would be lacking, and this theoretical position is very common in literature (Rondoy, Dibrell, Craig, 2009; Schulze et alii, 2001; Sciascia, Mazzola, 2008). An important conclusion drawn from all these studies is that the managerialization of family businesses would improve their performance: but, in spite of this, most family businesses fail to follow this requirement.

A family business can exclude the family from its management philosophy, but, obviously the opposite cannot happen, that is, the company of course can't be excluded from the managerial approach, otherwise it could no longer be called a family business. Family businesses, therefore, can, in fact, be divided into family businesses that pay attention only to business and family businesses that pay attention to both the business and the family. Therefore, a further step in the study of a family business is to measure its performance both from the family point of view (Sorenson, 1999; 2000) and the company (Gupta, Govindarajan, 1984) (see Table 1)

Furthermore, current literature in the area of knowledge and project management pays close attention to human involvement in project management. But, at the same time, this literature proposes a predominantly engineering and organizational approach, with limited contributions from other disciplines (occupational psychology, sociology). And this is a clear limitation of current literature, which this chapter wants to overcome. Indeed, all organizations, and family firms, in particular, have *non-economic goals*: this is a well-established fact in managerial literature (e.g. Astrachan Jsaskiewicz, 2008; Berrone et al., 2012; Chua, Chrisman, & Sharma, 1999; Zellweger & Astrachan, 2008). In pursuing these goals, organizations are particularly exposed to relationship conflicts (Davis & Harveston, 2001; Levinson,

Table 1. Business performance and family performance (source: Basco, Pèrez-Rodriguez, 2009)

Business performance indicators	Family performance indicators
Sales growth Market share Net profit Cash flow Profit:sales ratio Return on investment Product development Market development Adapting to client needs Reduction of costs Staff development Environmental protection Customer satisfaction Service quality	Money available for family Quality of life at work Enterprise generates family security Enterprise interest in the family Time to be with family Family loyalty and support Family unity Respected name in society Customer loyalty to family name Good reputation in the business community Family interest in the enterprise Development of children's skills Generate possibilities for the children

Table 2. Stereotypical differences between family and non-family businesses

	Nonfamily business	Family business	Representative citation
Ownership	Dispersed, non-kinship based No wedge between cash flow and ownership rights Well diversified	Concentrated, kinship based Wedge between cash flow and ownership rights Nondiversified	Achmad et al. (2009) Morck et al. (2005)
Governance	Ownership and control split External influences on board Transparency, disclosure	Ownership and control united Internal dominance of board Opaqueness, secrecy	Andres (2000) Slrrnen et al. (2008} Parada et al. (20 I O) Gedajlovk et ai. (2004)
Returns	Largely economically defined No private benefits Minority shareholders protected	Non-economic outcomes important Private benefits for family Minority shareholders exploited	Chrisman et al. (2010) Anderson and Reeb (2003a) Martinez et al. (2007)
Rewards	Achievement. merit based Employees: Based on performance Universalistic criteria	Ascription. nepotism based Family members: Indulged Particularistic criteria	Beehr et al. (1997) Ram (1994) Chua et al. {20(9)
Networks	External ties based on business Distinct business, family spheres Impersonal social responsibility	Embedded in kinship networks Role diffuseness Personalized social responsibility	Ingram and Lifschitz (2006} Lomnitz and Pèrez-Lizaur, (1987) Muntean (2009)
Leadership	High turnover with market discipline Formally educated Succession draws on large pool	Entrenched, long tenured Trained on the job Succession draws on kinship pool	Oswald et al. (2009} Jorissen et al, (200.5) Pèrez-Gonzàlez (2006)
Careers	Salariated managers Shorter term career horizons	Family members Longer term career horizons	Galambos (2010) Benedict(1968)
Management	Delegation to professionals Rational, analytical Innovative Formalized command and control	Autocratlc Emotional, intuitive Rent-seeking, stifling innovation Organic, mutual accommodation	Greenhalgh n994) Zellweger and Astrachan, (2008) Morck and Yeung (2003) Zhang and Ma (2009)

1971; Zellweger & Astrachan, 2008) and/or rivalry (Grote, 2003), which can negatively affect firm performance (Eddleston & Kellermanns, 2007) or group dynamics and business processes (Pelled et al., 1999). Not surprisingly, literature suggests that a project manager devotes up to 80% of his time in conflict resolution and negotiation activities, both internal (project team) and external (stakeholders), thus, generating non-financial costs (Astrachan & Jaskiewicz, 2008). These negative non-economic outcomes are often overlooked (Astrachan & Jaskiewicz, 2008) and, at least, are difficult to measure in spite of their potentially significant impact on project performance and business value (Demetz & Lehn, 1985).

Emblematic from this point of view, the empirical analysis (Basco, Pèrez-Rodriguez, 2009) that involved a sample of 732 companies through a telephone interview, with a closed questionnaire. This research showed that the performance of family businesses is improved when management's attention is placed equally on both the economic aspects and those closely related to the family as such. Unfortunately, however, these data are to be accepted with extreme caution, since they are self-reported by the interviewees, and do not derive from an analysis of objective results. In addition, the sample was divided into 3 + 3 clusters, according to the performance coefficient both company and family, and as many as 205 companies (28% of the sample) were defined by the authors as immature, as with low levels of attention both to the family aspects that corporate. Therefore, if there remains uncertainty regarding the best philosophy of approach to business management, the idea that the family business differs from the non-family business relative to the overall level of managerialization is well founded. A recent trend in research is stimulating family businesses to develop a more business-oriented and less family-oriented philosophy, developing a higher level of maturity regarding managerial practices used (Cagliano, Spina, 2000, Biasetti et alii, 2009).

Obstacles to Family Firms' Managerialization: Non-Economic Goals

Thesetopic have been studied using multiple organizational approaches, like Agency Theory, Stewardship Theory, Resource-Based View, and even from an evolutionary perspective. In particular, trust and commitment are two fundamental pillars upon which many of the positive approaches towards family business research are built (Eddelston, Morgan, 2014). These concepts are often used to describe distinct attributes of family businesses, like *familiness* (Frank, Lueger, Nose, & Suchy, 2010; Irava & Moores, 2010; Zellweger, Eddleston, & Kellermanns, 2010), *social capital* (Arregle, Hitt, Sirmon, & Very, 2007; Pearson, Carr, & Shaw, 2008; Sirmon & Hitt, 2003), *reciprocal altruism* (Eddleston, Kellermanns, & Sarathy, 2008; Lubatkin, Durand, & Ling, 2007), *family firm identity* (Zellweger, Kellermanns, Eddleston, & Memili, 2012) and *stewardship* (Davis, Allen, & Hayes, 2010; Dibrell & Moeller, 2011; Eddleston & Kellermanns, 2007).

More recently, a promising approach has been developed in the specific field of family business studies: the Socio-Emotional Wealth approach (SEW). This approach (Gomez-Mejia, Haynes,Nuñez-Nickel, Jacobson, and Moyano-Fuentes (2007); Gomez-Mejia, Makri, and Larraza Kintana (2010); Berrone, Cruz, Gomez-Mejia, and Larraza-Kintana (2010); Gomez-Mejia, Cruz, Berrone, and De Castro (2011); Berrone et al., 2012) suggests that "family firms are typically motivated by, and committed to, the preservation of their SEW (socio-emotional wealth), referring to non-financial aspects or "affective endowments" of family owners" Berrone et al., 2012, p. 259). The SEW construct has proven to be a good analytical lens for interpreting a wide variety of family firm phenomena. Berrone and colleagues (2012) suggest that SEW presents at least three dimensions: the *intention of handing the business down* to future generations, *the control and influence* of family members and the *close identification* of the family with the firm.

Given these considerations, it becomes necessary to investigate these organizational dynamics which could affect project management practices. Therefore, this chapter aims to develop the following theoretical propositions:

P1: In a family firm, the *intention to hand the business down to future generations* affect project and knowledge management practices.

P2: In a family firm, the *influence and control of family members over business* affect project and knowledge management practices.

P3 In a family firm, the *close identification of the family with the firm* affect project and knowledge management practices.

Of course, these are not research hypotheses, but propositions that, deductively, emerge from the analysis of the literature. They are logically derived from the empirical evidence available in the SEW field of study, and future research will submit them to empirical verification.

SPECIFIC ATTENTION POINTS FOR PROJECT AND KNOWLEDGE MANAGEMENT IN FAMILY FIRMS

In family business field of study, SEW is defined as 'the intrinsic and inextricable emotional endowment that all family businesses have i.e. the set of feelings, emotions, relationships and binding ties between members of the business family (Martinez-Romero et al., 2016, 3). Although scholars have suggested some criteria for considering these aspects in business value appraisals, (e.g. Astrachan and Jaskiewicz, 2008), a complete and reliable evaluation of these factors is far from being accomplished (see Martinez-Romero et al., 2016, for a recent review). In the field of project and knowledge management especially, literature has until now paid scarce attention to the impact of family firm context on managerial practices (Ferrari, *forthcoming;* Sadkowska, 2017; 2018).

Theoretically, considering the intention to hand the business down to future generations, the influence and control of family members over business and the close identification of the family with the firm, it's possible to identify several undesired consequences for the project and knowledge management and related attention points in project management body of knowledge (see Table 3 for a synopsis of this theoretical framework). By developing the theoretical propositions above presented, the following sections discuss the impact of each SEW factor on project and knowledge management practices.

The Intention to Hand the Business Down to Future Generations

Many authors (Berrone et al., 2012; Zellweger and Astrachan, 2008; Zellweger, Kellermanns, et al. (2011) suggest that trans-generational sustainability is one of the central aspects of SEW. Evidence shows that maintaining the business for future generations is commonly seen as a key goal for family firms (Kets de Vries, 1993; Zellweger, Kellermanns, et al., 2011) and that many family firms exhibit long term planning horizons (Miller & Le Breton-Miller, 2006b; Miller, Le Breton-Miller, & Scholnick, 2008; Sirmon & Hitt, 2003). The intention to hand the business down to future generations could lead to the intention to protect the next generation, and, in turn, lead to a lack of meritocracy. This affective dynamic is the foundation of the intention to protect the next generation, especially if the successor is low skilled or shows low or even no aptitudes for the specific business. These dynamics could have an impact on project management practices. The intention to hand the business down to future generations could lead to a discrimination versus non-family members in managerial jobs assignment, thus leading to a low level of organizational justice (both distributive and procedural). The next generation can be considered an investment for the future, so the non-family member project manager will be forced to integrate a family member into the project team, regardless of their skill level. This in order not to exclude a member of the family from the production and sharing of strategic organizational knowledge. Or again, the long-term time horizon could induce the owner family to invest in unattainable projects. Finally, the need to protect and value the following generations could lead to invest in wrong projects from a strategic point of view but consistent with the skills and / or attitudes of junior members. Moreover, in addition to the close identification of the family with the firm (see the following point below), the intention to hand the business down to future generations could result in a sort of mistrust of non-family members, perceived as 'the others', favoring a family member in a project team staffing process. Consistent with Agency Theory approach, an adverse selection problem arises due to the lack of fit between the skills required and the skills possessed by the young family member. This fact could result in a low level of organizational justice (OJ) as perceived by non-family members, thus undermining their work engage-

Table 3. Theoretical development of propositions

Family firms' features	Theoretical proposition	Undesired consequences for project and knowledge management	Attention points in project management body of knowledge (PMBOK®, 6th ed.)
Intention to hand the business down to future generations	P1: In a family firm, the *intention to hand the business down to future generations* affect project and knowledge management practices.	Protecting the next generation: lack of meritocracy **Discrimination versus non-family members in managerial job assignement** **Low level of organizational justice (both distributive and procedural).**	**ORGANIZATIONAL GOVERNANCE FRAMEWORK:** Rules Policies Procedures Norms Relationships Systems Processes **Objectives of the organization** **MANAGEMENT ELEMENTS:** Division of work using specialized skills and availability to perform work; Authority given to perform work; Responsibility to perform work appropriately assigned based on such attributes as skill and experience; Discipline of action (e.g., respect for authority, people, and rules); Unity of command (e.g., only one person gives orders for any action or activity to an individual); Unity of direction (e.g., one plan and one head for a group of activities with the same objective); General goals of the organization take precedence over individual goals; **Paid fairly for work performed;** Optimal use of resources; Clear communication channels; Right materials to the right person for the right job at the right time; Fair and equal treatment of people in the workplace; Open contribution to planning and execution by each person; and Optimal morale
Influence and control of family members over business	P2: In a family firm, the *influence and control of family members over business* affect project and knowledge management practices.	**Intergroup conflict vs. non family members** **High non-family member turn-over rate.** Skill shortage in managerial positions	
Close identification of the family with the firm.	P3 In a family firm, the *close identification of the family with the firm* affect project and knowledge management practices.	**Self-centred family** Low innovation rate Over-valuing of the current products, processes and managerial practices;	

ment which, in turn, affects their Organizational Citizenship Behaviours - OCB and can even result in a high non-family member turn-over rate (Colquitt et al., 2001; Ng et al., 2001; Bowen et al., 2007). Due to these theoretical considerations and the available empirical evidence, it's possible to posit the Proposition 1: „in a family firm, the *intention to hand the business down to future generations* affect project and knowledge management practices".

The Control and Influence of Family Members

Indeed, control and influence are an integral parts of SEW and are highly desired by family members (Zellweger, Kellermanns, et al., 2011). However, this could lead to intergroup conflict vs. non-family members. First of all, the control maintained by the family can hinder the autonomy of the Project Manager if he is an external manager, undermining their effectiveness, motivation, satisfaction and overall performance. The purpose to maintain the strategic control under the owner family is also often associated with the discrimination of out-groups, and, at the same time, in-group favouritism. This finding is consistent with social-psychological literature (Tajfel, Turner, 1986) and previous research in the family business field (Ferrari, 2014; De Massis et al., 2012). The *out-group discrimination* is associated with a high level of inter-group conflict versus non-family members and, more frequently, with low level of organizational justice (OJ) as perceived by non-family members, according to current literature (Colquitt et al., 2001). Furthermore, literature supports the impact of the OJ dimensions (distributive, procedural, informational and relational) on organizational performance (Cropanzano et al., 2007), especially non-family member turnover rate (Colquitt et al., 2001; Ng et al., 2001). Contrastingly, literature shows that *in-group favouritism* can explain dynamics such as nepotism (Salvato et al., 2012) and adverse selection (Schulze, Lubatkin, Dino, 2003; Chirsman, Chua, Litz, 2004), and more in general result in a lack of meritocracy, eventually resulting in severe skill shortage in managerial positions. This finding is consistent with literature: family SMEs firms have often showed a lack of *managerialisation* (Cagliano, Spina, 2000; Biasetti et al., 2009), with a lack of meritocracy in managerial selection and promotion (Bugamelli et al., 2012). Indeed, *Familism* and *Cronyism* appear to be the main causes of the delayed development of SMEs organizations (Pellegrino and Zingales, 2014).

Finally, discrimination versus non family members could easily lead to high non-family member turn-over rate in both managerial (e.g. project manager) and executive (eg expert/ project team member) positions. As a consequence, such discrimination could result in a skill shortage in managerial positions, in case of non-family member resignation from project team). Given this theoretical scenario, and the available empirical evidence, it's possible to posit the Proposition 2: "in a family firm, the *influence and control of family members over business* affect project and knowledge management practices".

The Close Identification of the Family with the Firm

The identity of a family firm's owner is inextricably tied to the organization: this causes the firm to be seen both by internal and external stakeholders as an extension of the family itself (Berrone et al., 2010; Dyer & Whetten, 2006). Often, it is maybe more correct to sustain that there is a total identification between the founder and the firm.

The close identification of the family with the firm could result in a low innovation rate and in over-valuing of the current products, processes and managerial practices. This fact is often associated with an affective over-investment in family business, resulting in over valuing of the current products, processes and managerial practices. Literature also shows biases in strategic decision-making, for example in launching new products or developing process innovation, pursuing the desire to preserve the current situation. Often, an innovation is perceived as a threat to the family firm past history. Literature highlights that the identity of a family firm's owner is inextricably tied to the organization: this causes the firm to be seen both by internal and external stakeholders as an extension of the family itself (Berrone et al., 2010; Dyer & Whetten, 2006). In particular, there is almost always the founder's identification with the firm.

The close identification of the family with the firm could result in a *self-centred family*. At *family group* level, the empirical evidence shows that a key factor in producing a positive outcome in family firms is the *kind of relationship* amongst *all* involved family members (Davis & Harveston, 1998; Dunn, 1999; Lansberg, Astrachan, 1994; Morris et al., 1997). Family members' involvement could lead to positive outcomes such as *reciprocal altruism* (Eddleston, Kellermanns, & Sarathy, 2008; Lubatkin,

Durand, & Ling, 2007), *family firm identity* (Zellweger, Kellermanns, Eddleston, & Memili, 2012) and *stewardship* (Davis, Allen, & Hayes, 2010; Dibrell & Moeller, 2011; Eddleston & Kellermanns, 2007), but also to a *self-centred* and *strict family*. In fact, literature suggests, (for a review and an assessment of these dimensions, see Beavers & Hampson, 1995, Olson, 2000; 2011; Michael-Tsabari, Lavee, 2012), that every situation experienced by a family group can be defined by assessing two different and variable family features: its *cohesiveness* (how self-oriented the family is) and its *flexibility* (how changeable and adaptable the relationships within the family are).

Following this approach, the so called *Circumplex Model* (Olson, 2000; 2011), each dimension/variable (both cohesiveness and flexibility) could be described in a curvilinear graph, where the extreme values are negative (dysfunctional). Families which are, at the same time, both connected *and* flexible/versatile show the highest likelihood of positive outcomes. Cohesiveness and flexibility are in some way correlated (Friedman, 1986; Gersick et al., 1997), and the former seems to play a prominent role in reducing variability around group/family norms (Schermerhorn et al., 2000).

The circumplex model has been previously applied in order to investigate several family features and family firms' performances (for a review, see Daspit et al., 2017). Lee (2006) investigated the impact of balanced cohesion and flexibility on job satisfaction; Nosé, Korunka, Frank, & Danes (2015) found that balanced family structures reduce relationship conflict within the family. Additionally, circumplex research shows that balanced levels of cohesion and flexibility help the family firm to survive through business transmission (Labaki, 2011). More generally, balanced family structures lead to positive outcomes in family performance (Zody, Sprenkle, MacDermid, & Schrank, 2006) and even ensure success over multiple generations (Michael-Tsabari & Lavee, 2012).

Often, the founder is unwilling to relinquish power and control, showing a lack of delegation, mainly to non-family members, thus exacerbating intergenerational conflict dynamics. Other research has highlighted how heirs feel intense pressure to join the family business despite their reservations or personal preferences (Freudenberger, Freedheim, & Kurtz, 1989). Finally, within a family business, also family cohesion has an impact on non-family members. As suggested by Daspit and colleagues (2017, 17), "if non-family members are indeed less likely to be hired and promoted, the organization is likely to suffer from a lack of specialized skills and diverse knowledge resources gained from the employment and advancement of non-family members". Given this theoretical scenario, and the available empirical evidence, it's possible to posit the Proposition 3: in a family firm, the *close identification of the family with the firm* affect project and knowledge management practices".

CONCLUSION

This chapter developing the above discussed theoretical propositions suggests striving for non-economic goals (the *intention to hand the business down to future generations*, the *influence and control of family members over business* and the *close identification of the family with the firm*) affect project and knowledge management practices. Therefore, family firms appear to be an environment for project management prone to several criticism. SEW factors have impact on, at least, on two general areas of project management knowledge (PMBOK, 6th Edition, Chapter 2): the *organizational governance framework* and *management elements*.

Regarding the former, all peculiar family-related factors shape rules, policies, procedures and norms. The family business is an environment in which the rules do not follow impersonal criteria, but are adapted to the involved people. In a family business, often, the law is not the same for everyone. The objective of transmitting the company to next generations, the influence and pervasive control of owners and the close identification of the family (or often, even of the founder) with the company can create problems of Organizational Justice easily, of a procedural, distributive, informative and relational nature (Colquitt et al., 2001; Ng et al., 2001; Bowen et al., 2007).

In addition, SEW factors have an impact on organizational relationships, between family and non-family members especially (Sieger, Bernhard, Frey, 2011). Non-family members are usually excluded from the succession, and, therefore, discouraged in long-term investments in the company, of which they will not become co-owners. If they are project managers, non-family members will usually have to deal with conflicting relationships with incompetent and de-motivated members of the family.

Further, SEW factors have an impact on organizational systems and processes. The pervasive influence and control of the owner and the close identification of the family with the enterprise could undermine process innovation, or the introduction of procedures aimed at preserving knowledge. For example, the poor managerialization of the family business easily has an impact on the formalization of procedures to create knowledge repositories.

Finally, the prevalence of non-economic objectives can change the nature of specific project objectives, ensuring that family objectives prevail over those potentially conflicting.

Regarding further management elements, striving for non-economic goals as transmitting the company to next generations, the influence and pervasive control of owners and the close identification of the family with the company could affect pivotal organizational mechanism in project management practices as division of work using specialized skills and availability to perform work, authority given to perform work, responsibility to perform work appropriately assigned based on such attributes as skill and experience.

Moreover, as discussed in developing theoretical propositions, SEW goals can easily have an impact on further organizational dynamics as discipline of action (e.g., respect for authority, people, and rules), unity of command (e.g., only one person gives orders for any action or activity to an individual) and unity of direction (e.g., one plan and one head for a group of activities with the same objective).

SEW goals could hinder the achievement of general goals of the organization and take precedence over non-family members individual goals, therefore, once more generating conflicts and undermining the overall Organizational Justice level (De Massis, 2012). Family firms show criticism in paid fairly for work performed (Carrasco, Sanchez, 2007).

Furthermore, striving for SEW could affect negatively clear communication channels with non-family members, and right materials to the right person for the right job at the right time if non-family members are involved. Finally, literature suggests a family firm doesn't ensure fair and equal treatment of people in the workplace (Barnett, Kellermanns, 2006).

REFERENCES

Achmad, T., Rusmin, R., Neilson, J., & Tower, G. (2009). The iniquitous influence of family ownership structures on corporate performance. *Journal of Global Business Issues*, *3*, 41–49.

Agarwal, R., & Hoetker, G. (2007). A faustian bargain? The growth of management and its relationship with related disciplines. *Academy of Management Journal*, *50*(6), 1304–1322. doi:10.5465/amj.2007.28165901

Albert, S., & Whetten, D. (1985). Organizational identity. In L. L. Cummings & B. M. Staw (Eds.), *Research in organizational behavior*, 7, pp. 263–295. Greenwich, CT: JAI Press.

Anderson, R. C., & Reeb, D. M. (2003a). Founding-family ownership and firm performance: Evidence from the S&P 500. *The Journal of Finance*, *58*(3), 1301–1328. doi:10.1111/1540-6261.00567

Andres, C. (2008). Large shareholders and firm performance: An empirical examination of founding-family ownership. *Journal of Corporate Finance*, *14*(4), 431–445. doi:10.1016/j.jcorpfin.2008.05.003

Ariely, D., Huber, J., & Wertenbroch, K. (2005). When do losses loom larger than gains? *JMR, Journal of Marketing Research*, *42*(2), 134–138. doi:10.1509/jmkr.42.2.134.62283

Arregle, J. L., Hitt, M. A., Sirmon, D. G., & Very, P. (2007). The development of organizational social capital: Attributes of family firms. *Journal of Management Studies*, *44*(1), 73–95. doi:10.1111/j.1467-6486.2007.00665.x

Astrachan, J. H., & Carey Shanker, M. (2006). Family businesses' contribution to the US economy: a closer look. In P. Poutziouris, K. X. Smyrnios, & S. B. Klein (Eds.), *Handbook of research in family business*. Cheltenham, UK: Edward Elgar. doi:10.4337/9781847204394.00011

Astrachan, J. H., & Jaskiewicz, P. (2008). Emotional returns and emotional costs in privately held family businesses: Advancing traditional business valuation. *Family Business Review*, *21*(2), 139–149. doi:10.1111/j.1741-6248.2008.00115.x

Barach, J. A., & Gantisky, J. B. (1995). Successful succession in family business. *Family Business Review*, *8*(2), 131–155. doi:10.1111/j.1741-6248.1995.00131.x

Barnes, L. B., & Hershon, S. A. (1976). Transferring power in family business. *Harvard Business Review*, (July-August), 105–114.

Beavers, W. R., & Hampson, R. B. (1995). Misurare la competenza famigliare: il modello sistemico di Beavers, tr. it. In *Ciclo vitale e dinamiche familiari, a cura di F*. Milano, Italy: Walsh, Franco Angeli.

Beckhard, R., & Dyer, W. G. Jr. (1983). Managing continuity in the family-owned business. *Organizational Dynamics*, *12*(1), 5–12. doi:10.1016/0090-2616(83)90022-0

Beehr, T. A., Drexler, J. A. Jr, & Faulkner, S. (1997). Working in small family businesses: Empirical comparisons to non-family businesses. *Journal of Organizational Behavior*, *18*(3), 297–312. doi:10.1002/(SICI)1099-1379(199705)18:3<297::AID-JOB805>3.0.CO;2-D

Benedict, B. (1968). Family firms and economic development. *Southwestern Journal of Anthropology*, *24*(1), 1–19. doi:10.1086outjanth.24.1.3629299

Berrone, P., Cruz, C., & Gomez-Mejia, L. R. (2012). Socioemotional wealth in family firms: Theoretical dimensions, assessment approaches, and agenda for future research. *Family Business Review*, *25*(3), 258–279. doi:10.1177/0894486511435355

Berrone, P., Cruz, C. C., Gomez-Mejia, L. R., & Larraza Kintana, M. (2010). Socioemotional wealth and corporate response to institutional pressures: Do family-controlled firms pollute less? *Administrative Science Quarterly*, *55*(1), 82–113. doi:10.2189/asqu.2010.55.1.82

Bryant, A., & Charmaz, K. (2007). *The Sage handbook of grounded theory*. London, UK: Sage. doi:10.4135/9781848607941

Cartwright, D. & Zander, A. (Eds.). (1984). Group dynamics: Research and theory (2nd ed., pp. 414-448). Evanston, IL: Row, Peterson, & Company.

Cascino, S., Pugliese, A., Mussolino, D., & Sansone, C. (2010). The influence of family ownership on the quality of accounting information. *Family Business Review*, *23*(3), 246–265. doi:10.1177/0894486510374302

Cater, J. J. III, & Justis, R. T. (2009). The development of successors from followers to leaders in small family firms: An exploratory study. *Family Business Review*, *22*(2), 109–124. doi:10.1177/0894486508327822

Chan, D. (1998). Functional relations among constructs in the same content domain at different levels of analysis: A typology of composition models. *The Journal of Applied Psychology*, *83*(2), 234–246. doi:10.1037/0021-9010.83.2.234

Charmaz, K. (1995). *Grounded theory. a practical guide through. qualitative analysis*. London, UK: Sage.

Chenail, R. (2009). Qualitative research like politics can also be local: A review of interdisciplinary standards for systematic qualitative research. *The Weekly Qualitative Report*, *2*(11), 61–65.

Chirico, F. (2008). Knowledge accumulation in family firms: evidence from four case studies. *International Small Business Journal*, *26*(4), 433–462. doi:10.1177/0266242608091173

Chirico, F., & Salvato, C. (2008). Knowledge integration and dynamic organizational adaptation in family firms. *Family Business Review*, *21*(1), 169–181. doi:10.1111/j.1741-6248.2008.00117.x

Chrisman, J. J., Chua, J. H., & Litz, R. (2003). A unified systems perspective of family firm performance: An extension and integration. *Journal of Business Venturing*, *18*(4), 467–472. doi:10.1016/S0883-9026(03)00055-7

Chrisman, J. J., Chua, J. H., & Litz, R. (2004). Comparing the agency costs of family and nonfamily firms: Conceptual issues and exploratory evidence. *Entrepreneurship Theory and Practice*, *28*(4), 335–354. doi:10.1111/j.1540-6520.2004.00049.x

Chrisman, J. J., Kellermanns, F. W., Chan, K. C., & Liano, K. (2010). Intellectual foundations of current research in family business: An identification and review of 25 influential articles. *Family Business Review*, *23*(1), 9–26. doi:10.1177/0894486509357920

Chrisman, J. J., Sharma, P., & Taggar, S. (2007). Family influences on firms: An introduction. *Journal of Business Research*, *60*(10), 1005–1011. doi:10.1016/j.jbusres.2007.02.016

Chua, J. H., Chrisman, J. J., & Bergiel, E. B. (2009). An agency theoretic analysis of the professionalized family firm. *Entrepreneurship Theory and Practice*, *33*(2), 355–372. doi:10.1111/j.1540-6520.2009.00294.x

Chua, J. H., Chrisman, J. J., & Sharma, P. (1999). Defining the family business by behavior. *Entrepreneurship Theory and Practice*, *23*(4), 19–39. doi:10.1177/104225879902300402

Corbetta, G., & Montemerlo, D. (1999). Ownership, governance and management issues in small and medium sized family businesses: a comparison of Italy and the United States. *Family Business Review*, *12*(4), 361–374. doi:10.1111/j.1741-6248.1999.00361.x

Corbin, J., & Strauss, A. (2008). *Basics of qualitative research* (3rd ed.). Thousand Oaks, CA: Sage.

Daspit, J. J., Madison, K., Barnett, T., & Long, R. G. (2017, in press). The emergence of bifurcation bias from unbalanced families: Examining HR practices in the family firm using Circumplex Theory, in Human Resource Management Review, Special Issue on "The Role of Family Science Theories for Human Resource Management in Family Firms".

Davis, J. H., Allen, M. R., & Hayes, H. D. (2010). Is blood thicker than water? A study of stewardship perceptions in family business. *Entrepreneurship Theory and Practice*, *34*(6), 1093–1116. doi:10.1111/j.1540-6520.2010.00415.x

Davis, P. S., & Harveston, P. D. (1998). The influence of family on the family business succession process: A multi-generational perspective. *Entrepreneurship Theory and Practice*, *22*(3), 31–53. doi:10.1177/104225879802200302

De Massis, A., Chua, J., & Chrisman, J. J. (2008). Factors preventing intra-family succession. *Family Business Review*, *21*(2), 183–199. doi:10.1111/j.1741-6248.2008.00118.x

Dibrell, C., & Moeller, M. (2011). The impact of a service-dominant focus strategy and stewardship culture on organizational innovativeness in family-owned businesses. *Journal of Family Business Strategy*, *2*(1), 43–51. doi:10.1016/j.jfbs.2011.01.004

Donckels, R., & Fröhlich, E. (1991). Are family businesses really different? European experiences from Stratos. *Family Business Review*, *4*(2), 149–160. doi:10.1111/j.1741-6248.1991.00149.x

Dukerich, J. M., Golden, B., & Shortell, S. M. (2002). Beauty is in the eye of the beholder: The impact of organizational identification, identity, and image on the cooperative behaviors of physicians. *Administrative Science Quarterly*, *47*(3), 507–533. doi:10.2307/3094849

Dunn, B. (1999). The family factor: The impact of family relationship dynamics on business-owning families during transitions. *Family Business Review*, *12*(1), 41–60. doi:10.1111/j.1741-6248.1999.00041.x

Dutton, J. E. & Dukerich, J. M. (1991). Keeping an eye on the mirror: image and identity in organizational adaptation, *Academy Of Management Journal, 34,* 3, 517-554.

Dyck, B., Mauws, M., Starke, F. A., & Mischke, G. A. (2002). Passing the baton: The importance of sequence, timing, technique and communication in executive succession. *Journal of Business Venturing*, *17*(2), 143–162. doi:10.1016/S0883-9026(00)00056-2

Eddleston, K. A., & Kellermanns, F. W. (2007). Destructive and productive family relationships: A stewardship theory perspective. *Journal of Business Venturing*, *22*(4), 545–565. doi:10.1016/j.jbusvent.2006.06.004

Eddleston, K. A., Kellermanns, F. W., & Sarathy, R. (2008). Resource configuration in family firms: Linking resources, strategic planning and technological opportunities to performance. *Journal of Management Studies*, *45*(1), 26–50.

Eddleston, K. A., Kellermanns, F. W., & Zellweger, T. M. (2012). Exploring the entrepreneurial behavior of family firms: Does the stewardship perspective explain differences? *Entrepreneurship Theory and Practice*, *36*(2), 347–367. doi:10.1111/j.1540-6520.2010.00402.x

Eddleston, K. A., & Kidwell, R. E. (2012). Parent–child relationships: Planting the seeds of deviant behavior in the family firm. *Entrepreneurship Theory and Practice*, *36*(2), 369–386. doi:10.1111/j.1540-6520.2010.00403.x

Eddleston, K. A., & Morgan, R. M. (2014). Trust, commitment and relationships in family business: Challenging conventional wisdom. *Journal of Family Business Strategy*, *5*(3), 213–216. doi:10.1016/j.jfbs.2014.08.003

Feltham, T. S., Feltham, G., & Barnett, J. J. (2005). The dependence of family businesses on a single decision-maker. *Journal of Small Business Management*, *43*(1), 1–15. doi:10.1111/j.1540-627X.2004.00122.x

Fendt, J., & Sachs, W. (2008). Grounded theory method in management research. Users' perspectives. *Organizational Research Methods*, *11*(3), 430–455. doi:10.1177/1094428106297812

Fiol, C. M., Pratt, G. M., & O'Connor, E. J. (2009). Managing intractable identity conflicts. *Academy of Management Review*, *34*(1), 32–55. doi:10.5465/amr.2009.35713276

Fletcher, D., De Massis, A., & Nordqvist, M. (2016). Qualitative research practices and family business scholarship: A review and future research agenda. *Journal of Family Business Strategy*, *7*(1), 8–25. doi:10.1016/j.jfbs.2015.08.001

Foo, M.-D. (2011). Emotions and entrepreneurial opportunity evaluation. *Entrepreneurship Theory and Practice*, *35*(2), 375–393. doi:10.1111/j.1540-6520.2009.00357.x

Frank, H., Lueger, M., Nosé, L., & Suchy, D. (2010). The concept of "Familiness": Literature review and systems theory-based reflections. *Journal of Family Business Strategy*, *1*(3), 119–130. doi:10.1016/j.jfbs.2010.08.001

Freudenberger, H. J., Freedheim, D. K., & Kurtz, T. S. (1989). Treatment of individuals in family business. *Psychotherapy (Chicago, Ill.)*, *26*(1), 47–53. doi:10.1037/h0085404

Galambos, L. (2010). The role of professionals in the Chandler paradigm. *Industrial and Corporate Change*, *19*(2), 377–398. doi:10.1093/icc/dtq009

Gedajlovic, E., Lubatkin, M. H., & Schulze, W. S. (2004). Crossing the threshold from founder management to professional management: A governance perspective. *Journal of Management Studies*, *41*(5), 899–912. doi:10.1111/j.1467-6486.2004.00459.x

Gómez-Mejía, L. R., Cruz, C., Berrone, P., & De Castro, J. (2011). The bind that ties: socioemotional wealth preservation in family firms. *The Academy of Management Annals*, *5*(1), 653–707. doi:10.1080/19416520.2011.593320

Gomez-Mejia, L. R., Haynes, K. T., Nunez-Nickel, M., Jacobson, K. J. L., & Moyano-Fuentes, J. (2007). Socioemotional wealth and business risks in family-controlled firms: Evidence from Spanish olive oil mills. *Administrative Science Quarterly*, *52*(1), 106–137. doi:10.2189/asqu.52.1.106

Gomez-Mejia, L. R., Larraza-Kintana, M., & Makri, M. (2003). The determinants of executive compensation in family-controlled public corporations. *Academy of Management Journal, 46*, 226–237.

Gomez-Mejia, L. R., Makri, M., & Larraza-Kintana, M. (2010). Diversification decisions in family-controlled firms. *Journal of Management Studies, 47*(2), 223–252. doi:10.1111/j.1467-6486.2009.00889.x

Greenhalgh, S. (1994). Deorientalizing the Chinese family firm. *American Ethnologist, 21*(4), 746–775. doi:10.1525/ae.1994.21.4.02a00050

Habbershon, T. G., Williams, M., & MacMillan, I. C. (2003). A unified systems perspective of family firm performance. *Journal of Business Venturing, 18*(4), 451–465. doi:10.1016/S0883-9026(03)00053-3

Haveman, H. A., & Khaire, M. V. (2004). Survival beyond succession? The contingent impact of founder succession on organizational failure. *Journal of Business Venturing, 19*(3), 437–463. doi:10.1016/S0883-9026(03)00039-9

Haynes, G. W., Walker, R., Rowe, B. R., & Hong, G. S. (1999). The intermingling of business and family finances in family-owned businesses. *Family Business Review, 12*(3), 225–239. doi:10.1111/j.1741-6248.1999.00225.x

Hitt, M. A., Beamish, P. W., Jackson, S. E., & Mathieu, J. E. (2007). Building theoretical and empirical bridges across levels: Multilevel research in management. *Academy of Management Journal, 50*(6), 1385–1399. doi:10.5465/amj.2007.28166219

House, R., Rousseau, D. M., & Thomas-Hunt, M. (1995). The meso paradigm: A framework for the integration of micro and macro organizational behavior. In L. L. Cummings & B. M. Staw (Eds.), *Research in organizational behavior*, 17, pp. 71–114. Greenwich, CT: JAI Press.

Ingram, P., & Lifschitz, A. (2006). Kinship in the shadow of the corporation: The interbuilder network in Clyde River shipbuilding, 1711-1990. *American Sociological Review, 71*(2), 334–352. doi:10.1177/000312240607100208

Irava, W. J., & Moores, K. (2010). Clarifying the strategic advantage of familiness: Unbundling its dimensions and highlighting its paradoxes. *Journal of Family Business Strategy, 1*(3), 131–144. doi:10.1016/j.jfbs.2010.08.002

Jorissen, A., Laveren, E., Martens, R., & Reheul, A.-M. (2005). Real versus sample-based differences in comparative family business research. *Family Business Review, 18*(3), 229–246. doi:10.1111/j.1741-6248.2005.00044.x

Kaye K. (1991). Penetrating the cycle of sustained conflict, *Family Business Review, 4*(1), 2, 1-44.

Kaye, K. (1992). The kid brother. *Family Business Review, 5*(3), 237–256. doi:10.1111/j.1741-6248.1992.00237.x

Kaye, K. (1996). When the family business is a sickness. *Family Business Review, 9*(4), 347–368. doi:10.1111/j.1741-6248.1996.00347.x

Kellermanns, F. W., Eddleston, K. A., & Zellweger, T. M. (2012). Extending the socioemotional wealth perspective: a look at the dark side, *Entrepreneurship Theory and Practice,* 36, Special Issue on Family Business, 1175–1182.

Kets de Vries, M. F. (1977). The entrepreneurial personality: A person at the crossroads. *Journal of Management Studies*, *14*(1), 34–57. doi:10.1111/j.1467-6486.1977.tb00616.x

Kets de Vries, M. F. (1993). The dynamics of family-controlled firms: The good and the bad news. *Organizational Dynamics*, *21*(3), 59–71. doi:10.1016/0090-2616(93)90071-8

Kidwell, R. E., Eddleston, K. A., Cater, J. J. III, & Kellermanns, F. W. (2013). How one bad family member can undermine a family firm: Preventing the Fredo effect. *Business Horizons*, *56*(1), 5–12. doi:10.1016/j.bushor.2012.08.004

Kidwell, R. E., Kellermanns, F. W., & Eddleston, K. A. (2012). Harmony, justice, confusion, and conflict in the family firm: Implications for ethical climate and the "Fredo effect.". *Journal of Business Ethics*, *106*(4), 503–517. doi:10.100710551-011-1014-7

Kozlowski, S. W. J., & Klein, K. J. (2000). A multilevel approach to theory and research in organizations contextual, temporal, and emergent processes. In K. Klein & S. Kozlowski (Eds.), Multilevel theory, research, and methods in organizations: Foundations, extensions, and new directions, San Francisco, CA: Jossey-Bass.

Kraatz, M. & Block, E. (2008). Organizational implications of institutional pluralism. In R. Greenwood, C. Oliver, R. Suddaby, & K. Sahlin-Andersson (Eds.), Handbook of organizational institutionalism, 243-275. London, UK: Sage. doi:10.4135/9781849200387.n10

Kucharska, W. & Kowalczyk, R. (2016). Trust, collaborative culture and tacit knowledge sharing in project management – a relationship model, In *Proceedings of the 13th International Conference on Intellectual Capital, Knowledge Management & Organisational Learning, ICICKM.* (pp. 159-166). doi:10.13140/RG.2.2.25908.04486

Lambrecht, J. (2005). Multigenerational transition in family businesses: A new explanatory model. *Family Business Review*, *28*(4), 267–282. doi:10.1111/j.1741-6248.2005.00048.x

Lambrecht, J., & Lievens, J. (2008). Pruning the family tree: An unexplored path to family business continuity and family harmony. *Family Business Review*, *21*(4), 295–313. doi:10.1177/08944865080210040103

Lansberg, I. (1988). The succession conspiracy. In C. E. Aronoff, J. H. Astrachan, & J. L. Ward (Eds.), *Family business sourcebook II* (pp. 70–86). Marietta, GA: Business Owner Resources.

Lansberg, I., & Astrachan, J. H. (1994). Influence of family relationships on succession planning and training: The importance of mediating factors. *Family Business Review*, *7*(1), 39–59. doi:10.1111/j.1741-6248.1994.00039.x

Le Breton-Miller, I., Miller, D., & Steier, L. P. (2004). Toward an integrative model of effective FOB succession. *Entrepreneurship Theory and Practice*, *28*(4), 305–328. doi:10.1111/j.1540-6520.2004.00047.x

Leibowitz, B. (1986). Resolving conflict in the family owned business. *Consultation*, *5*(3), 191–205.

Linstead, S., Marechal, G., & Griffin, R. W. (2014). Theorizing and researching the dark side of organization. *Organization Studies*, *35*(2), 165–188. doi:10.1177/0170840613515402

Lomnitz, L. A., & Pérez-Lizaur, M. (1987). *A Mexican elite family, 1820-1980: Kinship, class and culture*. Princeton, NJ: Princeton University Press.

Lubatkin, M. H., Durand, R., & Ling, Y. (2007). The missing lens in family firm governance theory: A self-other typology of parental altruism. *Journal of Business Research*, *60*(10), 1022–1029. doi:10.1016/j.jbusres.2006.12.019

Martínez, J. I., Stöhr, B. S., & Quiroga, B. F. (2007). Family ownership and firm performance: Evidence from public companies in Chile. *Family Business Review*, *20*(2), 83–94. doi:10.1111/j.1741-6248.2007.00087.x

McCollom, M. E. (1990). Problems and prospects in clinical research on family firms. *Family Business Review*, *3*(3), 245–262. doi:10.1111/j.1741-6248.1990.00245.x

McKenny, A. F., Payne, G. T., Zachary, M. A., & Short, J. C. (2014). Multilevel analysis in family business studies. In L. Melin, M. Nordqvist, & P. Sharma (Eds.), *The SAGE handbook of family business* (pp. 594–608). Thousand Oaks, CA: Sage. doi:10.4135/9781446247556.n30

Melin, L., & Nordqvist, M. (2007). The reflexive dynamics of institutionalization: The case of the family business. *Strategic Organization*, *5*(3), 321–333. doi:10.1177/1476127007079959

Miller, D., Steier, L., & Le Breton-Miller, I. (2003). Lost in time: Intergenerational succession, change, and failure in family business. *Journal of Business Venturing*, *18*(4), 513–531. doi:10.1016/S0883-9026(03)00058-2

Morck, R., Shleifer, A., & Vishny, R. W. (1988). Management ownership and market valuation: An empirical analysis. *Journal of Financial Economics*, *20*, 293–315. doi:10.1016/0304-405X(88)90048-7Morck, R., Wolfenzon, D., & Yeung, B. (2005). Corporate governance, economic entrenchment, and growth. *Journal of Economic Literature*, *43*(3), 655–720. doi:10.1257/002205105774431252

Morck, R., & Yeung, B. (2003). Agency problems in large family business groups. *Entrepreneurship Theory and Practice*, *27*(4), 367–382. doi:10.1111/1540-8520.t01-1-00015

Muntean, S. C. (2009). *A political theory of the firm: Why ownership matters*, (Unpublished doctoral dissertation).University of California, San Diego, CA.

Murphy, L., & Lambrechts, F. (2015). Investigating the actual career decisions of the next generation: The impact of family business involvement. *Journal of Family Business Strategy*, *6*(1), 33–44. doi:10.1016/j.jfbs.2014.10.003

Murray, B. (2003). The succession transition process: A longitudinal perspective. *Family Business Review*, *16*(1), 17–33. doi:10.1111/j.1741-6248.2003.00017.x

Nordqvist, M., Hall, A., & Melin, L. (2009). Qualitative research on family businesses: The relevance and usefulness of the interpretive approach. *Journal of Management & Organization*, *15*(3), 294–308. doi:10.1017/S1833367200002637

Olson, D. H. (2000). Circumplex model of marital and family systems. *Journal of Family Therapy, 22*(2), 144–167. doi:10.1111/1467-6427.00144 PMID:6840263

Oomen, O., & Ooztaso, A. (2008). Construction project network evaluation with correlated schedule risk analysis model. *Journal of Construction Engineering and Management*, (1), 49–63.

Oswald, S. L., Muse, L. A., & Rutherford, M. W. (2009). The influence of large stake family control on performance: Is it agency or entrenchment? *Journal of Small Business Management, 47*(1), 116–135. doi:10.1111/j.1540-627X.2008.00264.x

Pache, A.-C. & Santos, F. M. (2010). Inside the hybrid organization: an organizational level view of responses to conflicting institutional demands (August 2010). *Research Center ESSEC Working Paper 1101*. doi:10.2139srn

Parada, M. J., Nordqvist, M., & Gimeno, A. (2010). Institutionalizing the family business: The role of professional associations in fostering a change of values. *Family Business Review, 23*(4), 355–372. doi:10.1177/0894486510381756

Pearson, A. W., Carr, J. C., & Shaw, J. C. (2008). Toward a theory of familiness: A social capital perspective. *Entrepreneurship Theory and Practice, 32*(6), 949–969. doi:10.1111/j.1540-6520.2008.00265.x

Peay, R. T., & Dyer, G. (1989). Power orientations of entrepreneurs and succession planning. *Journal of Small Business Management, 27*(1), 47–52.

Pellegrino, B. & Zingales, L. (2014). Diagnosing the Italian disease. *Chicago Booth Working Paper*. Retrieved from http://faculty.chicagobooth.edu/luigi.zingales/papers/research/Diagnosing.pdf

Pérez-González, F. (2006). Inherited control and firm performance. *The American Economic Review, 96*(5), 1559–1588. doi:10.1257/aer.96.5.1559

Pratt, M. G., & Foreman, P. O. (2000). Classifying managerial responses to multiple organizational identities. *Academy of Management Review, 25*(1), 18–42. doi:10.5465/amr.2000.2791601

Ram, M. (1994). *Managing to survive: Working lives in small firms*. Oxford, UK: Blackwell.

Rousseau, D. M. (1985). Issues of level in organizational research: Multi-level and cross-level perspectives. In L. L. Cummings & B. M. Staw (Eds.), *Research in organizational behavior* (Vol. 7, pp. 1–37). Greenwich, CT: JAI Press.

Rubenson, G. C., & Gupta, A. K. (1996). The initial succession: A contingency model of founder tenure. *Entrepreneurship Theory and Practice, 21*(2), 21–32. doi:10.1177/104225879602100202

Sadkowska J. (2017). The impact of the project environment uncertainty on project management practices in family firms. *PM World Journal, 6*, 7, July 2017, 1-12.

Salovey, P., O'Leary, A., Stretton, M. S., Fishkin, S. A., & Drake, C. A. (1991). Influence of mood on judgments about health and illness. In J. P. Forgas (Ed.), *Emotion and social judgments* (pp. 241–262). Elmsford, NY: Pergamon Press.

Santiago, A. L. (2000). Succession experiences in Philippine family businesses. *Family Business Review, 13*(1), 15–35. doi:10.1111/j.1741-6248.2000.00015.x

Scabini, E. (1995). *Psicologia sociale della famiglia*. Torino, Italy: Bollati Boringhieri.

Scabini, E., & Iafrate, R. (2003). *Psicologia dei legami familiari*. Roma, Italy: Carocci.

Schulze, W. S., Lubatkin, M. H., & Dino, R. N. (2003). Toward a theory of agency and altruism in family firms. *Journal of Business Venturing, 18*(4), 473–490. doi:10.1016/S0883-9026(03)00054-5

Schulze, W. S., Lubatkin, M. H., Dino, R. N., & Buchholtz, A. K. (2001). Agency relationships in family firms: Theory and evidence. *Organization Science, 12*(2), 99–116. doi:10.1287/orsc.12.2.99.10114

Shanker, M. C., & Astrachan, J. H. (1996). Myths and realities: family businesses' contribution to the us economy—a framework for assessing family business statistics. *Family Business Review, 9*(2), 107–119. doi:10.1111/j.1741-6248.1996.00107.x

Shepherd, D., & Haynie, J. M. (2009). Family business, identità conflict, and an expedited entrepreneurial process: A process of resolving identity conflict. *Entrepreneurship Theory and Practice, 33*(6), 1245–1264. doi:10.1111/j.1540-6520.2009.00344.x

Sirmon, D. G., Arrègle, J.-L., Hitt, M. A., & Webb, J. W. (2008). The role of family influence in firms' strategic responses to threat of imitation. *Entrepreneurship Theory and Practice, 32*(6), 979–998. doi:10.1111/j.1540-6520.2008.00267.x

Sirmon, D. G., & Hitt, M. A. (2003). Managing resource: Linking unique resource, management and wealth creation in family firms. *Entrepreneurship Theory and Practice, 27*(4), 339–358. doi:10.1111/1540-8520.t01-1-00013

Sorenson, R. (1999). Conflict management strategies used by successful family businesses. *Family Business Review, 12*(4), 325–340. doi:10.1111/j.1741-6248.1999.00325.x

Sorenson, R. L., Goodpaster, K. E., Hedberg, P. R., & Yu, A. (2009). The family point of view, family social capital, and firm performance: An exploratory test. *Family Business Review, 22*(3), 239–253. doi:10.1177/0894486509332456

Spaltro, E., & de Vito Piscicelli, P. (1990). *Psicologia per le organizzazioni*. Roma, Italy: Carocci.

Steier, L. (2001). Family firms, plural forms of governance, and the evolving role of trust. *Family Business Review, 14*(4), 353–367. doi:10.1111/j.1741-6248.2001.00353.x

Steier, L., & Miller, D. (2010). Pre- and post-succession governance philosophies in entrepreneurial family firms. *Journal of Family Business Strategy, 1*(3), 145–154. doi:10.1016/j.jfbs.2010.07.001

Stewart, A., & Hitt, M. A. (2012). Why can't a family business be more like a nonfamily business? modes of professionalization in family firms. *Family Business Review, 25*(1), 58–86. doi:10.1177/0894486511421665

Strauss, A., & Corbin, J. (1998). *Basics of qualitative research: grounded theory procedures and techniques* (2nd ed.). Newbury Park, CA: Sage.

Stryker, S. (1968). Identity salience and role performance. *Journal of Marriage and the Family, 4*(4), 558–564. doi:10.2307/349494

Sundaramurthy, C. (2008). Sustaining trust within family businesses. *Family Business Review*, *21*(1), 89–102. doi:10.1111/j.1741-6248.2007.00110.x

Swogger, G. (1991). Assessing the successor generation in family businesses. *Family Business Review*, *4*(4), 397–411. doi:10.1111/j.1741-6248.1991.00397.x

Tajfel, H., & Turner, J. C. (1986). The social identity theory of intergroup behaviour. In S. Worchel & W. G. Austin (Eds.), *Psychology of intergroup relations* (pp. 7–24). Chicago, IL: Nelson-Hall Publishers.

Tsang, E. W. K. (2002). Learning from overseas venturing experience: The case of Chinese family business. *Journal of Business Venturing*, *17*(1), 21–40. doi:10.1016/S0883-9026(00)00052-5

Uhlaner, L. M. (2006). Business family as a team: Underlying force for sustained competitive advantage. In P. Z. Poutziouris, K. X. Smyrnios, & S. B. Klein (Eds.), *Handbook of research on family business* (pp. 125–144). Cheltenham, UK: Edward Elgar. doi:10.4337/9781847204394.00016

Villanueva, J., & Sapienza, H. J. (2009). Goal tolerance, outside investors, and family firm governance. *Entrepreneurship Theory and Practice*, *33*(6), 1193–1199. doi:10.1111/j.1540-6520.2009.00340.x

Westhead, P., & Howorth, C. (2007). "Types" of private family firms: An exploratory conceptual and empirical analysis. *Entrepreneurship and Regional Development*, *19*(5), 405–431. doi:10.1080/08985620701552405

Whetten, D., & Mackey, A. (2002). A social actor conception of organizational identity and its implications for the study of organizational reputation. *Business & Society*, *41*(4), 393–415. doi:10.1177/0007650302238775

Whetten, D. A., Foreman, P. & Dyer, W. G. (2014), Organizational identity and family business, In L. Melin, M. Nordqvist, & P. Sharma (Eds.), The SAGE handbook of family business (1st ed.), Thousand Oaks, CA: Sage. pp. 480–497. doi:10.4135/9781446247556.n24

Yin, R. K. (2003). *Case study research: Design and methods* (3rd ed., Vol. 5). Thousand Oaks, CA: Sage.

Zahra, S. A., Hayton, J. C., Neubaum, D. O., Dibrell, C., & Craig, J. (2008). Culture of family commitment and strategic flexibility: The moderating effect of stewardship. *Entrepreneurship Theory and Practice*, *32*(6), 1035–1054. doi:10.1111/j.1540-6520.2008.00271.x

Zahra, S. A., Hayton, J. C., & Salvato, C. (2004). Entrepreneurship in family vs. non-family firms: A resource-based analysis of the effect of organizational culture. *Entrepreneurship Theory and Practice*, *28*(4), 363–381. doi:10.1111/j.1540-6520.2004.00051.x

Zellweger, T. M., & Astrachan, J. H. (2008). On the emotional value of owning a firm. *Family Business Review*, *21*(4), 347–363. doi:10.1177/08944865080210040106

Zellweger, T. M., & Dehlen, T. (2012). Value is in the eye of the owner: affect infusion and socioemotional wealth among family firm owners. *Family Business Review*, *25*(3), 280–297. doi:10.1177/0894486511416648

Zellweger, T. M., Eddleston, K. A., & Kellermanns, F. W. (2010). Exploring the concept of familiness: Introducing family firm identity. *Journal of Family Business Strategy*, *1*(1), 54–63. doi:10.1016/j.jfbs.2009.12.003

Zellweger, T. M., Kellermanns, F. W., Chrisman, J. J., & Chua, J. H. (2011). Family control and family firm valuations by family CEOs: The importance of intentions for transgenerational control. *Organization Science*, *23*(3), 851–868. doi:10.1287/orsc.1110.0665

Zellweger, T. M., Kellermanns, F. W., Eddleston, K. A., & Memili, E. (2012). Building a family firm image: How family firms capitalize on their family ties. *Journal of Family Business Strategy*, *3*(4), 239–250. doi:10.1016/j.jfbs.2012.10.001

Zellweger, T. M., Nason, R. S., Nordqvist, M., & Brush, V. (2013). Why do family firms strive for nonfinancial goals? an organizational identity perspective. *Entrepreneurship Theory and Practice*, *37*(2), 229–248. doi:10.1111/j.1540-6520.2011.00466.x

Zhang, J. & Ma, H. (2009). Adoption of professional management in Chinese family business: A multilevel analysis of impetuses and impediments. *Asia Pacific Journal of Management*, 26, 119-139.

KEY TERMS AND DEFINITIONS

Circumplex Model: The Olson Circumplex Model (Olson, 2000) conceptualises flexibility, cohesion and communication skills as three central variables that define family interactions. Based on a conceptual clustering of many concepts designed to describe family and couple dynamics, the model "is specifically designed for clinical assessment, treatment planning and research on outcome effectiveness of marital and family therapy" (Olson, 2000, p. 144).

Dysfunctional Behaviour: The term dysfunctional is defined as "abnormal or impaired functioning" on the part of an individual person, between people in any sort of relationship, or amongst members of a family. Poor functioning refers to both behaviour and relationships that aren't working and have one or more negative, unhealthy aspects to them, such as poor communication or frequent conflict. This is a term used often by mental health professionals for interactions between people and is often used to describe any relationship in which there are significant problems or struggles.

Family Business: This chapter follows Chua et al. (1999, p. 25), by defining a family business as: "a business governed and/or managed with the intention to shape and pursue the vision of the business held by a dominant coalition controlled by members of the same family or a small number of families in a manner that is *potentially sustainable across generations* of the family or families".

Multilevel Analysis: The fundamental premise of this paradigm comes from the theory of organizational systems and resides in the fact that organizations are complex and dynamic systems with multiple levels. Thus, in any organizational system we find multiple levels such as industry, organization, group, individual etc.; with lower levels nested and integrated within the upper ones.

Organizational-Identity: Albert and Whetten (1985) argue that organizational identity is (a) what is taken by employees to be the central attributes of the organization; (b) what makes the organization distinctive and, therefore, unique from other organizations in the eyes of the employees; and (c) what is perceived by employees to be enduring or continuing, regardless of objective changes in the organizational environments. The three characteristics described above suggest that organizations with a strong identity have central attributes, are distinctive from other organizations and remain the same for longer periods.

Socio-Emotional Wealth: Conceptualized in broad terms to capture the stock-of-affect-related-value-that a family derives from its controlling position in a particular firm.

Chapter 7
Restructuring the Production Process:
Use of Technology and Value Creation for a Law Firm

Valéria Rocha Da Costa
Fundação Dom Cabral, Brazil

José Márcio Diniz Filho
Fundação Dom Cabral, Brazil

ABSTRACT

Process management, innovation, technology, and knowledge management are tools to achieve better results and create value for an organization, specifically for the law firm. This is why organizational processes, or business processes, have become fundamental structures for the management of modern organizations and to maintain the competitiveness of organizations. As a result, it was possible to identify that the use of process management techniques and tools is decisive for rational use of processes, increased productivity, and better customer service, presenting an ideal conceptual model.

INTRODUCTION

The importance of processes for organizations is growing. Due to increases in competitiveness and, consequently, in customer focus, organizations should be conscious of their processes. The focus on consumer and business processes has never been so intense. Therefore, it is important to improve the processes in order to improve the *performance* of an organization.

Increasing and continuous pressures for performance improvement make organizations rethink and change the way they operate their activities. In this context, law firms, as in so many other businesses, increasingly seek the rational use of processes, increased productivity, and better customer service. They are, therefore, facing the following issues: How would it be possible to achieve more operational results

DOI: 10.4018/978-1-5225-9993-7.ch007

with fewer resources and at the same time generate value for the customer? How should we improve the management model to gain with integration, efficiency and increased economic-financial results?

In this line, this article will address the necessity and usefulness of the redesign of production processes, with the use of technology for value generation, for a law firm in the tax area.

The Processes of Production, Process Management, and Technology

Increased competitiveness, increased customer focus, and offerings of customized services, specialized in providing legal advisory services, consulting, and litigation all make the processes in a company's organizational structure more important. The focus on consumer and business processes has never been so intense (Seethamraju, 2012). This is why it is important to improve processes in order to improve the performance of an organization (Damian, Borges, & Padua, 2015).

Organizational processes, also known as *business processes*, have become fundamental structures managing modern organizations (Campos, 2013). In recent decades, this concept has gained great popularity and has been disseminated broadly (Albuquerque, 2012). According to Gonçalves (2000), "There is no product or service offered by a company without a business process." Understanding operational workflow has become a necessary condition in developing products and services (Kluska, Lima, & Costa, 2015).

Areas, such as human resources, information technology, and management, have shown an increasing interest in organizational processes. This interest is largely due to the capacity of process management to horizontally transpose the functional structures of an organization. From this point of view, processes transcend the areas of an organization "in order to create flows from end to end" (Kluska, Lima, & Costa, 2015), providing savings, optimization, and control over the operation (ABPMP, 2014).

The constant and growing need for improved performance causes organizations to rethink and change its activities and to review and redesign its processes. Results are most effective with a focus on the management of business processes, and studies show that the alignment of processes is a "mediator who contributes to the involvement of people and organizational performance" (Damian, Borges, & Padua, 2015).

Because of this, companies tend toward a management approach focused on processes rather than functional performance. Process-focused management is called the *Business Process Management* (BPM). According to Damian, Borges, and Pádua, (2015):

Business Process Management-BPM is an approach to identifying, designing, executing, documenting, measuring, monitoring, controlling and improving business processes so that desired results can be achieved (ABPMP, 2009). BPM considers the process as imperative for the business and as a means of understanding and making explicit the business activities according to the needs and interests of the clients (Smart, Maddern, & Maull, 2009). In addition to being important for the transparency of the company in the internal environment, BPM is one of the ways to handle the challenge of improving the company's business processes to optimize its performance.

BPM is one of the most important practices in the business world. Companies should seek permanent performance improvement to provide greater strategic focus and align strategy with results (Hernaus, Bach, & Vuksic, 2012). BPM enables organizations to create dynamic, flexible collaborations to synergistically adapt to competitive market conditions by understanding, documenting, modeling, and analyzing business processes. With it, organizations can achieve visibility and transparency improvements, reduce costs, earn more efficiency in production, and achieve a better business performance result (Damian, Borges, & Pádua, 2015).

Certain factors improve process-focused management practices, through which routine activities are disseminated by those with knowledge of the process. These are called Critical Success Factors (Damian, Borges, & Pádua, 2015).

The *Critical Success Factors* can be classified into three groups, one relating to the fit between the business environment and business processes, one related to continuous improvement to ensure sustained benefits, and one related to adjustments between information technology (IT) and business processes, meaning process standardization, automation, and training. These factors should be a goal of any organization wanting to manage processes and find success.

Business Processes

According to Kluska, Lime, & Costa (2015), following a deep comparative study and a review of literature on business processes, reached the following definition:

Business processes are a representation of how a business, guided by a systemic model composed of processes, events, activities and tasks, whose purpose is to organize the flow and transactions between suppliers and customers.

Table 1 describes key research and theoretical literature on business processes.

The conceptual understanding of business processes is necessary to scale processes integrated to the operational reality, as well as to understand those that are closely linked with the greater productivity, efficiency, control, and transparency. Business process modeling is the set of activities involved in creating complete and accurate business processes. It creates models about activity routines observed in organizational reality. Through business process modeling, an organization can improve its processes, gain efficiency, and gain competitive advantages (Kluska, Lima, & Costa, 2015).

Process Management and Value Generation

Process management is with the increasing adoption of information technology; however, there is a growing consensus that an organization should be considered holistically, mainly because of the corporate culture that is allied to the practical implementation of BPM. This practical implementation is comprised of designing, developing, and executing business processes, as well as considering the interactions between these processes. In addition, BPM has the following basic foundations: (a) process mapping and documentation activities, (b) customer focus, (c) measurement activities to evaluate the performance of the processes, (d) optimizing processes, (e) best practices to improve competitive positioning, and (f) approaches to culture change in an organization (Damian, Borges, & Pádua, 2015) .

Organizations that want to apply process-based management need to focus on customers because businesses start and end in them (Gonçalves, 2000). Projects that involve value creation with a high level of customer engagement require a process management and modeling approach that integrates the perspectives of the client and the company. Processes to generate value must not be focused on the organization but rather on the customer. Because process management presupposes performance, through to delivering value to the client, processes must overcome traditional departmental barriers in an organization (ABPMP, 2009).

Table 1. Business process terms and definitions

Termo Term	Definição Definition
Processos (PLATTS et al. ,Process (Platts et al.1996), 1996)	São eventos sequenciados que descrevem modificaçõesSequenced events that describe changes aoover time, usually developed for a single goal.
Processo (JURAN,Process (Juran, 1992) 1992)	É uma série sistemática de ações direcionadasSSA systematic series of directed actions to achieve a goal. A definição genérica aplica-se The generic definition applies here, for all functions, related to amanufacturing or not. Também inclui as forças humanas, It also includes human forces,assim like facilities andfísicas. physics.
Processo (D'ACENÇÃO,Process (D'Acenção, 2001) 2001)	É um conjunto de causas, que provoca um ou maisAAA set of causes that result in one or moreefeitos. effects.
Processo (OLIVEIRA,Process (Oliveira, 2006) 2006)	UmAAA conjunto estruturadostructured sequende atividades sequenciaisce of related activitiesquetoto to atender e, preferencialmente, suplantar as necessidades emeet and preferably exceed the needs andexpectativas dos clientes externos e internos da expectations of external and internal customers.
ProcessoABNT Process (ABNT - Association BRASILEIRA DE Brazilian NORMAS Standards TÉCNICAS, Techniques, 2008) 2008)	Processos podem ser compreendidos como aThetransformação transformationde entradas e of inputs intosaídas. outputs.
ProcessoProcess de in negócio business (ABPMP, (ABPMP, 2014) 2014)	É uma agregação de atividades e componentesAnaAn aggregation of activities and components executadosexecutedpor humanos ou máquinas para alcançar um by humans or machines to achieveoumais results.
Processo de negócioBusiness process (ERIKSSON (Eriksson E PENKER, & Penker, 2000) 2000)	ÉAnAn abstraçãoabstraction dobusiness operations,composto por: objetivos, recursos, processos e composed of objectives, resources, processes, andregras. rules.
Processos de NegócioBusiness Processes (OMG, (OMG, 2011)	É qualquer atividade executada dentro de uma companhiaAny activity performed within a company ouororganização. organization.

Source: Kluska, Lime, & Costa, 2015.

Process Management Versus the Functional Approach

In the functional and hierarchical approach, processes are managed in isolation, separated by departments. The organization assumes characteristics of silos, with limited coordination capacity and low market orientation. In management by processes, however, there are changes in the integral elements of the organizational project, aiming at prioritizing the processes as a central managerial point of greater importance than the functional axis, orienting decisions predominantly by processes (Damian, Borges, & Padua, 2015).

A functional view focuses on dividing the process activities assigned to departments and areas with specific functions. The practice of this vision produces silos, or fiefdoms, in organizations and does not consider the process as a whole, allowing for some areas of greater influence and voids between functional areas which, in turn, causes opposing and conflicting goals in the organization. In this format, it is more difficult to perceive the contribution of each department in the generating value for the client, since the functions prove to be more important than the result of the process and the responsibilities for each role end up being lost in the voids between functional areas.

On the other hand, for Neubauer (2009), an organization focused on process management needs to have:

- A strategy systematically aligned with business;
- Mechanisms such as the *balanced scorecard* to support and measure alignment;

- Management methodologies such as *lean, six sigma,* or processes improvement;
- A *Chief Process Officer* (CPO) supported by processes;
- IT resources allocated according to the processes;
- Coverage of processes between organizations partners.
- Coverage of all organizational processes (end-to-end); and
- Mechanisms for risk management.

BPM maturity in organizations has not yet reached a significant level, and factors such as quality improvement, process standardization, and productivity should receive more emphasis (Damian, Borges, & Pádua, 2015).

The role of the process manager changes over time. In immature organizations, process managers seem to work alone and fight for resources. A successful BPM project, however, requires a well-organized team that is able to analyze, design, implement, and continually refine processes according to a business strategy (Neubauer, 2009).

Tasks for Process Management

In order to improve and implement BPM, the authoritative literature identifies twenty-four necessary tasks, grouping them under (a) designing processes, (b) managing day-to-day processes, and (c) promoting learning and development, as can be seen in Table 2.

Migration from the traditional management model to process management results in improved process performance and also promotes increased accountability for results, monitoring processes, and improving steadily. There are critical success factors in the migration from the traditional (functional) model to the process management model.

Table 2. Process management tasks

Group of tasks	Description
Designing processes	Understanding the internal and external environment; strategy and approach to change; securing support for change; understanding, selecting and prioritizing processes and tools; teams and their processes; understanding and modeling processes in the current situation; setting and prioritizing solutions for the current problem; defining management practices the execution of processes; understanding and modeling future processes; defining and implementing changes in processes.
Managing day-to-day processes	The tasks of this group relate to implementing processes and changes; Promote the realization of the processes; monitor and control the implementation of the processes; and make short-term changes.
Promoting learning and development	Process benchmarking, recording and controlling impact deviations, evaluating process performance, and documenting learning.

Source: Damian, Borges, & Paddle, 2015.

Restructuring the Production Process

Key Critical Success Factors in BPM

A predominant model has emerged based on functional management, with limitations to deal with the current reality. Organizations are managed by functional, segregated departments and still cannot discern the relationships between organizational processes. This doctrine emphasizes Critical Success Factors listed below that contribute to a change of management focus and provide a process management environment (Damian, Borges, & Pádua, 2015).

1. Executive sponsorship (Minonne & Turner, 2012) and governance mechanisms that support management with processes, beyond gives selection and The training project leaders (Antonucci et al., 2009);
2. Alignment of business strategy with operation operations (Jeston & Nelis, 2006);
3. Definition of goals for the processes (Antonucci et al., 2009);
4. Definition of process owners (Antonucci et al., 2009);
5. Monitoring business processes and maintaining the improvement program for the processes (Jeston & Nelis, 2006);
6. Focus on concepts and principles related to management (Jeston & Nelis, 2006);
7. Definition of the roles of process management with the systemic view of organization (McComack et al., 2009);
8. Changing the functional vision for process vision (Sentanin, Santos, & Jabbour, 2008) through a change in organizational culture (Brocke & Sinnl, 2011);
9. Development of improvement actions should be focused on the causes and not the effects (Ohtonen & Lainema, 2011);
10. Standardization of processes, since it is common for departments to work the same logical abstraction of the process, but with different denominations; With little integration between the methodologies, techniques and tools employed in the various areas; and different views of the set of activities, products, customers and other components of the process (Jeston; Nelis, 2006);
11. Direct involvement of the leadership and executive team (ABPMP, 2009); management of organizational change is important; people's involvement is key to BPM's success, and leaders are responsible for achieving this involvement (Jeston & Nelis, 2006);
12. Customer expectation for identification in relation to the expected results of the process (Burlton, 2010);
13. Proposing process improvements based on performance measurement (ABPMP, 2009);
14. Alignment between organization strategy and BPM projects (Bandara et al., 2007);
15. Continuity of measuring, monitoring, and controlling the process (Antonucci et al., 2009).

These critical success factors are important so that, internally, the organization can practice process management. It turns out that all the factors listed point toward the importance of continuous improvement to ensure the benefits offered by BPM.

It is possible to relate task groups, design processes, manage day-to-day processes and promote evolution and learning with Critical Success Factors, as shown in Table 3.

Table 3. Process management tasks connected with critical success factors

Group of tasks	Description	Critical Success Factors
Designing processes	Understanding internal and external environment; establishing the strategy and the approach to change; ensuring support for change; understanding, selecting, and prioritizing processes and tools; forming teams of diagnoses of processes; understanding and modeling processes in the current situation; defining and prioritizing solutions for current problems; defining management practices and the execution of processes; understanding and modeling future processes; defining and deploying changes in processes.	Executive sponsorship (Minonne & Turner, 2012) and mechanisms in governance (Antonucci et al., 2009); Alignment of strategy and operations (Jeston & Nelis, 2006); Setting targets for processes (Antonucci et al., 2009); Definition of owners of process (Antonucci et al., 2009); Domain of concepts and principles related to the management of processes (Jeston & Nelis, 2006); Definition of management roles in processes with systemic vision of organization (McComack et al., 2009); Change of functional vision to a processes view (Sentanin, Santos, & Jabbour, 2008) by means of change in organizational culture (Brocke & Sinnl, 2011); Standardization of processes. It is common that departments the same logical abstraction of a process but with many different denominations and with little integration between methodologies, techniques and tools (Jeston & Nelis, 2006); Direct involvement of the leadership and executive team (ABPMP, 2009); people's involvement is the key to BPM's success, and leaders are responsible for achieving this involvement (Nelson & Nelis, 2006); Identification of client expectations regarding the expected results of the process (Burlton, 2010); alignment between the organization's strategy and BPM projects (Bandara et al., 2007);
Managing day-to-day processes	Implementing processes and changes; promoting the execution of processes; monitoring and controlling the execution of processes; and making short-term changes.	Monitoring of business processes and maintenance of process improvement program (Jeston & Nelis, 2006); Continuity of measurement, monitoring, and control of the process (Antonucci et al., 2009).
Promoting learning and development	Recording the performance of the processes; performing process benchmarking; recording and controlling impact deviations; evaluating process performance; and recording learning about the processes.	Monitoring of business processes and maintaining a process improvement program (Jeston; Nelis, 2006); Proposition of process improvements based on performance measurement (ABPMP, 2009); Development of improvement actions should focus on causes and not effects (Ohtonen, Lainema).

Source: Damian, Borges, & Paddle, 2015.

Table 3 shows that on one hand, the migration from the traditional, functional management model to process management results in improvements in process performance, and on the other, it requires culture change, project management with process focus, management of the day-to-day process, and the promotion of learning and development.

PROCESS MANAGEMENT, INNOVATION, AND KNOWLEDGE MANAGEMENT

After having demonstrated the importance of good practices of Process Management to boost the business of organizations, especially due to the generation of value, gains with productivity and efficiency, it is worth analyzing how innovation and knowledge management can help this BPM environment.

Innovation can be described as a search process for renewal that is based on knowledge accumulated over a period of time in order to achieve a novel product, whether tangible or intangible (Mendrot et al., 2017). Another definition of innovation is the creation of sustainable value through a change in the internal paradigm of an organization. This paradigm break can be interpreted as, but not limited to, the commercialization of products/services, expansion/development of distribution channels, creation and expansion of new products/processes, and application of new marketing techniques, among others (Cavalcanti et al., 2012).

It follows that the greatest motivation for innovation is economic development and the creation of value. Thus, innovation assumes an important internal role in organizations that can be represented in their different faces, as shown in Table 4.

Innovation, depending on the scope of the organization's business model, may be radical or incremental. Innovation occurs in organizations for two reasons: economic and technological. Innovation is closely linked to BPM. Innovation culture will only achieve its objectives if the organization takes on the risks of changing its processes, supported by BPM best practices (De Bes & Kotler, 2011). The literature explains that one of the possible reasons for organizations not investing in product and process innovation lies in a resistance to taking the risks inherent in this process. For Sarkar (2010), the willingness to take risks and tolerate mistakes is indispensable for creating an environment that fosters innovation in organizations (Mendrot et al., 2017).

Table 4. Types of innovation

Classification of innovation	Definition
Product and services innovation	Develop and market new products and services, based on new technological concepts and driven by the needs and satisfaction of customers;
Innovation of processes	Develop new forms of production or new ways of relating with the client to provide services;
Innovation of business	Develop new business models that provide a competitive advantage in a sustainable way; and
Innovation of management	Develop new structures hierarchically.

Source: Mendrot et al., 2017.

There will always be limits to innovation in organizations, such as term, budgets, and demand. The way to overcome constraints is through creativity in order to innovate. The innovation process will not always represent a paradigm shift. Incremental innovations in established operational and management processes are more easily verified, while radical, disruptive innovations are rarer.

Regardless of the type of innovation (radical or incremental), investment is the biggest success factor in its execution is on creativity, knowledge, skills, and dedication of person participating in this innovative process (Drucker, 2007). The main source of innovation is an organization's process structure and employees (i.e., employees and partners), who live the day-to-day life of the company, providing greater knowledge of the company's innovation needs. Innovation and management processes are closely related, yielding an environment conducive to creativity in order to develop innovative processes and create value for the organization.

Process Automation with a Focus on Value Generation

Organizations are increasingly automating their processes. Automation can occur by deploying workflow tools or by using systems specific to certain needs. The goal of automation is to make processes more efficient and robust, supporting a workload and production that is greater and more secure. However, if processes are not well established and built, the information systems deployed from them will not be better. That is, the reason for the failure in automation is often not information technology but inconsistencies in the pre-established processes.

For Kluska, Lime, & Costa (2015), workflow tools usually include process development, management, and optimization systems based on best BPM practices:

The workflow tools are, for the most part, BPMS solutions, which are specialized suites for the development, management, and optimization of processes. Usually, these are software that associates graphic asset palettes with programming scripts in the format of lines of code. These tools within the organizations, due to its short- and long-term benefits (ABPMP, 2014; Gou et al., 2003).

Process Management and Knowledge Management

Organizations are like brains, managing a large number of data and information, compiling and processing them in systems, and then, through intelligent analysis, yielding new knowledge (Morgan, 2000). In organizations, one can consider any pure and simple record of organizational events or structural texts of transactions performed to be given. *Information* can be considered the result of data analysis with the purpose of modifying the perception, judgment, or attitude of its receiver about the context in which it intervenes (Lins, 2003). *Knowledge* can be defined as a justified and true belief (Takeuchi & Nonaka, 2008). More fully, Devanport and Prusak (1998) present knowledge as a condensed mass of experiences, values, contextual information, allowing the evaluation and incorporation of new experiences.

In organizations, knowledge is embedded in documents, operational routines, processes, practices, and organizational norms. Good organizational knowledge can be socially constructed by individuals within an organization while engaging in daily activities to achieve predetermined goals (Mendrot et al., 2017). Knowledge can be explicit or tacit (Takeuchi & Nonaka, 2008). *Tacit knowledge* is defined as personal and specific to its context, making it difficult to formalize and transmit. *Explicit*, or *codified*, knowledge refers to knowledge that is systematized, transmissible in formal language. It can be expressed in text and number, easy to communicate and share.

Institutional knowledge is generated by the transformation of tacit knowledge into explicit and vice versa. The transcription of the phases of tacit knowledge conversion into explicit is worthwhile (Takeuchi & Nonaka, 2008). The transformation of knowledge into knowledge socialize individual knowledge throughout the organization. Each individual is unique and able, through their knowledge, skills, and competencies, to contribute to the group and to the organization in general. The sum of individual knowledge provides an organization with a competitive advantage, creating the concept of *intellectual capital* (Stewart, 1998).

It is important to develop management by processes and culture with mechanisms for retention of such information and experiences lived by the organization's employees, making it the most productive and intelligent. In this sense, it is worth mentioning the main motivations for knowledge management in the view of Mendrot et al. (2017):

Alvarenga Drummond (2005) presents as main motivators for the adoption of management of knowledge by organizations: 1 - recognition of information as an important factor in commercial competitiveness; 2 - need for continuous innovation; 3 - relative problems in the dissemination of information, evidencing The lack in Add the same [sic]; 4 - lack in routines for sharing and protection of information; 5 - cultural change for collaborative culture; and 6 - to stimulate organizational learning.

Medrot et al. (2017) also highlights the main drivers for promoting and creating intellectual capital in organizations:

Takeuchi and Nonaka (2008) present five factors to promote the creation of the organization's intellectual capital:

1. Inculcate the vision of knowledge: communicate the need to transfer the mechanics of business strategy in order to create an overview of organizational knowledge. This information, if properly managed, will provide the strategic business plan, through the full knowledge of its competitive advantages, a solid strategy of advancement;
2. Management of the talks: manage communications between employees who carry the essence of organizational activities;
3. Mobilization of knowledge activists: correct allocation of professionals who stimulate and enjoy pre-established organizational knowledge and create in the activities performed new knowledge. These employees must be used through the organization with six purposes: focus and initialization of knowledge creation; reduction of time and costs required to create knowledge; leverage in initiatives for creation of knowledge; improvement of conditions for creation of knowledge; preparation in participating professionals of context in creation of knowledge for new tasks; and stimulation of discussion in group in factors in organizational transformation;
4. Creation of correct context: verification of the relationship between the organizational structure, the strategy and promotion of knowledge, in the end ensuring that the company develop conditions suitable for encouraging the creation in institutional knowledge;
5. Globalization of Knowledge: Ensure that the information generated in a local unit quickly and efficiently reaches the other units of the corporation, taking into account the regional particularities that must be analysed and adapted in compared to the scenario faced by each specific unit.

Therefore, an efficient program of intellectual capital management must include the retention of the intrinsic capital to the collaborators, through measures of reward and recognition, the support of information technology, and well defined processes, in order to store in a standardized way the mass of data generated with the implementation of a knowledge management program, or to manage the availability of this information for use in the organization.

Process Management and Reduction Costs

BPM is a structured approach to analyze and continuously improve the core activities of organizations such as production, marketing, communications, financial, accounting and other departments that are part and operation a company. In short, it is not wrong to say that BPM can be understood as an important methodology for measurement of key processes that analyzes what works and what does not work and needs to be improved in an organization. BPM is focused on eliminating waste and adding value to processes and organization.

The Toyota Production System (STP) can be understood in the same line of BPM as a system that aims at total loss elimination, i.e., a functional network of operations and processes, which can be a factory or an office, with the clear objective of elimination of activities that generate cost and that does not add surcharges value and the customer's point of view (Tegner et al. 2016).

Losses, widely discussed in the literature, are defined as seven kinds: defects, overproduction, waits, transport, processing itself, stock, and handling. Besides these, Shingo (1996) expresses two types of generic losses in the manufacture: a) *mura*: lack of regularity in production, leading to high variability; and b) *died*: overload in the people or equipment, leading to early wear due to the fast pace.

Tegner et al. (2016) explains that the designation STP for the term Lean Manufacturing was brought by Womack, Jones, and Roos (1990) after the search Massachusetts Institute of Technology (MIT). Tegner et al. (2016) also states that the concepts of Lean apply not only to manufacturing but also to various production areas such as construction, human resources, education, services, public service, and administrative areas and states:

The seminal work on the subject is written by Tapping and Shuker (2010), which presents a methodology for planning, mapping and sustaining Lean improvements through the systematic collection and analysis of data. The publication addresses the project in eight steps, which correspond to:

a) Commit to Lean: Alignment between management, management, and employees on their ongoing efforts with Lean initiatives, as well as the definition of the implementation team;
b) Choose Value Flow: Definition of all activities, including those that do not add value, that transforms the information and raw material into an end product that the customer is willing to pay;
c) Learn about Lean: Reviewing concepts and Lean tools that must be transmitted to those involved during training;
d) Map Current State: Expression of the flow of object of job and information per means of visual representation;
e) Identify Metrics Lean: Determining the metrics that will help achieve the goals Lean company, using them to help drive the improvement continuous and disposal waste;
f) Mapping the Future State: Understanding customer demands, establishing a continuous flow so that internal and external customers receive the object of the correct work, at the right moment and quantity, and distribute the work evenly;

Restructuring the Production Process

g) Create the Kaizen Plans: The creation of Kaizen plans to modify and improve the processes studied, as well as the planning of the phases in implementation of the Kaizen;
h) Implement the Kaizen Plans: Moment of performing the Lean transformation, implementing the previously planned Kaizen activities.

The analysis of the two methods management BPM and STP shows convergence points: a) focus on maximizing processes, b) increased productivity and efficiency, c) integration of the areas through well-defined and efficient processes, d) generating value for organizations, and e) reduction and waste of human resources, inputs and raw materials.

Conceptual Model Ideal of Management Processes Production in Law Firm

The ideal model of production process management in tax litigation is based on the BPM's best practices, the design of processes focused on generating value for the client and, consequently, for a law firm. Process management allows companies to adapt synergistically to competitive market conditions because, by understanding, documenting, modeling and analyzing business processes, organizations can achieve improvements in visibility and transparency, reduce costs, increase efficiency in production and achieve a better business performance result.

Organizations implementing process management should focus on the customer because it is in the perception of value that the organization is reaching its goal and achieving superior results. Above all, it is necessary to adopt functional management models. In Process Management, the manager must envisage the relationship between the organizational processes.

Process management must contribute to creating an environment of innovation and economic development and creating value for the organization. The processes should be well designed and defined and they should be automated through information technology systems and workflow whenever possible, making them more efficient and robust, supporting a larger and safer production workload. The generation of knowledge must be considered and supported in the management of processes, in a way that recognizes and develops the intellectual capital of the organization.

Implementation of process management requires the following resources:

1. Sufficient quantity of trained people;
2. Development of culture focused on process management;
3. Definition and design of processes in an organizational way, and
4. Implantation of technology systems to assist in production and control.

The main aspect for the implementation of a process management project in the production area of a law firm is in the best use of people for each activity, in a well-organized way, with well-designed processes and with the massive use of technology that can provide greater productivity, generation of knowledge and control for organization, generating value for the customer with a faster service and a superior quality to the market.

CONCLUSION

It is not difficult to conclude that management by processes is an important way to achieve value creation for the organization. Changing a process management system based on a functional structure to a BPM model reinforces the integration of production, financial, and accounting areas, generating value for the company. The implementation and adoption of success factors in process management contribute to process transparency, improved productivity, and efficiency in the production process of the law firm.

Innovation is important in law business because it makes possible the economic development and the creation of value for the organization. The law firm, being an intellectual capital intensive company, should always seek to retain this capital intrinsic to its employees, using measures of reward and recognition, with comprehensive use of information technology and well defined processes, in order to store the mass of data generated by the implementation of a knowledge management program in a normalized way, as well as manage the availability of this information for use in the organization.

The success of law firms is in the best use of people for each activity, in a well-organized way, with well-designed processes and with the full use of technology systems that can provide greater productivity, knowledge generation, and control for the organization, generating value for the customer with fast, transparent and effective customer service.

REFERENCES

ABPMP. (2014). *BPM CBOKTM V3.0. guide to the business process management common body of knowledge. 2*. Brasil: Association of Businees Process Management Professsionals Brasil.

ABPMP (Association of Business Process Management Professionals). (2009). *Guide to business process management: Common knowledge body*, Version 2.0, 2009.

Albuquerque, J. D. (2012). Flexibility and business process modeling: A multidimensional and national relationship. *RAE: Journal of Business Administration, 52*(3), 313–329.

Alvarenga, N., & Rivadávia, C. D. (2005). *Gestão do conhecimento em organizações: proposta de mapeamento conceitual integrativo*. São Paulo, Brazil.

Antonucci, Y. L., & ... . (2009). *Business process management common body of knowledge*. Terre Haute, IN: ABPMP.

Associação Brasileira de Normas Técnicas. (2008). *NBR ISO 9001:2008. 2*. ABNT - Associação Brasileira de Normas Técnicas. [s.l.]

Brocke, J. V., & Sinnl, T. (2011). Culture in business process management: A literature review. *Business Process Management Journal, 17*(2), 357–377. doi:10.1108/14637151111122383

Burlton, R. (2010). Delivering *business strategy through process management*. In J. Vom Brocke & M. Rosemann (Eds.), *Handbook on business process management: strategic alignment, governance, people and culture* , 2, 1, (pp. 5–37). Berlin, Germany: Springer. doi:10.1007/978-3-642-01982-1_1

Campos, A. (2013). *Modeling of processes with BPMN*. Rio de Janeiro, Brazil: Brasport Books and Multimedia.

Cavalcanti, A. M., Oliveira, M. R. G., Gracas Vieira, M., & Cavalcanti Filho, A. (2012). O característico de inovação setorial: uma métrica para avaliar potencial crescimento de inovação nas micro e pequenas empresas. In: Encontro Nacional de Engenharia de Produção, 32., Bento Gonçalves, 2012. Anais... ENEGEP.

D'Acenção, L. C. M. (2001). *Organização, sistema e método: análise, desenho e informatização de processos administrativo. 1*. São Paulo, Brazil: Editora Atlas.

Damian, I. P. M., Borges, L. S., & Pádua, S. I. D. (2015). The importance of tasks and critical success factors for business process management. *Journal of Management of UNIMEP 13*(2), p. 162. Retrieved from http://www.raunimep.com.br/ojs/index.php/regen/editor/submissionediting/899#scheduling

De Bes, F., & Kotler, P. (2011). *The innovation bible*. São Paulo, Brazil: Leya.

Drucker, P. F. (2007). *Management challenges for the 21st century*. New York, NY: Routledge.

Eriksson, H.-E., & Penker, M. (2000). *Business modeling with uml: business patterns at Work. 1*. USA: OMG PRESS.

Fleury, A. C. C., & Fleury, M. T. L. (2004). Entrepreneurial strategies and skills training: A kaleidoscopic puzzle of Brazilian industry (3rd ed.). São Paulo, Brazil: Atlas.

Gonçalves, J. E. L. (2000). *Companies are large collections of processes*. RAE: Revista dein Business Administration 40(1), p. 1.

Hernaus, T., Bach, M. P., & Vuksic, V. B. (2012). *Influence of strategic approach to BPM on financial and non-financial performance*, Baltic Journal of Management, 7(4), 376–396.

Jeston, J., & Nelis, J. (2006). *Business process management: practical guidelines to successful implementations*. Oxford, UK: Butterworth-Heinemann.

Juran, J. M. Juran Planejando para a qualidade. 2. ed. São Paulo - SP: Editora e livraria Pioneira, 1992.

Kluska, R. A., da Lima, E. P., & da Costa, S. E. G. (2015). *A proposal for structure and use of business process management (BPM)*. Revista Produção Online 15(3), 886–913. Retrieved from http://web.a.ebscohost.com/ehost/pdfviewer/pdfviewer?Vid=5&sid=e40d1be6-302b-4759-885b-ac4e11392bcc%40sdc-v-sessmgr05

Lins, S. (2003). *Transferring tacit knowledge: A constructivist approach*. Rio de Janeiro, Brazil: E-papers.

McComack, K., Willems, J., Bergh, J., Deschoolmeester, D., Willaert, P., Stemberger, M. I., ... Vlahovic, N. (2009). A global investigation of key turning points in business process maturity. *Business Process Management Journal, 15*(5), 792–815. doi:10.1108/14637150910987946

Mendrot, R. A., Oliveira, E. A. D. A., de Moraes, M. B., & Monteiro, R. (2017). The use of technical tools of project management and knowledge management to stimulate success in innovation projects. *Magazine Alcance—Eletrônica 24*(4).

Minonne, C., & Turner, G. (2012). Business process management – Are you ready for the future? *Knowledge and Process Management, 19*(3), 111–120. doi:10.1002/kpm.1388

Mintzberg, H. (1987). Crafting strategy. *Harvard Business Review, 65*(4), 66–75.

Morgan, G. (2002). *Images of the organization*. São Paulo, Brazil: Atlas.

Neubauer, T. (2009). An empirical study on the state of business process management. *Business Process Management Journal 15*(2), p. 2.

Nonaka, I., & Takeuchi, H. (1997). *Creation of knowledge in the company: How Japanese companies generate the dynamics of innovation* (4th ed.). Rio de Janeiro, Brazil: Campus.

Ohtonen, J. & Lainema, T. (2011). Critical success factors in business process management – a literature review. In Proceedings of IRIS (pp. 572-585). London, UK.

Oliveira, D. (2006). *Administração de processos: conceitos, metodologia e práticas*. São Paulo, Brazil: Editora Atlas.

OMG. (2011). *O. M. G. Business process model and notation (BPMN)*. Needham, MA: OMG.

Platts, K. W. (1993). A process approach to researching manufacturing strategy. *International Journal of Operations & Production Management, 13*(8), 4–17. doi:10.1108/01443579310039533

Sarkar, S. (2010). *Entrepreneurship and innovation. Forte da Casa*. School Publishing.

Seethamraju, R. & Marjanovic, O. (2009). Role of process knowledge in business process improvement methodology: A case study, *Business Process Management Journal 15*(6), 920–936.

Sentanin, O. F., Santos, F. C. A., & Jabbour, C. J. C. (2008). Business process management in a Brazilian public research centre. *Business Process Management Journal, 14*(4), 483–496. doi:10.1108/14637150810888037

Shingo, S. (1996). *The Toyota production system from the point of view of production* (2nd ed.). Porto Alegre, Brazil: Bookman.

Stewart, T. A. (1998). *Intellectual Capital* (3rd ed.). Rio de Janeiro, Brazil: Campus.

Takeuchi, H., & Nonaka, I. (2008). *Knowledge management*. Porto Alegre, Brazil. *The Bookman*.

Tapping, D., & Shuker, T. (2010). *Lean Office: gerenciamento do fluxo de valor para áreas administrativas - 8 passos para planejar, mapear e sustentar melhorias lean nas áreas administrativas*. São Paulo, Brazil: Leopardo Editora.

Tegner, M. G., de Lima, P. N., Veit, R. R., & Neto, S. L. H. C. (2016). Lean office and BPM: Proposition and application method for the reduction of waste in administrative areas. *Revista Produção Online, 16*(3), 1007–1032.

ADDITIONAL READING

Araújo, L., & Gava, R. (2014). *Proactive business strategies: The four keys to proactivity—Strategy, marketing, innovation, and people* (1st ed.). Rio de Janeiro: Elsevier.

Azevedo, D., Vaccaro, G. L., Lima, R. C., & da Silva, D. O. (n.d.). A computational simulation study for the analysis of organizational learning profiles. Retrieved from: http://www.scielo.br/pdf/prod/2010nahead/aop_t6_0008_0132.pdf. Accessed 10/15/2018.

Besanko, D., Dravone, D., Shanley, M., & Schaefer, S. (2006). The origins of competitive advantage: Innovation, evolution and the environment. In *The economics of strategy* (3rd ed., pp. 444–464). Porto Alegre: Bookman.

Besanko, D., Dravone, D., Shanley, M., & Schaefer, S. (2006). *Fundamentals: economic concepts for strategy. The Economics of Strategy* (3rd ed., pp. 33–64). Porto Alegre: Bookman.

Brickley, J., Smith, C. W., Zimmerman, J. L., & Willett, J. (2003). *Designing organizations to create value: From strategy to structure*. New York: McGraw-Hill.

Brickley, J. A., Smith, C. W. Jr, & Zimmerman, J. L. (1997). Management fads and organizational architecture. *Journal of Applied Corporate Finance*, *10*(2), 24–39. doi:10.1111/j.1745-6622.1997.tb00134.x

Cohen, W., & Levinthal, D. A. (1990). Absorptive capacity: A new perspective on learning and innovation. *Administrative Science Quarterly*, *35*(1), 128–152. doi:10.2307/2393553

Davenport, T., & Prusak, L. (1998). *Business Knowledge*. Rio de Janeiro: Campus.

Fama, E. F. (1980). Agency problems and the theory of the firm. *Journal of Political Economy*, *88*(2), 288–307. doi:10.1086/260866

Galbraith, J., & Downey, D. (2001). *Designing dynamic organizations: A hands-on guide for leaders at all levels*. New York: AMACOM.

Garrido, E., Gomes, J., Maicas, J., & Orcos, R. (2014). The institutional-based view: How to measure it. *Business Research Quarterly*, *17*(2), 82–101. doi:10.1016/j.brq.2013.11.001

Ghemawat, P., & Rivkin, J. (2006). *Creating competitive advantage* [Case study]. Boston: Harvard Business Review.

Jensen, M. C. (2002). A Theory of the firm: Governance, residual claims, and organizational forms. *Administrative Science Quarterly*, *47*(2), 387–389. doi:10.2307/3094817

Kavadias, S., Ladas, K., & Loch, C. (●●●). The transformative business model: How to tell if you have one. *Harvard Business Review*, *94*(10), 90–98.

Kerzner, H., & Saladis, F. P. (2011). *What executives need to know about project management*. Porto Alegre: Bookman.

Koehn, N. F., McNamara, K., Khan, N., & Legris, E. (2014). *Starbucks Coffee Company: Transformation and renewal*. Boston: Harvard Business School.

Ling, K. (2011). *Making a blue ocean strategic move that discourages imitation: The case of Wikipedia*. Paris: INSEAD.

Mendes, G. D. M. (2017). *Architecture and dynamics of organizations*. New Lima: FDC.

Nadler, D., Gerstein, M. S., & Shaw, R. B. (1992). *Organizational architecture: The key to business change*. Rio de Janeiro: Campus.

Peng, M. W. (2012). *An institution based on intellectual property rights (IPR)*. Dallas: Kelley School of Business.

Perin, M. G. (2002). *The relationship between market orientation, organizational learning and performance* (Doctoral thesis). Federal University of Rio Grande do Sul, Porto Alegre, RS, Brazil.

Piskorski, M. J., & Spandini, A. L. (2007). *Procter & Gamble: Organization 2005*. Boston: Harvard Business School.

Porter, M. E. (1996). What Is Strategy? *Harvard Business Review*, 74(6), 61–78. PMID:10158474

Porter, M. E. (2008). The five competitive forces that shape strategy. *Harvard Business Review*, 86(1), 78–93. PMID:18271320

Porter, M. E. (2017). From competitive advantage to corporate strategy. *Harvard Business Review*, 85(4), 53–56. PMID:17183795

Prahalad, C. K., & Hamel, G. (1990). The core competence of the corporation. *Harvard Business Review*, 68(3), 79–91.

Rabechini, R. Jr, & de Carvalho, M. M. (2005). *(Organizers). Project Management in Practice*. São Paulo: Atlas.

Schmidt, S. L., & Brauer, M. (2006). Strategic Governance: How to assess board effectiveness in guiding strategy execution. *Corporate Governance*, 14(1), 13–2. doi:10.1111/j.1467-8683.2006.00480.x

Stowell, D. P. (2011). *Best deal Gillette could get? Procter & Gamble's acquisition of Gillette*. Chicago: Kellogg School of Management.

Chapter 8
Project Management in Risk Analysis for Validation of Computer Systems in the Warehouse System

Jorge Lima de Magalhães
https://orcid.org/0000-0003-2219-5446
Instituto de Tecnologia em Fármacos Farmanguinhos, Brazil

Zulmira Hartz
Institute of Hygiene and Tropical Medicine (IHMT). GHTM, NOVA University of Lisbon, Portugal

Juliana Satie Oliveira Igarashi
Daudt Oliveira Pharmaceutical Laboratory, Brazil

Adelaide Maria de Souza Antunes
National Institute for Industrial Property & Chemical School, University of Rio de Janeiro, Brazil

Elizabeth Valverde Macedo
https://orcid.org/0000-0002-4815-6878
Federal Fluminense University (UFF), Brazil

ABSTRACT

The informational and digital era of Big Data presents a non-trivial and unprecedented way in history for data and information management in organizations. Thus, to manage, protect, and ensure the validation of this data, it is imperative to develop new technologies for project management and their respective implementation in organizations. This chapter shows a case study in a pharmaceutical industry with the proposition of a methodology for validation of emerging technologies in the computerized systems. Data validation and security for project management in the organization is increasingly in demand. So, this implies that time and human resources in organizations are not infinite. It is necessary to prioritize the activities and resources dedicated to maintaining the validated state of the system. Authors propose a risk analysis to help companies with validation. They also present a proposed methodology for risk analysis from the point of view of the validation of computerized systems in a Warehouse Management module in a validated SAP ERP.

DOI: 10.4018/978-1-5225-9993-7.ch008

BACKGROUND

In the late 19th and early 20th centuries, the first pharmaceutical industries emerged in the world. In Brazil, the development of this segment was linked to sanitary practices to prevent and combat contagious diseases. During the first part military government was not yet talked about globalization in the world and in Brazil the pharmaceutical industry grew with the Technological Development Company (CODETEC - Brazilian term). Thus, there were developments and productions for drug companies. In this way, the Government bought the drugs until the middle of the 80's. With the increase of inflation and with the beginning of the Fernando Collor government in 1990, the Brazilian economy opened. In this sense, the era of globalization began for any sector in Brazil. Since then, the internal dependence of drugs has started gradually in Brazil. With this scenario, the reduction of investments lasted from 1990 to 2000. In contrast, between 1980 and 2000, the lack of credibility of national products vis-à-vis the international ones, the patent law that enhanced the monopoly on products, as well as the increase of inspection with the creation of ANVISA in 1999 (Antunes; Mercado, 2000; Basil Achilladelis, 2001b; Chaves Et Al., 2007b; Jacobzone, 2000b; Magalhães; Quoniam; Boechat, 2013b)

The industries active in the pharmaceutical market are extremely regulated and are subject to national inspections periodically and internationally if they carry out exports of their products. In Brazil, the pharmaceutical industries are increasingly regulated by the National Health Surveillance Agency (ANVISA – Brazilian term) through various types of legal instruments such as the Resolutions of the Board of Directors (RDC – Brazilian term), Specific Resolutions (RE – Brazilian term), Normative Instructions (IN – Brazilian term) and Joint Ordinances with other actors of the federal government. In order to establish the minimum requirements to be followed for compliance with Good Manufacturing Practices (GMP) for Medicinal Products for human use, the Agency published RDC No. 17 on April 16, 2010, still in force, which included new requirements, among them the validation of computer systems (VCS) (ANVISA, 2019).

ANVISA released the Validation Guide for Computerized Systems (GVCS), still in force, which guides a set of criteria to perform the critical VCS that participates in the process of producing medicines for human use. A Computerized System is considered critical if it offers risk to the patient[1], product quality and data integrity. In this context of criticality, the computer system of the type Enterprise Resource Planning (ERP) is widely used in the industries (ANVISA, 2019; RICHMOND, L.; STEVENSON, J.; TURTON, A., 2003).

The ERP proposes to solve difficulties of integration, integrity and availability of information in a single Computerized System and to keep safe all the functionalities that support the activities of the various business processes of the companies (PINHEIRO, M. G.; DONAIRES, O. S.; FIGUEIREDO, L. R., 2011).

The German company SAP is one of the largest suppliers of ERP solutions on the market. Through ERP it is possible to place all the information in a single source of data, thus being able to carry out detailed searches of complete and updated data, attending to the needs of the clients more quickly.

The system is flexible and adjustable to meet the needs of the organization, directly assisting in decision making (VECCHIA, 2011). The modules of an ERP SAP system are related to the existing areas within the organization. Each module aggregates a set of different transactions[2] and shares data with other modules (ERP SAP, 2019). In this sense, any relevant change in Good Practices, i.e., BPx [3]

(Brazilian term) relevant in a critical system such as ERP SAP should be evaluated in relation to patient risk, product quality and data integrity (ISPE, 2019). Risk analysis (RA) is the process of understanding the nature of the risk and determining its level and by providing RA is the basis for risk assessment and treatment decisions. Risk management is characterized as coordinated activities to direct and control an organization in terms of risks (ABNT, 2019).

Risk management contributes to the demonstration of objectives and to the improvement of performance, for example, in the health and safety of persons, legal and regulatory compliance, public acceptance, environmental protection, product quality, project management, operations efficiency.

In the pharmaceutical segment, risk management is a way to address the risks present in the pharmaceutical production processes, and this type of management is one of the requirements established in RDC 17/2010[4]. Thus, the pharmaceutical industry must establish responsible for identifying, analyzing, evaluating, monitoring, treating and communicating risks or the set of risk management activities (ANVISA, 2019).

Although the validation of computer systems is mandatory, the resolutions published by ANVISA do not guide how to carry out the validation in practice, causing companies to hire consulting, invest in training, set up a team directed to the validation of computer systems to meet regulatory requirements.

This chapter aims to demonstrate the importance of project management in a computerized system validated in the pharmaceutical industry. Specifically, with the inclusion of a Warehouse Management (WM) module in a validated ERP SAP system.

In this context, the risk approach is fitted with the use of Failure Modes and Effects Analysis (FMEA) in the evaluation of potential failure modes for the processes and their likely effect on the results and / or performance of the product or process. Once failure modes are established, risk management can be used to eliminate, contain, reduce or control potential causes. The FMEA, when extended, incorporates, in addition to investigating the degree of severity of the consequences and their respective probabilities of occurrence, and their detectability, thus becoming the Criticality Analysis of Failure Mode and Effect (FMECA - Brazilian term) (IEC 60812:2006 GUIDANCE, 2006).

Thus, the identification of possible failure modes for a productive process proposed in this study passes within a validated ERP SAP system where a new relevant BPx module is being included as the WM module, as well as the assignment of values for Severity, Occurrence and Detection in conjunction with the application of an FMECA matrix contribute to the prioritization of risks and efforts and direction of the resources in the VCS are committed more assertively.

GOOD MANUFACTURING PRACTICES

The minimum requirements for Good Manufacturing Practices (GMP) were established in the mid-1960s, when the US government faced problems with the quality of personal care products and asked the Food and Drug Administration (FDA) available on the market. At the time, GMP had no legal but rather an orientate nature and were incorporated into the legal framework only after publication

Currently, in Brazil, the GMP standards are governed by ANVISA's Ordinance RDC 210/2003. This ordinance, in addition to issuing guidelines in GMP, establishes minimum requirements for process validation, cleaning procedures, analytical methodology and supplier qualification (CALARGE; SATOLO; SATOLO, 2007; SOUZA et al., 2014; VOGLER, M. et al., 2017).

According to the FDA Code of Federal Regulations (21 CFR), Section 820 (2018), GMP requirements apply to the manufacture, packaging, labeling, labeling, storage, installation and servicing of all end products made for human use. These requirements are proposed to ensure that the products are safe and effective. It is recommended that manufacturers establish their actions based on a quality system to help ensure that their products consistently meet applicable requirements and specifications.

Validation

The concept of the term validation is broad and there are numerous definitions for this that have been written almost 30 years since the appearance of the pharmaceutical industry (NUSIM, S., 2009). The first FDA guidelines on validation were in 1987, and included in the guidelines on General Principles of Process Validation. The 1987 recommendations have undergone revisions and the FDA states that current guidelines are consistent with the principles first introduced in the 1987 guidance. The 2011 review provides recommendations that reflect some of the goals of the FDA's GMP initiative in the 21st century that makes a risk-based approach, particularly regarding the use of technological advances in the manufacture of pharmaceuticals, as well as in the implementation of risk management and quality system tools and concepts ("FDA Guidance for Industry: Process Validation: General Principles and Practices - ECA Academy", 2011).

In the pharmaceutical industry, the concept of validation was introduced with the aim of establishing more strictly the sterility of pharmaceutical products, since the normal analytical methods did not serve the purpose. Over the years, the concept has been extended to several other operational aspects of the pharmaceutical industry: water systems, tablet and capsule manufacturing processes, environmental control, analytical methods and computer systems (NUSIM, S., 2009).

In this sense, the validation approach is necessary in the pharmaceutical industry to ensure quality. It is applied in several aspects in the production process, for example, in equipment, computer systems, process, cleaning their systems, and can thus affect the quality of the product if it was performed in the wrong way. In each case, the purpose of validation is to produce documented evidence that provides a high degree of assurance that all parts of the facility will function consistently when put into use. Therefore, validation is an important contribution to Quality Assurance. (HOFFMAN, 2011).

Validation of Computer Systems (VCS)

The pharmaceutical industries already have the culture to validate their production processes, cleaning, their systems for obtaining purified water and for injectable use, when applicable, etc. However, the concept of VCS can still be considered recent in the segment, since for years it was observed that the organizational computerization was restricted to administrative processes that did not have direct intervention with the quality, concentration, identity and purity of the medicines produced (FERREIRA, 2004).

For HOFFMANN et al. (1998), VCS efforts should use underlying principles that are used for process validation or analytical methods.

The VCS must assert with a high degree of security that computerized system analysis, controls and records are performed correctly and that the data processing complies with predetermined specifications (ANVISA, 2019).

According to RDC 17/2010, computer systems cover a wide range of systems, including but not limited to automated manufacturing equipment, automated laboratory equipment, process control, analytical process, manufacturing execution, laboratory information management, manufacturing resource planning, and document management and monitoring systems. In short, a computer system consists of hardware, software and network components, in addition to the controlled functions and related documentation.

System security should be based on Information Technology (IT) procedures and policies that should consider physical security, logical security, and network security. The adoption of system security measures and their data are recommended to ensure that the data is secure and protected against loss, intentional or accidental damage, or unauthorized changes. The measures should ensure the continuous control, integrity, availability and confidentiality of data linked to regulatory activities (ISPE, 2019).

The purpose of maintaining validation of computer systems is to ensure the accuracy and integrity of data created, modified, archived, retrieved or transmitted by the computer system (HUBER, 2005).

Faced with this, pharmaceutical and healthcare companies should validate all computer systems with BPx impact. Validation applies to Computerized Systems which, for example, monitor and / or control the manufacturing process, the malfunction of which may affect the safety, quality and effectiveness (during manufacture) or traceability of data, such as traceability during product distribution. Other computer system applications are also relevant for maintaining and distributing operational procedures, scheduling training, and / or documenting whether individuals can perform a function at work. From this perspective, it is noted that the list of potential applications of the computerized system that require validation is extensive. According to WINGATE (2003), some experts in the field even suggest that all computer systems used within a manufacturing environment, whatever their application, should be validated.

According to the Validation Guide for Computerized Systems (ANVISA, 2010), systems are classified into 3 classes, as described in Table 1.

ENTERPRISE RESOURCE PLANNING (ERP)

The entry of new technologies into administration systems is of great interest to both academics and executives. Increasing competition has pushed companies around the world to adopt ERPs in their production, sales, planning and finance processes, so that they can be closely supervised. Therefore, ERP

Table 1. Systems classifications

Classification	Product
1	Infrastructure Software: consists of connected infrastructure elements to form an integrated environment to run and support applications and services. Disregarded in validation. Example: operating systems, antivirus; network monitoring software;
2	Non-configurable products: They have well defined characteristics and are considered off-the-shelf software. They are standard software that cannot be changed. For system with classification 2, at least the protocol of tests and traceability matrix must be elaborated with reference to the requirement of the user.
3	Configurable or Custom Products: Software with functions that are configurable, developed and / or customized for specific uses. It involves the life cycle approach and supplier assessment. Example: ERP.

Source: Created by the authors from "Validation Guide for Computerized Systems".

system is a software platform developed to integrate several sectors or processes, composed of modules integrated with each other, to support the decision making in the various sectors, through the facilitation of communication of information and data of their processes.

In Brazil, the second half of the 1990s saw the beginning of the use of ERP in companies due to the economic stabilization of the Brazilian currency, with advances at the end of the century due to the "millennium bug" (MOREIRA; SANTANA; MIRANDA, 2014).

ERP can be applied, with minor adaptations, to any company. Scale gain brings cost advantage over custom solutions to suit a specific company. In theory, integrated systems can integrate the whole management of the company, streamlining the decision-making process. They also allow the company's performance to be monitored in real time (ROSS; WEILL; ROBERTSON, 2006; VECCHIA, 2011).

SAP is a German company founded in 1972. 56% of its base is in Europe, 20% in North America, 14% in Asia and 9% in Latin America (CARVALHO; VIANNA, 2018). With it, you can put all the information in a single data source, so you can perform detailed searches of complete and updated data, meeting the needs of customers faster. The system is flexible and adjustable to meet the needs of the organization, directly assisting in decision making (VECCHIA, 2011).

The classic ERP solution presented by SAP follows a modular pattern where modules designed for each type of process are offered, as shown in Figure 1. Each module modules different transactions and shares data with other modules. Transactions are programs that triggered direct to some process modeled within the SAP system (ERP SAP, 2019).

The main modules of the SAP system are Materials Management (MM), Warehouse Management (WM), Sales and Distribution or Sales and Distribution (SD), Production Planning (PP), Quality Management or Quality Management (QM), Finance or Finance (FI) and Controlling or Controlling (CO). In relation to the BPx relevance assessment and according to the GVSC (ANVISA, 2010) can be defined according to Table 2.

Figure 1. Modular proposal of the SAP Company
Source: SEIDOR (2019).

Table 2. SAP module and BPx relevance

Module	Main Approaches	BPx Relevance
MM	Procurement processes, master data records of suppliers and materials.	Yes
WM	Warehouse Management	Yes
SD	Registration of customer master data, invoicing and shipment of materials.	Yes
PP	Execution of Production Orders	Yes
QM	Quality control	Yes
FI e CO	Finance and Accounting	No

Source: Created by the authors.

About the classification for systems validation, the ERP SAP is classified as category 3. A category 3 system, by definition, consists of software's with functions that are configurable, developed and / or customized for specific uses. This classification involves the life cycle approach and supplier assessment. For systems classified as 3, at least: User Requirement; Risk analysis; Validation Plan; Functional Specification; Technical Specification (Hardware Design); Software Design; test protocols (installation, operation and performance); Traceability Matrix; Final Validation Report (ANVISA, 2010).

LOGISTICS AND WAREHOUSE MANAGEMENT SYSTEM

Logistics has always been present in companies due to the need for inventory, storage and transport. However, only after 1950, logistical activities began to receive more attention, with the introduction of the full cost approach and the need to improve and expand distribution channels. In the 1980s, the kanban and just-in-time (JIT) concepts, introduced by the Toyota® model, led to a greater integration of logistics functions into the production process (GUIDOLIN; MONTEIRO FILHA, 2010).

According to the Council of Supply Chain Management Professionals (CSCMP) logistics or logistics management is defined as part of the Supply Chain which is responsible for planning, implementing and controlling the operations of storage of goods, services and related information between the point of origin and the point of consumption, in order to meet the needs of customers. Thus, CCSMP defines "logistical activities" as inbound and outbound transport management, fleet management, warehousing management, materials management, inventory management, supply planning and management of logistics service providers (CARVALHO; VIANNA, 2018).

In relation to business costs, logistic activities are responsible for absorbing, on average, 25% of the total costs related to sales. For success in the logistics process, it is fundamental that companies have an information system that supports the organizational processes that make up their structure (PAOLESCHI, B., 2014).

According to GUARNIERI et al. (2006), companies must satisfy their client. Thus, they provide the support of the company in a competitive market. In this way, the development of business logistics has been exponential in recent years as it is an essential factor for the competitiveness of companies.

There are several factors that accelerate this development: pressure for greater turnover and reduction of stocks, service to distant markets, introduction of new technologies, short product life cycle, among others (GUARNIERI et al., 2006a).

In the globalized world and with a high degree of consumption, logistics plays a fundamental role within companies, because it is through good logistics management that the organization will be able to provide its services and products effectively, satisfying and retaining its customers (ANTUNES; PUGAS, 2018).

As for the pharmaceutical logistics chain, it is differentiated, because it is a health technology, the drug is subject to production, storage and distribution directives directed to the segment. The pharmaceutical sector is the junction of the health and industry sectors: the production, distribution and consumption of medicines. Thus, it is essential to incorporate more and more logistic tools in order to avoid drug shortages due to inadequate planning or the use of inefficient management tools (REIS; PERINI, 2008).

The Information Technology (IT) area plays a fundamental role in improving the logistics service. Logistic information systems are like links that integrate logistics activities into an integrated process, using the combination of hardware and software to measure, control and manage logistics operations. Thus, it has been observed over time that with the support of IT The quality of the information generated during the logistic processes is greater, that is, the information is intact, updated, accurate and at the right time for decision making (BESSA, M. J. C.; CARVALHO, T. M. X. B., 2005).

In this holistic view, there are the information systems for Warehouse Management System (WMS). It provides, stores and reports information relevant to the efficient management of the flow of products from receipt to shipment of materials. Benefits of using WMS include higher productivity, reduced inventory, optimized space, reduced operational errors, and increased operational safety (GRANT, D. B., 2013).

A Warehouse Management Systems (WMS) can be defined as the integration of software, hardware and peripheral equipment to manage inventory, space, equipment and manpower in warehouses and distribution centers. In general, a WMS is a Management System (software), which improves the operation of storage, through the efficient management of information and the resources of the same (GUARNIERI et al., 2006b)

A WMS system when well implemented can become a competitive differential for organizations (PINHEIRO, M. G.; DONAIRES, O. S.; FIGUEIREDO, L. R., 2011). When using a tool such as WMS, it is expected that the organization will have greater productivity since the tool saves time in operations such as loading and unloading, transport, storage and inventory control. The receipt of the materials also becomes optimized since the software shows the free positions in the warehouses to the operator in the moment of transit with the material for the processing. This function also acts in conjunction with all other areas since in the distribution center all goods will have exact and easily accessible addresses using WMS (HÉKIS et al., 2013; SISMEIRO, 2014).

The reduction of costs and improvement of the quality of the products is mainly due to the decrease of the handling of the materials, which preserves the packages during the movement, with the reduction of the time of search of materials when the storage in a place with known address (CORREIA; MARCELINO; PIZOLATO, 2018).

In the acronym "WMS" the letter "S" represents System and refers to the management of deposits in a decentralized way where there are interfaces that can be from SAP to SAP or to systems from other suppliers; the letters "WM" of the acronym are commonly referenced in the literature of the SAP library as a module that will increment the already active ERP SAP. The SAP WM module uses centralized warehouse systems where balances are managed in SAP's own warehouse system without communication with other systems.

In relation to the better performance of the WMS SAP system, it is observed that the solution is better adapted in companies that use ERP SAP in its entirety and not WMS independently (HÉKIS et al., 2013).

The Deposits Management module (WM) responsible for managing the flow of materials in the warehouse, transfers and labeling is also a module with relevance in BPx, but its implementation is considered optional and, in practice, it is observed that the implementation takes place post the maturation of the MM module, as shown in figure 2. This probably occurs because, despite the benefits of a WMS system, employees may have difficulties adapting to the new processes that will be implemented because the new process requires greater discipline to maintain the system to operate at maximum efficiency, therefore, it is necessary an evolution in the line of learning to use the modules that SAP offers.

The benefits derived from the industrial production of medicines are important for humanity, however, the risks are also present. Risks are not only associated with inadequate consumption of medicines, but also due to the greater probability of failure during a productive process and / or development (DEUS; SÁ, 2011). The relevance of the WM module in BPx makes SC validation with a risk analysis approach mandatory.

From the point of view of the pharmaceutical segment, risk management is a systematic process for risk assessment, control, communication and review. Regarding responsibilities, risk management is the responsibility of the company that holds the drug registration, however, these responsibilities can be attributed to a committee or internal team (ANVISA, 2019). Globally, risk management activity has been promoted by the European Medicines Agency (EMA) and the FDA as a relevance to GMP (SANDLE, 2012).

RISK ANALYSIS

Organizations of all types and sizes are subject and internal and external factors that make it uncertain whether and when they will achieve their goals. The effect that this uncertainty has on the organization's goals is called "risk," meaning that risk is the effect of uncertainty on objectives. A deviation from that expected is considered as an effect and may be positive and / or negative (ABNT, 2019).

Figure 2. Sequence of implementation of MM and WM modules
Source: (HERRANDO, M., 2019).

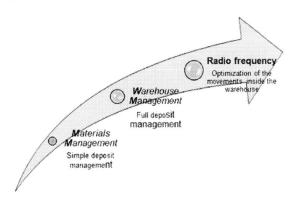

SUTER (2016) defines the risk as a probability of a specific harmful effect occurring, or as the relationship between the magnitude of the effect and its probability of occurrence which goes according to the definition of ISO 73 (ABNT, 2019). This Guide (ISO 73) defines risk with the combination of probability of occurrence of damage and the severity of damage to health, including damage that may occur due to loss of quality or product availability (HERRANDO, M., 2019; ERP SAP, 2019).

VCS is an interactive and complex process used throughout the life cycle of the computer system, including its discontinuity process. To initiate the process of risk analysis, a declaration of the intended use of the computerized system is required, that is, a User Requirements Specification (ERU) or equivalent document. As requirements may change during development or are redefined, this is the main reason why risk assessment needs to be reviewed and updated periodically (ANVISA, 2019).

The Failure Mode and Effects Analysis (FAA) risk analysis tool is a technique that identifies failure modes and mechanisms and their effects, provides an assessment of the potential failure modes for the processes and their probable effect on results and / or product performance, process, system, software, service. Once failure modes are established, risk reduction can be used to eliminate, contain, reduce or control potential causes.

In relation to non-governmental publications, the pioneering of international publications is also observed when compared to national publications, as can be seen in ICH publication Q9 on guidelines for Quality Risk Management (INTERNATIONAL COUNCIL FOR HARMONISATION OF TECHNICAL REQUIREMENTS FOR PHARMACEUTICALS FOR HUMAN USE, 2016). The ICH Guide Q9 presents options for risk analysis tools to assist the practical application of the concept in the pharmaceutical industry, as well as the PIC / S report on Good Practices for Computer Systems, published in September 2007, which, in addition to bringing the concepts on computer system validation provides considerations for the performance of audits by regulatory authorities, suggests checklists, makes reference to relevant points that should be remembered in audits and in computer system validation, leaving the practices in these processes in sight. Guidance on how the process should be viewed from the auditor's point of view generates benefits in relation to the suitability of the industry in meeting regulatory requirements.

As well as the Q9 Guide of ICH (2005) and PIC / S (2007), another relevant literature focused on the practice of the pharmaceutical industry is the 5th version of GAMP (Good Automated Manufacturing Practice), published by ISPE[5] in 2008. GAMP 5 proposes a methodology for risk analysis, categorizes the systems according to the degree of complexity and application in the industrial processes, as well as guiding the validation strategy according to each category of computerized system, thus guiding the aligned practice with the regulatory requirement that the validation extension should be proportional to the complexity of the system and based on a documented and justified risk analysis, which makes validation scalable.

In this sense, it observes the relevance of the pharmaceutical industrial segment and the urgent needed to adopt its production processes to the legislation in force in the country and its respective updates. It should be noted the constant alignment to the requirements of international standards. In the context of this chapter, the importance of the computerized systems of a pharmaceutical industry should be validated, from the point of view of knowledge management in consistent projects, in this case, joining the tool for risk analysis, applicable to VCS type ERP in module added to the ERP SAP system

AN MODEL FOR PROJECT MANAGEMENT

The checklist template should be generated based on the analysis of the production process in question. Failure modes should be listed in the checklist model as can be observed in table 3. In this model, the data contained in the "item" column do not represent failure modes, i.e., they refer to the process description that the failure modes are associated, for the following classification: "O" - Occurrence; "S" - Severity; "D" - Detectability and "NPR" - Priority Level (obtained by the formula NPR = O x S x D). It should be noted that each value that is assigned with the following values: 1 (low occurrence), 3 (average occurrence) and 5 (high occurrence) in each field "O", "S" and "D", will generate index NPR final. This index, the higher the result at the end of the fill, will be the one where there should be maximum priority for problem resolution.

As a good project management considering the pharmaceutical GMP standard, together with the risk analysis and recommendation of the FMECA matrix, it is recommended to prioritize the tests related to failure modes that have NPR index greater than number 9. This is observed, that average NPR and high, because they are more critical. In this way, it is recommended to direct all validation efforts to the most critical processes.

Therefore, the prioritization of the action with the validation of systems demonstrates the company's concern to act in advance to the risks making its processes are safe, reproducible and robust, this being the reason for being validation.

For better understanding, the classifications of the "O", "S" and "D" criteria are described in Tables 4, 5 and 6, respectively. Therefore, the parameterization of the "O" - occurrence of the fault, i.e. the probability of the failure occurring, can be classified by the descriptions in the Table 4.

Concerning the classification of Severity "S", it was classified as Table 5.

For the classification of the "D" detectability of the faults, the criteria were assigned according to Table 6.

it should be noted that the model for project management of risk analysis in computerized systems in the pharmaceutical industry can be applied in any company with the inclusion of the WM module in an ERP system already validated. However, adaptations may be necessary, so that the tool adapts to the peculiar process of each company.

CONCLUSION

Project management can be used in any area. Thus, for risk analysis in computer systems is no different. Applying this tool in a ERP SAP system validated in the pharmaceutical industry, one can observe the contribution to ensure the quality in the processes of computerized systems. In this way, considering the legislation in force nationally and internationally.

The use of computer systems in the relevant BPx processes in the pharmaceutical industry makes the validation of computer systems a broad field for study and discussion. The risk management approach to computer system validation has international and national benchmarks. Therefore, it evidences the tendency of the pharmaceutical segment to harmonize concepts and practices of the best forms of management in the present time.

Table 3. A model list proposal for project management in the fault mode check

Item	Failure Modes	O	S	D	NPR
1	Receipt and storage of inputs in the SAP system warehouse				
1.1	Material does not lie in the transitional deposit type 902 after the movement of the goods inbound movement 101 using transaction "MIGO"				
1.2	Printing material label printing (pdf mode) in transaction for example ZSXX0099 does not contain a readable bar code				
1.3	Printing UD label printing (pdf mode) in transaction for example ZSXX0099 does not contain a readable bar code				
1.4	No "Label already printed, use reprint" error message in transaction for example ZSXX0099 when trying to reprint media label when print mode is selected, not reprint				
1.5	Absence of error message "There are documents reversed" in transaction for example ZSXX0099 when trying to generate labels of materials that had the PC reversed				
1.6	When carrying out the data collector movement of the reservoir type material for example 902 to I01, it does not receive the message "Successful movement"				
1.7	System does not display "Limit Exceeded" message when attempting to transfer quantity greater than that which supports position.				
2	Material Sampling				
2.1	When moving the material to be sampled, via data collector, from type of tank I01 to type of tank Q01 does not receive the message "Handling carried out successfully".	.			
2.2	After transferring the material to be sampled from deposit type I01 to Q01, refer to the material in deposit type Q01 and do not display the transferred balance				
2.3	Print view of the sampled material label (pdf mode) in transaction for example ZSXX0098 does not contain a readable bar code				
2.4	When attempting to print sample material label via manifold, visualize on collector screen that system did not perform calculation in field "quantity generated" correctly				
2.5	Do not receive message "Transfer order created successfully" when returning with the sampled material to the shelf position				
2.6	When you perform transaction 900 in the MIGO transaction for material cost to cost center, you do not see the material stock output in MM and WM				
3	Production Supply				
3.1	After opening the OP, do not view NT in transaction LB10				
3.2	The materials listed in the OT generation by the transaction LB10 are not in agreement with the one determined in the OP				
3.3	In the generation of OT by LB10, inform the amount of material different from what is being separated to meet the demand of the PO and not receive the message "amount divergent"				
3.4	When completing the OT on LB10 does not receive the message "The OT generated successfully"				
3.5	Confirm the OT by LT12 and not receive the message "Confirmed OT"				
4	Provision of Finished Product to the Logistic Operator				
4.1	Finished Product label print view (pdf mode) in transaction for example ZSXX0097 does not contain a readable bar code				
4.2	After moving from PA to the storage of finished products in MM, do not display the item in storage type E01, "receive" position in WM.				
4.3	Palette label print preview (pdf mode) in transaction for example ZSXX0096 does not contain a readable bar code				
4.4	When moving the material, via the data collector, from the "receiving" position to the storage position in tank type E01, you do not receive the message "Handling carried out successfully".				
4.5	After transferring the material from the "receiving" position to the storage position in storage type E01, consult the material in the storage position and do not display the transferred balance				
4.6	When issuing Nota Fiscal transfer to OL by movement 931 in MM, do not view the inventory output in WM				

Source: Created by the authors.

Table 4. Occurrence classification – "O"

Classification of criterion "O" – occurrence of the fault	Description
High	Operation is performed in SAP with interface via data collector
Average	Operation is performed via desktop in SAP using custom transactions
Low	Operation is performed in desktop in SAP with standard transactions

Source: Created by the authors

Table 5. Severity rating

Classification of criterion "S" – severity	Description
High	Failure to use the SAP transaction generates direct impact on BPx (product quality, patient safety, data integrity).
Average	Failure to use SAP transaction generates direct impact on the routine of the operation and has no contingency by other means.
Low	Failure to use the SAP transaction generates direct impact on the routine of the operation and has contingency by other means.

Source: Created by the authors.

Table 6. Detection rating

Classification of criterion "D" – detectability	Description
High	Failure detected only through management report
Average	Failure perceived by the employee specialized in the routine of the operation
Low	Failure perceived by collaborator who is not specialized in the operation routine

Source: Created by the authors.

It can be observed that the major contributions to the practice in computer system validation came from non-governmental organizations that have as main source of information the contribution of members of the pharmaceutical industry, the scientific community, regulatory authorities and other interested parties, such as for example ISPE, SINDUSFARMA and ICH.

The validation of a computerized system cannot be superficial and must identify possible ways of failure and allow, in advance, to plan ways of acting on the problem causes.

The proposed checklist model for validation of computer systems containing the possible failure modes generated by including the WM module in SAP can be used in full or as a reference for future scenarios. In this way, it can direct validation teams on the limits of the process and the rational on what should be considered or not in order to calculate the NPR. This process can optimize the time to perform a computerized system validation.

The execution of the risk analysis using the FMECA tool in the process limits contributes to the provision of documentary evidence for the decision making regarding the computer system validation. This fact is evidenced by the fact that it demonstrates possible failure modes, effects, causes and control points that can be used as basis for other analyzes in computerized system validation.

REFERENCES

ABNT. ABNT - Associação Brasileira de Normas Técnicas. Disponível em <http://www.abnt.org.br/

Antunes, A., & Mercado, A. (2000). *A aprendizagem tecnológica no Brasil: a experiência da indústria química e petroquímica* [s.l.]. Editora E-papers.

Antunes, V. M., & Pugas, P. G. O. (2018, April). A Logística De Distribuição No Setor Moveleiro: Um Estudo De Caso Em Uma Empresa De Grande Porte. *Ciências Gerenciais em Foco*, *5*(2), 1.

ANVISA. Agência Nacional de Vigilância Sanitária. Disponível em http://portal.anvisa.gov.br/

Basil Achilladelis, N. A. (2001a). The dynamics of technological innovation: the case of the pharmaceutical industry. *Research Policy*, *30*(4), 535–588. doi:10.1016/S0048-7333(00)00093-7

Basil Achilladelis, N. A. (2001b). The dynamics of technological innovation: the case of the pharmaceutical industry. *Research Policy*, *30*(4), 535–588. doi:10.1016/S0048-7333(00)00093-7

Bessa, M. J. C., & Carvalho, T. M. X. B. (2005). *Tecnologia da informação aplicada à logística* , 11, pp. 120–127.

Calarge, F. A., Satolo, E. G., & Satolo, L. F. (2007). Aplicação do sistema de gestão da qualidade BPF (boas práticas de fabricação) na indústria de produtos farmacêuticos veterinários. Gestão &amp. *Produção*, *14*(2), 379–392.

carvalho, g. M.; vianna, n. M. C. Sistema de gerenciamento de armazéns: um estudo de caso sobre sua implementação no setor aeronáutico. Politécnico: ufrj, 2018.

Chaves, G. C., & .... (2007a, February). Evolution of the international intellectual property rights system: Patent protection for the pharmaceutical industry and access to medicines. *Cadernos de Saude Publica*, *23*(2), 257–267. doi:10.1590/S0102-311X2007000200002 PMID:17221075

Chaves, G. C., & .... (2007b, February). Evolution of the international intellectual property rights system: Patent protection for the pharmaceutical industry and access to medicines. *Cadernos de Saude Publica*, *23*(2), 257–267. doi:10.1590/S0102-311X2007000200002 PMID:17221075

correia, p.; marcelino, s.; pizolato, c. Análise das ferramentas de armazenagem e estocagem. **Anais da SEMCITEC - Semana de CiÃªncia, Tecnologia, InovaÃ§Ã£o e Desenvolvimento de Guarulhos**, v. 1, n. 1, 6 ago. 2018.

FDA guidance for industry: process validation: general principles and practices - ECA Academy. Disponível em https://www.gmp-compliance.org/guidelines/gmp-guideline/fda-guidance-for-industry-process-validation-general-principles-and-practices

Ferreira, H. P. (2004). Sistema de gestão da qualidade - estudo de caso: Far-Manguinhos.

Grant, D. B. Gestão de Logística e Cadeia de Suprimentos. [s.l: s.n.].

Guarnieri, P., & .... (2006b, April). WMS -Warehouse management system: Adaptation proposed for the management of the reverse logistics. *Production, 16*(1), 126–139. doi:10.1590/S0103-65132006000100011

Guarnieri, P., Chrusciack, D., Oliveira, I. L., Hatakeyama, K., & Scandelari, L. (2006a, April). WMS -Warehouse management system: Adaptação proposta para o gerenciamento da logística reversa. *Production, 16*(1), 126–139. doi:10.1590/S0103-65132006000100011

Guidolin, S. M.; Monteiro Filha, D. C. (2010). Cadeia de suprimentos: o papel dos provedores de serviços logísticos. set.

Hékis, H. R., Medeiros Araújo de Moura, L. C., Pires de Souza, R., & De Medeiros Valentim, R. A. (2013, Oct. 1). Sistema de informação: Benefícios auferidos com a implantação de um sistema WMS em um centro de distribuição do setor têxtil em Natal/RN. *RAI Revista de Administração e Inovação, 10*(4), 85–109. doi:10.5773/rai.v10i4.920

Herrando, M. Cicle optimització processos SAP amb MM, WM i RF. (2019). Disponível em http://www.marcherrando.com/2010/03/cicle-optimitzacio-processos-sap-amb-mm.html

Hoffman, S. G. (2011, Nov. 1). The new tools of the science trade: Contested knowledge production and the conceptual vocabularies of academic capitalism. *Social Anthropology, 19*(4), 439–462. doi:10.1111/j.1469-8676.2011.00180.x

Huber, L. (2005). Risk-based validation of commercial off-the-shelf. *Computer Systems, 2005*(6).

IEC. (2006). Guidance on failure modes and effects analyses (FMEAs). Disponível em https://webstore.iec.ch/publication/3571

International council for harmonisation of technical requirements for pharmaceuticals for human use. Disponível em http://www.ich.org/home.html

ISPE. Welcome to the Brazil Affiliate | ISPE. Disponível em http://www.ispe.org.br/ispe_about

Jacobzone, S. (2016a). Pharmaceutical policies in OECD countries. Disponível em http://www.oecdilibrary.org/social-issues-migration-health/pharmaceutical-policies-in-oecd-countries_323807375536

Jacobzone, S. 2016b. Pharmaceutical policies in OECD countries. Disponível em http://www.oecdilibrary.org/social-issues-migration-health/pharmaceutical-policies-in-oecd-countries_323807375536

Magalhães, J. L.; Quoniam, L.; Boechat, N. (2013a). Pharmaceutical market and opportunity in the 21st for generic drugs: a Brazilian case study of Olanzapine. *Problems of Management in the 21st Century, 6*.

Magalhães, J. L.; Quoniam, L.; Boechat, N. (2013b). Pharmaceutical market and opportunity in the 21st for generic drugs: a Brazilian case study of Olanzapine. *Problems of Management in the 21st Century, 6*.

Moreira, L. B.; Santana, A. A.; Miranda, A. R. A. (2014, Dec. 10). Os impactos da implementação do SAP R/3 em uma empresa do setor de laticínios. Revista Ciências Administrativas ou Journal of Administrative Sciences, 18(1).

Nusim, S. (2016). *Active pharmaceutical ingredients: development, manufacturing, and regulation* (2nd ed.). Boca Raton, FL: CRC Press.

Paoleschi, B. (2018). *Estoques e armazenagem*. Editora Saraiva.

Pinheiro, M. G., Donaires, O. S., & Figueiredo, L. R. (2011). Aplicação da Visão Sistêmica na implantação de Sistemas Integrados de Gestão ERP. Anais do 7o Congresso Brasileiro de Sistemas, 7, p. 409–421.

Reis, A. M. M., & Perini, E. (2008, April). Desabastecimento de medicamentos: Determinantes, conseqüências e gerenciamento. Ciência &amp. *Saúde Coletiva, 13*(suppl), 603–610. doi:10.1590/S1413-81232008000700009

Richmond, L., Stevenson, J., & Turton, A. (2003). Essay review the pharmaceutical industry: a guide to historical records. Aldershot.

Ross, J. W., Weill, P., & Robertson, D. (2006). *Enterprise architecture as strategy: creating a foundation for business execution*. Boston, MA: Harvard Business Review Press.

Sandle, T. (2012, December). Application of quality risk management to set viable environmental monitoring frequencies in biotechnology processing and support areas. *PDA Journal of Pharmaceutical Science and Technology, 66*(6), 560–579. doi:10.5731/pdajpst.2012.00891 PMID:23183652

SAP ERP. Seidor BRASIL - SAP Business Suite (ERP). Disponível em http://www.seidorbrasil.com.br/solucoes/saperp

Sismeiro, L. F. L. (2014, Dec. 18). Projectos de consultoria em SAP e tecnologias microsoft: análise e desenvolvimento de soluções de software à medida.

Souza, C., Oliveira, J., & Kligerman, D. (2014, September). Advances and challenges in standardization of free samples of drugs in Brazil. *Physis (Rio de Janeiro, Brazil), 24*(3), 871–883. doi:10.1590/S0103-73312014000300011

Vecchia, A. F. D. (2011, March 23). Sistemas Erp: A Gestão Do Processo De Implantação Em Universidade Pública.

Vogler, M., Gratieri, T., Gelfuso, G. M., & Cunha Filho, M. S. S. (2017). As boas práticas de fabricação de medicamentos e suas determinantes. Vigilância Sanitária em Debate: Sociedade, Ciência & Tecnologia, 5(2), 34-41.

ENDNOTES

[1] The medicinal product is a pharmaceutical product, technically obtained, or processed, containing one or more drugs and other substances, for prophylactic purposes; curative; palliative; or for diagnostic purposes. According to the World Health Organization (WHO), the proper use of medicines occurs when patients receive the medicines appropriate to their health condition, in doses appropriate to their individual needs, for an appropriate period and at the lowest cost possible for them and

2 SAP transactions are "links" typed in a specific area of SAP, where, in turn, the transaction, ie a transaction is an easy reminder to SAP You can find any program within SAP. A default SAP ratio is identified by a 4-function code that can be used directly in the command field in the presentation interface or in the corresponding menu option. When a transaction is not SAP standard, but custom is the title of the letter Z and may be longer than 4 characters.
3 BPx are compliance requirements for all best practice approaches that encompass the regulation of the pharmaceutical supply chain from discovery to post-marketing. These may be Good Distribution Practices (BPD), Good Laboratory Practice (GLP), Good Manufacturing Practice (GMP)).
4 Brazilian sanitary regulation on Good Manufacturing Practices.
5 Founded in 1980, ISPE is a non-profit organization focused on the development and dissemination of knowledge and GMP for professionals engaged in the research, production and distribution of quality medicines for human consumption. The activities of the association are supported through the voluntary work of professionals from the Life Sciences, Pharmaceuticals, Veterinary, Cosmetics, Medical Devices and related areas: regulatory agencies, academia and class entities. It currently has more than 18,000 members in 90 Countries. In Brazil, it has about 220 members (ISPE 2019).

(Note: item 1 continues from previous page: "their community (WHO, 1987). WHO has announced as a priority for patient safety measures to minimize errors in the use of drugs, which may even result in death (WHO, 2017).")

Chapter 9
Canvas Marketing Plan:
How to Structure a Marketing Plan With Interactive Value?

Miguel Magalhães
Universidade Portucalense, Portugal

Frederico D´Orey
Universidade Portucalense, Portugal

Manuel Pereira
https://orcid.org/0000-0002-6238-181X
Universidade Portucalense, Portugal

António Cardoso
Universidade Fernando Pessoa, Portugal

Alvaro Cairrão
Polytechnic Institute of Viana do Castelo, Portugal

Jorge Figueiredo
Universidade Lusiada, Portugal

ABSTRACT

Canvas Marketing Plan is a design thinking tool to help companies build a marketing plan that allows them to make better decisions. It provides a simple structure that allows the user to visualize the dynamics and interaction of the different stages of the marketing plan and adapt the products and services to the needs of their clients, thus, "finding" the best position in relation to their competitors. This chapter presents a methodology of marketing that aligns the marketing plan with a highly connected and constantly changing market, but also online interaction vs. offline interaction, thus facilitating marketer planning. The canvas marketing model is validated by 146 marketeers from 17 distinct sectors of activity, allowing authors to gauge the timeliness and usefulness of this Framework.

DOI: 10.4018/978-1-5225-9993-7.ch009

INTRODUCTION

Digitization is destroying old business models and creating room for new models, in addition to emerging new ecosystems. As the boundaries between sectors become dematerialized, markets redistribute and startups take the place of companies already established in the market. What causes organizations to wonder if their business model is under threat? This paradigm change requires that the Marketing Plan becomes a working tool that contemplates the systematic changes of the external and internal surroundings. Taking into account this imperative, the Canvas Marketing Plan was inspired by Osterwalder, A., in the Business Model Ontology (2004), which was the conceptual basis of the publication of the book Business Model (2009) by Osterwalder, A. and Pigneur, Y. The Business Canvas Plan is a summary map of the "nine key blocks" of the business plan, but does not, in itself, exclude the preparation of the business plan. For those who are not yet familiar with the Canvas model, there is a chronological order to fill in the nine blocks of the canvas. First, we must identify the customer segments (the targets) to whom we wish to sell the products and services. Second, what are the products and services that we wish to sell. The value proposition represents the "value" for a given specific target. Also, we must describe how an organization differentiates itself from its competitors. Which is why customers buy from one company and not from another, Osterwalder, A. (2004). Third, we must identify the distribution channels that we will use to get products and services to customers, including marketing and distribution strategy. Fourth, how will the organization relate to the chosen targets? The company establishes links between itself and the different targets. Fifth, what will be the income model? It will identify how the organization "makes money". The business model should describe the logic of creation, delivery and value capture by an organization, Osterwalder, A. (2004). Sixth, what are the key features needed to enable you to create "value" for the target. They list the assets that are needed to maintain and support the business model. Resources can be human, financial, physical or intellectual. Seventh, what are the key activities that will enable you to execute the organization's value proposition. Eighth, what are the key partners and strategic alliances that complement the business model. Finally, in ninth place, we must quantify what is the structure of spending that the business model translates. The Canvas Marketing Plan screen, presented in this article, proposes an interactive framework that allows structuring the marketing plan of an organization, analogical and digital, structured in "16 key blocks" grouped into three "P's" value proposition "," Value research "(" market offer ")," Value proposition "(what and how to offer?) and Value production (planning the offer?), but does not exclude the elaboration of the marketing plan. (The **marketing** mix is most commonly executed through the 4 **P's of marketing**: Price, Product, Promotion, and Place?)

MARKETING BACKGROUND

Throughout the twentieth century, business activity takes on three distinct perspectives before becoming a marketing science in the social sciences. Until the 1930s, there was only one production perspective, in which the organizations' objective was to produce the best product, based on an organization that would allow the lowest cost and the lowest prices. In the mid-30s, the perspective of the sale was developed, in which it is not enough to produce a good and at the lowest cost. It has become necessary to work the sales force and the promotional activities (push strategy), but it remains to ignore that the market is a set of customers who have needs, requirements and desires; who share values and cultures;

who know, choose, decide and buy that which gives the best satisfaction for their needs (pull strategy). Post World War II, the notion began to appear that it is no longer sufficient and effective to produce and to dispose of according to the best techniques and methods hitherto defended: it is necessary to be able to produce at the lowest cost, to sell at the best price and to know how to promote a product or service. As such, techniques of analysis, evaluation and market research began to be developed, giving rise to the appearance of the market perspective. Until then, markets were understood as homogenous realities, for which it is produced, depending on the financial interest of the organization. In the later phase, the market begins to gain importance in defining the strategies of organizations, although the concepts of segmentation and positioning do not produce consequences in business activities. The market studies only served the purpose of sales forecasting and analysis of needs, not providing tools for the treatment of the results obtained a posteriori, namely, as regards forms of evaluation of satisfaction of these same purposes, serving before, to define promotional messages and advertising. Nowadays, marketing begins to be included as a science of the social sciences, with a body of knowledge properly autonomized and endowed with its own scientific methodologies. In it there are several independent variables and concepts developed, applicable to different scopes. The current marketing philosophy, such as highly successful management and income, positions the client and the market, as priorities and strategic guidelines of basic entrepreneurship based on a philosophy of customer service. The sales perspective, very characteristic in SMEs, bets on a strong commercial department, losing on innovation, entrepreneurship and continuous customer satisfaction. The orientation to the market and the marketing optics have as their common denominator, the primacy of entrepreneurship and the client, on the basis of the organization's action, together with the interpretation of all the surrounding reality. In fact, marketing in Portugal emerged in the 1920s as an economic application with Pessoa F. (1935) and became a management activity in the 1940s. However, it was not until the 1950s that scientific development developed as a quantitative science and later qualitative, with Namora F. (1969), through the publication of the book entitled Marketing. It can be said that marketing entered Portugal more quickly through culture, than through the management function, which was only truly implemented since the 1970s. Only in this decade did concepts such as "Marketing Mix" by Neil B. (1964), the "Product Life Cycle" by Dean J. (1950), the "Brand Image" by authors Sidney L. and Ogilvy D (1955), "Market Segmentation" by Wendell S. (1956), "Marketing Concept" by McKitterick John (1957), and "Marketing Audit" by Shuchman A. (1959). In the 1960s, marketing came to be seen as the science of behavior. McCarty J. (1960) introduced the notion of the 4 P's - Product, Place, Promotion and Price - later developed by Levitt T. (1980); Howard J. and Sheth J. (1969) introduced the "Theory of Consumer Behavior"; and William L. (1987) introduced the "Lifestyles" adapted from sociological concepts. It was in the 1960s that in the U.S. banned tobacco advertising in magazines, television and radio, in which IBM achieved market leadership with the computers of the 360 line and Intel began marketing the stereo cassette decks system. In the 1970s, marketing was the science of decision making. Zaltman G. and Kotler P. (1971) introduced the concept of "Social Marketing". Ries Al and Trout J. (1970) identified the concept of "Positioning", as a result of the articulation of the four variables of "Marketing Mix". The consulting firm, Boston Consulting Group (B.C.G.) developed "Strategic Marketing" as a decision system. With Shostack L. (1972), "Service Marketing" gained prominence, with the concept of "Macro Marketing" emerging as a response to the increase in social problems due to unbridled consumption and indebtedness. It was also in the 1970s, that Sony introduced the Walkman and Magnavox introduced the first video game – Odyssey to the US market. In 1972 e-mail was born and the software for sending and receiving electronic mail was created. In the 1980s, concepts such as "War Marketing" emerged, by Singh Ravi and Kotler P. (1980); the "Internal Marketing", by Gronroos C.

(1981); the "Global Marketing", by Levitt T. (1925); o "Relational Marketing", by Jackson B. (1985); and "Megamarketing," by Kotler P. (1986). Marketing is therefore "popularized," constantly being "recreated" and differently positioned in every corner of the "global village." Marketing has grown and matured in the twentieth century alongside the information society, occupying a dimension of knowledge worthy of autonomy in the scientific universe of economic sciences and entrepreneurship. The contributions of Kotler and Levitt, among other authors already mentioned, were remarkable for the systematization of this "discipline". According to Oliveira (1999), both authors helped to define the true personality of the term "Marketing" and "Entrepreneurship" as an organization's philosophy, based on the premise that the orientation of a business should address the needs and requirements of the client, for the purpose of the exchange of a benefit, to achieve their satisfaction. Levitt T. (1986) writes in "The Marketing Imagination," "I realize a constant that defines what is best." It tells us that "there can be no effective entrepreneurship strategy if it is not marketing-oriented." With the changing form of communication and production over the past 60 years, the marketing plan had to reinvent itself. Unleashed by the industrial revolution comes the concept of Marketing 1.0, where there is a high demand, reduced differentiation, focused only on the product. An example that depicts this era, in the launch phase of the Ford model in the market, there was much demand for the product and little supply. The companies did not care about product differentiation (design, color, ...), their only objective being the production of large quantities to try to satisfy a high demand. Recall, Henry Ford's famous phrase, which perfectly illustrates the concept of Marketing 1.0 Any customer can have a car painted any color that he wants so long as it is black. Customer interaction at this stage was a relationship of "one" (company) to "many" (consumers) and the value proposition was purely functional. The management of the organizations did not feel the need to elaborate a marketing plan, but only to plan the production. The perception of consumption was interpreted as "mass buying." With the increase of the competition and the technological evolution, the concept of Marketing 2.0 appears, the customers have a greater variety in the purchase options for the same product, which forces companies to seek differentiation of the product.

The focus now is on the customer. This new paradigm forces us to rethink the business model of organizations. And, at this stage, they begin to outline the first plans of Marketing. It is necessary to invest in the positioning of the product and the brand, in the differentiation and the conquest by the retention of the customers. It was necessary to create a value relationship between the brand and the customer. Now, the interaction with the client changes from "one" (company) to "one" (target). In the third "wave", marketing becomes focused on the human being and the concept of Marketing 3.0 emerges. According to Kotler P., Hermawan K. and Setiawan I. (2010), clients become complete human beings, with minds, hearts and spirits. Marketing starts to reside in the creation of products, services and corporate cultures that embrace and reflect human values. This was possible with massification of the internet. The communication of the companies stops being vertical, static, by the disclosure of a catalog of product, and becomes horizontal, dynamic, by the interactivity between the company and the customer, where the customer and the brand talk among themselves. The interaction with the consumer changes from "many" (companies and other consumers) to "many" (other companies and consumers). The differentiation of the product and the brand becomes emotional and not only functional, but also spiritual, the customer is seen as a person with mind, heart and spirit. To the point of building the Museum of Marketing 3.0 in Ubud, Bali. This museum highlights inspiring cases of professionals and marketing companies who adore the human spirit. The contents are organized in a modern multi-screen scheme. Recently, the museum has been gradually updated with state-of-the-art technology, such as augmented Virtual Reality. In fact, technology has been heavily affecting marketing practices around the world, bringing new trends to the forefront of, among others, the "sharing" economy, the "now" economy, global integration, content and relationship management with the customer (social CRM). There is an increasingly greater convergence between analogue and digital marketing through the use of high technology, allowing a "closer" contact between people. The more social we become, the more we desire things made to suit us. For example, big data analysis allows more personalized products and more personal services. Arriving in a transition phase, a new marketing approach was needed as a natural consequence of Marketing 3.0. Today, we live in the Marketing 4.0 phase. The phase of humanistic and collaborative marketing. The focus now is social. It is the marketing phase focused on customer information, covering it in all its needs. The barrier between digital and analog (traditional) is broken and communication becomes multi-channel (omnicanal), and to guide the client in their journey from the attention to the advocacy is essential that the organizations have a plan dynamic and interactive marketing.

MARKETING FRAMEWORK CANVAS

Today, in order to manage and monitor business, organizations must have a dynamic and interactive marketing plan. Markets are extremely volatile as a result of strategic changes and the systematic tactics of competitors, where they are increasingly looking for the best positioning with customers and consumer preference. This paradigm of "restlessness" naturally requires a constant adaptation of strategic and tactical decisions of the company. In anticipation of or even at the precise moment (anticipate or react to a competitor's campaign). Therefore, this requires that the marketing plan has an agile effectiveness in the business management process. Thus, allowing the decision-making process to be "oiled" by becoming a decision support tool responding to "what is true today" may "be mismatched tomorrow". Given these premises, Dubois B (1994) already considered that all customers are chameleons. The same applies to stakeholders. Similarly, Business Canvas Model, which allows the design of an organization's business model, it is opportune, the emergence of other frameworks that allow monitoring the course of the stakeholders in order to systematically rethink the status quo of marketing management. Even though there are innumerable options for drawing up marketing plans. Therefore, it is often difficult to choose and opt for the best model marketing plan that meets the real needs of an organization. It should be noted that the marketing plan must systematically monitor market changes, in particular competitors' 'inves- tments', as well as improve the experience of the targets chosen by the organizations. It is important to mention, first and foremost, the marketing plan has to be a nimble work tool, it can not be a static document and it must be a decision support tool. Part of the genesis of the Business Canvas Model, we propose a framework with the same philosophy and complement to the design of the business model.

The Canvas Marketing Plan will enable marketers to define the three "Value P's" for a business.

The first pillar is called "Value Research" (Figure 1), where the Six O's are identified: Objects, Objectives, Operations, Occasions, Organization and Occupants:

1. **The Objects**: What do you buy? It is intended to identify the goods or services that are marketed, or that may be. Besides, to identify what is the typology of goods: whether they are durable or not, current, thought, specialty or impulse purchase?
2. **The Objectives**: Why buy? What is the type of need? Physiological, safety, belonging, esteem and achievement)?

Figure 1. Value search
Source: Own elaboration

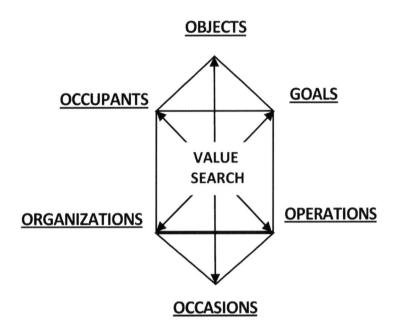

3. **The Operations**: How do they buy? Understand the awakening of desire, the process of information, the conviction of buying and repeating the purchase)?
4. **The Occasions**: When they buy? It will be necessary to measure the frequency, the rhythm, the seasonality of the purchase. But also, if there are events that may induce the purchase or if there may be conjuncture?
5. **The Organization**: In the B2B segment, who decides a purchase, is not always a single person, there are several players in the buying process: the initiator, the prescriber, the decision maker, the buyer (who pays), the opponent, the continuator, the destroyer and the user. Just as in the B2C segment, so do ambassadors, influencers, ... and the consumer.
6. **Occupants**: Which stakeholders are part of the market? The target chosen? Identification of direct, indirect and generic competitors?

The second pillar translates the "Production of value" (Figure 2) of an organization where the "4 P's" strategy must be planned: People, Programs, Processes and the Performace of the organization:

7. **The People**: Assessing an organization's competency score is critical. The score is obtained by summing up and valuing the knowledge, skills and attitudes of the people who constitute it. Being this, the main asset of an organization, the management and training of the marketing and sales team is crucial to the performance of the business.
8. Do the Programs identify the market opportunities for the definition of the marketing and sales strategy in order to define the strategic marketing and sales plan?

Figure 2. Value production
Source: Own elaboration

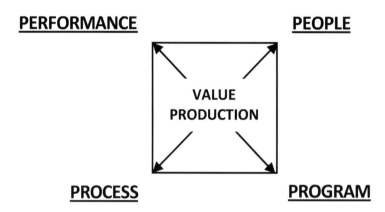

9. In the Processes, what is the marketing and sales action plan defined? In the traditional, is Attracting, Selling, Satisfy and Fidelize, while in the Digital is Attention, Attraction, Counseling, Action and Advocacy?
10. Finally, in Performance, the evaluation and performance measures are chosen to capture results that have financial and non-financial implications.

In the third pillar is defined with the tactic of Value Proposition (Figure 3), it refers to "what if?" (Product and Price), and "how are you going to propose" to the market (Promotion and " Channel).

11. Regarding Product / Co-creation, we must take into account the following aspects: brand coding, product characteristics (design, quality, 3D printing, range and sizes), packaging, service, post warranty -production, profitability and positioning. Even so, given the specifics of the business, we can add or remove some of the factors considered above.
12. In the Price we must consider the preparation of the price list of products, based on cost, competition and value for the customer), discounts (market fluctuation), subsidies, payment term and credit conditions.
13. In Sales Promotions we should take into account advertising budget, sales force spending, public relations, and direct marketing. The system of consumer classifications and connection, social networks (relationship policy), and privacy policy (solicit consent).

Figure 3. Value offer
Souce: Own elaboration

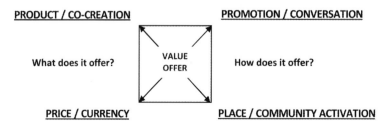

Figure 4. Value generation
Source: Own elaboration

VALUE PRODUCTION	VALUE OFFER	VALUE SEARCH
15. EXPENSES / INVESTMENTS What are the fixed and variable investments and structural expenses? What are the deviations from the budgeted?	16. INCOME What is the income of the organization? And what is the amount of invoicing compared to the budgeted?	

14. The choice of the distribution channel is essential to get the products to the market, due to their location, diversity and transportation, allowing the coverage rate to be evaluated later.

From research (1 to 6), production (7 to 10) up to the value proposition (11 to 14) will result in investments, expenses (15) and income (16) that will allow the organization to capture and generate value Figure 6). The Canvas Marketing Framework will also be a marketing intelligence process, and not just the simple verbalization of intentions.

Methodology

In order to validate the Canvas Marketing Framework, an empirical study was carried out with a population of 146 marketers from 17 different sectors of activity that are part of an important economic group in Portugal. The content of Figures 1 and 6 was addressed to the marketeers, in order to verify the practical applicability of this Framework. We obtained 95 valid surveys and the statistical treatment of the responses in the SPSS program (version 22).

BUSINESS CASE CANVAS MARKETING FRAMEWORK

The response rate obtained was high, obtaining 65% of the valid questionnaires, characterizing the sample as representative. In view of the conclusions obtained, we can say that about 97% of marketeers consider this framework intuitive, as is the Business Canvas Model. In addition, to enable an integrated view of an organization's strategic, operational and marketing marketing (97%). Many marketers recognize that this framework is complementary to Business Canvas Model (88%). It is a dynamic and interactive framework, even those who consider it disruptive (57%) because it allows to quickly make adjustments to the organization's value proposition. Approximately 86% of marketers recognize that Canvas Marketing Framework allows the integration of analogue and digital marketing (88%) and represent an organization's omnicanal presence (97%) in the same framework. Finally, almost all marketers considered the analysis of the 6 O's as particularly useful and relevant in building the marketing plan. (99%).

Figure 5. Canvas marketing model. How to structure a value marketing plan?
Source: Own elaboration.

Pillar	Marketing Plan	Framework	Element	Description
I	VALUE SEARCH The 6 O's Market research Connectivity "Know and Understand?"	1	Objects	What does the market buy?
		2	Goals	Why buy
		3	Operations	How to buy
		4	Ocasions	When to buy
		5	Organizations	Who is involved in the buying process?
		6	Occupants	Who is part of the market?
II	VALUE PRODUCTION 4 P's / 4 C's Marketing strategy "Proposal, Budget and Control?"	7	People	What are the skills (sum of Knowledge + Skills + Attitudes) of the people in the organization? Management and training of the marketing team?
		8	Programs	What are the market opportunities? Definition of marketing and sales strategy?
		9	Process	What is the marketing and sales action plan? Traditional: Attract, Sell, Satisfy and Loyalty Digital: Attention, Attraction, Counseling, Action and Advocacy?
		10	Performance	What evaluation and performance measures to capture results that have financial and non-financial implications.
III	VALUE OFFER 4 P's / 4 C's Marketing Tactic "Implementation and Adjustments?"	11	Product / Co-creation	Brand Coding, Features (Design, Quality, 3D Printing, Range, Sizes), Packaging, Service, Warranty and Profitability. Positioning.
		12	Price / Currency	Pricing table (cost basis, competition and customer value), Discounts (market fluctuation), Subsidies, Payment Period and Credit Conditions.
		13	Promotion / Conversation	Sales promotion, Advertising, Sales force, Public relations, Direct marketing, Consumer and connection classification systems, Social networks (relationship networks) and Privacy policy (requesting consent)
		14	Place / Community activation	Channels, Peer-to-peer Distribution, Coverage Rate, Channel Diversity, Location and Mode of Transport.
IV	PROFITABILITY VALUE OFFER	15	Income	What is the income of the organization? And what is the amount of invoicing compared to the budgeted?
		16	outgoing	What are the fixed and variable investments and structural expenses? What are the deviations from the budgeted?

Figure 6. Canvas marketing model
Source: Own elaboration.

VALUE PRODUCTION (4 P's / 4 C's) STRATEGY "Plan the offer?"		VALUE OFFER (4 P's / 4 C's) TACTICS "What does it offer? and How does it offer?"		VALUE SEARCH (6 O's) MARKET STUDY / CONNECTIVITY "Who's in the market?"	
PEOPLE People reflect internal marketing and the fact that employees are critical to marketing success. The competence of the people is sum of Knowledge + Skills + Attitudes.	PROGRAM The programs reflect all the activities directed to the consumer of the company. Paradoxing complementary offline and online interaction? What is the sales budget? Traditional: Attract, Sell, Satisfy and Fidelize. Digital: Attention, Attraction, Counseling, Action and Advocacy (Word-of-mouth)?	PRODUCT / CO-CREATION Mark / Position Brand Design / Coding Quality and Characteristics 3D printing Range / Sizes Packing Service Warranties Profitability	PROMOTION / CONVERSATION Sales Promotions publicity Sales force Public relations Direct marketing Consumer classification and connection systems Social networks (relationship networks) Privacy Policy (request consent)	GOALS Why buy? By type of need: physiological, safety, belonging, esteem and achievement.	OCCUPANTS Who is part of the market? What is the target? What are the direct, indirect and generic competitors?
				OPERATIONS How do they buy? Awakening the desire, information process, conviction of purchase, repeat purchase.	OBJECTS What does the market buy? Goods or services Durable or non-durable Current purchase, thought, specialty and impulse.
PROCESS The processes reflect all the creativity, discipline and structure brought to the marketing management. What is the raison d'être of the Brand? What is the connectivity strategy: screen technology and the internet? What is the positioning and differentiation of the brand and / or clarification and codification of the brand?	PERFORMANCE Performance is a holistic marketing to capture the range of possible outcomes and measures that have financial and non-financial implications and implications beyond the company itself. Big Data Analyzes? What is the level of appreciation, experience and involvement with the client?	PRICE / CURRENCY Pricing table (cost basis, competition and customer value) Discounts (market fluctuation) Subsidies Payment Term Credit Terms	PLACE / COMMUNITY ACTIVATION Channels Peer-to-peer distribution Coverage Rate Channel diversity Location Mode of transport	OCCASIONS When do they buy? Frequency, rhythm, seasonality, events and conjuncture.	ORGANIZATIONS Actors in the buying process? Initiator, prescriber, decision maker, buyer (who pays), opponent, perpetrator, product destroyer and user (consumer).
9. EXPENSES / INVESTMENTS				5. INCOME	

CONCLUSION

If we look at the recent past, we have innumerable cases, from several companies that failed to identify in a timely manner the first signs of change, and which eventually went completely unnoticed or were considered irrelevant by them, so that many of them disappeared or affected the threatened business model. Would Kodak believe that today photos are taken with a phone? The taxi industry would be worried when it appeared in 2009 to Uber? Did you believe a few years ago that it might be possible to buy a car online? Did any hotel, 20 years ago, think that it was possible to rent rooms by Airbnb? Five years ago would the crypto-coins banks speak? AND.... Will the pizzas be delivered by drones? Today, Uber is the largest taxi company without taxis. Airbnb is the largest booking company and has no houses. WhatsApp and Skype are the largest telecommunications companies and have no infrastructure and Alibaba is one of the largest retailers in the world and has no inventory. In this article, we present the new framework for marketers that will allow them to structure a strategic and operational marketing plan, in a simple and integrated way, through the Canvas Marketing Framework. It is a framework that will also allow to reconcile analog and digital marketing, as well as being a complementary tool to others, namely, with the conceptual model, the Business Canvas Model.

In summary, marketers consider this framework intuitive, allowing an integrated view of the search marketing, strategic and operational of an organization. In this sense, it is dynamic, interactive and even disruptive because it allows quick adjustments to the organization's value proposition. The Canvas Marketing Framework also allows for the integration of analogue and digital marketing, representing the most important factor today and for the future. Lastly, almost all marketers (99%) considered the analysis of the 6 O's as particularly useful and relevant in the construction of the marketing plan.

REFERENCES

Adolpho, C. (2011). *Os 8Ps do marketing digital*. Novatec Editora.

American Marketing Association. (2016). Definition of Marketing. Retrieved from https://www.ama.org/AboutAMA/Pages/Definition-of-Marketing.aspx

Anderson, E. (2015). Do the top U.S. corporations often use the same words in their vision, mission and value statements? *Journal of Marketing Management*, *6*(1), 1–15.

Arndt, J. (1978). How broad should the marketing concept be? *Journal of Marketing*, *42*(1), 101–103.

Bagozzi, R. (1975). Marketing as Exchange. *Journal of Marketing*, *39*(4), 32–39. doi:10.1177/002224297503900405

Berry, D. (1988). The marketing concept revisited: It's setting goals, not making a mad dash for profits. *Marketing News*, *22*(15), 26–28.

Cousins, L. (1991). Marketing plans or marketing planning? *Business Strategy Review*, *2*(2), 35–54. doi:10.1111/j.1467-8616.1991.tb00151.x

Eteokleous, P., Leonidou, C., & Katsikeas, C. (2016). *Corporate social responsibility in international marketing*: Review, assessment, and future research. International Marketing Review, *33*(4), 580–624. doi:10.1108/IMR-04-2014-0120

HBR. (2018). *Marketing estratégico. 10 artigos essenciais*. Actual Editora.

Kannan, P. K., & Li, H. A. (2017). Digital marketing: A framework, review and research agenda. *International Journal of Research in Marketing*, *34*(1), 22–45. doi:10.1016/j.ijresmar.2016.11.006

Khan, I., & Rahman, Z. (2015). A review and future directions of brand experience research. *International Strategic Management Review*, *3*(1), 1–14. doi:10.1016/j.ism.2015.09.003

Kienzler, M., & Kowalkowski, C. (2017). Pricing strategy: A review of 22 years of marketing research. *Journal of Business Research*, *78*, 101–110. doi:10.1016/j.jbusres.2017.05.005

Kotler, P., Keller, L., & Lane, K. (2015). Marketing management, Global Edition, 15/E, 2016, Pearson.

Kotler, P., Hermawan, K., & Setiawan, I. (2017). *Marketing 4.0: Moving from tradition to digital*. Hoboken, NJ: John Wiley & Sons.

Kotler P., Keller L., Tuck A., Ang H., Tan T., & Leong S. (2017. *Marketing Management*, An Asian Perspective, 7/E, Pearson.

Marques, V. (2015). *Vídeo marketing 360*. Actual Editora.

Marques, V. (2016). *Redes sociais 360. Como comunicar online*. Actual Editora.

Marques, V. (2018). *Marketing digital 360, 2.ª Edição*. Actual Editora.

Meffert J., Mendonça, P., McKinsey & Co. (2017). Eins oder Null, Planeta.

Mooij, M. (2015). Cross-cultural research in international marketing: Clearing up some of the confusion. *International Marketing Review*, *32*(6), 646–662. doi:10.1108/IMR-12-2014-0376

Osterwalder, A., Bernarda, G., Pigneur, Y., & Smith, A. (2015). *Criar Propostas de Valor*. Dom Quixote.

Pigneur, Y., & Osterwalder, A. (2015). *Criar Modelos de Negócio*. Dom Quixote.

Doyle, P. & Stern, P. (2006). *Marketing Management and Strategy*, 4/E, 2006, Financial Times Press.

Trout, J. (1969). Industrial Marketing Magazine, June, and then popularized by Ries, A. & Trout J. (1981). Positioning - The Battle for Your Mind. McGraw-Hill.

Van der Grinten, J., & Riezebos, R. (2011). *Positioning the brand*. Routledge.

Chapter 10
Sustainable Innovation Projects From Patent Information to Leverage Economic Development

Sérgio Maravilhas-Lopes
https://orcid.org/0000-0002-3824-2828
IES-ICS, Federal University of Bahia, Brazil

ABSTRACT

Patent information can provide a growing competitiveness through the technology transfer it fosters, and be economically important because of the innovation it leverages. Organizations are not monetizing their potential related to the use of patent information that could encourage more innovation and the largest number of patent applications, resulting in more businesses and greater economic growth. This chapter sustains that a coherent and effective use of patent information, containing information from research and development (R&D) activities with industrial application, can contribute to solving problems, fostering innovation through the resulting products and processes. Sustainable solutions can be realized, using unexploited inventions, as by the formulation of new products based on R&D that can be adapted to new global needs, creating jobs and protecting the environment and its resources.

INTRODUCTION

The concept of sustainable development has its origins in the attempt of integrating environmental conceptions in economic policy, bringing the ideas of environmentalists to the central area of world politics that currently focuses on Economy (Dresner, 2008, p. 69).

Sustainable development seeks to carefully balance environmental concerns with economic development, difficult task when the immediate concern focuses ever more on economic aspects than in preserving the environment and natural resources.

DOI: 10.4018/978-1-5225-9993-7.ch010

This concern is of importance in economic models to adopt in developing countries, mostly located in the southern hemisphere, where most of the natural resources to be exploited can be found and should be protected and used for the sake of improving the lives of the people who hold them.

Another concern, directly related to the economic issue, relates to the difficulty in accessing the most efficient technologies for the populations of these countries, often being the place where companies seek to sell the stocks of the products that are outdated and no longer have demand in developed markets, contributing for the availability in these countries of the most polluting products and environmental problem generators when it should be the opposite.

Patent information search and consultation helps prevent the waste of material and financial resources, because it avoids reinventing what already exists and has consumed resources to have been invented (Jolly & Philpott, 2009).

It allows the realization of sustainable solutions, both by the use of unexploited inventions, as by the formulation of new products based on R&D already done that can be adapted to new global needs, creating jobs and protecting the environment and its resources.

This source of information covers all scientific and technical activities of human creativity and is coded to permit its easy recovery and use.

Patent information repositories, in the form of databases and digital libraries are the largest source of scientific and technical information, available for free via the Web, globally.

The analysis of this type of information allows the free exploitation of certain inventions, without the obligation to pay any license fees if the patent is in public domain and free to be used (Petroski, 2008).

Such is the case of generic drugs that are the free use of the active substances of certain medicines that reached their protection limit and are free to be explored, what has been done successfully by various national and international companies.

It is described and sustained in this research project that a coherent and effective use of patent information, containing information resulting from R&D activities with industrial application, may contribute to the increase of creativity and support in solving research problems, fostering innovation through new products and procedures based thereon, with benefits for the sustainability of the countries due to the use of resources more efficiently by reducing costs and resources spent.

Several 'environmentally friendly' technologies are available to be explored, many at no cost, which may allow the use of clean energy for free and with economic advantages for those who want to implement them.

This research work aims to contribute to increase the use of patent information in R&D activities to stimulate creativity, contribute to solving research and manufacturing problems, minimize the costs related to these projects and maximize the results from this activity and their related investments, which may contribute to the creation of products and environmentally friendly businesses with benefits for all stakeholders.

ADVANTAGES OF PATENTS FOR DEVELOPING COUNTRIES

Sherwood says that "the protection of innovation has been the yeast of the economic development of many countries." This can be seen because "countries with advanced economies tend to be those who have property protection systems in which the public deposits a certain degree of confidence" (1992, p. 11).

With this, it warns us that we need to adopt effective measures of intellectual property (IP) protection in all its aspects (patents, brands, utility models, designs, etc.), at the risk of the developing countries watch escape to other countries its greatest asset, the Intellectual Capital, because not having the means to protect their inventions in their own countries make inventors seek other countries where they can get its protection, making the resultant wealth grow in those countries, and not in their own where it was most needed.

Thus, developed countries continue to develop themselves and emergent countries 'stagnate' without new ideas that can be channeled into sources of wealth to exploit their resources.

Also for the same reason, multinational companies do not invest heavily in these countries for fear of being dispossessed of their source of income due to the illegal appropriation of their discoveries and inventions, motivated by the weak property protection of intangible assets that characterize IP.

This position is also shared by Idris (2003), who defends the need to implement strict IP laws in developing countries in order to stimulate the creation of innovative companies, local or international, fixing themselves in these countries because they feel safe and secure, thereby promoting technological development that will condition economic growth by the competition that they motivate.

In this sense, Barbara Hansen (1980) in her analysis of the economic aspects resulting from technology transfer to developing countries, introduces the idea of 'absorption' of technology, defining this as the induction of technical progress based on a 'transferred' technology, noting that for this to occur, it is required the presence of several factors, such as adaptation, improvement and further development of the transferred technology, according to the conditions of the Economy that receives and integrates this technology, such as weather, production factors, resources, etc.

That research, whose core idea is that achieving goals aimed at the development is closely dependent on the correct adaptation of the mechanisms of absorption of new knowledge of a given Economy, stresses that the capacity to absorb such scientific and technical expertise contains three closely interrelated aspects, which are: i) The ability to recognize possibilities of adaptation of more advanced foreign technologies; ii) The ability to adapt the technology to the physical, social and economic contexts of the country; iii) The ability to adapt the social and economic conditions to the requirements of these new technologies.

Ullrich (1989) stresses that IP, especially patents and utility models, represent economic and industrial policy tools in the role they play as a stimulus to innovation. This stimulation is achieved by the temporary protection granted to the commercial exploitation of the results of R&D.

IP has, as its main function, the dissemination of technical and economic information that stimulates the economic performance of a country, but also the individual business units. IP also has the merit of supporting technology transfer processes so important for developing countries, establishing the link between R&D University and business centers, and the economic and financial structure characterized by market architecture.

SUSTAINABLE DEVELOPMENT AND ECONOMIC GROWTH

According to a study held together by the European Patent Office (EPO), United Nations Environment Programme (UNEP) and the International Centre for Trade and Sustainable Development (ICTSD), in 2010 (http://www.epo.org/news-issues/technology/sustainable-technologies/clean-energy/patents-clean-energy.html), six countries are the origin of about 80% of the innovations developed worldwide in the field of clean energy technologies.

This study, which was based on the analysis of 400,000 patents from *Espacenet* database, focused on the design and dissemination of such technologies worldwide.

It was analyzed the effect of patents on the transfer of Clean Energy Technologies (CET), being the first time that the importance of licensing technologies in this area was found, which allows to know the practices of the holders of these technologies in areas such as solar photovoltaic, geothermal, wind, carbon capture, among others.

The study clearly shows that the increasing number of CET patents coincides with the adoption of the Kyoto Protocol in 1997, making it clear that political decisions can be a major factor in stimulating the development of key technologies to combat global warming and climate change.

Through its statistical analysis, the study shows that the number of patents in the mentioned technology areas increased by about 20% per year since 1997, bypassing patents of traditional fossil energy sources and nuclear energy.

Among the six countries leading the area, Japan is the one who developed and patented most technologies, followed by the United States of America (USA), Germany, South Korea, France and the United Kingdom (UK). China is fast approaching the South Korean figures with regard to patenting in the solar photovoltaic area.

With regard to technology licensing, there is a reduced licensing activity between organizations from developing countries, being these activities limited to countries like China, India and Brazil.

According to analysis made by the company *Eloqua* entitled "Do 'Green' Companies Grow Faster?" of May 27, 2012 (http://blog.eloqua.com/do-green-companies-grow-faster/), based on a report by *Newsweek* in 2011 (http://www.thedailybeast.com/newsweek/2011/10/16/green-rankings2011.html), as well as generating a good reputation and positive image for brands and companies that adopt environmentally sustainable attitudes, what works well in aspects related to advertising and public relations for the firms that hold them, it appears that these promote further economic growth, with better performances in the markets where they operate.

Because they are committed to the mission of making the world a better place through social and environmental management policies, they acquire an intrinsic advantage gained by the respect and trust that their reputation generates among informed and responsible consumers.

They analyzed the 100 'greener' companies, according to the said *Newsweek* study, and compared the cumulative growth of the net income of those companies between 2005 and 2010, with the 500 largest companies in the Standard & Poor's (S&P) index for the same period.

The results show a performance advantage of 13.7% higher growth for the best companies regarding social responsibility.

Interestingly, even the 25 'less green' companies of the study have outperformed the S&P 500, showing that seems to compensate the bet of the company's investment in environmental programs.

Other authors had already drawn attention to the competitive advantages that sustainable environmental strategies (Esty & Winston, 2008) and clean energies (Krupp & Horn, 2009) allow to obtain. In addition to allowing lower costs with raw materials and energy necessary for the operation of some industrial activities, the use of clean and renewable energy sources reduces environmental degradation at the same time, imposing a positive image in public opinion, capturing a greater number of customers and consumers without spending on advertising. It's the Economy immersed in sustainability (Dresner, 2008, p. 81).

Figure 1. Performance comparison: Top 100 'green' companies X S&P 500 X 25 'less green' companies (Source: http://blog.eloqua.com/do-green-companies-grow-faster/?utm_source=feedburner&utm_medium=email&utm_campaign=Feed%3A+ItsAllAboutRevenue+%28It%27s+All+About+Revenue%29)

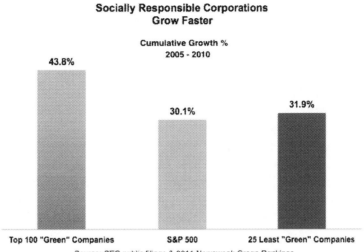

There are several natural energy resources that are lost daily without being used, like the energy of the sun that illuminates and warms us, the wind blowing, the sea waves and tides that succeed without our intervention, the abundant geothermal energy in some places and that the ancient Romans knew very well how to enjoy in its famous 'baths' (Krupp & Horn, 2009).

There are several patents that seek to use these resources, some of them not implemented for lack of financial investors to bet on these technologies because the markets are dominated by the coal and oil industries (Yeomans, 2006), two of the major causes of environmental problems of the planet (Esty & Winston, 2008; Krupp & Horn, 2009).

Currently with the start of oil scarcity and the environmental costs of unsustainable fossil materials, new solutions must be found to meet the energy needs of humanity (Dresner, 2008), whereby hydrogen is a candidate to consider (Yeomans, 2006).

However, other power solutions are already being explored, such as the methanol fuel cells, invented 15 years ago, which are the first commercial product on the market, surpassing the technical deficiencies that hydrogen presents for now (http://pt.euronews.com/2012/06/05/celulas-de-combustivel-para-dispositivos-portateis/).

Instead of wasting precious time and waste even more resources and cause more environmental damages, the ideal would be to look if there is already some solution available or adapt an existing solution to solve the current problems.

Patent information may contain the solution to adopt.

ORIGIN AND ADVANTAGES OF PATENT INFORMATION

During the registration and grant process of the patent, the Patent Offices will generate one or more legal documents that are designated by Patent Literature.

The information that these documents contain is called Patent Information.

After the publication of the patent application, normally 18 months after having been entered in the respective office (the granting of protection is only given, usually three years after application), this information becomes publicly available to those who wish to consult it.

Thus, analyzing patent information may allow the development of products, but by diametrically different processes, resulting in cost reductions that can drive their holders to competitive advantages (Petroski, 2008).

Thus, in the patent documents, we can find the following information: i) The 'state of the art', the technical knowledge available to date in the area in which operates the invention carried out; ii) Type and nature of the technical problems that the invention will solve; iii) Detailed description of the invention and how it works; iv) Illustrations, diagrams and drawings of the constituent parts of the invention for easier understanding of it, where necessary and appropriate; v) "Furthermore, patent information can clarify and supplement articles published by the inventor" (Macedo & Barbosa, 2000, p. 58).

One of the prerogatives, so that the patent can be granted, is that the information in the patent application is in such detail that a person skilled in the area will be able to perform the invention (product or process).

The patent document discloses the invention, necessarily liable of industrial application, but also defines the scope of protection required if the respective patent is obtained, granted by the responsible Office (Jolly & Philpott, 2009).

The disclosure of the technical secrets contained in the documents resulting from a patent application disseminates valuable information to the public about the 'state of the art' in a given area by promoting, through that knowledge, technological development.

As a result of this disclosure, constituent parts of this invention can even be used provided they do not incur in any violation of the claims contained and described in the patent. This procedure can lead to the attempt to develop competing products to those found in patent documents, by inventing more lucrative or more effective alternatives (Rivette & Kline, 2000).

Patent information, in addition to providing an excellent source of information to generate ideas, also has the advantage of being used as a source of inspiration when there is need to find the solution to assist in solving technology-related problems.

Where to Search Information for Sustainable Patents and Inventions

The Internet has brought many changes and advantages in the access to information.

The amount, extent and speed with which we can access the required information make this a privileged resource for the search and analysis of information.

According to Idris (2003), this possibility of access to the information available is the factor enabling the creation of knowledge and increasing the growth of wealth.

It relies on the generation and management of what he means by 3 "i's", namely, **I**nnovation, **I**nformation and **I**deas, supported by a fourth "i" which stands for **I**nternet.

For this author, these are the fuel that feeds the incredible current technological progress and the ownership or access to such vehicle and the information conveyed by them are vital for any company that wants to keep on top of their area of expertise, as this will allow them to create innovative products or find innovative ways of producing existing products with cost effectiveness.

According to Maia (1996), patent databases allow in a quick and efficient manner: i) Be certain of the originality of planned research programs; ii) Search inventions useful for further innovations; iii) Get an overview of new trends in R&D activities in a particular area of technological development; iv) Monitor the Marketing strategies of competitors, discovering the countries where they required patent protection.

The possibilities that modern technology have to offer regarding patent information search can prove to be very useful for allowing discovering inventions with high economic potential that are not being properly exploited by their holders, allowing to establish technology licensing agreements that lead to the exploitation of the invention by those concerned and interested in doing so.

Many inventions with environmental benefits have not yet been exploited because of the economic dominance of companies based on energy from coal, oil and nuclear.

Some digital platforms that allow access to patents and the information contained therein, as well as unprotected inventions (Open Innovation) with environmental and sustainable importance, are: i) Ecopatent Commons; ii) WIPO GREEN; iii) The GreenXchange, from *Nike*.

Ecopatent Commons (http://ecopatentcommons.org/), launched in 2008 by four companies (IBM, Sony, Nokia e Pitney Bowes) in collaboration with *The World Business Council for Sustainable Development* (WBCSD), aims the commitment to all those seeking to develop innovative products that respect the environment and the sustainability of the planet. To do this, these companies make available to those who want to explore their technologies in these conditions, a set of patents from ecofriendly inventions, promoting partnerships and collaborations between the companies who own the patents and entrepreneurs who have projects to use them. Besides the four founders, nine companies were part of this consortium (in a total of 13) that already contributed with more than 100 patents (103 in 01-05-2011) for this fund (HP, Bosch, Dow, DuPont, Fuji-Xerox, Ricoh, Taisei, Hitachi e Xerox). Nowadays, DuPont and Hitachi left the consortium. A partnership with the World Intellectual Property Organization (WIPO) green patents database (WIPO GREEN) was established and, in 2013, *The Environmental Law Institute* became the host organization. These initiatives must be cherished because lots of companies have several non-used patents that can be explored, in a useful and ecofriendly way, benefiting all humanity.

Similarly, WIPO created a platform for access to information of patents relating to 'green' technologies (https://webaccess.wipo.int/green/ - WIPO GREEN - *The Sustainable Technology Marketplace*), environmentally sustainable, to allow all stakeholders the access to information and mediation with transparency and trust, guaranteed by the organization itself. Several new index classes were created with the letter 'Y' to designate, classify and ease the search for green patents. It is addressed primarily to developing countries and emerging economies, intending to increase and accelerate the adoption, adaptation and diffusion of these technologies in these countries. It aims to contribute to make the world and the economy 'low-carbon' and, thus, reduce climate changes plaguing those countries, already very fragile and needy.

The GreenXchange platform from *Nike* (http://www.greenxchange.cc/), in partnership with *CreativeCommons* and *BestBuy*, was announced in Davos in the World Economic Forum in January 2011, and the network of partnerships now includes companies like *Ideo, nGenera, SalesForce, Yahoo!*, etc. Currently, they have 463 assets available for exploration, held by three organizations: the founders *Nike* and *BestBuy*, with 444 and 15 assets respectively, and the *University of California at Berkeley* with 4. There are three types of licenses available and there may be restrictions (geographical, for example) in the license to be granted for the invention, or even a payment for the use of it.

Figure 2. Ecopatent commons
(Source: http://www.wbcsd.org/work-program/capacity-building/eco-patentcommons.aspx)

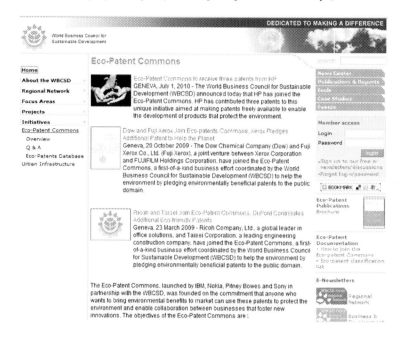

Figure 3. WIPO GREEN - The sustainable technology marketplace
(Source: https://www3.wipo.int/green/green-technology/techOverview)

Besides these specific information resources, directed to the promotion of sustainable inventions, other solutions may arise from the analysis of patent information in general allowing, as already mentioned, the use of available inventions with other purposes that can be adapted for the solution of problems related with sustainable development and the good manage of natural resources.

Currently, almost all industrialized countries offer via Internet their patent collections for easy access and consultation of this precious resource.

Figure 4. The GreenXchange from Nike
(Source: http://www.greenxchange.cc/)

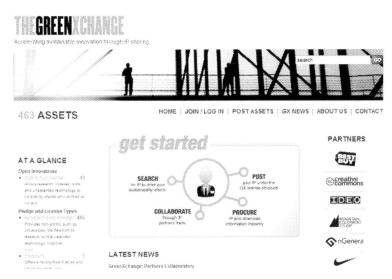

Usually, the documents are in the country's official language, which is not always easy for those who perform the research.

Hence access to information provided by WIPO and EPO are indispensable resources because they have large amounts of summaries (abstracts) in English and, in addition, they provide support for research in key Asian offices.

The advantage of all the resources presented in this work is that, besides they are all free, simply needing a computer with an internet connection to reach them, these allow to query patent information records of the major industrialized countries, with more inventions and, consequently, more patents awarded.

All these resources are of vital importance to obtain information to reduce costs and waste while increasing the probability of creating viable solutions to global problems.

USE OF PATENT INFORMATION FOR CREATING SUSTAINABLE BUSINESSES

The 'Solar Oven' developed by Prof. Manuel Colares Pereira was inspired in a patent already expired, in public domain, of a similar invention called the '*Phyreheliophoro*' but used for different functions, like melting metals and make fertilizers.

The consultation of this patent, from a Portuguese priest, Father Manuel António Gomes, better known as MAG Himalaya, that at the beginning of the XXth century, in 1904, won the 1st prize in a science competition in the Universal Fair in St. Louis, Missouri, USA (Rodrigues, 1999), motivated the development of an oven (http://www.sun-cook.com) for the preparation of food using the sun's energy (http://solarcooking.wikia.com/wiki/Sun_Co).

This invention uses only the sun energy, is non-polluting, eliminates the need to cut trees for firewood for cooking, reducing deforestation, release time available for women for other functions, such as studying and taking care of the family, which is impossible if they have to travel long distances to collect firewood, as is the case in less developed countries, and allows the creation of family businesses, like

Figure 5. The 'Phyreheliophoro' in Portugal and in St. Louis, Missouri, USA, in 1904 (Source: Rodrigues, 1999)

Figure 6. The sun cook: Solar oven by Sun Cº. – Comp.ª de Energia Solar S.A.
(Source: http://www.sun-cook.com)

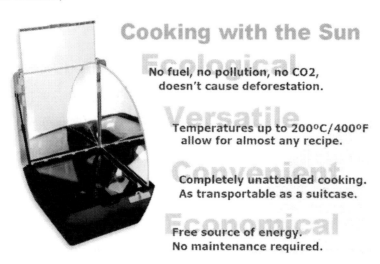

cooking to sell meals, enabling the realization of capital for the creation of other businesses and improve the quality of life. It is, therefore, an ecological product that allows sustainable businesses with inherent economic and environmental benefits.

This new invention and innovation, inspired in a 1900 technology, is patentable itself, since the solution found is new, it is not contained in the technical state of the art and is not intended for the same use of the previous invention.

Also, Dou (2004) proves all these advantages to us, through projects initiated in low technological development countries, where there have been notable changes in self-employment, social and economic improvements, diversification of supply and increased exports after the introduction of the surveillance

Figure 7. Solar oven patent
(Source: http://pt.espacenet.com/)

of markets and technologies practice, with particular emphasis on technological surveillance using information from patents through the Internet. In one of these projects the author tells us the situation experienced in an Indonesian province, in a remote island without large economic resources, but with great abundance of coconuts.

The fruit was eaten and its juice drunk (coconut water), but the shells resulted in a serious environmental problem because they were not given any use and constituted debris piles scattered around the island, stripping it of its beauty and sending tourists away. Training was given to the people, so that they could use the computers in the public library to conduct research in the *Espacenet* of patent technologies involving the use of coconuts, 'Kelapa' in their own language. Thanks to these documents they found out that they could use the fruit and its juice in many different applications, such as jams, sweets, cakes, liqueurs, etc., but that their bark could also have unknown uses that made possible to wash away this debris in a cost-effective manner. Thus, the shells or bark may be used as a fertilizer, material for the construction of furniture, building materials and insulation, toys and decorative articles (raw material for craft products), and if processed as material for filling pillows and mattresses. Such knowledge made possible a huge number of successful family businesses, transforming a small island, without economic resources in a prosperous place, with greater social equality and head of small business owners who have expanded their business across borders. We must highlight that most of the technologies and patents discovered were of Brazilian origin, not protected in Indonesia, and that could be used without any hindrance or payment of fees.

The Use of Natural Products to Generate Innovations With Economic Sustainability and Environmental Concerns

If analyzing an invention created with a certain purpose and intention, like melting metals, made possible the cognitive jump to the creation of a solar oven for food confection, visualizing patent information which already have available inventions using natural products, more easily may allow the insight necessary to create new businesses based on these inventions.

The following presents a few examples, to better realize the value of information in these repositories and, also, to emphasize the idea of using natural products and resources for solving problems in a sustainable way, eliminating waste that would otherwise be a problem to solve.

The WIPO GREEN platform provides a set of patented inventions, which can be exploited by those interested in taking advantage of available solutions.

Figure 8. Use of pineapple to produce plastic
(Source: https://www3.wipo.int/green/green-technology/withoutLogin)

Figure 9. Use of pineapple to produce paper
(Source: https://www3.wipo.int/green/green-technology/withoutLogin)

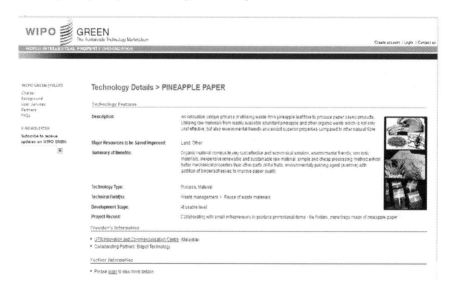

Among the available inventions we can find a way to produce plastic and paper from pineapple, which can be a solution for the recovery of waste, bark, and leaves, that are not used in the food industry.

Like the example of coconuts, this not only can allow a form of monetization of the surplus that was not used, but also how to avoid the ecological problem of waste disposal.

Again, as already pointed out, everyone wins with the situation, businesses, the environment and, ultimately, the general public (all of us).

Other wastes from the food industry, in this case banana peels (http://www.cienciahoje.pt/index.php?oid=46955&op=all), can have an extremely advantageous use in removing heavy metals, problem that industrial filters available on the market doesn't solve satisfactorily and whose cost is very high.

Other natural resources already used in this activity involve the use of clay, shellfish shells, corn cobs (after removing the grain) and algae (Agar).

Finally, we present an example developed in the Chemistry engineering department at the University of Coimbra, aimed at the use of other waste with advantages in the context of sustainability, the use of eggshells (http://www.cienciahoje.pt/index.php?oid=54436&op=all) as additive for soil correction.

Not only eliminates the environmental problem of deposition of eggshells as they become useful and, probably, profitable in the resolution of an ecological problem, eliminating the use of chemical fertilizers with undeniable environmental benefits.

For all the reasons presented, the analysis of patent information is a task to be considered by entrepreneurs and researchers because a monetization opportunity can be discovered for some invention that can be used or adapted for cost-effective sustainable practices.

FUTURE RESEARCH DIRECTIONS

Future projects should consider training entrepreneurs in developing countries how to search for patent information, and how to start a business using that knowledge, so they can explore inventions with sustainable focus and improve the quality of life in their countries, with economic results.

These measures aim to improve the number of people that effectively search patent information with benefits for the whole society.

CONCLUSION

The relation between IP and economic development was shown, based in the economic growth verified in the countries that have developed IP protection mechanisms to the inventive and creative work and the consequent technological progress.

Also, was demonstrated that the companies that adopt 'green' practices, environmentally friendly, can get a higher growing than the average of the top 500 companies in S&P index, holding a reputation and positive image that lead to a greater number of customers.

Patent information importance was analyzed for the economic growth and the obtainment of competitive advantages, easily accessible through the Internet that helps to disseminate it with very low costs.

Several examples of its strategic use were mentioned, where the highlight of the use of products invented for a function, but if adapted can serve to another distinct task, with the satisfaction of basic needs in less developed markets was done, and the use of inventions in public domain for the realization of sustainable products and businesses, environmentally responsible, to improve the lives of people (like the solar oven and the use of coconuts as raw material for several ecological products).

It was possible to realize that there are organizations that promote the use of patented solutions for those who want to start their operation without a high initial cost of R&D and that, sometimes, natural resources can be used to solve complex problems.

Although this information is available to all, not everybody benefits from it and it's not sufficient just to have it, it is necessary to build something new with it, to innovate.

Additionally, you can enable advances in the intersection (Johansson, 2007), allowing to create new knowledge from different scientific areas, or a new science based in others, like Bioinformatics, Biomedicine, Bioengineering, Genetic Engineering, and so forth.

To highlight the issue of social advantage and the circular economy, much more important than simple economic benefits, that may also be present in the system, particularly if we invest in technologies that induce an economy based in reusable practices that are economically sustainable and environmentally viable.

Patent information can be used to the development of clean technologies and solutions to save or improve lives.

Eco Patents Commons and WIPO GREEN may contain several solutions to explore for that purpose.

If some technology developed with a certain purpose allows its adaptation for a different one with cost advantage benefits for the environment, so it should be used, and everyone will gain with that adoption, as long as that invention is in public domain and free to use. Researchers should go back to those inventions that, although genial, never were made because there were no feasible and financially affordable techniques to perform them cost efficiently (Petroski, 2008). With the current technological development, it may be possible to build them successfully and allow some technological advancement for the preservation of nature and the planet.

The secret of success is not only to know what others do not.

It is to act on this information and produce advantages and benefits to its holder.

REFERENCES

Andrew, J., & Sirkin, H. (2008). *Payback: Como conquistar o retorno financeiro da inovação* (1st ed.). Lisboa, Portugal: Actual.

Ashton, W. B., & Klavans, R. A. (1997). *Keeping abreast of science and technology: technical intelligence for business*. Columbus, OH: Battelle Press.

Berle, G. (1992). *O Empreendedor do Verde: Oportunidade de negócios em que você pode ajudar a salvar a Terra e ainda ganhar dinheiro*. San Paulo, Brazil: Makron Books.

Butler, J. T. (1995). Patent searching using commercial databases. In Lechter (Ed.), Successful Patents and Patenting for Engineers and Scientists. New York, NY: The Institute of Electrical and Electronics Engineers (IEEE) Press.

Dantas, J. (2001). *Gestão da Inovação*. Porto, Portugal: Vida Económica.

Dantas, J. & Carrizo Moreira, A. (2011). O Processo de Inovação: Como Potenciar a Criatividade Organizacional Visando uma Competitividade Sustentável (1ª ed.). Lousã: Lidel.

Debackere, K., Luwel, M., & Veugelers, R. (1999). Can technology lead to a competitive advantage? A case study of Flanders using European patent data. *Scientometrics, 44*(3), 379-400.

Dou, H. (2004). Benchmarking R&D and companies through patent analysis using free databases and special software: A tool to improve innovative thinking. *World Patent Information*, *26*(4), 297–309. doi:10.1016/j.wpi.2004.03.001

Dresner, S. (2008). *The principles of sustainability* (2nd ed.). Chippenham, UK: Earthscan.

Esty, D. & Winston, A. (2008). Do Verde ao Ouro: Como Empresas Inteligentes usam a Estratégia Ambiental para Inovar, Criar Valor e Construir uma Vantagem Competitiva (1ª ed.). Cruz Quebrada: Casa das Letras.

Haberman, M. (2001). The role of intellectual property and patent information in successful innovation, production and marketing. Case study 1: The non-spill drinking vessel. *World Patent Information*, 23(1), 71–73. doi:10.1016/S0172-2190(00)00105-8

Hansen, B. (1980). Economic aspects of technology transfer to developing countries. *International Review of Industrial Property and Copyright Law*, 11, 430–440.

Holyoak, J., & Torremans, P. (1995). *Intellectual property law*. London, UK: Butterworths.

Idris, K. (2003). *Intellectual property: a power tool for economic growth*. Geneva, Switzerland: World Intellectual Property Organization.

Jegorov, A., Husak, M., Kratochvil, B., & Cisarova, I. (2003). How many "new" entities can be created from one active substance? the case of Cyclosporin A. *Crystal Growth & Design*, 3(4), 441–444. doi:10.1021/cg0300127

Johansson, F. (2007). O Efeito Medici: O que nos podem ensinar os Elefantes e as Epidemias acerca da Inovação (1ª ed.). Cruz Quebrada: Casa das Letras.

Jolly, A., & Philpott, J. (2009). *The handbook of European intellectual property management: developing, managing & protecting your company's intellectual property* (2nd ed.). Glasgow, Scotland: Kogan Page.

Krupp, F. & Horn, M. (2009). Reinventar a Energia: Estratégias para o Futuro Energético do Planeta (1ª ed.). Alfragide: Estrela Polar.

Macedo, M., & Barbosa, A. (2000). *Patentes, pesquisa & desenvolvimento: um manual de propriedade intelectual*. Rio de Janeiro, Brazil: Fiocruz. doi:10.7476/9788575412725

Maia, J. M. (1996). Propriedade Industrial: Comunicações e Artigos do Presidente do INPI. Lisboa: Instituto Nacional da Propriedade Industrial (INPI).

Marcovitch, J. (1983). *Administração em ciência e tecnologia*. São Paulo, Brazil: Edgard Blücher.

Marcus, D. (1995). Benefits of using patent databases as a source of information. In Lechter (Ed.), Successful Patents and Patenting for Engineers and Scientists. New York, NY: The Institute of Electrical and Electronics Engineers (IEEE) Press.

Petroski, H. (2008). *Inovação: da Idéia ao Produto*. São Paulo, Brazil: Edgard Blücher.

Rivette, K., & Kline, D. (2000). *Rembrandts in the attic: unlocking the hidden value of patents* (1st ed.). Boston, MA: Harvard Business School Press.

Rodrigues, J. (1999). *A Conspiração Solar do Padre Himalaya*. Porto, Portugal: Árvore - Cooperativa de Actividades Artísticas.

Sherwood, R. E. (1992). *Propriedade intelectual e desenvolvimento econômico*. S. Paulo, Brazil: EdUSP.

Tachinardi, M. H. (1993). *A Guerra das Patentes: O conflito Brasil x EUA sobre propriedade intelectual*. Rio de Janeiro, Brazil: Paz e Terra.

Ullrich, H. (1989). The importance of industrial property law and other legal measures in the promotion of technological innovation. *Industrial Property, 28,* 102–112.

Yeomans, M. (2006). *Oil - Petróleo: Guia Conciso para o Produto mais Importante do Mundo* (1st ed.). Lisboa, Portugal: D. Quixote.

KEY TERMS AND DEFINITIONS

Creativity: Creativity is based on reasoning that produces imaginative new ideas and new ways of looking at reality. Creativity is an individual process, arises from the idea that popped into someone's head. Relates facts or ideas without previous relationship and is discontinuous and divergent. No Creative Process exists if there is no intention or purpose. The essence of the Creative Process is to seek new combinations.

Innovation: The application of new knowledge, resulting in new products, processes or services or significant improvements in some of its attributes. When a new solution is brought to the market to solve a problem in a new or better way than the existent solutions.

Invention: The creation or discovery of a new idea, including the concept, design, model creation or improvement of a particular piece, product or system. Even though an invention may allow a patent application, in most cases it will not give rise to an innovation.

Patent Information: During the process of registration and grant of a patent, the official entities like the USPTO, EPO or WIPO, will generate one or more legal documents that are called patent literature. These documents contain information that is referred to as patent information.

Serendipity: Serendipity is the ability to make important discoveries by accident. Not all the ideas for new products or processes appear voluntarily and intentionally. Sometimes a mixture of luck and preparation provides valuable discoveries. A serendipitous discovery results from the combination of a happy coincidence with perspicacity.

Sustainability: The ability of producing goods and conduct business without exhausting nature's resources and polluting the environment or, if not totally possible, do the less harm and take measures to compensate the harm done. Can also be used to designate the ability of an organization of being capable of maintain itself on operation, generating profits and doing the best that it can for every stakeholder and shareholder.

Chapter 11
Change Management Projects in Information Systems:
The Impact of the Methodology Information Technology Infrastructure Library (ITIL)

Nuno Geada
https://orcid.org/0000-0003-3755-0711
College of Business Administration, Polytechnic Institute of Setúbal, Portugal

Pedro Anunciação
https://orcid.org/0000-0001-7116-5249
Research Center in Business Science, College of Business Administration, Polytechnic Institute of Setúbal, Portugal

ABSTRACT

In the current economic and social context, management of change should not be framed by managers on a passive perspective and only when there are clear signs of changes in organizational or market factors. The management of change must be framed in a perspective of continuous improvement, which justifies the development of capacities of economic and social vision associated with the sector in which they are positioned. The information society and the impact that new IT technology has on the functioning of economic organizations and the modus operandi of the market and the economy have been evident. The IT competitiveness potential of companies has attracted managers to the increasing inclusion of more technology in organizations, challenging them in managing the implicit changes.

DOI: 10.4018/978-1-5225-9993-7.ch011

INTRODUCTION

The increasing automation of industry (Hammer & Champy, 1994) and services seeks to improve the competitive position of companies in each investment. This strategic objective, well evident in the current economic context, has led many companies to assume the centrality of their competitive development in Information Systems and Technologies (IS/IT). In fact, the IS/IT has been gradually supporting the processes of automation in most activities and the modernization of economic organizations. Although this modernization is related to the levels of innovation of the various economic sectors, in each innovation organizations challenge their operating dynamics and interactions with the market.

As in the past, the introduction of technological innovations in organizational functioning continues to require structural, organizational, functional, informational and decision-making adaptations. The systemic perspective, under which organizational functioning and dynamics are due, must make it possible to question, in each innovation and consequent need for adaptation, the existing *status quo*. This logic of approach must allow management to open up competitive opportunities for change. These competitive opportunities may be internally driven by the fact that innovations provide reductions in cost structure or efficiency gains in organizational dynamics. And on the external economic level, the differentiation of products or services or new dynamics of interface with customers. However, when addressing the issue of change management, it should be stressed that these benefits cannot be achieved if economic organizations fail to match the potential of innovation and the achieved by its competitors.

Change management includes, among other things, the identification and management of the gap between a given organizational context and the desired one. The latter, the desired context, must be seen in the possibility of introducing innovation in processes, in the integration of systems, in the development of new products and services, in cultural change, in the introduction of new models and instruments, etc. In the field of management, innovation must be assimilated at the strategic level and, at this level, the frequent need to rethink the logics and organizational dynamics towards improving economic performance, the adequacy of communication with the market, the differentiation of the commercial offer, among several other examples.

In the current economic and social context, management of change should not be framed by managers in a passive perspective, and only when there are clear signs of changes in organizational or market factors. The management of change must be framed in a perspective of continuous improvement, which justifies the development of capacities of economic and social vision associated with the sector in which they are positioned. It is this capacity of vision, above all the reflexes that innovations can have on the market, which allows managers to identify business opportunities and, simultaneously, action on organizational weaknesses and constraints. The information society and the impact that new IT technology has on the functioning of economic organizations and on the *modus operandi* of the market and the economy have been evident. The IT competitiveness potential of companies has attracted managers to the increasing inclusion of more technology in organizations, challenging them in managing the implicit changes. This technological attraction has often been justified by the illusion of immediate and automatic generation of benefits in terms of efficiency, effectiveness and competitiveness.

However, many companies are subsequently faced with more fragile workings and with increased functional and organizational complexity as, in view of the increasing dependence on technologies, there are several shortcomings in IS/IT, for example, lack of adequate architecture models, lack of contingency plans, lack of risk management models, lack of definition of policies and safety plans, among other examples.

IS/IT should be understood in a systemic framework of economic and business functioning, which presupposes a coherent alignment between the business, the organization and IS/IT. Unless one understands the impact of the relationship between the market, the organization and IS/IT and manages its evolution, in order to provide new levels of efficiency in operation and effectiveness in relation to the market, it will be difficult to maintain the levels of competitiveness that the market currently demands. When we consider IS/IT as the pillars of current economic functioning, it is through them and it is through them that the management must seek to strengthen organizational performance, not only in support of operational activities, but especially organizational strategies. In this sense, IS/IT must be adequately designed (IS Architecture), integrated (Organizational Urbanism) (Anunciação, 2006) and managed (Governance) (ISGec, 2011) (ISGec, 2010) (ISGec, 2009).

It should be noted that, according to Oliveira (2010) (2009) (2004), because the economy develops in an information context, information is an economic resource. There is no Economy or Society without information, there are no economic organizations without IS/IT and there is no management without information, Governance practices are required for IS/IT in line with Corporate Governance, regardless of the type of organizations, profit and nonprofit organizations. All these arguments express the current dependence of organizations on IS/IT and the centrality they assume in the competitiveness and sustainability of developed economic activities. This centrality is even more evident when addressing current issues associated with new paradigms, such as Artificial Intelligence, Cloud Computing, Big Data, Digital Transformation, among others.

In this context, the adequacy of IS/IT to the challenges of the economy and digital innovation are central and essential to the value they can provide to companies. This understanding passes, among others, by two dimensions: the size of the architecture and the management dimension. Regarding the architecture dimension, although there are several methodologies, such as TOGAF, we can say that this dimension has not always been given due attention in the context of IS/IT. In a study carried out by the European Club for Governance of Information Systems in Portugal (Anunciação, 2013, 2012a, 2012b, 2011). It was possible to verify that, although most companies mention the existence of SI Architecture specification models, in terms of objectives, morphology, functionalities, level of integration of applications, relations between users and stakeholders, the centrality of the architecture in information systems is not expressive, as can be seen in Table 1.

Regarding the management dimension, although there are several proposed methodologies, such as the Information Technology Infrastructure Library (ITIL), we can see that, from a historical perspective, IS/IT was built for many years by adding layers of complexity that difficult and inefficient management processes. Considering also the previously mentioned study, it is possible to verify, through the results associated to some dimensions of the management of the IS, values that are not very expressive within the scope of some variables of the responsibility of the management. For example, most companies as-

Table 1. Existence of an SI architecture model

	1° Inquiry - 2011 -	2° Inquiry - 2012 -
• Yes	56,5%	60,5%
• No	39,1%	37,0%
• Do not know / Do not respond	4,4%	2,5%

sume that this evaluation was not carried out in 2011, although in 2012 a change of attitude in this area is evident. Despite this inversion, the value is not expressive, as evidenced in Table 2.

If we consider the evaluation of IS quality, for example, the study shows that most companies assume that this evaluation is performed, although a decrease is observed between 2011 and 2012, as can be seen in Table 3.

These results reveal the reality of managing some of the dimensions associated with IS in large companies in Portugal. Given this reality, we have proposed to evaluate the management of change associated with IS. We have taken as a reference, as evidenced previously, that in each technological innovation opportunities for improvement are opened, that these opportunities must be managed, and that this management must obey a project logic. The possibility of new forms of activity development, the search for efficiency and effectiveness gains, or even the implementation of new dynamics of operation through IS/IT should be sought from change projects. These projects must be managed as a way to achieve the desired benefits. Change management projects are justified insofar as IS/IT is an integral part of organizational systems, and it is important to understand how adequacy of organizational functioning should be made in relation to the introduction of new technologies or changes to existing ones. As mentioned above, the adoption of new IT may not correspond to the immediate and automatic generation of efficiency, effectiveness and competitiveness benefits. These should be appropriately introduced and managed in the context of IS and business. In this sense, in the domain of management, we will try to highlight, using one of the most well-known instruments/methodology in the IS/IT domain, ITIL, if organizations adopt change practices and if these practices are managed. In the study, we will analyse and highlight three central dimensions: the conceptual approach to the issue of change management, change management practices and the assumption of change management as an integral part of IS/IT management.

Table 2. Systematic evaluation of productivity performance, efficiency and efficacy of IS

	1º Inquiry - 2011 -	2º Inquiry - 2012 -
• Yes	34,8%	54,3%
• No	63,0%	43,2%
• Do not know / Do not respond	2,2%	2,5%

Table 3. Systematic evaluation of productivity performance, efficiency and efficacy of SI

	1º Inquiry - 2011 -	2º Inquiry - 2012 -
• Yes	71,7%	58,0%
• No	26,1%	38,3%
• Do not know / Do not respond	2,2%	3,7%

ITIL AND CHANGE MANAGEMENT PROJECTS

The choice of ITIL, as an inducing factor for change projects, was due to two main reasons. The first is due to the fact that it is a widely used methodology in the way of IT management. Its implementation has been applied in many organizations as a differentiating element in order to achieve efficiency gains and efficiency in the management of IT, for example, by reducing costs or optimizing investments. The second is due to the fact that this methodology proposes a life cycle for the management of IT services, with an impact on the performance of the Si and the organization itself. This life cycle, which starts with the strategy and ends in continuous improvement, proposes a set of steps that provide opportunities for the framing of the change in function of the design of the services and their availability.

ITIL, initially referred to as CCTA – *Central Computer and Telecommunications Agency,* is a methodology created in the late 1980s, which is based on IT service management, applying best practices to help companies achieve their business objectives, (re) focusing IS Services (Cartlidge et al., 2007). Associates the cycle of adequacy of IT services with the business divided by 5 central stages, as evidenced in Figure 1:

- **Service Strategy:** Definition of strategy for service through alignment with organizational strategies, policies, resources and business requirements and needs;
- **Service Design:** design and design of the solution to be adopted, through the design of the service according to the recommended solutions, the architecture (s) of IT, Standards, SDP;
- **Service Transition:** transition to new services by defining transition plans, test solutions, SKMS updated;

Figure 1. Inputs and outputs of the service cycles (Cartlidge, et al., 2007)

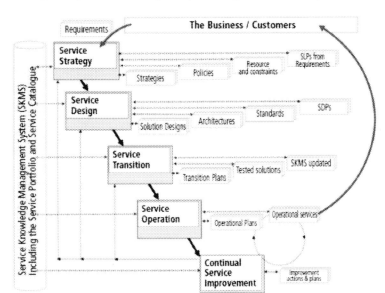

- **Service Operation:** assurance that services are managed based on Service Level Agreement (SLAs), identifying operational plans and operational services;
- **Continual Service Improvement**: maintenance of services provided and development of actions for continuous improvement of services based on the PDCA cycle (Plan, Do, Check, Act).

Taking as reference the respective steps, it is possible to verify the following benefits of the proposed methodology (Cartlidge, et al., 2007):

- Improved IT service and risk management with respect to business needs;
- More Flexible Service Levels;
- More consistent and predictable processes;
- Greater efficiency in the delivery of services;
- measurable and measurable Services and Processes;
- Optimization of customer experience;
- A common and universal language.

The adoption of this methodology, by allowing an alignment of the company's services with IT, enables the creation of a driving force for continuous improvement. Continuous improvement in IT services is naturally reflected in organizational services. It should be noted that the projects and the processes start with the service strategy, which means that, with the definition of the requirements and needs of the business, begin a path that leads us to the improvement of the existing service. However, the scope of this step presupposes capacity and success in the design and design of the service, that is, in the identification of the solution to be adopted. But you assume, above all, the ability to make the transition to a new service.

Note that ITIL (Van Haren - Publishing, 2011) does not exactly mention the solution, and what should be done. This methodology provides the identification of ways to adopt differentiating practices that aim at the excellence of IT services. With these practices at the level of IT Service Management, organizations can better their levels of efficiency and effectiveness in supporting IT to organizational activities, be they internal and external. Good practices, unlike standards and methodologies, provide flexibility and adaptability to internal competitiveness requirements. It is in the transition phase of the service that change is most evident. It is the service transition that depends on the ability to maximize the results of the redesign of the IT service (s). It is the phase where a service is tested and implemented and activated and changed in the production environment. The redesign of IT services provides guidance for the development and improvement of organizational and business capabilities, resulting from the introduction of new and modified services. It is also at this stage that it is ensured that the value identified in the service strategy becomes evident through the service design, and that it transits in the change that is implicit to it, which justifies the adoption of a set of good practices in several domains, according to Figure 2.

Considering that investments in IS/IT are normally critical to the operation and competitiveness of companies and that the management of associated projects is demanding, it is important that managers identify the critical orientations for managing the associated complexity, making evident also the set of areas parallel to the central focus of change in which the impacts of the innovations introduced (iceberg syndrome) will be reflected.

Figure 2. Schematic of the variables that constitute the good practices in the Transition of Services

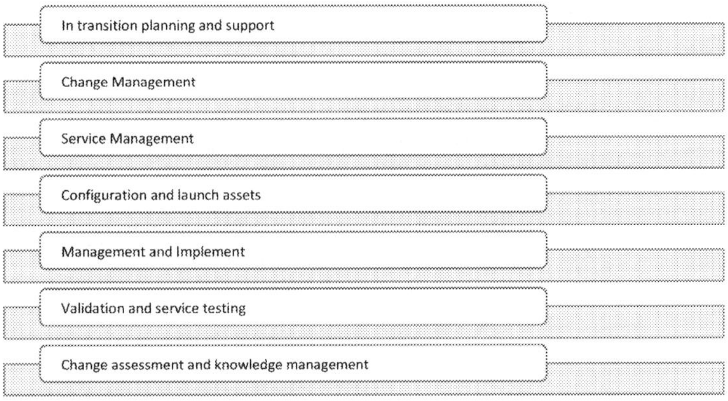

- In transition planning and support
- Change Management
- Service Management
- Configuration and launch assets
- Management and Implement
- Validation and service testing
- Change assessment and knowledge management

Project management provides and increases greater objectivity and assertiveness in managing the diversity of factors, be they human, financial, organizational, etc. Especially in situations of change associated with rapidly changing contexts, such as the contexts associated with IS/IT, the specification of the scope of projects, the hierarchization of the project phases, the identification and allocation of the necessary resources, the identification of risks, the analysis of the financial viability, among other relevant factors, provide management with a pragmatic assessment of the benefits and results and a possible correction of the strategy outlined.

In essence, the ITIL methodology foresees the execution of stages of project management: identification and validation of the needs, planning of the work, execution and revision after the implementation.

THE RELEVANCE OF CHANGE MANAGEMENT PROJECTS

The operationalization of change must be preceded by the identification and evaluation of the factors associated with it. In addition to the set of steps presented in the various models for change management, it is important to bear in mind that these projects or processes have a very strong cultural aspect. Understanding of existing culture, with respect to formal and informal rules, policies, norms and habits, is a critical success factor in change management. Galpin (1996) presents the following phases:

- Establish the need for change;
- Define the purpose of planned change;
- Diagnose and analyse the current situation;
- Create recommendations;
- Detail the recommendations;
- Create a pilot based on these recommendations;
- Prepare recommendations for roll-out;
- Distribute the recommendations;
- Measure, reinforce and redefine change.

Figure 3. PDCA versus SDCA (adaptado (Imai, 2006)

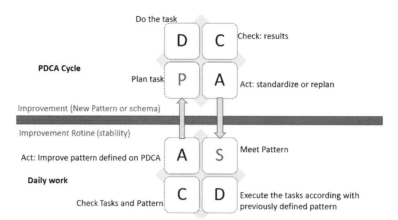

After these phases have been completed, the benefits of the change must be monitored and consolidated. In this sense, we can use the PDCA + SDCA cycle, or Deming cycle (Requeijo & Pereira, 2008).

The PDCA allows, within the scope of the management, to plan how to do, to develop or to execute, to validate what is done and finally to maintain everything that was executed in the previous phases with recourse to the continuous improvement, ensuring that the applied change is maintained. In order to ensure that there are no deviations or loss of effectiveness with regard to the benefits achieved, the Standardize-Do-Check-Act (SDCA) methodology must be adopted. This methodology allows to guarantee the uniformity of the applied change and the minimization of the associated impacts. The success of IS/IT project implementation depends essentially on the commitment and commitment of top managers and users.

Top management commitment, evidence of benefits, and expectations and positive reactions from end users to changes in their daily work routines are critical success factors for any change project as it is supported by users and the managers' commitment that improvements can occur, and the success of the change can be achieved. When the opposite situation occurs, known as resistance to change, due to the lack of credibility in it, fear of technologies, lack of involvement of the parties, among other examples, the central premises are committed to the success of any process or project associated with adaptations that are essential to the evolution of organizations or to their adaptation to the market.

However, it is normal for the introduction of new approaches or management instruments to be resistant to change. According to de Ven & Hargrave (2000), the resistance is verified when the change:

- It is not understood;
- It is imposed;
- It is incompatible with the existing environment;
- Introducing presents costs that outweigh the benefits;
- Does not have well defined processes.

The difficulties in managing change are usually difficult to identify the various variables sensitive to the new processes and the confrontation of the interests of all that are directly or indirectly involved, which is why there is a natural and tendential leverage to always make all the more difficult or complicated in order to maintain the existing status quo and to discourage management from achieving change.

Methodology and Objectives of the Study

As ITIL is one of the most used and most relevant methodologies in the IT management context, we have sought to analyse the extent to which this methodology is adopted in economic organizations and, second, to understand the extent to which practices are adopted associated with project management in IS/IT related change situations.

Four objectives were set out for this study, namely:

- analyse conceptual perception on the subject of change;
- to highlight the practices adopted regarding the management of change;
- evidence the knowledge and adoption of the ITIL methodology;
- to analyse the possibility of a relationship between the ITIL methodology and the main variables of change.

To achieve it there are some objectives, that we have tried to understand:

- Conceptual perception - the relationship between projects in IS or the introduction of IT and change, the need for mechanisms and management tools in change projects, the relationship between change management and organizational performance and competitiveness, and the need assessment tools and practices in change projects;
- Practices adopted in the management of change - the relationship between the management of change and investments or developments in IS/IT, the nature (internal or external) of the management of change projects, the level of management responsibility (top management, middle management or operational management) in change management projects, and the adoption of evaluation practices for change projects);
- Knowledge and adoption of the ITIL methodology - knowledge of the ITIL methodology, its adoption in IT management, and the relationship with change;
- Relevant dimensions in change management - identifying the most relevant dimensions for change management (leadership, vision, knowledge, evaluation, commitment, architecture, impact, strategy, planning and organization, implementation, celebration, learning);
- Identify the relationship between ITIL and change - the relationship between ITIL and change management phases.

As mentioned previously, the structure of the study contemplated two different dimensions: the validation of the relevance of the themes/strands under study, and the characterization of the practices adopted by economic organizations. The evaluation of each of the dimensions presented was done from a scale, in which the classification 1 expressed a total disagreement with the item in analysis, the classification 2 expressed some disagreement, the 3 expressed the nonconformity or disagreement, 4 showed concordance and score 5 showed total agreement. It was also given the possibility that respondents may not know how to respond or do not want to respond (Ns/Nr). The methodology adopted was based on the Focus Group technique and focused on the nature and depth intended for the study. The choice of this methodology was due to the specificity of the analysis and evaluation intended in the study, which presupposed a solid base of knowledge and experience that would characterize the situation and management of the information systems. The focus group methodology and the assumptions underlying its adoption seem to us to be relevant for a further extension of the study to other economic organizations and sectors.

The adoption of the focus group methodology resulted from the search for objectivity in the context under analysis and the topic under study. It was considered that the meeting of a group of experts, with different level of responsibility and knowledge, in various sectors of economic activity, would be the best way to achieve the objective, effective and pragmatic results with value for the assessment. Furthermore, it is also sought with this methodological option to understanding, knowledge and feeling about the relevance of the change management projects and information system. The possibility of interaction and sharing of participants' views, knowledge and experiences (Berg, 2001; Morgan, 1996; Queirós & Lacerda, 2013; Ivanoff & Hultberg, 2006; Gaiser, 2008); the provision of psychological and sociocultural characteristics (Berg, 2001); the generation of a consensus on ideas, issues, themes and solutions due to the synergistic effect (Berg, 2001); or the generation of comparative data between experiences and points of view rather than individual data (Morgan, 1996), were some of the benefits sought with this methodology and that justified the adoption.

Even though, a choice of elements for the focus group was based on experience, level of responsibility and knowledge of the theme, seeking a diversified contribution, regarding the identification of the most significant variables to consider.

Through this methodology, it was possible to collect, analyse and understand the situation in the organization through the involvement of different "actors" from different enterprises.

In this case, we sought actors who would play roles of responsibility in their organizations, namely at the intermediate or top level of the hierarchical pyramid. We also sought to diversify the economic sector present in the study, having sought to inquire specialists in the Banking and Insurance, IT, Commerce and Distribution, Public Administration and Services sectors.

Analysis of Results

Regarding the characterization of the focus group, it was possible to gather 11 specialists representing three sectors of activity: companies providing services in IT, Public Administration and service companies, as can be seen in Table 4.

Regarding the level of responsibility, the area of professional performance of respondents is mostly associated with IS/IT, according to Table 5, being the majority affected to functions at the level of intermediate management, as can be seen in the Table 6.

Regarding the conceptual characterization of change management associated with information systems, as shown in Table 7, we can see that the great majority considers that there should be practices associated with change management, such as the association of the change process with projects in IS or the introduction of IT, and management tools associated with change, such as evaluation tools.

However, the results are not so enlightening when respondents were confronted with the practices adopted in their organizations. The results allow to show that almost a third of the answers present a low performance in the domains under analysis. In this context, it can be seen that about 9% do not know or respond when faced with the existence of change management projects associated with IS/IT investments, or with the adoption of instruments and mechanisms for evaluating change projects, when they exist.

Table 4. Sector of activity of the focus group

	Activity Sector						
	Industry	Banking and Insurance	IT	Trade and distribution	Public administration	Services	Others
	0	0	4	0	6	1	0
TOTAL	0,00%	0,00%	36,36%	0,00%	54,55%	9,09%	0,00%

Table 5. Area / function performed by focus group

	Area / function performed by the respondent		
	Management / Administration	SI/TI	Others
	3	7	1
TOTAL	27,27%	63,64%	9,09%

Table 6. The hierarchical level of the focus group

	Hierarchical level of the respondent	
	Top Management	Medium
	2	9
TOTAL	18,18%	81,82%

Table 7. Conceptual and real dimension of SI projects changes

2 – INFORMATION SYSTEMS AND CHANGE PROJECTS	1	2	3	4	5	Ns/Nr
Projects in IS or the introduction of ICT must have associated processes of change	0,00%	27,27%	0,00%	36,36%	36,36%	0,00%
Change projects should be managed	0,00%	0,00%	0,00%	27,27%	72,73%	0,00%
Managing change is essential for improving performance and organizational competitiveness.	0,00%	0,00%	9,09%	36,36%	54,55%	0,00%
Change projects should always be evaluated	0,00%	0,00%	9,09%	18,18%	72,73%	0,00%
Company:						
Change management projects are developed when investments are made or developments in IS / ICT	9,09%	27,27%	18,18%	36,36%	0,00%	9,09%
Change projects are managed internally by the company	9,09%	18,18%	0,00%	45,45%	27,27%	0,00%
The responsibility of the projects of Management of the change are of the Top Management	9,09%	27,27%	0,00%	27,27%	36,36%	0,00%
Change projects are always evaluated	9,09%	27,27%	18,18%	27,27%	9,09%	9,09%

Regarding the knowledge and adoption of the ITIL methodology, its adoption in IT management (Laudon & Laudon, 2014) (O'Brien & Marakas, 2011), and the relationship with change, the results show that all the professionals know the methodology. As a reference methodology in the field of IT, the results are excellent in that they show the possibility of using this methodology in the field of IS/IT in their organizations.

However, it is observed that only about half of the organizations represented by specialists adopt this methodology. Apparently, it seems counter-intuitive, and it was not possible to determine the reasons for these results. However, most (81.82%) consider that this methodology has associated a process of change when it is adopted, as can be seen in Table 8.

With regard to the stages of ITIL that best evidence or are associated with the need for change, the answers do not show a uniformity of understanding. Considering only the answers that show agreement

Table 8. ITIL and changes

3 – ITIL (Information Technology Infrastructure Library)	TOTAL(SIM)	TOTAL(NÃO)
Do you know the ITIL methodology	100,00%	0,00%
Your organization adopts practices associated with ITIL	54,55%	45,45%
Do you consider that ITIL has associated a process of change	81,82%	18,18%

In the phases of ITIL presented, in your opinion, in which phase(s) can / should (s) be framed the change (s):	1	2	3	4	5	Ns/Nr
1 - Service Strategy	9,09%	9,09%	18,18%	27,27%	36,36%	0,00%
2 - Service Design	9,09%	18,18%	27,27%	18,18%	27,27%	0,00%
3 - Service Transition	0,00%	0,00%	9,09%	54,55%	36,36%	0,00%
4 - Service Operation	9,09%	0,00%	18,18%	45,45%	27,27%	0,00%
5 - Continuous Service Improvement	9,09%	0,00%	0,00%	36,36%	54,55%	0,00%

or total agreement, insofar as they are unequivocal in their confidence in the given response, the phases that can or must be framed in a change management process are the Service Transition phase (90, 91%) and the Continuous Service Improvement phase (90.91%). The Service Operation phase collected only 72.7% of the answers in grades 4 and 5 of the scale, as can be seen in Table 9. As it was not possible to confront the respondents with these results in the direction of their clarification, this situation provides an interesting opportunity to continue this study with a view to its deepening.

The next analysed dimension corresponded to the classification of the relevance of several relevant dimensions in the management of the change. The dimensions presented were as follows: leadership, vision, knowledge (for example, the current situation, the desired state and the existing gap), evaluation (e.g. current situation, desired state and existing gap), commitment, architecture, impact, strategy, planning and organization, implementation, celebration, learning. The results show that the majority of respondents consider that the stated dimensions are relevant in the management of change, as shown in Table 9.

However, it is noteworthy that, generally, about 10% of the experts do not agree with the relevance of the dimensions presented in the scope of the change. This is also an opportunity for further study in future studies.

Next, the experts were confronted with a proposal for a relationship between the ITIL methodology and the phases of change management. The proposal presented is that shown in Table 10, and has been affected by nature (strategic, design, transition, operation and continuous improvement). In this case, as we can see, the answers are even more ambiguous.

Table 9. Dimensions of change management

Relevant Dimensions in Change Projects	1	2	3	4	5	Ns/Nr
1 – Leadership	9,09%	0,00%	36,36%	18,18%	36,36%	0,00%
2 – Vision	9,09%	0,00%	18,18%	45,45%	27,27%	0,00%
3 – Knowledge (current situation, desired state, gap)	9,09%	9,09%	0,00%	45,45%	36,36%	0,00%
4 – Evaluation (eg current situation, desired state, gap)	9,09%	0,00%	9,09%	36,36%	45,45%	0,00%
5 – Commitment	9,09%	0,00%	18,18%	36,36%	36,36%	0,00%
6 – Arquitecture	9,09%	9,09%	9,09%	45,45%	27,27%	0,00%
7 – Impact	9,09%	0,00%	9,09%	36,36%	45,45%	0,00%
8 – Strategy	9,09%	9,09%	9,09%	18,18%	54,55%	0,00%
9 – Planning e organization	9,09%	0,00%	9,09%	54,55%	27,27%	0,00%
10 – Implementation	9,09%	0,00%	0,00%	72,73%	18,18%	0,00%
11 – Commemoration	9,09%	18,18%	18,18%	36,36%	18,18%	0,00%
12 – Learning	9,09%	9,09%	9,09%	27,27%	45,45%	0,00%

Table 10. Relationship between the phases of ITIL and change management

Mudança	1	2	3	4	5	Ns/Nr
1 - Identification of leadership for change	0,00%	27,27%	18,18%	27,27%	27,27%	0,00%
2 - Identification of vision, definition of commitment and definition of organizational capacities	0,00%	9,09%	18,18%	45,45%	27,27%	0,00%
3 - Assessment of the current situation and identification of the existing gap	0,00%	9,09%	18,18%	45,45%	27,27%	0,00%
4 - Desired state architecture	9,09%	9,09%	18,18%	45,45%	18,18%	0,00%
5 - Impact analysis	0,00%	27,27%	9,09%	27,27%	36,36%	0,00%
6 - Definition of change strategy	0,00%	18,18%	18,18%	27,27%	36,36%	0,00%
7 - Planning and organization of implementation	0,00%	9,09%	9,09%	54,55%	27,27%	0,00%
8 - Implementation of change	9,09%	0,00%	18,18%	54,55%	18,18%	0,00%
9 - "Commemoration" and integration of the new state	9,09%	9,09%	36,36%	36,36%	9,09%	0,00%
10 - Learning, Correction and Consolidation	9,09%	0,00%	18,18%	27,27%	45,45%	0,00%

Although, in general, a complete agreement or agreement with the presented affectation is evident, the dispersion of the responses is significant, and some cases of disagreement or indifference regarding this affectation are evident. This situation requires a differentiated analysis, in the sense of analysis and confirmation if the answers are undifferentiated, or if there is any relation with the professional nature of the specialists who participated in the focus group. This is one more aspect that can be developed and deepened in a future work.

CONCLUSION

The main evidence from this study shows that there is a clear assumption between systems and information technologies and change. The impact of investments, or the introduction of technological innovations in the context of the functioning of organizations, leads to the need to adapt the modus operandi of economic institutions. However, although this relationship is evident for most professionals, this awareness does not always follow organizational practices. The reasons for the difference in the values obtained are not clear, but the literature usually points to managers' lack of time, technological and market pressure for innovation, lack of specific knowledge about change management, among other things. The choice of the ITIL methodology for this study and its association with the change projects was due to the fact that it is a well-known methodology, as evidenced by the results obtained, and especially because it integrates a life cycle, which naturally presupposes adjustments associated with the new IT services provided. The results show a comfortable majority in this relationship between ITIL and change, but this evidence is not general as expected in the knowledge of the methodology.

The ITIL methodology is very dynamic and can adapt very easily to any reality, so we only need to know what services we want from IT, how we want them and how they should be made available. It should be noted that the topic of change does not have general "revenues" for any type of organization or any economic sector. It is necessary to define a road map, which should include a roadmap that presents the stages considered as nuclear for the management of the change, which should be managed from a project perspective. On the other hand, change does not have to be radical. It is necessary to consolidate what is well done or is considered appropriate to the organizational functioning and the requirements of the market, pursuing a path of continuous improvement towards the adaptation to the market evolution and the needs of the customers, since much of the products and services information technology. The advantage of applying these methodologies lies in the possibility of developing an integrated vision, at several levels, seeking to achieve greater competitive efficiency in the business and efficiency in the organizational functioning, thus ensuring a strategy of quality assurance in the delivery of services, optimization of processes, productivity, customer satisfaction, and growth and contributing to the stability and survival of the organization.

REFERENCES

Anunciação, P. F. (2011). Results of the 1st study of information systems governance in the large companies in Portugal. In *3rd International Conference of ceGSI – Governance of Information Systems: Models, Ethics & Performances*, INA – National Institute of Administration, Oeiras, Oct. 11 *(In Portuguese)*

Anunciação, P. F. (2012 a)). *Results of the 1st study of information systems governance in the large companies in Portugal*, Calouste Gulbenkian Foundation, Feb. 14 *(In Portuguese)*

Anunciação, P. F. (2012 b)). Results of the 2nd study of information systems governance in the large companies in Portugal, In *4th International Conference of ceGSI – Information Systems Governance: Economy & Security*, National Security Office, Lisbon, Portugal, Oct. 9 *(In Portuguese)*

Anunciação, P. F. (2013). *Results of the 2nd study of Information Systems Governance in the large companies in Portugal*, National Parliament – New Auditorium, March 19 *(In Portuguese)*

Anunciação, P. F., & Zorrinho, C. (2006). *Organizational Urbanism – How to manage technological shock*. Lisboa, Portugal: Sílabo Publishing. (In Portuguese)

Berg, B. L. (2001). Focus group interviewing. In B. L. Berg (Ed.), *Qualitative research methods for the Social Sciences*, 4, pp. 111–132. Needham Heights, MA: Pearson.

Cartlidge, A.; Hanna, A.; Rudd, C.; Macfarlane, I.; Windebank, J., & Rance, S. (2007). *An introductory overview of ITIL*, The UK Chapter of the ITSMF.

de Ven, A. V., & Hargrave, T. (2000). *Social, technical, and institutional change*. Oxford, UK: Oxford University Pr.

Galpin, T. J. (1996). The human side of change: a practical guide to organization redesign. San Francisco, CA: Jossey-Bass.

Hammer, M., & Champy, J. (1994). *Reengineering the Corporation: A manifesto for business revolution*. Londres, UK: Nicholas Brealy.

Imai, M. (2006). *Gemba Kaizen*. Warszawa, Poland: MT Biznes.

ISGec – Information System Governance European Club. (2009). Information Systems Governance Manifest. Retrieved from http://www.cegsi.org/index.php/qui-sommes-nous/o-que-e-o-cegsi

ISGec – Information System Governance European Club. (2010). Why the corporations are asking for an information systems governance? Retrieved from http://www.cegsi.org/index.php/documents/telechargement-du-document-la-gouvernance-des-systemes-d-information-pourquoi/la-gouvernance-des-systemes-d-information-pourquoi

ISGec – Information System Governance European Club. (2011). The importance of the information systems approach in the governance of organizations. Retrieved from http://www.cegsi.org/index.php/documents/l-importance-de-l-approche-par-les-systemes-d-information/the-importance-of-the-information-systems-approach-for-governance-organizations (In French)

Ivanoff, S. D. & Hultberg, J. (2006). Understanding the multiple realities of everyday life: basic assumptions in focus group methodology, *Scandanavian Journal of Occupational Therapy*, 13.

Laudon, K. C., & Laudon, J. P. (2014). *Management information systems - managing the digital firm* (13th ed.). Harlow, Essex: Pearson.

Morgan, D. L. (1996). Focus groups. *Annual Review of Sociology*, 22. Retrieved from http://www.jstor.org/stable/2083427

O'Brien, J., & Marakas, G. M. (2011). *Management information systems.* Mcgraw-Hill.

Oliveira, A. (2004). *Analysis of investments in information and communication technologies and systems.* Sílabo Publishing. (In Portuguese)

Oliveira, A. (2009). *Information & information systemas – Facts, myths, mystifications, dangerous half-truths & lack of common sense.* Refertelecom Publishing. (In Portuguese)

Oliveira, A. (2010). Governance of information systems – Why, ISGec Conference, Madrid, Spain.

Queirós, P., & Lacerda, T. (2013). The importance of interview in qualitative research. In I. Mesquita & A. Graça (Eds.), *Qualitative research in sport, 2*. Porto, Portugal: Center for Research, Training, Innovation, and Intervention in Sport, Faculty of Sport, Porto University. (In Portuguese)

Requeijo, J. F., & Pereira, Z. L. (2008). *QUALITY: Planning and statistical process control.* Prefácio Publishing. (In Portuguese)

Van Haren - Publishing. (2011). *ITIL Foundations - Best Practice.*

Chapter 12
Information Management for the University–Enterprise Interaction:
Considerations From the Research Groups Directory of the CNPQ in Brazil

Morjane Armstrong Santos de Miranda
Federal University of Bahia, Brazil

Sérgio Maravilhas
https://orcid.org/0000-0002-3824-2828
IES-ICS, Federal University of Bahia, Brazil

Ernani Marques dos Santos
Federal University of Bahia, Brazil

Antonio Eduardo de Albuquerque Junior
Oswaldo Cruz Foundation, Gonçalo Moniz Institute, Brazil

Daniella Barbosa Silva
Faculdade de Tecnologia e Ciências, Brazil

Platini Fonseca
https://orcid.org/0000-0003-4422-3671
Federal University of Bahia, Brazil

ABSTRACT

This chapter analyzes the importance of Information Management for the phenomenon of University-Enterprise (U-E) interaction, based on the Directory of Research Groups (DGP) in Brazil, of the National Council for Scientific and Technological Development (CNPq). The methodology used consisted in analyzing, by the empirical-analytic research and descriptive-analytical approach, the data available on this database. The data is about the activities of the research groups of the Federal University of Bahia (UFBA), interacting with companies from 2002 to 2010. Results show information management is important for this occurrence because it contributes to the recognition of interest and the conditions of interaction of the actors, enhancing the transfer of knowledge and technologies.

DOI: 10.4018/978-1-5225-9993-7.ch012

INTRODUCTION

Information Management plays a strategic role in the decision-making process in innovative organizations, covering from innovation management to competitive intelligence.

Based on the understanding of Reis (1993), it is perceived that, as important as having access to information, it is its quality and rapid acquisition that will allow the achievement of objectives.

In the Information and Knowledge Society (QUEVEDO, 2007), the generation and diffusion of these inputs has become increasingly intense. The rapid, systematized and qualified management of information is important for the effective transfer of technologies, reflecting the phenomenon of University-Enterprise (U-E) interaction.

The Directory of Research Groups in Brazil (DGP), of the National Council for Scientific and Technological Development (CNPq), is a database that contains information about research groups that are active in the country, and can contribute to the understanding of phenomena on U-E interaction by providing information on interaction between universities and the productive sector, serving both the scientific and technological community as well as political-administrative organizations and professionals working with innovations in general.

This paper aims to analyze the role of information management for University-Enterprise interaction, since there is, in this phenomenon, the need for a good communication channel to enable the exchange of knowledge and technology between the parties involved. For this purpose, the DGP is used as the object of study to map and analyze the interactions between research groups and companies.

BACKGROUND

Because of economic globalization and, consequently, the diffusion of technological advances, the economy and the social structure underwent a structural reorganization (BORGES, 2008). In this sense, one of the aspects that have had most influence on the world framework was the relative capacity to innovate, spread and apply new knowledge, and, secondarily, capital, natural resources or cheap labor force, with the purpose of improving competitiveness (QUANDT, 2004).

These changes demanded of man the expansion of freedom and human capacity, constituting a new society that was called the Information and Knowledge Society. Although the terms information and knowledge are used interchangeably, Carvalho (2000) understands that the Information Society directs the information object as a product or input, whereas, in speaking of the Knowledge Society, the focus is on the use of information by the individual as part of the process of knowledge formation.

For Albagli (2007), the Information Society is usually associated with the development and diffusion of Information and Communication Technologies (ICTs), providing applications and innovations in various fields of economic, political and social life, particularly the formation of networks of all kinds, connected by electronic and digital means. The Knowledge Society, for its part, refers to the capacity to generate and use knowledge relevant to innovation and development.

It is necessary here to distinguish information and knowledge. Information is a message, which may be in the form of a written or spoken document or communication. Knowledge, however, is designed to shape the person who receives it in the sense of making some difference in their perspective or insight (DAVENPORT & PRUSAK, 1998).

Takeuchi e Nonaka (2008) present information and knowledge as distinct and interrelated elements in a process of knowledge transmission. According to the authors, knowledge is a new know-how, resulting from analyzes and reflections on information and according to mental values and models, which increases the ability to adapt to real world circumstances. Information is an interpretation of a data set according to a relevant purpose and consensus for a target audience, while data is a collection of relevant evidence about a given fact. Vieira (1998) adds intelligence to these three elements, which stands out as a set of information analyzed and contextualized for decision-making, with political or market purposes. The result, according to the author, serves as a subsidy to the decision-making process.

In these conditions, it is worth stressing the importance of the information management process. To this end, the information management process requires mechanisms for obtaining and using human and technological resources (MARCHIORI, 2002). The author complements that the information manager has the role of mapping the points of use of information, identifying the needs and requirements indicated / negotiated with their applicants.

In addition to being identified and organized, information must be accessed quickly and accurately (BEAL, 2012). Thus, information technologies are also useful for reducing the time spent in the search for information, presenting itself as a potential instrument of approximation between the components involved in a process of University-Enterprise interaction.

University-Enterprise Interaction

Despite the relevance of the current innovation theme, Moreira and Queiroz (2007) emphasize that literature is not unanimous in its definition. The Oslo Manual of the Organization for Economic Co-operation and Development (OCDE, 2005, p. 55), which has a well-defined definition, defines innovation as "[...] the implantation of a new or significantly improved product (good or service), or a process, or a new marketing method, or a new organizational method in business practices, workplace organization or external relations".

Technological innovation has increasingly been used as a strategy to redeem companies, regions and nations from their economic problems and to promote their development. According to Plonski (2005), for this reason, since the 1990s, the implementation of effective policies to stimulate technological innovation has become one of the structuring axes of the OECD, which covers more than 30 countries. According to this author, innovation has been gaining ground also in Brazil, especially since 2001, as a result of the mobilization associated with the II National Conference on Science, Technology and Innovation held that year, and the Industrial, Technological and Trade Policy (PITCE), implemented in 2003, which resulted in the enactment of Law no. 10.973/2004, known as the "Innovation Law", which aims to encourage innovation and scientific research in the productive environment.

Investment in scientific and technological development policies is associated with the fact that knowledge is the most important resource for competitiveness in the modern world economy (LUNDVALL & JOHNSON, 1994). The intensification of technological change, characterized as one of the most striking aspects of capitalism in the last decades, is fundamentally related to the knowledge production process and the relations of this process with economic activity.

According to Rapini and Righi (2007), the new role of information and knowledge in economies and in the productive process has led to a repositioning of the role played by universities. As a result, they are no longer responsible only for training, but they have given crucial knowledge to the evolution of some industrial sectors.

Etzkowitz (2009), in turn, addresses the issue of innovation in contemporary society through the concept of Triple Helix, that is, a dynamic of cooperation between University-Enterprise-Government (U-E-G), inserting itself as an important institutional arrangement. In this perspective, the university is the propeller of societies based on knowledge, just as government and enterprise are representative institutions in industrial society. The company then takes on the role of key actor and locus of production, while the government serves as the source of contractual relations that ensure stable interactions and the exchange.

In Brazil, the development of the University-Enterprise interaction started from the difficulties that arose in the industrial and technological development of the country, which, until the 1970s, privileged the entry of foreign technology and the establishment of multinationals in the most sophisticated sectors (VELHO, 1996). In the meantime, according to Velho (1996, p. 49), the failure of the connection between the business and academic community in Brazil led the State to "transform itself into a demander of research and technology", opening space and stimulating different mechanisms to intensify the interaction between University and Enterprises.

In the context of this interaction, technology transfer is also a fundamental factor to be analyzed. Therefore, Parker and Zilberman (1993, p. 89) understand this activity as: "[...] any process by which basic knowledge, information, and innovations move from a university, institute or governmental laboratory to an individual or to enterprises in the private and semi- private sectors".

RESEARCH METHOD

The research consisted in the search and analysis of the secondary data published in the DGP portal (http://dgp.CNPq.br/planotabular/) about research groups in activity in Brazil. To describe the role of information management in the University-Enterprise interaction, data were analyzed on the activities of the research groups of the Federal University of Bahia (UFBA), from 2002 to 2010.

The DGP data are collected continuously by CNPq through a standardized electronic questionnaire, filled out by the leaders of the groups authorized and previously registered in the system by the Research Officers of the participating Institutions, with biannual updates, when quantitative information is presented about the groups in their various dimensions. Access to the DGP is public and the electronic portal offers textual search capabilities using user-specified filters, with the possibility of exporting the query data in various file formats.

The data sources are of two types: (1) Sources of bibliographic information, characterizing the interaction phenomenon U-E-G and its elements; and (2) Secondary information sources, by the Directory of Research Groups in Brazil, of CNPq, regarding the identification and characterization of the research groups, the companies with which they interact and their types of interactions, categorized by the platform itself.

PRESENTATION AND DISCUSSION OF RESULTS

Information on the research groups present in the DGP is certified by the entities to which they belong, among universities, isolated institutions of higher education, institutes of scientific research, technological institutes, laboratories of research and development (R&D) of state or ex-state companies, besides of some non-governmental organizations that perform research.

DGP provides information on the members of the groups, the lines of research in progress, the specialties of knowledge, the sectors of activity involved, the scientific, technological and artistic production of the members and the patterns of interaction with the productive sector.

The Directory was started at CNPq in 1992 and since then, on an almost biennial frequency, the Agency makes available to the public a qualitative and quantitative census of the country's installed capacity for research, as measured by the active groups in each period. From 2002, the Directory made the system available for continuous updating of the database, denominated Current Base, but maintaining the biennial frequency for the realization of the censuses, which became "photographs" of this current base.

In this website, the results of the Census carried out from 2000 onwards, and in interaction with companies from 2002 onwards are available, comprising in the meantime seven independent modules, consisting of General Information, Historical Series, Statistical Summary, Tabular Plan, Textual Search, Group Stratification and Annexes.

The most important methodological definition in the constitution of the database is that of its unit of analysis. Thus, the research group was defined as a set of individuals organized hierarchically and with some basic characteristics, namely: experience, prominence and leadership in the scientific or technological terrain as an organizing foundation; a professional and permanent involvement with research activities; works organized around common lines of research; and sharing, to some degree, of facilities and equipment's.

Each research group should, therefore, organize itself around a leadership (possibly two), which is the source of the information in the database. The concept of a group admits one constituted of only one researcher. In almost all cases, these groups are constituted by the researcher and his students.

In addition to the information collected on the forms of the research groups, other information's that complements the census databases of the Directory are imported from the Curriculum Lattes database, the CNPq Promotion System and the DATACAPES database of the Coordination for the Improvement of Higher Education Personnel (CAPES), and "frozen" for the preparation of censuses.

Information Management and University-Enterprise Interaction: Perceiving the Relationship Starting from the UFBA Research Groups

According to Moura (1999), the information's to be treated for interaction between the business and academic sectors are numerous, and it has become increasingly important to have access to all these data. The author reinforces this understanding by presenting the "paradox of University-Enterprise interaction", which states that although the results of this relationship can bring benefits to both parties involved in the process, there are aspects of resistance on the part of these actors.

These barriers would reveal, from the company's point of view, that universities are little interested in the common problems of companies and are concerned with developing research on topics of their own interest, and often of personal interest to the researchers (MOURA, 1999). Velho (1996) points out

Information Management for the University-Enterprise Interaction

Figure 1. Sequence of events in DGP / CNPq data acquisition
Source: Own elaboration, based on the DGP / CNPq database.

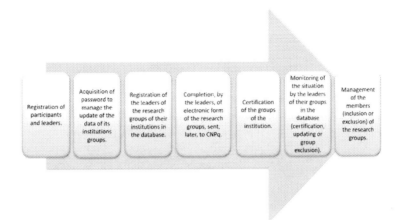

that there is some resistance from the companies to interact with universities since they often do not believe or do not know the scientific capacity of the researchers, or even the peculiar characteristics of Brazilian businessmen who are not used to innovation.

These findings demonstrate how much the knowledge about the peculiar characteristics of each actor is relevant, so that the process of University-Enterprise interaction leverages towards the effectiveness of innovation in these environments. In this sense, the DGP presents itself as an important tool for disseminating information about what universities and companies propose in the interaction question. In the case of universities, the use by the scientific and technological community in the day-to-day practice of the professional exercise and the wide source of information for the institutions and for the Brazilian and international scientific-technological activity are the greatest gains provided by the census database.

In the case of companies, information and its adequate management are critical success factors, whose importance is based, according to Moura (1999), in the formation of a strategy of competitive advantage and decision making; and the enhancement, in quantity and quality, of the activities inherent to the university, whether in the scope of research or teaching.

In this database, which is the focus of analysis for the present study, the available information is collected through a standardized electronic questionnaire that the CNPq makes available to group leaders previously registered in the system by the Research Officers of the participating institutions. These managers are also in charge of certifying the research groups after sending the data by the leaders, which contributes to the process of confirming the veracity of the information submitted.

The sequence of events in data acquisition from the Directory, performed through the site called Data Collection, is as follows:

Thus, the registration of the leaders of the groups and the certification of these are the responsibilities of the research leaders of the participating institutions. Information about the group (researchers, students, technical support staff and research lines) is the responsibility of the group leaders. Some personal data about researchers and students, such as degree, student training, age and sex, and those related to scientific, technological and artistic production are the responsibility of each leader, researcher and student, who inform them in their Lattes Resumes platform.

The Directory is, therefore, an activity shared among several actors that have access to a same database physically located in CNPq. This database, continuously updated, is called the Current Base, and its "photographs", made periodically, correspond to the Censuses.

From the DGP data, there was an increase in the number of institutions, groups and researchers with doctorates in the registered groups (Table 1).

Table 2 shows the distribution of the groups by regions of Brazil, from which it is possible to note the distance between the Southeast and South regions in relation to the others, since they contain more than 69% of the active research groups in the Country.

Table 3 compares the evolution of research groups in activity in Brazil between 2002 and 2010 with the UFBA groups in the same period. The data shows that the number of UFBA research groups increased more than in the same period the rest of Brazil groups did.

Although the University-Enterprise interaction brings advantages to the actors involved in the process, there is already the mentioned resistance to the effectiveness of this cooperation (VELHO, 1996; MOURA, 1999), which may be due to a lack of knowledge of one of the parties about what is being demanded by the other party.

For Leonard-Barton (1998), the main knowledge-generating activities that managers are responsible for guiding, controlling and encouraging are shared problem solving, experimentation and prototyping, integration of new processes and technical tools, and the importation of know-how from outside the company. This last element justifies and strengthens the importance of knowing and managing informa-

Table 1. Evolution of the number of institutions, groups, researchers with doctorates (PhD) - 1993-2008

Main dimensions	1993	1995	1997	2000	2002	2004	2006	2008	2010
Institutions	99	158	181	224	268	335	403	422	452
Groups	4.402	7.271	8.632	11.760	15.158	19.470	21.024	22.797	27.523
Researchers (P)	21.541	26.779	33.980	48.781	56.891	77.649	90.320	104.018	128.892
Researchers with Doctorates (PhD) (D)	10.994	14.308	18.724	27.662	34.349	47.973	57.586	66.785	81.726
(D)/(P) in %	51	53	55	57	60	62	64	64	63

Source: Directory of CNPq Research Groups in Brazil.

Table 2. Distribution of research groups by geographic region (2010)

Region	Groups	%
Southeast	12.877	46,8
South	6.204	22,5
Northeast	5.044	18,3
Center-West	1.965	7,1
North	1.433	5,2
Brazil	27.523	100

Source: Directory of CNPq Research Groups in Brazil.

Table 3. Number of research groups in Brazil and UFBA

Number of Research Groups		
Year	Brazil	UFBA
2000	*11.760*	*200*
2002	15.158	225
2004	19.470	348
2006	21.024	401
2008	22.797	406
2010	*27.523*	*484*

Source: Own elaboration, based on the DGP / CNPq database.

tion related to the interactions carried out by the research groups with companies, represented by the database in 14 relationships (Table 4).

As for the interactions among the UFBA research groups, the data show that among the 112 research groups of this university that interact with companies, seven had interaction for five biennia, which shows that there was little interaction in the analyzed period. There are also 16 groups that have interacted with companies for four biennia, 22 groups that have interacted for three biennia, 20 groups that have interacted for two biennia and 47 groups with only one year of interaction. It is noticed that U-E

Table 4. Types of CNPq DGP interactions

RELATIONSHIPS	TYPES OF INTERACTIONS
Rel1	Scientific research without considerations of immediate use of the results
Rel2	Scientific research with considerations of immediate use of results
Rel3	Non-routine engineering activities including the development of prototype head or pilot plant for the partner
Rel4	Non-routine engineering activities including the development / manufacture of equipment for the group
Rel5	Development of non-routine software for the group by the partner
Rel6	Software development for the partner by the group
Rel7	Transfer of technology developed by the group to the partner
Rel8	Transfer of technology developed by the partner to the group
Rel9	Technical consultancy activities not included in other types of interaction
Rel10	Supply by the partner of material inputs for the group's research activities without linking to a specific project of mutual interest
Rel11	Supply, by the group, of material inputs for the activities of the partner without linking to a specific project of mutual interest
Rel12	Training of staff of the partner by the group including courses and training "on-the-job"
Rel13	Training of group staff by the partner including "on-the-job" training and courses
Rel14	Other predominant types of relationships that do not fit into any of the above

Source: Own elaboration, based on the DGP / CNPq database.

interaction is still incipient in UFBA research groups, although it is not a completely new phenomenon in this scenario, which makes it possible to study and encourage these practices.

As for the companies with which UFBA research groups have been interacting over these five biennia's, it has been noticed that there has been an increase in this relationship, as well as a variety of types of companies. UFBA's research groups related to 33 companies in 2002, 64 companies in 2004, 40 companies in 2006, also 40 companies in 2008 and 70 companies in 2010. It is worth mentioning the strong presence of the government, with emphasis on the work of CNPq, the Foundation for Support to Research and Extension (FAPEX), the Foundation for Research Support in the State of Bahia (FAPESB) and the Secretariat of Science, Technology and Innovation of Bahia (SECTI).

Table 4 shows that 2004 was the year in which there was a greater number of interactions with companies, which leads to investigations into the influence of the Innovation Law, created in this period. More recent data needs to be analyzed, as soon as it becomes available and gets public for consultation.

Among the scientific areas present in Table 4, Engineering is the one that shows more interactions with companies, with 423 interactions declared in the period 2002-2010. The area of Exact and Earth Sciences has the second largest number of interactions, with 196 cases. The Health Sciences, with 141 cases of interaction, are in third position. Next, the Applied Social Sciences, with 128, Biological Sciences, with 118, Agricultural Sciences, with 61, Humanities, with 40 interactions and, closing the table, the area of Linguistics, Literature and Arts, with only 3 interactions. Figure 2 represents these interactions by scientific area.

In this same period, the types of relationships of the groups with companies that stand out most, focus on technological innovations with and without considerations of immediate use, but whose production and exchange of knowledge between partners is a constant, as Rel. 1 and 2. Next to this relationship is the transfer of technology by the group to the partner (Rel. 7), which can be characterized by the purchase of a technological package developed in the group (such as patent licensing), or by simply buying products developed by the research group.

In this sequence, there are also: Other predominant types that do not fit those listed in the DGP / CNPq (Rel. 14); Personnel training (Rel. 12), which may involve participation in courses, or the development of dissertations and theses (in this case is observed the exchange and production of knowledge useful for both the group and the company); and Consulting activities (Rel. 9), which can be understood as the

Table 5. Interactions of UFBA research groups per biennium

Scientific Area	2002	2004	2006	2008	2010	Total
Agricultural Sciences	15	21	12	10	3	61
Biological Sciences	15	36	25	33	9	118
Health Sciences	3	28	31	32	47	141
Exact and Earth Sciences	40	48	38	37	33	196
Humanities	0	4	12	8	16	40
Applied Social Sciences	8	30	32	34	24	128
Engineering	43	123	99	80	78	423
Linguistics, Literature and Arts	0	1	1	0	1	3
TOTAL	124	291	250	234	211	1383

Source: Own elaboration, based on the DGP / CNPq database.

Figure 2. Interactions of UFBA research groups with companies by scientific area (2002-2010)
Source: Own elaboration, based on the DGP database.

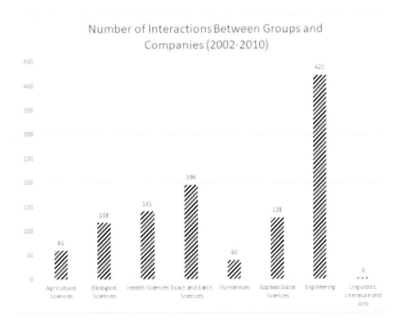

hiring of the research group to carry out a consultancy activity, such as solving a practical problem or diagnosing problems and bottlenecks that hinder the growth of the company. Table 6 presents, systematically, these interactions.

FUTURE RESEARCH DIRECTIONS

As suggestions for future research, one can glimpse the proposition of a methodology to measure the U-E interaction, being later applied to the universe of Brazilian institutions. Such methodology would serve as an important indicator of the economic environment in which the universities are inserted, of the interest and potential of these to interact with the business environment and of the knowledge potential of the researchers. In the meantime, studies should also be applied analyzing the information management for the interactions by area of knowledge and degree of complexity, as well as by types of relationships between universities and companies. It would also be appropriate to continue the analysis of the subsequent biennia, once all the information is available, to make it possible to see if the continuation of these relations between the U-E continues to exist, and if the growth of the research groups remained the same as in the period analyzed, linking this growth with the performance of the Brazilian economy.

Table 6. Number of relationship types per biennium

Interaction Type (Relationship)	YEARS					
	2002	2004	2006	2008	2010	Total
Rel 1	33	53	34	31	37	151
Rel 2	41	91	68	68	56	277
Rel 3	2	7	6	5	5	22
Rel 4	0	1	2	2	4	9
Rel 5	0	1	1	2	4	8
Rel 6	4	6	6	5	5	21
Rel 7	11	39	33	31	23	123
Rel 8	4	9	9	7	5	29
Rel 9	5	17	21	20	16	72
Rel 10	4	13	10	11	9	42
Rel 11	0	3	2	3	2	10
Rel 12	8	20	17	20	18	73
Rel 13	5	10	9	5	4	28
Rel 14	7	21	20	24	23	91
TOTAL	124	291	244	234	211	

Source: Own elaboration, based on the DGP / CNPq database.

FINAL CONSIDERATIONS

This work shows that the information management from research assumes a preponderant role for the phenomenon of University-Enterprise interaction, especially regarding to the structuring and storage of information from research projects and research groups, as well as its accessibility and diffusion.

Zorrinho (1995, p. 146) corroborates this statement when he quotes that: "Managing information is thus deciding what to do based on information and deciding what to do about information. It is to be able to select from an available repository of information that is relevant to a given decision and, also, to construct the structure and design of that repository".

When analyzing the Directory of Research Groups, it can be seen, at the outset, its scope of action, evidenced by the three main purposes, aimed at attending to the scientific and technological community in the professional day-to-day, to the scientific institutions, and to the science-technology activity in Brazil.

Another evidence that confirms the importance of this database, and consequently the information management from its research projects, is the strong growth of registered research groups, both in Brazil (81.57% growth) and UFBA (115.11%). In addition to this information, several types of interactions that the research groups carry out with companies are also brought forward, demonstrating the flow of information that serves as a subsidy for the actions of these actors, regarding the production, appropriation and diffusion of knowledge and technologies.

In the case of UFBA's research groups, it can be observed that, of the universe of 484 research groups listed in the Directory in the period from 2002 to 2010, 112 interact with companies, a number that signals the existence of this phenomenon, although in a shy way still.

Regarding the interaction with companies by scientific area, it was noticed that the most outstanding in interactions at UFBA are the Engineering research groups, whose innovation tends to be more technological, but this did not prevent the area of Applied Social Sciences (fourth), to appear in front of the Biological Sciences area.

When identifying the types of interactions listed in the Directory of Research Groups under analysis, the type of interaction that has more occurrence is the interaction that deals with technological innovations with and without considerations for immediate use. In other words, technological innovations with considerations of immediate use, namely the solution of a problem of the company together with its researchers, converges well with the marked interaction in the large area of Engineering, of a more technological nature, unlike the innovations without immediate use, which deals with groups and companies that develop new lines of research. However, it is highlighted that the production and exchange of knowledge between partners is a constant, in both types of relationship.

Finally, it should be emphasized that the paradox of University-Enterprise interaction still needs to be overcome, since there remains a certain strangeness of an actor regarding the contributions that the other can give, reinforced by the misalignment between actors with times of production and delivery of knowledge so different. It is believed that improving the understanding of the role of each side by the other can converge to achieve this goal. In addition to the creation of this communication channel, information management allows the formulation by the government of public policies aimed at supporting the interaction between different actors, encouraging more and more the collaboration that leads to innovation.

Throughout this work, some limitations were identified, such as completing the data in the database is optional, which may imply outdated or incomplete information, and the lack of a more detailed description of the types of relationships used in the mapping of interactions.

REFERENCES

Albagli, S. (2007). Tecnologias da Informação, Inovação e Desenvolvimento. In: VII Cinform Encontro Nacional de Ciência da Informação, Salvador, 16.

Beal, A. (2012). *Gestão estratégica da informação: como transformar a informação e a tecnologia da informação em fatores de crescimento e de alto desempenho nas organizações*. São Paulo, Brazil: Atlas.

Borges, M. A. G. (2008). A informação e o conhecimento como insumo ao processo de desenvolvimento. [RICI]. *Revista Ibero-americana de Ciência da Informação*, 1(1), 175–196.

Davenport, T. H., & Prusak, L. (1998). *Conhecimento Empresarial*. Rio de Janeiro, Brazil: Campus.

de Carvalho, H. G. (2000). Inteligência Competitiva Tecnológica para PMEs Através da Cooperação Escola-Empresa. Tese de Doutorado apresentada ao Programa de Pós-Graduação em Engenharia de Produção da Universidade Federal de Santa Catarina - UFSC. Florianópolis.

Departamento Intersindical de Estatística e Estudos Socioeconômicos (DIEESE). (2008). Política de Desenvolvimento Produtivo: nova política industrial do governo de 2008. n. 67. São Paulo, Brazil.

DGP/CNPq - Diretório dos Grupos de Pesquisa no Brasil, do CNPq (2014). *Censo dos Grupos de pesquisa no Brasil, do CNPq*. Retrieved from http://dgp.CNPq.br/planotabular/

Diretrizes de Política Industrial, Tecnológica e de Comércio Exterior de 2003. (2003). UNICAMP. Retrieved from http://www.inovacao.unicamp.br/politicact/diretrizes-pi-031212.pdf

Etzkowitz, H. (2009). *Hélice Tríplice: Universidade-Empresa-Governo, Inovação em movimento.* Porto Alegre, Brazil: EDIPUCRS.

Leonard-Barton, D. (1998). *Nascentes do saber: criando e sustentando as fontes de inovação.* Rio de Janeiro, Brazil: Fundação Getúlio Vargas.

Lundvall, B. A., & Johnson, B. (1994). The learning economy. *Journal of Industry Studies, 1*(2), 23–42. doi:10.1080/13662719400000002

Marchiori, P. Z. (2002). A Ciência e a Gestão da Informação: compatibilidades no espaço profissional. *Ciência da Informação. Brasília, 31*(2), maio/ago., 72-79.

Moreira, D. A., & Queiroz, A. C. S. (2007). *Inovação Organizacional e Tecnológica.* São Paulo, Brazil: Thomson Learning.

Moura, L. R. (1999). Gestão e Tecnologia da Informação como instrumento de interação Universidade-Empresa. In *IBICT. Interação Universidade-Empresa II.* Brasília: Instituto Brasileiro de Informação em Ciência e Tecnologia.

Organização para Cooperação e Desenvolvimento Econômico (OCDE). (2005). Manual de Oslo: diretrizes para coleta e interpretação de dados sobre inovação de 2005. (Traduzido pela FINEP - Financiadora de Estudos e Projetos, 3ed.)

Parker, D. P. & Zilberman, D. (1993). University technology transfers: impacts on local and U.S. economies. *Contemporary Policy Issues. XI*, 87-99.

Plonski, G. A. (2005). Bases para um movimento pela inovação tecnológica no Brasil. *Revista São Paulo em Perspectiva, 19*(1), 5–33.

Quandt, C. (2004). Inovação em clusters emergentes. *Revista Com. Ciência, 57*, ago, 1-5.

Quevedo, L. A. (2007). Conhecer para participar da sociedade do conhecimento. In Maria Lucia Maciel, Sarita Albagli (Org.). Informação e Desenvolvimento: conhecimento, inovação e apropriação social. Brasília: IBICT, UNESCO.

Rapini, M. S., & Righi, H. M. (2007). Interação Universidade-Empresa no Brasil em 2002 e 2004: Uma aproximação a partir dos grupos de pesquisa do CNPq. *Revista Economia (Brasília), 8*, 263–284.

Reis, C. (1993). *Planejamento Estratégico de Sistemas de informação.* Lisboa, Portugal: Presença.

Takeuchi, H., & Nonaka, I. (2008). *Gestão do Conhecimento, tradução Ana Thorell.* Porto Alegre, Brazil: Bookman.

Velho, S. (1996). *Relações Universidade-Empresa: desvelando mitos.* Campinas, SP: Autores Associados.

Vieira, A. da S. (1998). *Monitoração da competitividade científica e tecnológica dos estados brasileiros a partir do SEICT.* Brasília: Ibict.

Zorrinho, C. (1995). *Gestão da Informação.* Condição para Vencer. Iapmei.

KEY TERMS AND DEFINITIONS

Directory of Research Groups in Brazil of CNPq (DGP / CNPq): Inventory of scientific and technological research groups active in the country, maintained by the National Council for Scientific and Technological Development (CNPq).

Information and Knowledge Society: A model of organization of societies that has information as the central element of wealth production, source of power and guarantee of the well-being of people, supported by technology and characterized by easy access and exchange of information and their use for knowledge construction and decision making.

Information Management: A set of strategies and actions that comprise a process that involves searching, identifying, processing, storing and disseminating information to support decision making.

Innovation: Process of creating or perfecting a product or process to make it more efficient or introduce it to the market.

Research Groups: Group of researchers and students who organize themselves around one or more lines of research in a field of knowledge, to develop scientific research.

University-Enterprise-Government Interaction (Triple Helix): A model that considers the establishment of partnerships between universities, companies and government as a means to promote innovation in a country and serves as an approach to understanding and developing innovation strategies.

Section 3
Case Studies About Emerging Technologies for Project Management

Chapter 13
Information Technology Study Cases

Rabia Imtiaz Durrani
Institute of Management Sciences, Pakistan

Zainab Durrani
https://orcid.org/0000-0002-9292-9954
Institute of Management Sciences, Pakistan

ABSTRACT

This chapter focuses on the use of project management tools, techniques, and software in projects. The chapter includes a detailed discussion on the use of information communication technology within projects and provides a tour of the software and project management methodologies used to deploy projects. To contextualize the discussion, a case study of four startup projects hosted by two different incubation centers is presented. The case study discussion is structured around four themes: financial aspects, family support, legal perspective, and project success and failure. Findings from the cases are then compared against the literature reviewed; finally, the chapter concludes by providing recommendations. However, the result divulges there is no proper mechanism that encompasses the use of project management software.

INTRODUCTION

Project Management

Projects have been around since time immemorial. However, there are considerable differences between how projects were managed vs how they are managed now. Examples of some popular old projects include Egyptian pyramids, the Great Wall of China and the ancient Roman projects. Formally, it was not until the 20th century that the field of project management originated. Since then, different developments have taken place in the discipline, such are Gantt charts, PERT/CPM, EVM, statistical quality control, AHP, etc. Presently, the discipline of project management provides an ample amount of opportunities

DOI: 10.4018/978-1-5225-9993-7.ch013

and challenges for the organizations running them. However, the progress of both the field of project management and information technology (IT) has been considered revolutionary in many ways. Richard Nolan states that information technology has developed through three main eras; the electronic data processing era (EDP), the microcomputer era, and the network era. EDP era began in the 1960s, during which companies focused on organizational operations, such as inventory management, accounting task production, and scheduling management. This era includes the use of mainframe computers which has an impact on businesses. The main aim of this technology was to improve efficiency and lower the costs of the manual tasks carried out by individuals. However, IT projects in this era were generally structured and formalized, this was because things were quite stable in business operation and making changes was not a problem. However, in some cases, these systems created a communication gap between the departments, which hampered data exchange. It was the early 1980s, when the IBM Company launched personal computers (PC), ushering the beginning of the microcomputer era. The PC's eventually gained favor over centralized computers, but this transition did not take place immediately and without conflict. Management faced problems as they challenged the centralized management of data. The IT resources of organizations were split between a collection of decentralized PC's hosting information and a centralized computer with a centralized store of information. It was essential for companies to regain control of their IT resources. Therefore, IT was not only viewed as just a tool for lowering the cost of manual work but more of a tool for supporting knowledge management. Moreover, the managers needed an IT solution that would help manage and share information faster. IBM Company, in 1990 launched Windows 3.0 that was a breakthrough innovation in the tech world that made the use of PC much easier. Leading to eventually allowing people to own their digital content and share their content easily with everyone The Internet was born in the 1960s, as a project of the Defense Advanced Research Project Agency (DARPA). In the mid-1980s, as sharing information over networks gained popularity the Internet was the way chosen to connect distributed computers spread across the country. The Internet evolved to host the World Wide Web in 1995. The era of networks let the IT projects to focus on difficulties in building a software infrastructure that would assist the businesses, industries, vendors, employees, and customers to manage their day-to-day operations smoothly. The digital age has begun for providing numerous opportunities for innovative ways of delivering products and services to customers worldwide. Hence, the microcomputing era projects focused on internal organization networks while network-era projects focused on the external communication network (Marchewka, 2002). Moreover, 1990s, leads to the concept of globalization which made things more reachable anywhere in the world. According to Thomas L. Friedman, the world has become flattered which means that individuals and organizations can connect or can be bought on the same platform conveniently any time, any place with the help of IT e.g. virtual teams. In the present time and the fast-growing market, any business whether it is small or big the managers need to implement every step wisely and strategically due to the increasing competition of securing the market-leading position. They need to use and implement the right tools of IT and project management for successful business results of projects. As the increase in competition, globalization, communication management, outsourcing and speed of service along with intangible assets, such as knowledge management factors are the emerging trends or value drivers for different companies. In this tech-world, the projects and business environment are highly complex and there is a critical need to employ tools of PM and technology. The reasons for investing in projects are not only to gain operational efficiency but also to achieve strategic objectives that will lead to profitable gains. Additionally, the IT tools and project management techniques may help optimize new operations or processes, business expansion and time management in order to achieve timely delivery of new products in the market (Anantatmula, 2008;

Birk, 1990).Therefore, projects need to be designed to cater to the expectations of the project managers and project stakeholders. Information technology has made things a lot easier in operating projects, but to what extent IT has affected project management is a question to be addressed. To some extent, IT has changed businesses in every possible way one can imagine through advancement in both software and hardware and as well as communication technology. It could be argued that IT has thus, transformed the world of project management (Anantatmula, 2008).

A general overview of the origin and modern developments in the history of project management are in the table below. The table will give an idea about the progression of the project management field and its existence since the 20th century and its relationship with information technology.

Information Technology in Project Management

This section explains the importance of information technology in project management by focusing on the startup life cycle in projects along with the appropriate tools of information technology for project management. Moreover, this section explains the perceptual context of success and failure in projects.

Globalization has resulted in the interconnectedness of individuals and businesses using information technology (Fox & Hundley, 2011). In the entire process of globalization, the utmost important driving factor is information technology ("Information Technology," 2013).IT has taken hold all over the world and is changing every aspect of not only how individuals perceive their lives but the organization's way of survival in the market (Fox & Hundley, 2011). In the early 1990, telecommunication sector improved and resulted in an increased potential of the people accessing information ("Information Technology," 2013). Moreover, from the past ten years, the progress in Internet-based tools has resulted in social networking websites, such as Facebook, Twitter, Instagram, WhatsApp, Viber, LinkedIn etc. and appli-

Table 1.

Year	The history of Project Management development: A Timeline
1911	"The Principle of scientific management" publication of Frederick Taylor.
1917	The Gantt chart created by Henry Gantt
1956	The American Association of Cost Engineering (AACE International) was founded.
1957	The Critical Path technique developed by DuPont.
1958	The Program Evaluation Review (PERT) developed by the United State Department of US Navy for its Polaris Project.
1962	The Work Breakout Structure was mandated.
1965	International Project management Association was founded.
1969	The nonprofit Project management institution was founded.
1984	The management philosophy theory of constraints developed by Dr. Eliyahu Goldratt.
1986	Scrum was named as a project management style.
1989	Earn Value Management (EVM) came to prominence as a technique for project management.
1989	PRINCE was created by UK government for the projects in a controlled environment.
1997	Critical chain project management based on the methods and algorithm of constraint theory by Dr. Eliyahu Goldratt.
2001	Agile as a project style was introduced with a creation of the agile manifesto.
2008	ProjectManager.com is released bringing project management into the cloud.

cations that are sculpturing new ways of sharing and using data for personal, commercial and political purposes. IT communication has made it easy for companies to connect all over the world to smoothly conduct businesses and varies projects worldwide. In short, IT is reshaping society and economy regardless of the geographic location ("Information Technology," 2013). Today, in nearly every corner of the world people and enterprises are showing immense interest in project management. Up until the 1980s, project management elementally focused was on providing resource data and schedule data to the senior management of various institutions such as military, construction, and tech industries. Today's' project management has more to contend, including how to involve people or organizations working on a project anywhere in the world (Schwalbe, 2016). Therefore, both the fields of project management and information technology are interlinked. Moreover, the IT projects provide a vast range of business activities that involve maintaining existing systems and creating unique ideas taking inspiration from emerging technologies like Google Drive, Cloud, 3D printing and mobile tech. IT projects can be used for developing a website or upgrading a network or applications, like Enterprise resource planning (ERP) etc. Nowadays, companies are linking technology with customer products such as Apple smartphones, smartwatch, Google glasses. What does it all mean for a project manager? it means being a project manager you will be involved in dynamic nature, geographically scattered and more culturally different or ethnically diverse projects than ever before. Therefore, not only the risk the potential has increased but also the rewards are greater than before. However, to smartly deal with a fast-changing environment, project managers need a solid set of hard and soft skills to manage IT projects that will help them to code with emerging trends (Marchewka, 2002). IT projects are considered as organizational investments because when a new IT product or service is implemented it means the organization is committing in time, resource, money, and nature or lifecycle of a project with an expectation of value in return. Such are creditors or investors expects certain financial benefits and risk to avail an opportunity similarly, companies consider all the possible elements attached to a project life cycle. These factors help project managers make effective business decisions. Thus, projects are the vital source of the organization as they have a major effect on its growth. Particularly, IT projects are the ones that integrate technology in new product or services that have a direct impact on the relationships between customers, suppliers, and employees of that organization. However, IT-based projects can result in two outcomes, either it proves to be beneficial or best solution for an entity, or it may be a failure if not managed properly and fails to meet the required goals of the end users.

Technological Startup Life Cycle

This section will explain the startup life cycle of a project. According to Aidin (2015), there are three phases that each start-up goes through whether it is a tech-related startup or non-tech startup. While there is neither a predefined standard, nor do they have any principles exist for a start-up to follow. However, there is a specific life cycle that every company or startup goes through. To understand each phase of the startup, the life cycle will be explained in detail.

The 21st century is the age of technology, and technology has led to empowering the masses. There is huge potential that exists in the tech lucrative industry. Therefore, there are numerous opportunities available for start-ups and that keeps on increasing with every right set of circumstances (Mazhar, 2016). A tech start-up is a small venture seeking new avenues in the market to generate profits and to gain market presence or share. Weich (1979) defined "new venture creation as the organizing of new organizations, to organize is to assemble ongoing interdependent actions into sensible sequences that generate sensible

Figure 1. **The three stages of startup lifecycle** *(Adapted from Salamzadeh & Kawamorita Kesim, 2015)*

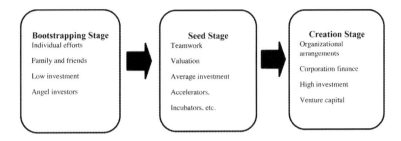

outcomes". Various researchers or scholars have characterized this phase in an enterprise life cycle as the "startup" phase (Vesper, 1990) Pre-organization(Katz & Gartner, 1988),Prelaunch (McMullan & Long, 1990), Gestation (Reynolds & Miller, 1992;Whetten, 1987),Organizational Emergence (Evers, 2003). Moreover, according to Steve Blank (2010) "A startup is a company, a partnership or temporary organization designed to search for a repeatable and scalable business model". Start-ups represent new ideas in a market that are remodeled into enterprises or firms. The basic purpose behind these new firms are to transforms the unique idea into profitable gains (Spender, 2014). Generally, newly formed businesses tend to be fast growing to meet the needs of the customer demands by offering unique or innovative product and services. There are various steps in the start-up creation process of a business.

Start-ups are different and complex in nature having a set of diversifying activities. The activities vary among different newly formed start-ups. However, here, it will present a holistic approach in order to better understand the initial phase of a business.

Bootstrapping Stage

It is well researched small ventures face multiple problems while acquiring funds for their business from outside parties such are: trust issues, the capacity to pay back a loan, information asymmetries, and high risk are involved because they are new in the market. To deal with these constraints, the startup businesses opt for creative or unique ways to reduce the overall cost of business activities or face the problems through a technique known as bootstrapping (Ebben & Johnson, 2006). This is the first stage of a startup lifecycle in which the entrepreneurs present the unique idea to establish it into a successful business. Bootstrapping is a general technique used in establishing a business by spending frugally and utilizing the business assets carefully in the initial stages of the startup. New ventures generally begin on a small scale but entail high-risk profile. In the early stages of the business, many startups rely on the personal source for funding the business such as personal savings, borrowing from family and friends. According to Freear, 'bootstrapping' is defined as "highly creative ways of acquiring the use of resources without borrowing money or raising equity financing from traditional sources" (Freear et al., 1995b, 1995c). There are two offshoots of the bootstrapping stage, which is the business development and product development. The most common techniques used in business development are founders' saving, home equity loans, deferred compensation for the owners, credit card debt, and low rent space. Whereas, the frequently used techniques in product development bootstrapping are prepaid licenses, royalties, special offers on hardware access, customer-finance research and development, and commercial projects (Freear, Sohl, & Wetzel, 2002). The purpose of this stage is to enable the business executive to manage the overall business in an ingenious and resourceful manner (Bhide, 1999). Owners who learn how to effectively apply the bootstrapping technique often gain trust in the eyes of the potential stakeholders who are directly or indirectly being affected by them (Freear et al., 2002). However, bootstrapping also helps in building knowledge, technical resource, and human resource etc. These factors lead entrepreneurs to adopt a 'lean' methodology to minimize waste without affecting business productivity (Harrison, Mason, & Girling, 2004). Moreover, bootstrapping helps in directing the startups towards growth by exhibiting capital expenditure management, product feasibility, customer acceptance and retention, and team management (Brush, Carter, Gatewood, Greene, & Hart, 2006).

Seed Stage

After the bootstrapping stage, the startup companies are ready to set foot into the seed stage. Whereas, at this stage of the startup the business is still considered an idea or a notion (Manchanda & Muralidharn, 2014).The seed stage includes the collaborative work of teams, the establishment of the prototype, evaluation of the venture, access to the market, look for assistance mechanisms, such as incubators and accelerators and the requirement of necessary funds to grow the startup (Salamzadeh, 2015 a). The primary funding is maneuvered in startups and used for the development of product and service, market research surveys, business recruitments, and patents registration. However, the startups in seed stage face numerous challenges to work on their business idea and to introduce it successfully in the commercial market. The most difficult thing for the startups at this phase is to bring in the huge amount of capital to convert their initiative into a successful business (Manchanda & Muralidharn, 2014). Therefore, the owners are in need to strive for support systems such as incubators, accelerators, business centers, and hatcheries to mentor or groom them to sustain in the market. However, numerous startups or new ventures often fail in this stage. When startups, cannot find the right platform to support them, they end up in a situation that turns their business into a low profit generating company with a slow growth rate success. On the contrary, those startups that receive support to sustain in the market would have possible higher chances of becoming successful entities (Aidin, 2015).Until now, there are few business giants or investors that are financially strong enough and ready to invest in the new innovative ideas known as angel investors. However, the angel investor and the venture capitalists have always been considered as the mentors of the creative world. They always support those selective startups that deemed to have high potential growth (Manchanda & Muralidharn, 2014).

Creation Stage

The startups enter this phase is depicting the picture when they enter into a real market world. Such as the startups start selling their products in the market, access to the different market and hiring of employees begins etc. (Salamzadeh & Kawamorita Kesim, 2015). Moreover, the entrepreneurial process remains only at the beginning and conception phase of a new business or venture. When the creation stage ends the entrepreneurship even stops, and the business starts focusing on the growth and development of the venture that is then mainly decided by the management.(Ogorelc, 1999).Therefore, as the creation stage ends, it signals an organization forms and they mostly rely on the corporate finance mechanism for financing the business. Another source of financing is the venture capitalist that facilitates this creation stage, by providing funds to the ventures.

PROJECT LIFE CYCLE

The life cycle a project provides a framework for the managers to manage any type of project and successfully deliver it to the clients. It is essential to understand that project life cycle may vary for different projects. Project management literature explains different project life cycle models. For example, technological software project comprises of five stages: definition, design, code, integration, and maintenance (Larson & Gray, 2010).It is important to use project life cycle model in a proper manner keeping nature of the project work in mind and which can be split into manageable phases. Every project consists of a definite beginning period, a middle phase that comprises of multiple activities that directs the projects towards its completion phase. It is not necessary the project will be successful or not (Watt, 2014). The life cycle acknowledges that every project has a limited lifespan and each phase of the life cycle should be effectively dealt to face any problems over the lifetime of the project. Generally, project life cycle has to go through four stages: defining, planning, executing and delivering. The project starts the moment when it is given the authority to proceed. The project starts at a slow pace, reaches to a peak position, and then declines when the project is delivered to the final user (Larson & Gray, 2010)

Defining Stage

In the inceptive phase, the objective or need of a project is recognized. It can be related to a business problem or new opportunity available. A relevant response to the demand is properly documented along with recommended possible solutions. Before starting or selecting a project a feasibility test is must to investigate a few questions such as "can we do the project?" or "should we do the project?"(Watt, 2014). Once the project is declared feasible, project objectives are established, team building begins, and major roles and responsibilities are assigned. This is the point where the project manager can move forward to the next detailed planning stage of the project.

Planning Stage

This is the next stage in the project lifecycle where a project plan is created in detail to meet the project goals systematically. Here, the project manager along with the team members identifies what work to be done and how? That includes detailed planning, when the project will be scheduled, what activities it will entail, the task and timeline management, what level of quality should be maintained? who will be the end beneficiaries, and prepare strategies, etc. (Larson & Gray, 2010). Moreover, the project manager prepares a detailed budget of the project by providing estimates for the equipment, human resource, and materials costs etc. The budget estimation will help project manager to closely monitor the cost expenditures during the implementation of a project period. Once the project group or team members has identified the basic work, done with schedule working, and determined the project cost means the three essential elements or components the process of planning is completed. This is the right time or period to recognize and manage to deal with any potential threats that can create hurdles in the process of a project to complete successfully. This process is known as risk management. During the risk management process, possible problems are identified and the practical solution for dealing with it. It either will reduce the chance of occurrence of that problem or will reduce the effect on project performance if it does happen. On the other hand, it is also a perfect time to identify all the direct or indirect stakeholders of a project and project manager should build a communication plan mechanism to describe the potential information required and the delivery tools and methods to be utilized in the procedure of keeping the stakeholders well informed. Finally, the project manager wants to document a good quality plan, imparting achievable or quality targets, assurance of work, and control measures arranging the criteria list to be fulfilled to gain final customer acceptance. At this stage, the project would have been discussed and planned in every possible detail to be execute (Watt, 2014).

Execution Stage

The execution phase of a project is the third stage of the project life cycle. During this phase, the project plan is put into action. A sizable portion of the work takes place in this stage both mentally and physically. It is essential to maintain control over the work and communicate only that amount of information that is required during the implementation. The progress of the project is continuously monitored, and relevant tweaks are made and recorded if deviates from the original project plan. During this phase, the team members are carrying out multiple tasks and through regular meetings, progress information reports are discussed. The project manager utilizes the information to manage the overall management of the project by comparing the progress document with the original project plan to measure the performance of the project tasks and take timely action as required. The basic job of the project manager in this phase is always to keep the project on the track, means; the project is direct according to the initial plan. If the project cannot be brought back on track, the project manager or the team members should record the deviations from the original project plan. Throughout this stage, all the stakeholders should be kept well informed about the status of the project. The timely reports should be focused on project cost, schedule, and deliverables. Concurrently the project manager should compare each deliverable as produced with the required quality and against the approved criteria. Once all the required deliverables are manufacture and the customer has agreed upon the final solution means the project is all set for closure (Watt, 2014).

Closing Stage

This is the fourth and last stage of the project life cycle during which you formally close your project. The major emphasis in the closure stage is on providing the final deliverables, terminating the contracts of the suppliers, release the remaining project resources, analyzing the project success and informing the stakeholders about the closure of the project. Yet the work of the project manager is not done here although the ample work of the project is finished. The last thing to do is the evaluation of the project or the post project review that includes not only the performance assessment but also what things worked out and what did not for the project. Through this investigation will indeed help the organization in future projects work ("The 4 Phases of the Project Management Life Cycle," 2017)

However, the project life cycle is a complete mechanism that a whole project goes through, on the other hand, the startup life cycle only focuses on the initial phase of a project. A startup life cycle is a part of the project lifecycle.

OVERVIEW OF PROJECT MANAGEMENT SOFTWARE, TOOLS AND TECHNIQUES

There are multiple software applications, tools, and techniques that are developed to help project managers to smoothly conduct business affairs and there is few other software that are not developed specifically for the project management but still project managers use them. In the following paragraph, the project management application, tools, and techniques are explained with the little description. Firstly, that software will be discussed which are not designed for project management but are useful and then those will be explained which are designed and used for project management.

SOFTWARE NOT DESIGNED FOR PROJECT MANAGEMENT

Microsoft Office Access

It is the unit of the Microsoft Office package that is used for the creation of the database. The main function of this tool is to store data and categorize the information as required that makes it easier to use. This software is mostly used in both small and large firms. The latest version comprises a lot of new elements or features. This software is very simple and easy to use whereas, its basic characteristic is to establish databases, forms, tables, reports, queries, etc.

Microsoft Office Accounting

It is used to easily manage the accounting work of the businesses. This software was developed for the small businesses which comprised of one to fifteen employees. That would help them save time, work efficiently, and organize or support accounting functions such as formulating budget, preparing the financial statements, and balance sheets. This software smoothly integrates with other Microsoft Office tools to exchange data information easily.

Microsoft Office Excel

It is important software that supports the project manager. This program of Microsoft is developed to arrange, calculate and format data along with formulas using a spreadsheet mechanism. This software has improved sharing of data functions and provides a high level of data safety. Moreover, Ms. Excel enables to perform not only the basic calculations, formatting spreadsheets but creates diagrams, graphing tools, and pivot tables etc. ("Microsoft Excel," n.d.)

Microsoft Office Visio

The software that helps project managers making their professional tasks easy to manage. Microsoft Visio is the part of the Microsoft family and its primary function is to create diagrams and visualizations. It facilitates those users that are dealing with managing data processing and organization activities to easily develop examine and share visual descriptions of complex information, processes, and systems.

Microsoft Office Word

It is one of the popular text editor's software that assists the project manager. The basic purpose of this tool is to allow the end users to type or save documents. It helps prepare and share documents easily by using a variety of writing tools.

SOFTWARE DESIGNED FOR PROJECT MANAGEMENT

Microsoft SharePoint Workspace

The Microsoft SharePoint workspace previously named as Microsoft Office Groove facilitates carrying out a vast number of operations linked with the given project. By using this application, it becomes easy to bring, all the team members, all tools, and the information on one platform or place. Moreover, with the help of this software any team of a project can work any time, from anywhere, and with anybody.

Microsoft Office Project

This software is a utility for managing projects, management of IT conditions, risk and resource management, forming of products, management of quality using six sigma setup, automation of professional service etc. Microsoft Project entitles creation of schedules, the establishment of work assigning structure, use of various financial functions (for example, project finances management), emphasize on changes, highlighting of cell background, tracking of budget, maintaining visual reports.

Microsoft Project Portfolio Server

A continuation of the previous software for organizations to use to obtain information within the scope of projects, programs, and portfolios that includes; selection and identification of portfolios with a potential to manage high on risk matters of business operations to be strategically solved on time. This software is an important unit or component of EPM (Microsoft Office Enterprise Project Management) solutions tool for project operations.

Primavera Risk Analysis Versions v8.6.

The software is similar to MS project is created by Oracle company. The main functionality of this software is to improve risk analysis, cost, and schedules. This software assists project managers to create a model and prepare an analysis of project costs, schedules, resources and also provide a situational and conditional logic for example weather modeling.

Ganttproject.biz Version 1.x and 2x

Created on java-based technology. It is an excellent planning software on which the project managers are planning the whole project activities, breaking tasks into subtasks, depicting project dependency factors and management of the project resources ensuring the start and end date and keeping a track on the progress report. These functions are done by using Gantt charts.

Open Workbench Version 1.1x

Another planning software that has the potential to help the project manager to plan tasks using in the form of Gantt charts. It can also support the management of calendars, project resources, data sorting and filtering them accordingly. It functions comprises critical path, time estimation for task completion and project progress reports.

Web Timesheet

The software was created by Replicon Company for managing projects at different levels of organizational level from regular employees to management staff by giving reliable up to date data for decision making in different management areas like budgeting, which includes profitability analysis, scheduling, invoice. Other than that, the software gives a project manager tracking opportunity of project budget, expenditures, calculation of total project receivables, employees working hour's accountability and project configuration.

Ace Project Version 1.x- 4.x

Developed by Websystem Inc. embedded with top performance and a wide range of tools, offered in two versions of software, allowing users to manage this software via different ways of access. The first version of this software can be accessed online on the internet directly. This software has the ability to manage large scale projects, multiple task management, dependencies of tasks, project reports, integrated activities and data transfer to MS Excel.

Achievo

The software is manufactured by an achievo company based on network access for the purpose of resource management in projects. This software is considered the best solution for those firms that want to record their progress, task scheduling, and resources in project implementation. The basic feature of the achievo software are human resource management, project management, customer relationship management, sharing schedule data with other users of the software tool and comparing schedules.

Artemis Software Version 7.0

Manufactured by Artemis international solutions corporation. The artemis software tool specifically developed for portfolio management which can be accessed on the internet easily. It consists of various basic features: portfolio management, financial planning, resource management, project planning, strategic planning, requirement management, statistics, and analyses.

InventX Version 2.5, 3.0 and 4.0

Also developed for the purpose of project management operations specifically for the reasons of strategic portfolio management, strategic planning, resource management, task management, and reporting.

Planisware Version 5

The software manufacture by Planisware Corporation is specifically designed to help institutional customers. This software is most helpful for the companies that are manufacturing new products and making strategic decisions for its basic features consist of project portfolio management, resource management, project management, scheduling management, budgeting and controlling management etc.

Planning Force Version 2.3

This software is manufactured by Planning Force Company and is a tool of project management. This software is design for the assessment of feasibility reports, task identification and investigating the level of risk involved in projects. Moreover, it helps project-related decisions. It mainly focuses on the project simulations, project monitoring, and the process of decision making etc.

Vert Abase Version 3.x,4.x

A project management tool for tracking project progress, analysis, reporting, scheduling, budgeting, task creation, automatic email notifications, controlling and can access the activities of the project online from anywhere.

Primavera P6

It is the most user-friendly and strongest software in the project management field, and it is manufactured by Oracle. It's developed for project management operations that include; planning, cost management, budgeting, scheduling, analysis, resource management, reporting, portfolio management, and project progress measurement. However, this software is for large and complicated projects. Mainly construction, manufacturing industry, finances, power engineering, public sector projects, and R&D department. This software can be accessed at anytime and anywhere in the world (Gorecki, 2015).

Agile Software

Agile as a concept is the ability of a project to respond to change in an uncertain environment. The software was developed in the year of 2001 for project management. The Agile manifesto software built on four core values and twelve principles. This approach enables the employee to use their innovative ideas, increase the productivity of individuals and improves the quality of work. Agile project management is an iterative process that value communication and feedback of every individual that is flexible in adapting change and producing business results.

Information Technology Study Cases

Scrum

Scrum is a concept that is used to explain the breaking down big tasks into small tasks and it is an agile project management framework that renders assistance for teams to work together. Also, it encourages teams to learn from their experience to be organized while sorting out the problems and continue to improve their work. Scrum is widely used to address the complexity of projects and initiates the flow of information transparent so that an individual can work smoothly on the project.

Lean Project Management Technique

The main concept of the lean method is to cut down the waste and produce more value within the business process. The lean project management emphasizes on quality of work, empowering the team, improving the data analysis, focusing on the business value-added, project planning, project documentation, project analysis, process improvement, and most importantly eliminating waste by efficient resources management.

SUCCESS AND FAILURE OF THE PROJECT

The field of project management has progressed over the decades focusing on project related challenges, problems and reasons for project failure rather than project success. (Alotaibi & Nufei, 2014; Balachandra & Raelin, 1984; Bedell, 1983; Hall, 1980; Morgan, H. and Soden, 1979).

However, different researchers argue that several projects are not able to achieve their end goals and thus, deemed unsuccessful. (Tumi, Omran, & Pakir, 2009; Doloi, 2013; Gündüz, Nielsen, & Özdemir, 2013). The nature of projects is considered unique in every aspect because of their areas, dimensions, industries, and difficulty levels. Due to the uniqueness of each project, it makes it difficult for project managers to list down the failure factors for future guidelines. According to the research, the factors on which projects are listed as failure projects are still not clear. To provide a list of factors that can prove to be guiding lines for future project managers to minimize the failure impact in projects are overlooked repeatedly. From Westerveld's (2003) perspective, the possibility of establishing and maintaining a set of standards for project success is repeatedly changing due to high failure rate. The success criteria of a project are said to be the characteristics involve time, cost and scope (Cooke-Davies, 2002).Projects often fail when they do not meet their deadlines, scope, and budget in a given opportunity of time for delivering the required work to the end users. In the same context, (Ika, Diallo, & Thuillier, 2012) adds a new perspective that the projects completed within the given cost, scope, and deadline are still considered to be unsuccessful. For this reason, it is compulsory for project managers to dig in deep and investigate failure factors beyond the mentioned criteria by different scholars. Also, other supporting factors involved in project operations such as interests of community, stakeholders, and organization play a vital role in the project's failure and success. Nelson, (2005) favors the above approach and encouraged the project managers to proactively identify problems and incorporate the best solutions for their projects. Whereas, according to the chaos definition, the projects that are functionally brilliant and over the budget and schedule is a fail project e.g. In 1994 American Airlines Early (AAE) sort out their legal proceedings with Hilton hotels, Marriott corps and budget rent a car system company for the confirmed car rental and hotel reservation costing $165 million system project failed into chaos. Furthermore, the project does not fail because of one single factor to be blamed for but due to multiple problems and challenges. However, the researcher argued that the success rate of projects is getting better with the passage of time. There are multiple factors for a project to strategically analyze before it is commercialized; one of the factors is overpromising. e.g. In 1983 Apple Macintosh company advertised commercial of first desktop Lisa featuring Kevin Costner in its ads costing $10000 and has a 1 MB of RAM. This project was considered as the biggest failure because the company could not fulfill the promise due to high financial costs. The whole scenario embarrassed the apple company of not being transparent enough about Lisa desktop. The article of work and project management (2018) discussed that most startup companies and various entrepreneurs are still hesitant and failing to appreciate the effective use of project management software, tools, and techniques.

THE STRUGGLE OF STARTUP COMPANIES IN THEIR INITIAL PHASES

Startup companies in their initial phases are strategically in search for multiple sources of finance from family, friends, banks, government loans, venture capitalists and angel financiers to invest for their growth and business activities. (Berger & Udell, 2006;Beck & Demirguc-Kunt, 2006).The financial market system is imperfect due to which the flow of information is not systematic and costs on transactions and contracts are huge that de-motivates or restrict the small business to gain the confidence of financial providers to acquire monetary support that results into a loss of credit history, guarantee, and reputation. (Beck, Demirguc-Kunt, & Martinez Peria, 2007). For these reasons, financial institutions and banks may find it difficult and costly to monitor the small business and their activities for their loan payback (Korosteleva & Mickiewicz, 2011). Family support is also another struggling factor for startups in their initial phase. The scholar such as Shapero,(1982) sheds light on family support and explains that father and mother are the two influential people for bringing out the entrepreneurial features in the family. However, the new venture still does not have enough trust developed on stakeholders that is why they need support from family to start their business ideas successfully (Ferina, Khurotul, & Sukowidyanti, 2018). Similarly, another factor for startup success is to get legally secured. According to the Agrawal, (2016) when entrepreneurs are busy and are excited to launch their startup business often have a legal aspect of the startup in the back of their mind. But because it is a small business that doesn't mean they do not require the registration process for their startup.

STUDY CASES

This section will give us an overview of the organizational background of the institution, centers, and startup business. For a better understanding of the cases, different themes are extracted from the interviews such as financial, family support, legal perspective, and success and failure. Moreover, the case study findings will be explained in detail with author recommendations in the end.

Information Technology Study Cases

Case Study Background

Institute of management sciences Peshawar (IMSciences) is quite renowned and leading management school in the province of Khyber Pakhtunkhwa. It is not only providing professional degrees to students in various disciplines, but to encourage and polish their professional skills and facilitate them to remain competitive in the marketplace. At present IMSciences has developed various centers and is involved in multiple projects such as human resource development centre, Centre for Excellence in Islamic Finance (CEIF), Micro Finance, Technology and Business Incubation center (TBIC), Center for Public Policy Research (CPPR), Research and development Division, Entrepreneurship Development Center (EDC),Lincoln Reading Lounge, Business Incubation Centre (BIC) and IM|Durshal. The case study will focus on two startups from IM-durshal and two from Business incubation center. IM-durshal is the joint collaboration project of the Institute of Management Sciences and Khyber Pakhtunkhwa Information Technology Board (KBITB). KBITB has been working to develop a digital strategy for the province: with activities aimed at improving digital skills, digital governance, digital access and promoting digital markets. A key component of this strategy is the development of a complete ecosystem for the startup culture that not only provides an enabling environment for early-stage technology startups but also ensures the sustainability of such tech-startups that are in the expansion stage. Establishing a network of **"Durshals"** is a step forward towards the Digital Transformation of the province. Durshals are physical spaces that can be used to host training programs and house coworkers, and tech-startups in order to promote the development of a digital economy across the province. To this end, the KP IT Board intends to collaborate with existing Public sector Universities and degree awarding institutes. This collaboration is based on an Incubation program. The program is built around the concept of efficient utilization of existing resources and expertise available at reputable public sector universities through a sustainable partnership for the benefit of the local tech community. The aim is to produce successful businesses that can create jobs and strengthen the economy. Whereas, the Business incubation center (BIC) project was established at the Institute of management sciences with the support of the Higher Education Commission (HEC) Pakistan. HEC established BIC platform at IMS to nurture the unique ideas of entrepreneurs. The main aim is to polish their skills and make them sustainable in the market moreover, to encourage an entrepreneurial ecosystem in the province of KP.

A brief description of the chosen startups is given below in a tabulated form.

To make things easier we will do the coding for the startups: Bikeon=B1, Innovation Limited = B2, Khawateen Rozgar = D3, and Smart agriculture D4. And for Manager BIC= BM and Manager IM-Durshal = DM.

Themes

The following themes are extracted from the interviews:

Financial Aspect

Finance is depicted as the lifeblood running through any business without which survival is impossible. All the four startups adopted "bootstrapping" strategy to finance their business operations. For the initial investment in the business B1 and B2 startups used personal savings whereas, D3 and D4 along with personal savings get a monthly stipend of 30,000 Rs (187.21 $) from the KBITB.

Table 2.

Incubation Center Name	Name of startup	Description	Coding for Startups	Coding for Managers
Business Incubation Centre	Bikeon	It is software developed for food delivery service system that will replace the existing system of fast food delivery rider.	B1	BM
	Innovation Limited	A mechanism for self-service printing that will reduce time and cost.	B2	
IM-durshal	Khawateen Rozgar	Services providing for creating job opportunities by linking women (khawateen) to organizations and industries.	D3	DM
	Smart Agriculture	Automatic plant watering device.	D4	

Family Support

In Asian culture society, the consent of the family members in every important decision of life is essential. Therefore, family support is indispensable for the members of all four startups. Nevertheless, B1 and B2 startups family members were not really agreeing with their unique ideas initially but, with the passing time they understood and showed support towards their business idea.

Legal Perspective

The registration process for startups is to obtain legal authorization to manage their business. Moreover, it's protection under the law. Business registration is one of the important steps for startups to form an entity. As all the four startups are in the initial phase of their business but D3 startup is one step ahead of the other three by registering their business.

Success and Failures

The phenomena of success and failure are subjective that keeps on changing with the project's scenario. The reason for the failure of one project will not be the same for the other projects similar is the case with success factors of a business. Initially, all the four startups have in one way or other faced small setbacks while working on their unique business idea. On the contrary, the startups are gradually moving towards their goal by achieving their set targets.

DISCUSSION

The case study involves two incubation centers at Institute of management sciences, among which two cases are selected from BIC and two from IM-Durshal. Six interviews are conducted in total to extract information, four from the startups and the rest of two from the managers of incubation centers. According to the DM (IM-durshal manager), IM-Durshal is purely based on the tech-startups to boost the entrepreneurial culture and promote ICT based startups in Khyber Pakhtunkhwa. Beside the project manager, there are two more members in the team one is coordinator marketing and outreach and the other is training coordinator.DM mentioned in the interview "we've been given a set of 20 modules by the IT board related to the different functional areas of business" according to which we provide training sessions to startups. An important aspect was revealed by the manager regarding how they are managing the project work "Well frankly speaking this whole entrepreneurial ecosystem in the region is in its budding phase and so are we. So, we actually have no such software which can guide us that this is how we need to do things. These are the things which can be saved and documented in a particular manner or some software which can reduce the reporting time for the project as such; In short, there is no mechanism in place and we are doing everything manually". Moreover, the Institute of management sciences is not charging any equity from the startups but charging KPITB for the services they are providing. However, the manager of BIC stated: it's a collaboration project between HEC and IMsciences to create a whole ecosystem for young entrepreneurs. It is a platform that will groom startups to achieve their desired goals. HEC has finalized twelve modules based on which BIC provides training lectures and provides subject specialist sessions per requirement. This incubation center provides opportunities for both tech and non-tech startups. BIC manager statement was similar to the IM-durshal manager regarding work management. The BM stated: "The incubation center is in the initial phase in itself, just got operational in January 2019. Presently, we are not using any project management tools and software to manage the day to day work". They are using Microsoft Word for documentation purpose, WhatsApp and Google drive for communication and checklist mechanism for completing a specific task etc.

After conducting interviews with all the four startup groups, we found out all of them have identified the market gap and initially pitched their idea at incubation centers where they got selected. In the beginning, all four startups adopted a bootstrapping strategy to finance their ideas. Moreover, IM-durshal incubatees are receiving a stipend of 30,000rs (187.21$) per team member and each team comprises of four members hence, a total of 120,000rs (789.83$) is being given to them by the KPITB per month. On the other hand, the literature supports the concept of bootstrapping strategy to be adopted for the initial investment in the project (Freear et al., 2002)& (Salamzadeh & Kawamorita Kesim, 2015). However, the startups have fulfilled the seed stage criteria that are characterized by Manchanda & Muralidharan, in their paper published in 2014. Furthermore, startups are in the initial phase of the lifecycle and not fully commercialized to legally secure themselves, though only D3 startup has registered their services.

Last but not least; family support plays an important role in encouraging one's achieving their dreams. D3 and D4 startups mention in their interviews that they were fully supported by their families. It is one of the factors that D3 is doing well as compare to the other startups in IM-durshal. Whereas B1 and B2 faced little resistance from family and friends regarding their unique idea that had an emotional impact on them but with passing time everything has to settle down. However, it is proved from literature as well that family members and friends play an important role in supporting the business idea of the entrepreneur (Hendieh & Osta, 2019; Shapero,1982).Finally, we are going to shed light on the success and problem factors in the case study. It is a fact; individuals face numerous challenges on the way of achieving their goals. Success cannot attain without hard work and dedication khawateen Rozgar (D3) is a case in point here. They have entered the market, signed MOU with different institutions and placed forty females in various organizations in the province of Khyber Pakhtunkhwa. But, they are still facing the problem of data management and working on a solution for it. The smart agriculture startup is still working on the prototype and has not entered the market yet. Similar is the case with the bikeon startup they are still not fully prepared to enter the market because they are doing some modifications to make software user-friendly. On the other hand, innovation limited has successfully launched its Air print product at IMsciences and earned 30,000rs (187.21$) revenue in three months. However, they are still dealing with the problem of product maintenance, due to security reasons in Khyber Pakhtunkhwa, the Institute of management sciences is not allowing the owner to keep an outsider to maintain the product on a regular basis. But is allowing whenever a problem encounters, they can come and rectify it but, this whole process consumes time.

On the contrary, all four startups are not using any project management software tool to carry out work. They are using different software's and non-project management techniques such as B1 is using self-developed software bike on and manually maintaining activities, B2 is using PaperCut and tally ERP software, D3 is using Sublime editor, Bootstrap Studio, Google Sheets, and Google forms. When they were using the sublime editor software, they faced security issues then they shifted the data to Google sheets and Google forms because of their security modules.

UNKNOWINGLY DOING THINGS KNOWINGLY

All the startups and both the incubators members are using project management methods such as agile, scrum and lean in their business operations unknowingly due to lack of knowledge. For instance, some of the startups meet on a daily basis discuss and divide their work this is known as a scrum method. Whereas, both the incubation center managers coordinate work themselves and divide it among internees regularly, is an agile method. Moreover, all the startups and the incubation centers are in the budding phase of the lifecycle that's why avoids wastage of resource and utilizing it in the best possible way this technique is known as a lean method. Therefore, not only startups but the incubation center managers should be given training regarding project management tools, techniques, and software in order to plan, execute and control all aspects of the project process efficiently and effectively. Moreover, due to the complex nature of the project tasks requires implementing project management software to streamline the whole process.

CONCLUSION

This chapter provides a quick and thorough explanation of the concepts and the underlying discussions on Information Technology software for project management. This chapter has covered detailed literature on project management and information technology. Then technological startup lifecycle stages vs project lifecycle stages were addressed in detail. Following the explanation of the IT software providing assistance to project manager were elaborated. Moreover, the success, failure factors of the projects and struggle of startups in the initial phase of the business lifecycle were comprehended. A case study methodology was applied in order to find the required results. Four cases of startups from two incubation centers at IMsciences were selected for analysis. To extract data interviews were conducted with the incubatees and the managers of both incubation center. Furthermore, four themes: financial aspect, family support, legal perspective, and success and failure were excerpted. The result revealed there is no proper mechanism that involves the use of project management software to conduct their day to day operations.

REFERENCES

Agrawal, A. (2016). The dumbest legal mistakes early startups make. Retrieved from https://www.inc.com/aj-agrawal/the-dumbest-legal-mistakes-early-startups-make.html

Aidin, S. (2015). New venture creation: controversial perspectives and theories. *Economic Analysis, 48*(1984), 101–109. Retrieved from http://ssrn.com/abstract=2711731

Alotaibi, A. B., & Al Nufei, E. A. F. (2014). Critical success factors (CSFS) in project management: critical review of secondary data. *International Journal of Scientific & Engineering Research, 5*(6), 325–331. Retrieved from http://www.ijser.org/researchpaper%5CCRITICAL-SUCCESS-FACTORS-CSFs-IN-PROJECT-MANAGEMENT-CRITICAL-REVIEW-OF-SECONDARY-DATA.pdf

Anantatmula, V. S. (2008). The role of technology in the project manager performance model. *Project Management Journal, 39*(1), 34–48. doi:10.1002/pmj.20038

Balachandra, R., & Raelin, J. A. (1984). When to kill that R & D project. *Research Management, 27*(4), 30–34. doi:10.1080/00345334.1984.11756846

Beck, T., & Demirguc-Kunt, A. (2006). Small and medium-size enterprises: Access to finance as a growth constraint. *Journal of Banking & Finance, 30*(11), 2931–2943. doi:10.1016/j.jbankfin.2006.05.009

Beck, T., Demirguc-Kunt, A., & Martinez Peria, M. S. (2007). Reaching out: Access to and use of banking services across countries. *Journal of Financial Economics, 85*(1), 234–266. doi:10.1016/j.jfineco.2006.07.002

Bedell, R. I. (1983). Terminating R & D projects prematurely. *Research Management, 26*(4), 32–35. doi:10.1080/00345334.1983.11756785

Berger, A. N., & Udell, G. F. (2006). A more complete conceptual framework for SME finance. *Journal of Banking & Finance, 30*(11), 2945–2966. doi:10.1016/j.jbankfin.2006.05.008

Bhide, A. (1999). *The origin and evolution of new businesses*. The Oxford University Press.

Birk, J. (1990, October). A corporate project management council. *Proceedings of The IEEE International Engineering Management Conference, Gaining the Competitive Advantage* (pp. 180-181). IEEE.

Brush, C. G., Carter, N. M., Gatewood, E. J., Greene, P. G., & Hart, M. M. (2006). The use of bootstrapping by women entrepreneurs in positioning for growth. *Venture Capital, 8*(1), 15–31. doi:10.1080/13691060500433975

Cooke-Davies, T. (2002). The "real" success factors on projects. *International Journal of Project Management, 20*(3), 185–190. doi:10.1016/S0263-7863(01)00067-9

Doloi, H. (2013). Cost overruns and failure in project management: Understanding the roles of key stakeholders in construction projects. *Journal of Construction Engineering and Management, 139*(3), 267–279. doi:10.1061/(ASCE)CO.1943-7862.0000621

Ebben, J., & Johnson, A. (2006). Bootstrapping in small firms: An empirical analysis of change over time. *Journal of Business Venturing, 21*(6), 851–865. doi:10.1016/j.jbusvent.2005.06.007

Evers, N. (2003). The process and problems of business start-ups. *The ITB Journal*, *4*(1). doi:10.21427/D7WT8K

Ferina, N., Khurotul, A. E., & Sukowidyanti, A. P. (2018). Does family social support affect startup business activities? *2*, 41–54. DOES

Fox, P. & Hundley, S. (2011). The importance of globalization in higher education. In New Knowledge in a New Era of Globalization. doi:10.5772/17972

Freear, J., Sohl, J. E., & Wetzel, W. (2002). Angles on angels: Financing technology-based ventures-a historical perspective. *Venture Capital*, *4*(4), 275–287. doi:10.1080/1369106022000024923

Freear, J., Sohl, J. E., & Wetzel, W. E. J. (n.d.-a). *Early stage software ventures: what is working and what is not*. Report for the Massachusetts Software Council.

Freear, J., Sohl, J. E., & Wetzel, W. E. J. (n.d.-b). Who bankrolls software entrepreneurs. In Frontiers of Entrepreneurship Research.

Gorecki, J. (2015). Information technology in project management, *77*.

Gündüz, M., Nielsen, Y., & Özdemir, M. (2013). Quantification of delay factors using the relative importance index method for construction projects in Turkey. *Journal of Management in Engineering*, *29*(2), 133–139. doi:10.1061/(ASCE)ME.1943-5479.0000129

Hall, P. (1980). *Great planning disasters*. London, UK: Weidenfeld and Nicolson. doi:10.1016/S0016-3287(80)80006-1

Harrison, R. T., Mason, C. M., & Girling, P. (2004). Financial bootstrapping and venture development the software industry. *Entrepreneurship and Regional Development*, *16*(4), 307–333. doi:10.1080/0898562042000263276

Hendieh, J., & Osta, A. (2019). Students' attitudes toward entrepreneurship at the Arab Open University-Lebanon. *Journal of Entrepreneurship Education*, *22*(2).

Ika, L. A., Diallo, A., & Thuillier, D. (2012). Critical success factors for World Bank projects: An empirical investigation. *International Journal of Project Management*, *30*(1), 105–116. doi:10.1016/j.ijproman.2011.03.005

Information Technology. (2013). Retrieved from http://www.globalization101.org/information-technology/

Katz, G., & Gartner, W. B. (1988). Properties of emerging organisation. *Academy of Management Review*, *13*(3), 429–441. doi:10.5465/amr.1988.4306967

Korosteleva, J., & Mickiewicz, T. (2011). Start-up financing in the age of globalization. *Emerging Markets Finance & Trade*, *47*(3), 23–49. doi:10.2753/REE1540-496X470302

Larson, E. W., & Gray, C. F. (2010). *Project management the managerial process* (5th ed.).

Manchanda, K. & Muralidharan, P. (2014, January). Crowdfunding: a new paradigm in start-up financing. In *Global Conference on Business & Finance Proceedings, 9*(1), p. 369. Institute for Business & Finance Research.

Marchewka, J. T. (2002). *Information technology project management: providing measurable organizational value*. Retrieved from /citations?view_op=view_citation&continue=/scholar%3Fhl%3Den%26start%3D10%26as_sdt%3D0,5%26scilib%3D1%26scioq%3Dentrepreneurship%2Bin%2BEgypt&citilm=1&citation_for_view=EJ3GnDgAAAAJ:SP6oXDckpogC&hl=en&oi=p

Mazhar, M. (2016). *Challenges faced by startup companies in software project management*.

McMullan, W. E. & Long, W. A. (1990). *Developing new ventures: the entrepreneurial option*.

Microsoft Excel. (n.d.). Retrieved from https://www.techopedia.com/definition/5430/microsoft-excel

Morgan, H., & Soden, J. (1979). Understanding MIS failures. *Database*, (5): 157–171.

Nelson, R. R. (2005). Project retrospectives: Evaluating project success, failure, and everything in between. *MIS Quarterly Executive*, *4*(3), 361–372.

Ogorelc, A. (1999). Higher education in tourism: An entrepreneurial approach. *Tourism Review*, *54*(1), 51–60.

Reynolds, P., & Miller, B. (1992). New firm gestation: conception, birth and implications for research. *Journal of Business Venturing*, *7*(5), 405–417. doi:10.1016/0883-9026(92)90016-K

Salamzadeh, A. (2015, June). Innovation accelerators: emergence of startup companies in Iran. In *Proceedings 60th Annual ICSB World Conference* (pp. 6–9).

Salamzadeh, A., & Kawamorita Kesim, H. (2015). *Startup companies: life cycle and challenges*. doi:10.2139srn.2628861

Schwalbe, K. (2016). Information technology - project management.

Shapero, A. (1982). *Inventors and entrepreneurs: Their roles in innovation*. College of Administrative Science, Ohio State University.

Spender, J. (2014). *Business strategy: managing uncertainty, opportunity, and enterprise*. Oxford University Press. doi:10.1093/acprof:oso/9780199686544.001.0001

The 4 phases of the project management life cycle. (2017). Retrieved from https://www.lucidchart.com/blog/the-4-phases-of-the-project-management-life cycle

Tumi, S. A. H., Omran, A., & Pakir, A. H. K. (2009). Causes of delay in construction industry in Libya. In *The International Conference on Economics and Administration* (pp. 265–272).

Vesper, K. H. (1990). *New venture strategies*. Prentice-Hall.

Watt, A. (2014). *Project management*.

Whetten, D. A. (1987). Organisational growth and decline processes. *Annual Review of Sociology*, *13*(1), 335–358. doi:10.1146/annurev.so.13.080187.002003

APPENDIX

Table 3. The startup and incubation centre member's detail

Name	Designation
Dr. Shahwali Khan	Manager, IM-durshal
Ahsan Zia	Manager, BIC
Fizza Majeed	Manager, Smart Agriculture
Faaiz Majeed	Hardware Support, Smart Agriculture
Aiman Abbas	CIO, Smart Agriculture
Asma Razaq	CTO, Smart Agriculture
Zarmina Orakzai	Co-founder and Director, Khawateen Rozgar Service
Misbah Faiz	Founder Khawateen Rozgar Service
Moneeba Anwar	Director Web Development Khawateen Rozgar Service
Junaid Ahmed	Founder, Innovation Limited
Shezad Afridi	Founder, Bikeon

Chapter 14
Pharmaceuticals and Life Sciences:
Role of Competitive Intelligence in Innovation

António Pesqueira
Takeda, Zurich, Switzerland

Maria José Sousa
 https://orcid.org/0000-0001-8633-4199
ISCTE, Instituto Universitário de Lisboa, Portugal

ABSTRACT

This chapter analyzes innovation, knowledge, and competitive intelligence (CI). Besides these concepts, the focus will be on the role of innovation profiles. The innovation profiles include the creation, capture, organization, and integration of knowledge into the innovation process. The CI variable will be analyzed, demonstrating the potential for creating a context of competition for companies. A case study is presented about the pharmaceutical (pharma) industry with the application of the concepts of competitive intelligence, knowledge, and innovation to a real context.

INTRODUCTION

Most of the pharmaceutical companies continue to face crescent competition in the operating markets from different sources. Some of the solutions to overcome the current difficulties and barriers may be a constant search of new R&D methods and new operations mechanisms in order to obtain economic scale and gain new commercial capabilities that can drive or improve effectiveness from R&D investments.

Currently, Pharmaceutical companies are searching for new disruptive strategic methodologies through innovative methods of business models, in a quest to reduce the business risks and prepare the commercial models to a more ready technological approach.

DOI: 10.4018/978-1-5225-9993-7.ch014

There is no doubt that the pharmaceutical landscape and drivers for growth are changing, with new patterns around a crescent aging worldwide population, technological advancements, products innovation, new standards of living and transformed health care access systems.

While pharmaceutical companies continue to face substantial difficulties resulted from a crescent competition, decreasing revenues, product patents expiring and limited access to physicians, most of the competitors such as Medical technology companies, personalized medicine, diagnostic technologies, medical devices, and biotechnology companies continue on maintaining their good momentum, with strong predictors of continued success.

Increasing the complexity of pharmaceutical effectiveness operations it's evident the constant pressure and increased regulatory scrutiny from the official authorities with severe impacts from health care reforms and fewer drugs approvals across all world but particularly in Europe, where the escalating costs of products forces the global authorities and local governments to impose new pricing rules, fair trade conditions and price decreases. Therefore is not only crucial for today's pharmaceutical companies to continue leveraging and growing geographic scale, creating strong positions in dominant markets and acting as global serving points on global needs, is also crucial and critical for those same companies to have long-term supporting processes and monitoring capabilities to face the constant business influential factors.

Although global life sciences sector persists in exhibiting some resilience in such market conditions and crescent constraints, the need for reinvention of the current business models continues to be one of the most significant market demands and patients expectations for more efficient products.

The way that global life sciences sector can project new research and developments (R&D) efforts and step back in some ongoing projects to add fragmented pieces of scientific innovation is a capability that not all the companies can adopt.

The chapter includes a brief study of innovation fundamentals and theories. Its contribution is to review basic ideas about innovation process in the pharma industry, allowing future discussion about the main issues to potentiate the competitive intelligence in the pharma industry. Innovation allows managers to implement successful organizational practices and processes, resulting in efficient new business models (Kalakota & Robinson, 2001; Kearns & Lederer, 2003; Takeuchi & Nonaka, 2004). This can deliver results in complex scenarios with different strategic design and execution pathways.

LITERATURE REVIEW

The Concept of Innovation

The innovation theory literature gives the idea that innovations occur mostly within the national system of innovation (Freeman 1987; Lundvall 1992; Nelson 1993; Edquist 1997). However, another perspective was studied by organizational studies in innovation in organizational microsystems (Van de Ven 1986; Aldrich & Fiol 1994; Van de Ven et al. 1999; den Hertog & Huizenga 2000).

Literature shows that the concept of innovation is very complex, which makes it difficult to have a single definition. The Green Book on Innovation from the European Commission (1996) defines innovation as "the successful production, assimilation, and exploration of something new." More recently,

Pharmaceuticals and Life Sciences

Mulgan and Albury (2003) made their contribution to the concept pointing out the importance of the innovation implementation results: "new processes, products, services and methods of delivery which result in significant improvements in outcomes efficiency, effectiveness or quality."

Leadbeater (2003) exposes the complexity of the concept including the interactive and social dimensions: he argues that "the process of innovation is lengthy, interactive and social; many people with different talents, skills and resources have to come together."

On the other hand, the literature assumes various categorizations of innovation. OECD (2002) structures the concept around three areas: the renewal and broadening of the range of products and services and associated markets; the creation of the production, procurement, and distribution methods; and the introduction of changes to management, work organization and workers' qualifications.

Baker's (2002) typology also differentiates three types of innovation: Process; product/service; and strategy/business.

Process innovation (i.e., work organization, new internal procedures, policies and organizational forms) and the strategic and new business models (i.e., new missions, objectives, and strategies) are called organizational innovation.

Following OECD (2002), organizational innovation includes three broad streams: 1) the restructuring of production and efficiency processes, which include business re-engineering, downsizing, flexible work arrangements, outsourcing, greater integration among functional lines, and decentralization; 2) human resource management (HRM) practices, which include performance-based pay, flexible job design and employee involvement, improving employees' skills, and institutional structures affecting the labor-management relations; and 3) product/service quality-related practices emphasizing total quality management (TQM) and improving coordination with customers/suppliers (Table 1).

The analyses of organizational and innovation literature point out innovation as one of the most critical strategic/management dilemmas. Organizations survival requires that they became more and more competitive, and organizational innovation can be a key solution. Currently, the organizations invest in a very consistent way in an innovation strategy. The answer for this phenomenon is itself a fundamental and complex dilemma because the importance of innovation for competitiveness is not explicit and the choice between investing in technology and investing in people always raises some questions about short- and long-term survival of the organizations. In this context Sousa (2009, 2013) has created several profiles intending to understand the potentialities of each profile as a tool to help employees develop their competencies and become more skillful along the innovation process. The set of competencies associated with each innovation profile are identified in the following Table 2, according to Sousa (2009; 2010; 2013; 2016) methodology:

Table 1. Types of organizational innovation

Production and efficiency practices	**Human resources management practices**	**Product/service quality**
• Business re-engineering • Downsizing • Flexible work • Outsourcing • Greater integration among functional areas • Decrease degree of centralization	• Performance-based pay • Flexible job design and employee involvement • Developing skills • Labor-management cooperation	• Total quality management (TQM) • Improving coordination with customers/suppliers • Improving customer satisfaction

Source: Wulong Gu & Surendra Gera (2004)

Table 2. Knowledge profiles competencies

Profile	Competencies
Innovator	• Ability to use creative techniques • Ability to use schematization and simulation techniques • Ability to use content analysis • Ability to create new knowledge • Ability to innovate
Integrator	• Ability to apply the accumulated technical knowledge into new projects • Ability to apply organizational knowledge • Ability to use individual knowledge in problem-solving • Ability to work in a team
Organizer	• Ability to create and organize organizational memories • Ability to create and manage knowledge centers • The ability for knowledge mapping • Ability to create and manage knowledge networks
Facilitator	• Ability to organize learning processes • Ability to share best practices • Ability to organize spaces of share, like seminars or workshops • Ability to develop young talents • Ability and knowledge to shape behavior • Ability to encourage subordinates and co-workers to innovate and change • Ability to help subordinates and co-workers to participate and accept change

Source: Sousa (2009, 2013, 2016)

According to Sousa (2009, 2013, 2016), these four knowledge profiles (defined as sets of attitudes and behaviors) can be defined as the following:

1. *The Innovator* is an organizational actor that focuses on experiments to develop new knowledge and new solutions. He or she makes things happen and creates results using existing knowledge through experimentation.
2. *The Organizer* is an organizational actor who prefers to create structures that make explicit, collect, combine, and analyses knowledge. He or she creates mechanisms that transform tacit knowledge into explicit knowledge for future application.
3. *The Integrator* is an organizational actor that uses and integrates the knowledge developed and shared by all organizational actors, including him or herself.
4. *The Facilitator* is an organizational actor that promotes reflection, learning and tacit knowledge sharing processes. He or she makes sure that the right competencies are present when knowledge is applied in a controlled process.

Types and Sources of Innovation

In the last years, several types of research has produced new concepts, and benefit analysis to innovation contributed significantly to academic and corporate knowledge. It is a concept of great relevance that influences not just the corporate procedures but also various products and services obtained from business dynamics and models.

Pharmaceuticals and Life Sciences

It is a process that includes techniques, conceptions, development and work tools, which originate improved products or procedures, which are then readily available for promoting and selling.

Innovation can be identified through 4 distinct levels:

1. **Incremental**: Considers the improvement in products and procedures that upgrade the quality and still can reduce costs and increase productivity. Identifying the role that technology plays, specifically in equipment compatibility.
2. **Radical**: Discontinuous result of actions regarding R&D in corporate, academic or state environments organizations.
3. **Modification of the "Technological System"**: Technological changes that affect certain economic sections and originate completely new sectors.
4. **A Shift in the "Tech-Economic Paradigm" (Tech Revolution)**: Result from the evolution of technological systems, such as new products, procedures, changes in the economic and social structure and the behaviors of the leading economic agents.

Radical innovations can frequently create or destroy specific market niches and client segments, or even profoundly change a market or sector, resulting in new paradigms, skills, abilities, and knowledge in the same market.

Reversely to the incremental model, radical innovation usually follows along technology-push, which is to say that new concepts and paradigms are created inside the corporations, many times being led by researchers or professionals dedicated to research and development of new products. On the other hand, radical innovations are very rarely altered or changed while in their development stage, mostly to protect the main idea, which does not give way to accepting new concepts or developments in this stage.

Presenting radical innovations to the market complies a great deal of effort, which is exponentially higher than in incremental innovation, we are not always positive outcomes are obtained, regardless of the quality of the innovation.

Innovations is a new way to do things, which can be marketed. The various types of innovation are Product Innovation & Technological Innovation; Innovation of Procedures; Innovation in the organizations; Marketing Innovation.

Innovation models can be presented the following way: Push-pull Model; Funnel Model; Disruptive Innovation Model; Closed Innovation & Open Innovation.

We can also perceive innovation through four types of innovation: management innovation; strategic innovation; product/service innovation and Operational innovation. Each of the innovation levels results in asymmetrical levels of competitive value outcome, all though not all the levels necessarily originate competitive advantage. Some innovations can be incremental, where the competitive advantage is strongly dependent on the changes made.

Innovation has in its core different new concepts and ideas, which can come from numerous places. Several innovation sources can play a significant role in creating new ideas and exploring innovation opportunities. These sources can be internal and external to the organization and also formal and informal.

The external sources are:

- Internet and the Media;
- Specific studies;
- Client Surveys;

- Market Surveys;
- Opponent companies benchmarking;
- R&D entities or design.

The informal external source is:

- Internet and the Media;
- General surveys;
- Customers Feedback;
- Competitor's products and services;
- Distributors and Partners advice;

For formal internal sources:

- Innovation programs;
- Informal brainstorming,
- Internal processes;
- R&D Departments;

As far as Informal internal sources we have:

- Associates ideas;
- Products;
- Procedures;
- Services;

It is therefore essential that the organization arranges its activities and procedures to identify market tendencies and opportunities, in order to reinforce its competitive advantage. Having the ability to develop tools to monitor the competitor's competitive advantage, clients and social context surrounding is very relevant. This surrounding can be defined as contextual, transactional and intracompany.

The contextual surrounding incorporates economic factors, technological, political, social and cultural factors. The Transactional surrounding includes the clients, opponents, suppliers and the community. Relating to the Intracompany surrounding is the knowledge of the organization when compared to its opponents and its sector.

Innovation is a result of the investigation and development procedures, focused mainly on development. We will consider however that innovation is the center of these interactions where R&D is considered a source of procedures directed at innovation and as one of the processes that itself catalyzes innovation.

As far as influencing innovation outcomes, we have Organization location, a specific region, pertaining sector, dimension of the organization, globalization level and relation in environments with catalyzing abilities. Continuous innovation in a company has as crucial factors of influence the external and internal sources of data and information, along with client, supplier and partner relationships.

Innovation sources are miscellaneous, but consumers, products, and suppliers are the most relevant. It is crucial to emphasize the useful sources of innovation, where the role of each source is actually inside innovation and is related to the implementation goals. The functional sources of innovation include the organization or the people that will directly benefit from the products innovation, procedures or services. These same organizations or people are not passive bystanders of innovation; they vary according to the innovations analysis, due to the functional relationship between the innovator, the user, and the innovation. Given the right conditions, any functional class can be a potential source for innovation. The manufactures can explore many innovation sources, be it in an internal or external context, having a strong influence by the economic market, cultural and social factors, clients, suppliers, opponents, stakeholders or partners. One of the biggest challenges that life sciences companies faces are the methods to efficiently and productively turn innovation into commercial products. Many of the solutions can be explained through the use of multidisciplinary activities around collaboration and cooperation between internal and external teams. The main characteristics of these life sciences companies are the risk-sharing activity models.

Life sciences sector currently needs to support the innovation efforts with a multi-prolonged strategy to cope with the current market environment and constraints, preventing the future decline in sales revenues by a decreasing potential market or directly with patent expirations of some of the drugs available in a pipeline.

Knowledge as a Driver for Innovation

The analyses of organizational and innovation literature point out innovation as one of the most essential strategic/management dilemmas. Organizations survival requires that they became more and more competitive, and organizational innovation can be a key solution. However, very few organizations invest in a very consistent way in an organizational innovation strategy. The answer for this phenomenon is itself a fundamental and complex dilemma because the importance of organizational innovation for competitiveness is not explicit and the choice between investing in technology and investing in people always raises some questions about short- and long-term survival of the organizations.

In microanalysis, there are some dilemmas arise concerning interactions between organizational actors and whether their knowledge affects the organization's dimensions.

This research analyses knowledge management dilemmas that emerged from the literature review, and will conceptualize them in order to identify situations where organizations continuously face dilemmas, determine their responses to these situations, and, over time, how they succeed:

1st Dilemma: "Literature emerges the idea that the use of individual knowledge accumulated through life and professional experiences is a competitive advantage for the organizations' success. However, sharing and transferring inexpressible knowledge is almost an impossible task to accomplish."

Knowledge sharing and transference requires specific competencies of interaction. One of the main factors of successful knowledge sharing is a trusting climate among workers. This makes them more participative and more involved.

Concerning workers' interactions, the assumption is that the individual learns and then affects the group with the new knowledge acquired, but needs to be inserted in an organization whose purpose is developing individuals and producing skills and innovation for the organization (Jacobs & Washington, 2003).

On the other hand, transforming tacit knowledge to explicit knowledge, namely life and work experiences and all the knowledge that workers develop and store along the years, seems to be a challenging activity because it represents knowledge that people possess, but which is inexpressible because it incorporates both physical skills and cognitive frameworks.

However, when the knowledge becomes explicit, it can be passed on and acquired by another person (Morris & Beckett, 2004). Several research works about workplace learning also imply the assumption that individuals acquire knowledge, for example, by listening to information presentation and when this becomes standard practice, they become more open to sharing it with other colleagues.

2nd Dilemma: "The use and sharing of employees' knowledge is an essential factor in solving problems and strengthening performance. However, several organizational and individual barriers condition the process."

Organizations use particular processes in order to solve problems - testing new and different ideas on how to achieve success is one of these processes and employee's knowledge can perform a relevant role in it.

However, employees cannot always use their knowledge to help their organization to solve problems and respond to challenges because organizations do not always give them space to think, act, make informal contacts, gain experience, experiment and take risks. In many situations, employees and even managers find it difficult to use the knowledge that they have developed in other working experiences just because it was not requested.

To stimulate the use of individual knowledge and strength the core competencies of an organization, top management can promote a learning attitude, intensive knowledge exchange, and internal entrepreneurship. It is also possible to use an approach to problems through precise routines, procedures and methods like brainstorming, problem-solving cycles and risk management.

Jashapara (2007) suggests that "organizational routines provide the contingent condition or `spark' to activate organizational knowledge processes." The processes can be initiated and guided by existing or expected problems that are seen as a chance to learn or innovate.

Managers can focus themselves on developing and mobilizing employee's knowledge to innovate and introduce new practices using tools like mapping out the individual competencies of each employee - it will help to understand which employees have valuable knowledge and what the existing knowledge gaps are in order to take some measures to narrow and eliminate them. They can also create more "communicative knowledge-accomplishing activities, which frame and respond to various problematic situations" (Kuhn & Jackson, 2008).

Nevertheless, organizations need to have a high level of openness to risk and tolerance to mistakes and failures instead of penalizing employees for them. Only this perspective allows organizations to create a culture of innovation.

3rd Dilemma: "Using and sharing individual knowledge is crucial to organizational innovation processes, but organizational culture and management resistance makes it very difficult to promote employee's involvement and participation."

Organizations can promote and invest in a learning environment characterized by positive thinking, self-esteem, mutual trust, willingness to intervene preventively, taking responsibility for business performances and rewarding the employees who continually study their work and give ideas to better it when needed.

Skilled workers are more open to innovation and change because accepting new work practices is easier when the skill level of the workforce is higher. A skilled labor force will accelerate the introduction of organizational changes because skilled workers are more able to analyze and synthesize new pieces of knowledge (Caroli & Reenen 2001).

However, knowledge and learning competencies need to be a part of every employee's competence profile, and organizations can have an essential role in stimulating employees to think about, identify and solve common problems; to let go of traditional ways of thinking; to continually develop their own skills, and let them acquire experience and feel responsible for organization and team performances.

From the employee's perspective, it is interesting to analyze their position concerning the balance between personal ambition and the shared ambition of the organization. In the literature, there are two kinds of workers: a) Individuals who care about the organization and what it stands for; those with the vision, competences, and resources to apply what they have learned to make the company and themselves the best they can be; b) Individuals who would be satisfied with the fact that their manager takes all the responsibility, and they just do what they tell them to do.

Finally, it is essential to take a look at the leadership style - it is crucial for leaders to coach, help, inspire, motivate and stimulate; to be action-oriented and that give feedback about improvement actions undertaken. A participative style can be used as an advantage to the decentralization of decisions and the communication process to involve employees. Leaders should become facilitators instead of barriers to organizational innovation and change.

4th Dilemma: "Organizations need to promote individual knowledge sharing among all organizational actors, but organizations do not see the need to create mechanisms to promote this sharing."

Top management can have an essential role in the promotion of dialogue, creating conditions whereby people are willing to apply their knowledge, share it and exchange it with each other. Developed knowledge can be continually documented through reports, images or even metaphors, and made available to everyone in the organization.

Informal contacts, internal lectures, conferences, problem-solving and project review meetings, dialogue sessions, internal rapports, and memos are an essential means to share knowledge. Organizations can also use some mechanisms that facilitate knowledge sharing: the Internet, the Intranet, the library, comfortable meeting rooms, an auditorium, an electronic archive, and even a documentation system.

To reinforce the dialogue, organizations can develop a proactive competence policy, which may include internal and external training, courses, working conferences, symposium, seminars, and informal employee contacts.

The organization can also create networks of knowledge with workers with different backgrounds for developing new knowledge using several processes to develop and share knowledge like using images, metaphors, and intuitions.

Not only do the internal actors perform a relevant role in the process of organizational innovation and change, but also external actors, like universities, consultant companies, trade unions, and others. As innovation agents, their involvement can be significant for the organizational development itself.

5th Dilemma: "Researchers and practitioners recognize knowledge as a fundamental asset to an organizations survival. However, organizations do not integrate and effectively use new knowledge created or developed by employees."

In some organizations, knowledge is continually being implemented and incorporated into new products, services, and processes. For instance, processes like benchmarking are done systematically to gain new knowledge and develop new practices or new business models.

The organization itself promotes critical thinking development and applies it in the workplace, constantly developing employees' knowledge through training, coaching, and talent development programs.

However, some organizations have difficulties in integrating and effectively using new knowledge in the job description. Even workers and managers rarely use knowledge from training courses or self-development processes.

Also, an essential dimension for knowledge integration is the need for a coherent company-wide social identity instead of a multiple community or group based social identity in order to promote useful knowledge integration in organizations (Willem et al., 2008).

Competitive Intelligence Definition

Competitive Intelligence (CI) is an area of investigation and corporate applicability that is expanding, reaching a clear peak in knowledge, as can be seen in worldwide research and many of the more recent corporate procedures.

Competitive intelligence is a part of the everyday economic competition of various subjects (companies, national economies, multinational concerns, economic integrations, etc.) at both domestic and world markets. For many analysts, the critical feature of competitive intelligence is that it functions are based on strict ethical codes and standards, in other words, that it uses legal devices for collecting and analyzing data and turning them into the economic knowledge of a company or country.

Today's society is built on a complex system of information and competitiveness. It is therefore crucial for an organization survival to be able to process quickly and systematically, large volumes of data regarding the surrounding context and converting these into corporate knowledge, allowing some room to anticipate external, internal and market changes.

CI is based on gathering, analysis and processing data and information regarding the economic information about the market, competitors, current economic development, consumers, customers, suppliers, government, regulators, partners, and all the surrounding entities or factors, in order to obtain a competitive advantage in a specific organizational context.

The concept of Competitive Intelligence began drawing more attention in the 1960s, when it was mostly looked at as a corporate procedure of gathering and processing information about internal and external data, to obtain a strategic advantage to benefit the overall strategic plan. Basically, during the '60s and '70s, CI activities were associated with data gathering, informal and tactical activities.

After the 80's all the analysis around the competitors and industries became popular where CI converted from informal activities to marketing and strategic functions.

After the 1990s CI assumed a more strategic position than other functional areas such as Market Research to most of the sectors but especially to Life sciences. Market Research activities were not able to provide more strategic and decision matter intelligence, providing content with a lack of context and lack of follow up strategic items. CI receives moderate attention from top management and is often a valuable contributor to strategic decision-making.

CI concept and scope has been frequently studied in the last years in developed countries, being considered a subject of investigation before the organizational performance. Therefore, CI is the primary influence in the strategic planning and competitive advantage in organizations, not acceptably inserted in other information models or strategic management, but, a systemic and cyclic process of corporate intelligence, being classified as a product and as a procedure.

If knowledge is the source in competitive advantage, then the access to information is used to create knowledge, and the process used to retain and transfer that same knowledge is vital for the institution. In an efficient organization, since the moment that knowledge is absorbed and processed, it originates entirely new knowledge, and it is a force to create intelligence. This same intelligence is the result of the collective cognitive process inside the organization. The culture, society or each situation inside

Pharmaceuticals and Life Sciences

the culture and society, determine an individual's intelligence, which in its turn is affected by ones the values and believes and the interaction between all these factors. With the constant evolution of science, intelligence is still at the heart of many research and investigation; still, there is not a common ground or understanding as to what is intelligence.

The primary or internal sources are those that can be obtained by personal contact with people and specialist (Analysts, consultants, journalists, and others), customers, suppliers, and employees. These sources are prone to direct contact and create a competitive advantage, which makes them intuitive and informal. Primary sources account for almost 90% of the analyzed information in the CI procedure. The secondary sources or external pertain to the information that is widely available publicly, such as databases, publications, legislation, radio, television, interviews, technical reports, patents, among others. These sources account for almost 10% of the information analyzed in the CI processes.

The CI system allows close monitoring of both the external and internal environment. By monitoring the opponent's external environment, it analyzes their potential, suppliers, negotiation potential with clients, new threats by new players, products or services. CI also incorporates macro environmental factors, such as political, economic and social, that directly affect the company in all the industrial segment and services. The uses of CI are very broaden: in marketing, where the constant search for new products and opponents is frequent; in the production departments, where there is a constant quest for competitive costs and procedures; in human resources, where the institution's HR policies are compared to those of the market.

Most of the pharmaceutical companies are already understanding the overlap and scope among Market Research (MR), Market Access (MA) and Competitive Intelligence activities, improving as well all the understanding around CI and how it can be integrated with the other functions available for companies.

So the action to proceed with an understandable integration between MR, MA, and CI for some pharmaceutical companies should come in a structured way and having formal boundaries for each of the scopes. Nevertheless, the full understanding of each function scope and operations should be perceived as the main factor to have an efficient combination between all.

Currently, in several pharmaceutical companies, CI plays already a fundamental paper in the operating strategy, where for several cases of M&A or R&D complex processes, is CI the main responsible for the positive and effective strategic impact on those operations.

Competitive Intelligence and Its Influence in Innovation

Innovation is the mechanism through which new products, services, and systems arise, that are necessary to keep up with a regular market, technology, and competitiveness.

Many of the worldwide organizations have developed sophisticated CI abilities, that represents a constant search for opportunities and threats, which allows for greater corporate knowledge and promotes innovation that accompanies the organizations' strategic planning.

Many organizations support their development and their business's importance in a constant effort to differentiate, adapting to change and trying to obtain the most return on investment. Another solution is innovating procedures, products, and services, with the help of CI, to allow for a better understanding of the competitive surroundings. CI is used as an objective factor in the competitive advantage of many current organizations, where through an adequate perception of the surrounding context, better knowledge, and understanding of the various influences around and the market risks are understood (Figure 1).

Figure 1. Dynamic tree influential level between innovation, competitive intelligence and strategy

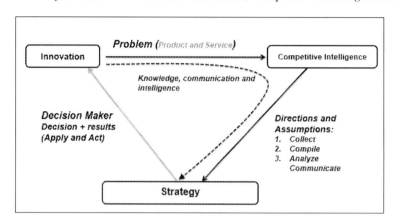

As a result of the more excellent knowledge of the variables that surround the organization, also the internal processes regarding competitive advantage are helped by this new knowledge offered by CI. In the last few years, several corporate systems have appeared to aid the decision centers, in data analysis, market behavior and tendencies statistic models, gathering and processing of data. CI is a transversal way to answer all of them, and it also completes the decision and analysis center, almost simultaneously as the market and its external variables. Therefore, CI can be considered as knowledge generating tool to the innovation process, market observation, customer's strategic behavior analysis, suppliers and opponents, as well as all the external environment and market necessities. Since CI is focused on gathering and processing information to gain a strategic advantage to the leading corporate decision makers, it facilitates the creation of strategies and solutions towards the critical stages of the innovation process, with a financial and strategic perspective of the investigations and development assumptions. The innovation is evaluated, monitored and controlled employing using the knowledge acquired by the organization, for innovation to reach its higher potential, the market must have a necessity to obtain this product or final service. This necessarily implies a duty for innovation to have the maximum possible knowledge about the market, to create these necessities in the customers or the market.

In the model above, the search for market knowledge with high potential, creates an internal and external corporate quest that allows new products to be developed, that match the markets necessities.

Even though organizations create internal market necessities, through innovated products or services, another necessity still exists as a result of the competitive markets nature. This necessity is based on the competition towards the opponents and towards the customers' needs. CI can also aid a pharmaceutical organization in helping to understand the market, as an opponent develops their unique capacities and strength in the market. Another advantage in using CI for innovation in products or services is its abilities to gather and analyze consumers' perceptions and opinions towards specific products and services available. This ability can directly change the way innovation is made and developed, allowing for information to be collected through CI in response to the consumers' specific innovation. Studies conducted regarding CI shows that organizations that have developed CI systems are better prepared for competition than others, also creating a competitive advantage through innovation. It is therefore sensible to conclude, that after this review on CI and its influence on innovation that is does play a crucial role in innovation, in organizations that intend to gain a competitive advantage in the market and to gain added value to their products and services.

Pharmaceuticals and Life Sciences

Organizations that have CI procedures are capable of gaining considerable competitive gains in innovation, as compared to their opponents. This advantage results in lower costs in the overall business procedures to improve the general perspective on the market and create sustainable innovation.

Case Study: Omega Pharma and Competitive Intelligence Context

As already described, the current global life sciences sector constraints and difficulties are increasing the importance of new commercial models, new sales and marketing analytics solutions, new business effectiveness methodologies, competitive and scalable systems, more efficient business processes and more flexible and agile tools.

Although it is not always clear what the best commercial model is, many pharmaceutical companies are looking at a range of options, from total re-organisation, to the use of alternative selling channels (such as e-detailing, Closed Loop Marketing (CLM), digital platforms, telemarketing or merely a full re-organisation of the entire commercial systems) in order to engage more successfully with all their customers, physicians, payers and partners.

Currently, several pharmaceutical companies are being able to collect in a very efficient way several details from a different type of important sources by simply using Closed Loop Marketing (CLM) or digital service portals into their commercial strategies. However, the gaps exist between the collecting phase up to the decision-making phase, where the focus of the decision problem falls in the analysis phase.

As a consequence, marketing teams are getting isolated from the senior management levels, IT departments, external analytical influencers, and internal or external commercial players. Is being asked to the marketing teams to react as intelligence connecting bridges with the other involved departments, being not only the brand and products management decision makers but also the intelligence and market insights experts. However, the real scenarios do not place Marketing as the strategic internal partner inside an organization.

Many pharma companies are not being fast enough in the data analysis to be able to make meaningful decisions, effectively failing to close the loop. Nowadays in some of the cases, the sector is relying on CRM solutions to capture different types of market indicators, opportunities, customer segmentation, targeting and profiling, organizational understanding and market access conditions on the various levels of activity.

Holistic CRM solutions have as main premises the joining combination of business process management (BPM), Competitive Intelligence (CI), Business Intelligence (CI), Multi-channel Marketing (MCM) and as well social CRM. There is an evident need for holistic solutions that can support the business services integrating all the offering aspects into robust customer experience and customer-centricity vision. This is particularly true when the business owner (Marketing, Sales or Market Access) is the primary driver of any solution and essential part of an initiative to shift how the enterprise interacts with customers.

From another perspective, we have today a more meaningful perspective about data information, where new forms of analysis have appeared in order to use the concept of information science with Big Data new capabilities and data science mighty analytical and statistical power.

The combination of new methods of analysis for decision support system such as Big Data brought another type of value to innovation and technological development for any pharmaceutical company. Today's pharma companies can collect several patterns and data sets from the marketplace using the most recent technology available to improve traditional business methods, such as iPad platforms, where new mobile sales force automation (SFA) is having an incisive effect on overall pharma sales and marketing activities.

While new systems and new IT paradigms are bringing dramatically new capabilities into commercial and marketing interactions, there are visible signs that the ability of the sector performing meaningful analysis and create strategic outcomes that can bring value to the decision makers is not being useful and productive.

Although the main job of competitive intelligence is to support management decision making, having a formalized competitive intelligence system in place can help any pharmaceutical company address several different issues.

Currently, pharmaceutical organizations not only have technology supporting the main critical needs but also have human resources knowledge around new powerful concepts like Big Data and Data Science to bring intelligence and insights through data and information. However, the gap resides in the connecting bridge between the Information technology systems and the human resources, where is fundamental to have processes and organizational structures to support and build consistent frameworks and methodologies to have on a systemic pattern critical insights, decision resources, analysis, and answers to support the business strategy and decision committees. Thus CI processes are strongly related and from a technologic perspective to business intelligence (BI) and from a business perspective to knowledge management (KM). BI and KM can be both perceived as critical inputs to competitive intelligence processes.

A systemic CI process allows to anticipate changes in the marketplace, anticipate actions of competitors, relocate in an efficient matter investment from R&D operations and initiatives, monitoring of new technologies, products, and processes that impacts the business and monitoring the political, legislative or regulatory changes.

CI outcomes can be described as cutting-edge data collection sets and ethical and professional human analysis that allows insightful competitor analysis to the next corporate decision-makers at a reasonable strategic point. Therefore, CI should not be considered as a function, but as a process, that should appear in several aspects of the business and as one seamless process not relegated to one area, division or unit.

Omega Pharma is well positioned to evolve already existing processes with new CI capabilities or support the integration of CI with another system, allowing the organization to move to a single and unified CI strategy.

Omega Pharma understands that in order to successfully realign the business model to respond to these pressures, Life Sciences marketers must find more effective ways to reach and understand their audiences. Dynamically addressing customer needs is difficult in highly regulated industries, presenting challenges never previously encountered. Marketing success in this environment requires different thinking and different capabilities. In order to provide real value to the customer, marketing requires internal collaboration with the full spectrum of stakeholders and external collaboration with health systems, biotech, academia, payers and governments.

Omega Pharma Services Include:

- **Marketing Center of Excellence Strategy:** Analyzing and refining business process and technology to deliver a complete view of key influencers in the healthcare network; Utilizing new sources of information and evolving technologies to deliver on the promise of a Marketing COE

- **Advanced Analytics:** Helping teams more rapidly integrate disparate sets of data utilizing novel technologies to provide actionable insights in very short timeframes
- **Multi-Channel Marketing (MCM):** Identifying and leveraging new channels to communicate with patients, advocates, and health care practitioners, and using knowledge gained through those interactions to evolve the customer experience to become a "human" experience
- **Data Harmonization and Standardization**: Evaluating the acquisition and integration of data assets to ensure data is purchased and integrated once thereby minimizing cost and increasing speed to deliver

FUTURE RESEARCH DIRECTIONS

New expectations of CI and knowledge sharing are emerging as this research field is becoming strategically important for business organizations. In this context, a consistent framework needs to be developed. Further research could also be undertaken:

- Studies on CI integration across organizational functions and in other types of organizations.
- Studies that develop and test a theoretical framework that relates CI mechanisms, situational characteristics, and organizational outcomes.
- Studies that analyze the capabilities of employees' in order to achieve efficient integration of CI into work practices.

Finally, more research is needed on the facilitation of the different types of CI tools and systems and their impacts on the innovation process of organizations.

CONCLUSION

CI and knowledge are fundamental for organizations, particularly, in encouraging the innovation process and the implementation of new practices and processes.

In this context, two elements need to be managed together: people and knowledge. Assuming that people are the source of knowledge, practices such as communication, skills development and recognition are core to promote the sharing of knowledge.

It is essential to implement mechanisms for systematic involvement of employees, either through meetings, technological platforms allowing discussion forums or specific systems of knowledge management.

Also, it is necessary to highlight the importance of CI as support for problem-solving and decision-making, identifying new solutions and routines in order to develop conditions to implement new management practices and organizational changes for better and more competitive strategies.

REFERENCES

Abernathy, W., & Clark, K. B. (1985). Innovation: Mapping the winds of creative destruction. *Research Policy*, *14*(1), 3–22. doi:10.1016/0048-7333(85)90021-6

Aldrich, H., & Fiol, C. M. (1994). Fools rush in. The institutional context of industry creation. *Academy of Management Review*, *19*(4), 645–670. doi:10.5465/amr.1994.9412190214

Baker, W. E., & Sinkula, J. M. (2002). Market orientation, learning orientation and product innovation: Delving into the organization's black box. *Journal of Market Focused Management*, *5*(1), 5–23. doi:10.1023/A:1012543911149

Calof, J. (2006). The SCIP06 academic program - Reporting on the state of the art. *Journal of Competitive Intelligence and Management*, *3*(4).

Calof, J. L., & Beamish, P. W. (1999). The right attitude for international success. *Business Quarterly*, *59*(1), 105–110.

Caroli, E., & Van Reenen, J. (2001). Skill-biased organisational change. Evidence from a panel of British and French establishments. *The Quarterly Journal of Economics*, *CXVI*(4), 1449–1492. doi:10.1162/003355301753265624

den Hertog, J. F., & Huizenga, E. (2000). *The knowledge enterprise. Implementation of intelligent business strategies*. London, UK: Imperial College Press. doi:10.1142/p117

Drucker, P. F. (1974). *Management: tasks, responsibilities, practices* (p. 65). Oxford, UK: Butterworth.

Edquist, C. (Ed.). (1997). Systems of innovation; technologies, institutions, and organisations. London, UK: Pinter European Commission. 1996. The Green Book on Innovation. Brussels, Belgium: EU.

Fahey, L., & Prusak, L. (1988). The eleven deadliest sins of knowledge management. *California Management Review*, *40*(3), 265–276. doi:10.2307/41165954

Freeman, C. (1987). *Technology policy and economic performance: lessons from Japan*. London, UK: Pinter.

Fuld, L. M. (1995). *The new competitor intelligence: the complete resource for finding, analysing, and using information about your competitors*. New York, NY: Wiley & Sons.

Fuld, L. M. (1995). *What competitive intelligence is and is not. Your competitors*. New York, NY: Wiley & Sons.

Fuld, L. M. (2006). The secret language of competitive intelligence. New York, NY: Crown Publishing, Random House.

Hamel, G. & Prahalad, C. K. (1989). Strategic intent (3rd ed.). Harvard Business Review.

G, Hamel,. & Prahalad, C. K. (1990). The core competence of the corporation. *Harvard Business Review*, *68*(3), 79–91.

Wulong, G. & Gera, S. (2004). *The effect of organizational innovation and information technology on firm performance.* Statistics Canada, Catalogue No. 11-622-MIE No. 007.

Jacobs, R. L., & Washington, C. (2003). Employee development and job performance: A review of literature and directions for future research. *Human Resource Development International, 6*(3), 343–355. doi:10.1080/13678860110096211

Jashapara, A. (2007). Moving beyond tacit and explicit distinctions: A realist theory of organisational knowledge. *Journal of Information Science, 33*(6), 752–766. doi:10.1177/0165551506078404

Kalakota, R., & Robinson, M. (2001). *M-Business: the race to mobility.* New York, NY: McGraw-Hill.

Kearns, G. S., & Lederer, A. L. (2003). A resource-based view of strategic IT alignment: How knowledge sharing creates a competitive advantage. *Decision Sciences, 34*(1), 1–29. doi:10.1111/1540-5915.02289

Kuhn, T., & Jackson, M. H. (2008). Accomplishing knowledge: A framework for investigating knowing in organisations. *Management Communication Quarterly, 21*(4), 454–485. doi:10.1177/0893318907313710

Leadbeater, C. 2003. Open innovation in public services. The Adaptive State - Strategies for personalizing the public realm. In T. Bentley & J. Wilsdon (Eds.), *Demos Collection,* 37-49. Retrieved from http://www.demos.co.uk/files/HPAPft.pdf

Lundvall, B. A. (Ed.). (1992). *National systems of innovation: towards a theory of innovation and interactive learning.* London, UK: Pinter.

Morris, G., & Beckett, D. (2004). Performing identities: the new focus on embodied adults learning. In P. Kell, S. Shore, & M. Singh (Eds.), *Adult education @ 21st century. Studies in the postmodern theory of education.* New York, NY: Peter Lang.

Mulgan, G., & Albury, D. (2003). *Innovation in the public sector. Strategy Unit, Cabinet Office, 1,* 40.

Murphy, M. (2002). Organisational Change and Firm Performance, DSTI/DOC (2002)14. Paris, France: OECD.

Nelson, R. R. (Ed.). (1993). *National innovation systems: a comparative analysis.* Oxford, UK: Oxford University Press.

Nonaka, I., Von Krogh, G., & Voelpel, S. (2006). Organizational knowledge creation theory: Evolutionary paths and future advances. *Organization Studies, 27*(8), 1179–1208. doi:10.1177/0170840606066312

Sousa, M. J. (2009). *Knowledge dilemmas: the case of two Portuguese organizations.* (Doctoral thesis, University of Aveiro).

Sousa, M. J. (2010). Dynamic knowledge: An action research project. *The International Journal of Knowledge, Culture and Change Management, 10*(1), 317–331.

Sousa, M. J. (2013). Knowledge profiles boosting innovation. *Knowledge Management, 12*(4), 35–46.

Sousa, M. J., & González-Loureiro, M. (2016). Employee knowledge profiles – a mixed-research methods approach. *Information Systems Frontiers, 18*(6), 1103–1117. doi:10.100710796-016-9626-1

Takeuchi, H., & Nonaka, I. (2004). *Hitotsubashi on knowledge management.* Singapore: John Wiley & Sons.

Van de Ven, A. H. (1986). Central problems in the management of innovation. *Management Science, 32*(5), 590–607. doi:10.1287/mnsc.32.5.590

Van de Ven, A. H., Policy, D. E., Garud, R., & Venkataraman, S. (1999). *The innovation journey.* Oxford, UK: Oxford University Press.

Von Krogh, G., Ichijō, K., & Nonaka, I. (2000). *Enabling knowledge creation: how to unlock the mystery of tacit knowledge and release the power of innovation.* New York, NY: Oxford University Press. doi:10.1007/978-1-349-62753-0

Willem, A., Scarborough, H., & Buelens, M. (2008). Impact of coherent versus multiple identities on knowledge integration. *Journal of Information Science, 34*(3), 370–386. doi:10.1177/0165551507086259

Chapter 15
A Business Intelligence Maturity Evaluation Model for Management Information Systems Departments

Josenildo Almeida
Universidade Salvador, Brazil

Manoel Joaquim Barros
https://orcid.org/0000-0003-0719-2802
Universidade Salvador, Brazil

Sérgio Maravilhas-Lopes
https://orcid.org/0000-0002-3824-2828
IES-ICS, Federal University of Bahia, Brazil

ABSTRACT

One of the biggest challenges for managers today is decision making. The adoption of technological solutions to obtain information more easily and intuitively is increasing, so decisions are taken with greater coherence. In this aspect, Business Intelligence (BI) appears as a tool that extracts, transforms, and enables data to be crossed to assist managers in making decisions. This chapter proposes a BI maturity assessment model to assess the level of this phenomenon in the management of the Information Technology (IT) area to verify the main reasons why the IT managers of a company from the private sector in the city of Salvador, Bahia, Brazil, do not use BI tools in their management practices whereas their clients implemented such processes in the last two years. As a result, the level of maturity reached was 01, denominated empirical management or without maturity.

DOI: 10.4018/978-1-5225-9993-7.ch015

INTRODUCTION

The adoption of an agile management model, currently, is needed for any organization, which wishes to promote a management with focus on results, in an effective and efficient way. This is because the demands and implementation of solutions happen in a brief time.

In the competitive scenario, where one of the business differentials is the speed in managerial actions, decision-making has become fundamental for organizations. In this regard, the area of information technology (IT) contributes by providing a solution that seeks to integrate all information available in enterprise management systems and other systems of the company, Business Intelligence (BI) is a concept of tools that aims to group and present information from the most varied areas, allowing analysis through reports and screens created by the users themselves according to their specific needs.

Since the advent of BI in organizations, managers have a set of tools that extracts, transforms and enables the cross-referencing of data to assist them in their strategic management and decision support. Such tooling is developed and kept by the IT sector and, in some cases, IT managers do not use BI in their management practices.

In the IT sector of organizations, the reality is no different. The fact that it is a strategic sector, where the control of the productivity and resources used directly affects the strategic aims of the organization, imposes that the IT sector needs, more than ever, tools to monitor and control its operations, besides decision-making process. Another point to highlight is the great difficulty for IT managers to deal with the decision-making process, which justifies the need to adopt a technological solution, such as BI, where it is possible to obtain information more easily and intuitively, to make decisions more consistently. This environment forces the manager, because he does not have a set of technological solutions to help him in the decision-making process, to make decisions based only on historical data or individual experiences.

To understand the reasons why IT managers do not use BI in their management practices, it is first necessary to evaluate the level of maturity of use of this tool. Talking about maturity implies an evolutionary process to achieve specific skills or a final goal, that is, it is a measuring instrument between the current state and the intended state. Such measurement is possible through maturity models.

The scenario, presented above, is a common reality in several organizations. The locus of study of this research, the Municipal Institute of Public Administration (IMAP), presents these characteristics. The organization supplies technological solutions for management and, within these solutions, BI modules are offered to city halls, to assist public managers in the decision-making process.

The starting point of this research is based on the fact that IMAP's IT industry supplies and supports BI solutions for these customers and some internal industries; however, IT managers themselves do not make use of these solutions in their management work. Faced with this prerogative, some assumptions were raised through literature consulted and expert assessment, such as: lack of investments; culture in decision-making; lack of training of professionals in the sector; insufficient time to plan their decisions / actions; IT managers evaluate which BI tools are targeted to both lay and end users.

The general objective of this study is to understand the main reasons why IMAP IT managers do not use BI in their management practices after their implementation. Together with the central aim, the specific objectives are: to propose a model of BI maturity focused specifically on the IT area; measure the BI maturity level of the IT industry of IMAP; to raise the possible reasons why IMAP IT managers do not use BI in their management practices.

BUSINESS INTELLIGENCE

The term Business Intelligence (BI) was created in the 1980s, coined by the Gartner Group, and its main features are: extracting and integrating data from multiple sources, making use of experience, analyzing contextualized data, working with hypotheses, searching for cause and effect relationships, transforming records and data in useful information for business knowledge. BI was already foreseen in the old information engineering, that is, as a company, over time, accumulate operational or transactional data in its databases, there will come a time when this data should be used to support the tasks to be performed (Rosini & Palmisiano, 1998). In this paper, we present the results of the analysis of the strategic decision-making process at the strategic level of the company.

For Inmon (1997), the concept of BI was used since the time of ancient peoples, who used the principles of this concept, crossing information from nature to improve the life of their communities. Using BI, you can format your own information and also connect it to others in order to get a better analysis and a better result with your use, that is, you can become more independent in the search for appropriate information, without requiring separate reports (Mcgeever, 2000).

Turning to the company's business environment, Geiger (2001) states that BI is the complete set of processes and data structure with the aim of supporting strategic analysis and decision making. Kimball and Ross (2002) reinforce the concept, saying that the term is related to the collection of information relevant to managers, aiming at making better business decisions.

For Barbieri (2011), the BI concept is even broader. The author understands it as the use of various sources of information to define strategies of competitiveness in the business of the company. In this definition, the concepts of data structures, represented by traditional databases, data warehousing (DW), and data marts, can be included in this definition, aiming at the relational and dimensional treatment of information.

Still within the perspective presented above, the idea of BI is to assist the end user to access and analyze diverse sources of information, structured and unstructured, that must be organized so that they can be centralized and made available to users at any time and in any location (Serra, 2002).

For Ferraz (2009), the main components of BI are: (a) ETL (extract, transform and load) tools: set of tools responsible for extracting data from transactional systems and / or any data source that contains contextualized information for decision making, transformation of these extracted data according to the indicators defined by the managers and data load, already transformed into the data repository; (b) Data repository (DW): large data repository, responsible for receiving the transformed data, storing the history of several sources, generating integrated information; (c) OLAP (online analytical processing) tools: tools used for the exploration and manipulation of DW data.

Several benefits are attributed to BI, such as: anticipating changes in the market, anticipating actions of competitors, discovering new or potential competitors, learning from successes and failures of others, getting to know their potential acquisitions or partners, getting to know new technologies, products or processes that impact your business, enter in new business, review your own business practices and assist in the implementation of new management tools (Larson, 2009).

For the purposes of this study, based on the contributions of Inmon (1997), Kimball and Ross (2002) and Barbieri (2011), we conceptualized BI as a set of technological tools for data collection and transformation to assist managers in decision-making.

To characterize the existence / presence of BI, we will consider an environment that has tools for extracting, processing and loading operational data, storing the data contextualized in a data repository (DW) and analyzing indicators through reports and dashboards, resulting from the crossing of these data.

BI MATURITY LEVELS

Talking about maturity implies an evolutionary process to achieve specific skills or a final goal, that is, it is a valuable tool to find gaps between the current state and the intended state of the person or thing. The purpose for creating maturity models is to evaluate and be a guide in an evolutionary process for companies to reach certain levels, be they methodological, procedural, or technological. They are based on the premise that people, organizations, functional areas, processes, etc. evolve through a process of development or growth towards maturity, crossing several distinct stages (Barbieri, 2011).

To be able to balance the BI investment and the value added by it, it is especially important to understand the maturity of an economic agent through a technology, process and organizational level assessment (Logica, 2009).

In conceptual terms, maturity is assumed to be a more descriptive rather than normative definition, since we cannot define an ideal state for maturity. The basic idea is that organizational maturity, related to information systems and their components, must be understood in a context (Santos, 2014, p. 41).

There are several reasons why BI maturity is assessed. Getz (2013) considers that conducting this type of evaluation is especially important and should be performed on a regular basis, since only then can it detect failures and make improvements. Rocha and Vasconcelos (2004) point out two essential reasons for measuring this maturity: the need to justify investments and the management of BI initiatives to verify user satisfaction.

Currently, there are some BI maturity models proposed, and the vast majority do not have enough information for analysis, since they are proprietary constructs and made available exclusively to the clients of the models' companies. Among the models identified in the market, we can highlight: SAS (Hatcher & Prentice, 2004); TDWI (Eckerson, 2009); HP (Henschen, 2007); Gartner (Rayner & Schlegel, 2008); AMR Model (AMR Research, 2006); Hierarchical BI Maturity Model (Deng, 2007); Enterprise BI maturity model (Tan, Sim, & Yeoh, 2011).

For the purposes of this work, a BI maturity level evaluation model was proposed based on the following models: TDWI (Eckerson, 2009), HP (Henschen, 2007) and Gartner (Rayner & Schlegel, 2008). The SAS models (Hatcher & Prentice, 2004) and BI hierarchical maturity (Deng, 2007) will not be used because they have as weaknesses the lack of deepening of the criteria to define in which stage the maturity level management is. The AMR model (AMR Research, 2006) was also not used because its focus is only on performance management and indicators, rather than BI. However, the Enterprise BI maturity model (Silva, Santos, & Gonçalo, 2017) will not be used because its focus is only on information management, leaving aside other important criteria to be analyzed in BI maturity. The strengths of the models that will not be used are like those of the models that will be used as the basis for the construction of the proposed for this study.

Most mature maturity models address BI, as does Eckerson (2009), which includes six levels and seven dimensions. This model, known as The Data Warehouse Institute (TDWI), was built in 2004, but was improved in 2007, with five growth stages: baby, child, teenager, adult, and "mature." Besides the stages, the model has two obstacles, "gulf" and "abyss", and, at the same time, aims to study maturity in three dimensions: people, processes, and technology.

The "Baby" phase is a "previous - Data Warehouse (DW)" environment, where the organization relies primarily on operational reports, which are typically static and inflexible, showing a limited number of data and processes. In addition, the visualization of the information is not dynamic, causing the end users to spend much time in their analysis. Right after the "Baby" phase, organizations face the first hurdle, called the "Gulf." Eckerson (2009) proposes the following challenges to overcome this obstacle: the feeling of executives, the adequacy of funding for BI resources, the change of culture in decision-making and the improvement of the quality of data to be analyzed.

In the "Child" phase, the first initiatives between DW and BI solutions are carried out at the departmental level, and there are some attempts to align these projects with other initiatives at the organizational level. The first BI solutions are implemented, with improved decision-making and business understanding.

In the "Teen" phase, each department creates its DW, which results in a set of repositories that do not have consistent data definitions, nor common rules and data elements. The biggest feature of this phase is that there is a greater use of BI applications by the organization. After this, organizations usually go through the second obstacle, called the "Abyss". At this stage, organizations are faced with the difficulty of managing information in a global and unique way. The challenge is to combine departmental data repositories to deliver a more consistent set of information for decision making.

Once the second obstacle is overcome, the "Adult" stage comes, where the BI team supplies adequate and targeted tools, allowing the organization to achieve its aims. It is characterized by centralized management of information, business-driven applications and a flexible DW architecture. At this stage, BI assumes the strategic resource character, based on a single DW with rules, metrics and semantics common to the entire organization.

In the last phase, called "Mature", BI is seen as a strategic resource fully coupled with the processes, applications, and strategies of an organization. In this phase, interactive dashboard reports (dashboards with charts and tables having the indicators to be analyzed) and other information services are made available on the organization's intranets.

The HP model (Henschen, 2007) aims to outline the way in which companies work to get closer to IT organizations. Its maturity model is based on Hewlett-Packard (HP) experiences with customers from various industries, differentiating five levels (with generic labels e.g. improvement or transformation) and three dimensions to assess BI maturity, namely: a) business strategy: this dimension concerns the behavior of managers in the decision process, which would be the indicators used by this management and the strategies adopted in the business; (b) program management: evaluates the portfolio management capability of tools for extracting, processing and viewing strategic data, and how they are supported by transactional systems. (c) information management: evaluates how information is classified and made available to strategic users, who will use decision support tools.

The measure used by the maturity model is targeted to (potential) HP customers and is not available for free (Côrte-Real, 2010).

At the first level, business needs are focused on improving baseline reporting and analytical action. Only the executives and managers use the analysis tools, where the focus is to consume the information of the departmental data marts. Another feature of this level is the size of the projects, which are generally small and intradepartmental (HP, 2009).

The second level is characterized by the implementation of basic dashboards and scorecards, as well as planning, budgeting and forecasting applications. The users are the same as the first level but have more autonomy to customize their reports, reducing the large amount of manual work existing in the earlier level.

At the third level, there is greater capacity for integration and alignment, which increases the value of the business. Metrics and processes are well defined. The number of users of the tool is much higher compared to earlier levels. In addition, from now on, a specific BI sector appears with well-defined programs and projects.

Because of the earlier level, the fourth level emerges, marked by the capacity of BI to transform the way processes are designed and the way of working of organizational elements. There is full integration of information and processes, and the organization begins to invest even more in analytical technologies. A single version of truth is created around the organization, that is, there will be no more than one argument for a fact.

At level five, the last level, analytical capacity is seen not only as an increment of value but as a factor of differentiation. Information management capabilities are at the highest level as well as their availability.

Rayner and Schlegel (2008) take an approach to assessing the maturity of the organization's efforts in BI and process modeling, and how these need to be changed to achieve business aims. The GARTNER model defines five levels and three dimensions, namely: processes and metrics, people, and technologies.

Level 1, called "no perception", is characterized by poor data quality, limited use of reports to support decision making, processes are not specified, and no metrics are defined for performance analysis. In an organization with this level, information management is done by the IT sector (Côrte-Real, 2010).

At level 2, called "tactical", the first investments in BI projects appear. Processes and metrics are defined, but at the departmental level, without integration. Processes are repeated and are set up by business units, but there is little consistency in the approach used at the organizational level.

After the "tactical" level, level 3, called "focused", appears, where BI projects succeed, but are still focused on perfecting the efficiency of departments and business units. From now on, there is a set of defined and documented processes for each application, but the data is not integrated.

At Level 4, called "strategic," organizations have a clear business strategy for BI development, especially in an interorganizational framework. From this level, management policies and data quality are implemented. In addition, there is a set of measures that show the performance of the processes and the need to improve them.

At the last level, called "infiltration", BI is already present in all business areas and around the organizational culture. Organizations are more flexible to change, more mature processes and fully integrated information. At present, there is confidence in the information available and users have easy access to analytics tools, generating value for the organization.

Moreno and Carvalho (2017) also point out that the models try to explain maturity in a partial way, focusing on points that they assume as critics. None of them explains maturity holistically. Another crucial point, mentioned by the author, concerns the variables presented in most of the models, which are not restricted only to technology, but also to organizational factors, processes, people and the quality of information used. Finally, the author states that the TDWI model is the most complete, not only in terms of documentation, but also in the perspectives it considers, such as: organizational aspects, through variables such as executive feeling; culture of information, financing and sponsorship; and technical and functional aspects, with the data variable.

Becker and Knackstedt (2009) consider the subjective maturity models, being later changed like any other theoretical model. These need to be completed and adjusted to the new market trends to keep their true value. In addition, they draw attention to the speed with which models become obsolete, given the technological progress, and changing business conditions.

Rajteric (2010) proved that many models studied supplied an inadequate basis of comparison, since they did not cover the entire BI field, concentrating on specific areas of analysis. On the other hand, Chuah and Wong (2011) show that there is a great chasm between the models proposed in the literature, especially with regard to a greater emphasis on technical aspects in one part of the models and, in business aspects, in the other set of models. In this sense, an analysis effort was made by Lahrman et al (2011), seeking to determine theoretical elements that approach the models from a perspective more associated to the value added by BI to the business.

Tan, Sim and Yeoh (2011) show the need for systematic evaluations of maturity measurement models to ensure an implementation that is increasingly costly, complex and comprehensive involving large segments of the organization. In turn, Raber, Winter and Wortman (2016) suggest that such an evaluation should be developed to find the level of updating of the model with emerging BI practices, as well as its level of transparency.

PROPOSAL FOR A MODEL FOR THE EVALUATION OF BI LEVELS OF MATURITY

Considering the elements collected in the proposed analysis, an important and common point, to be highlighted, is that all presented models focus on the organization, not on the analysis of a specific sector. The proposed model aims to measure the level of BI maturity in IT management practices. Therefore, the necessary elements will be extracted from the models and adapted according to the reality of the IT sector. For this, the proposed model has 04 levels, and, at each level, 03 dimensions will be evaluated, namely: people, processes, and technology. In the people dimension, the cultural aspects, degree of investment in BI solutions and the behavior of managers in the decision process will be evaluated. In the second dimension, it will be evaluated how the processes are defined and their alignment with the strategic aims. In the third dimension, the resources that the IT sector supplies to access the information needed by managers in the decision-making process and their structure to manage those resources will be verified. Table 1 is a summary of the elements extracted from each model used, divided by analysis dimension.

At each maturity level, these dimensions behave as follows:

At level 01, characterized by more empirical management - (a) the people dimension would be characterized by a culture of decision-making based on personal experience, relying in some cases on static and inflexible operational reports; the end users take a long time in the analysis, since the information visualization is not dynamic; there is no concern about the quality of data used in decision making; there is no intention on the part of the executives in investments in the area of BI; (b) the process dimension would be recognized by sector actions, that is, each sector defines its processes individually and there is no link with the other sectors of the organization; there is no concern about the quality of the data used in the decision-making process; (c) Finally, in the technology dimension, the organization relies essentially on operational reports, which are generally static and inflexible; absence of BI tools and their components.

Table 1. Conception of business intelligence maturity model

Base Model	People Dimension	Process Dimension	Technology Dimension
TDWI (Eckerson, 2009)	Cultural aspects; financing and sponsorship.	Variable of information management.	Technical aspects of applied technologies.
Gartner (Rayner; Schlegel, 2008)	Investments made.	Process integration; use of information and data quality.	Technology applied in management.
HP (Henschen, 2007)	Business strategy, which refers to the behavior of managers in the decision-making process.	Information management evaluates how information is classified and made available to strategic users.	Program management that evaluates the portfolio management capability of tools for extracting, processing, and viewing strategic data, and how they are supported by transactional systems.

(Source: Authors)

At level 02, basic - (a) the people dimension is characterized by the interest in using appropriate tools to support managers in decision making, even though culture in decision-making is still predominantly based on personal experience; there is no concern about the quality of data used in decision making; (b) in the process dimension, the characteristics remain unchanged from the previous level; (c) Regarding the technology dimension, the first initiatives to implement BI tools at the departmental level emerge.

For level 03, considered intermediate - (a) in the people dimension, managers understand that they need to rely on historical facts to improve the quality of decision making; there is a significant increase in investments in BI solutions; each sector has an individualized DW, making it difficult to manage information in a global and unique way; however, there is still no concern about the quality of data used in decision making; (b) in the process dimension, the characteristics remain unchanged from the previous level; (c) Finally, in the technology dimension, there is the creation of an environment for implementation and maintenance of BI projects, where the BI team provides appropriate and targeted tools.

At the last level, 04, the most advanced - (a) in the people dimension, BI is seen as a strategic resource; there is a single data repository (DW) for the organization and managers make decisions based on unique and global information; the quality of the data becomes trivial for continuous improvement of the decision-making process; (b) in the processes dimension, there is a unification of the departmental processes, allowing an alignment with the business objectives of the organization; in this case, data quality policies are implemented; (c) Finally, the technology dimension is characterized by a time when the information technology (IT) team manages the portfolio of tools for extracting, processing and viewing strategic data, and how they are supported by transactional systems . In addition, it evaluates how information is classified and made available to strategic users, who will use decision support tools.

METHODOLOGY

To achieve the proposed goals, a particular case study was conducted through semi-structured interviews with IT managers of the selected organization. The research had quantitative and qualitative questions, where the levels of BI maturity in the organization's IT sector were analyzed and the practices of the IT managers in the use of BI tools were deepened, trying to understand the reasons why the degree of maturity was not reached.

In addition to the semi-structured interviews with the managers, surveys were also carried out with the analysts of the IT sector of the particular case on-screen and external market experts, with the aim of triangulating the managers' discourse. This procedure presented the results of the qualitative-quantitative research with the managers for the subjects of the triangulation, so that they could confirm the revealed reasons. Like the interviewed managers, the subjects of triangulation, too, were exposed to the research presuppositions.

Starting from one of the foundations of Yin (2001), what characterizes the uniqueness of the application of the single case study, for this work, is the fact that the locus represents a rare and / or extreme case, that is, the IT organization of the selected organization provides and supports BI solutions for its clients, but does not use this tool in its management practices.

The analysis model aims to define a strategy to verify the assumptions made. Therefore, the purpose of this work is to verify the possible reasons that IMAP IT managers do not use BI tools in their management practices after their implementation in the organization. The proposed model is based on the proposition of a maturity level model to characterize the existence or not of BI solutions and instruments to verify the level of maturity of the management practices in the IT area of the organization, evaluating three dimensions: people, processes, and technology.

For the dimensions evaluated, a maturity level was assigned, according to the criteria defined in the proposed BI maturity model, and could be from level one, "empirical management", to level four, called "advanced". The next step in the strategy was to define the indicators / variables to be used in the data collection instrument, to verify the assumptions made.

After the definition of the analysis model, semi-structured interviews were carried out, which verified all the issues defined in the analysis dimensions, namely, the existence of BI solutions in management practices to understand the possible reasons for their non-existence. The field research had a quantitative / qualitative character, with affirmations of a single answer, being "I agree," or "I do not agree". The purpose of these statements, quantitatively, was to "gauge" the level of BI maturity in the organization's IT sector. If the respondent labeled "I do not agree" as an answer, he would justify the possible reasons why such a proposed condition would not meet his reality. In the moment of the justification, of qualitative nature, it is expected to understand the possible reasons why IT managers do not use BI in their management practices.

To determine the level of BI maturity, through the collected data, a vertical and horizontal scale was assigned. The vertical scale has the function of determining the percentage of the level of maturity reached, while the horizontal scale handles determining the percentage of requirements met within the level of maturity. In the vertical scale, the percentage of maturity reached can vary from "empirical management" (level 01) to 100% of maturity (level 04). The criteria used for classification were based on the requirements set up for each level. Table 3 summarizes the verification scale, which determines the degree of maturity reached.

At the level 01, called empirical management, the percentage of maturity can reach up to 25% because, even if the IT managers understand that BI processes are important for their management, the culture in decision-making is based on personal experiences, relying, in some cases, on static operational reports without the use of BI tools in management practices. In addition, their processes are sectoral, that is, they are defined in an individualized way.

At level 02, the percentage of maturity may reach the level of up to 50%, since, from this level, the investments and use of solutions that support managers in the decision-making process are started, despite the culture in decision-making still be based on personal experiences.

Table 2. Analysis model

Indicators / Analysis Variables	Maturity Level			
	01	02	03	04
People Dimension				
Use of BI tools in the decision-making process		X	X	X
There are investments in BI solutions		X	X	X
Managers understand the importance of adopting BI processes to improve decision making, but culture in decision-making is based on subjective experiences, relying, in some cases, on static and inflexible operational reports	X	X		
Process Dimension				
There is concern about the quality of data used in decision-making				X
Formalized data quality policy				X
Each sector defines its processes individually	X	X	X	
Unification of departmental processes, allowing alignment with the organization's business aims				X
Technology Dimension				
Existence of individualized data repository (DW)		X	X	
Existence of a single data repository (DW) for the organization				X
Existence of a BI team, who supplies and supports the appropriate tools			X	X
Portfolio management of BI tools and classification of information made available to strategic users				X

(Source: Authors)

Table 3. BI maturity scale

Level	Quantity of Requirements	Value per Requirement (%)	Total (%)
01	02	12,5	Until 25
02	05	10,0	Until 50
03	05	15,0	Until 75
04	08	12,5	Until 100

(Source: Authors)

The level 03 can reach the level of up to 75% of maturity, as there is an evolution with the formation of a BI team, which supplies and supports the proper tools, which assist in the decision-making process.

At the last level, called advanced, the maturity can reach 100% if, in addition to meeting the requirements of the previous level, there is a unification of the organizational processes, facilitating access to information through a single data repository (DW), besides implementation of a data quality policy, used in the decision-making process.

The horizontal scale is a complement to the vertical scale, since it has the function of determining the percentage of maturity reached for each level, from the established requirements. This is an important scale because, even if the assessed entity is classified as a specific maturity level, it is possible that not all the requirements of this level are met, determining a variation of the percentage of the BI maturity level reached. Thus, the criterion used to set up the score of each requirement was to divide the maximum percentage of the level with the number of requirements established.

For the analysis of the interviews, to understand the reasons that IMAP IT managers do not use BI in their management practices, data analysis was structured in the following steps: horizontal, vertical, and diagonal analysis. To compare the starting point with the arguments of the researched facts, the horizontal analysis looks to find the agreement and disagreement of the studied subject, based on the information obtained in the interviews. In the vertical analysis, the collected material is organized, cataloging it from the ideas of each interviewee. And, as the last stage of the process, the diagonal analysis seeks to find ideas, information, and positions that appeared during the interviews and which are related to the central hypothesis of the study (Bardin, 2002).

SEARCH RESULTS

The data collected in the interviews were used for the following purposes: to determine the level of BI maturity of the IT sector of IMAP and to understand the reasons why its managers do not use BI in their management practices. The IT managers interviewed have a background in technology and more than ten years of experience in software development. They always acted as service providers and, until then, had not held management positions. Although they always act as consultants, their profile is technical, with a focus on execution demands, not control and / or management. Currently, due to the complexities of the sector, attributed to the management position, this profile has been changed, giving greater importance to management aspects, to the detriment of technical aspects.

Maturity Level Achieved

After applying the collection instrument, based on the assumptions of the proposed maturity model, performing a quantitative analysis, it was verified that the IT sector of IMAP met only two requirements, namely: I) Managers understand the importance of adopting the BI processes to improve decision-making, but culture in decision-making is based on personal experiences, relying, in some cases, on static and inflexible operational reports; and II) The processes are sectoral, that is, each sector defines its processes individually. Therefore, in the vertical scale, responsible for determining the level of maturity, the maturity of the IT processes of the IT sector of IMAP was classified as level one (01), also known as empirical management. In the horizontal scale, which handles verifying the percentage of requirements met within a classified level, the study locus obtained a 10% maturity of BI processes, since it reached the two requirements of level one (01). The classification was determined by the following identified ratios, divided by size:

- Size People: (a) IT managers do not use BI tools in the decision-making process; (b) There are no investments in BI solutions; (c) Culture in decision-making is based exclusively on personal experience.
- Dimension Processes: (a) There is no data quality policy implemented in the sector; (b) The IT sector has looked to adopt processes that meet exclusively internal needs without worrying about the other sectors of the organization.

- Dimension Technology: (a) There is no industry data repository (DW) where managers can rely on historical data to make decisions; (b) There is no IT team in the IT industry that provides and supports the appropriate tools, nor portfolio management of BI tools, and classification of information made available to strategic users. The BI solutions deployed to the organization's customers are supported by partner companies and managed by IMAP IT managers.

Key Reasons Why IT Managers Do Not Use Business Intelligence in Their Management Practices

Analyzing the interviewees' speeches, from a qualitative perspective, based on the affirmative answers as "I do not agree", the main reasons why IT managers of IMAP do not use BI in their management practices could be extracted. The analysis took the following perspectives: I) horizontal, II) vertical and III) diagonal.

In the horizontal analysis, the interviewed managers defend the use of BI in management practices, as this could be more correct in the facts that come from the demands of the sector to make decisions. They were categorical in considering that they do not use such tooling because the organization's culture is to invest in solutions that can have a direct financial return, such as customer service, and / or in internal sectors that have strategic indicators for the organization. Allied to the cited question, another common reason found was the cost of implementation. They believe that the prohibitive costs of existing BI solutions in the marketplace are a critical barrier to negotiating with the board for defining new investments. Coordinator A also points out that it is always difficult to conquer new investments for the sector, since, in the view of the board, IT costs are high, and its results are not tangible to verify the return on such investments.

In the vertical analysis, a peculiar reason raised by coordinator B is that, even if BI is an important management tool, not having enough time to plan its actions could also be considered a barrier for their use. Currently, the priorities of the sector's demands are often determined by criteria set up by the organization's board, neglecting deadlines, resources to be used and level of complexity. Once again, the cultural issue of the organization was clear in maintaining a reactive management, that is, a focus on "putting out fires", where there is a direct interference in all actions of the IT sector, which, in turn, do not have subsidies to justify some possible planning of the actions by the managers of the area.

Another important account of coordinator B, who drew a great deal of attention, was that he pointed out, as one of the possible causes of the lack of time available to plan his actions / decisions, the fact that he had his busy time performing operational routines in the sector, or in addition to managing the team and its assets, the coordinators also effectively participate in software development activities, among other technical activities in the sector.

In the diagonal analysis, due to the lack of incentives in financial investments for the implantation of BI solutions in the sector, by the management of the organization, it was also verified the lack of training of the professionals of the sector. BI solutions deployed on clients are delivered and supported by partner vendors.

With the results of the vertical, horizontal and diagonal analysis of the interviews, it was possible to extract from IT managers the main reasons for not using BI solutions in their management practices since their implementation in IMAP in the last two years. As foreseen in the data collection procedures, in addition to the interviews with the managers, a survey of information was also carried out with the systems analysts of the sector and to two external experts from the study area. The aim of this survey is to make a triangulation between the collected discourses, to confirm the arguments pointed out by the managers and to find new arguments that possibly were not listed for diverse reasons.

By triangulating managers' discourses with industry analysts and industry experts, the following conclusion was reached: the IT managers' arguments, the lack of investment by the board, the prohibitive costs of deploying BI and the lack of time available to plan their actions, have been proven by the team of analysts. The analyst team also adds new arguments in relation to those presented by managers about the size of the team and the positioning of the IT sector as operational rather than strategic for the organization. The specialists in the field add, as a motivating element, the cultural question of IT managers, who position the sector as operational rather than strategic. Table 4 shows the triangulation between the research participants, and the possible reasons for not using BI in management practices.

In the analysis presented, we can highlight three isolated issues: lack of training of professionals in the sector; the IT sector is operational rather than strategic; and culture of IT managers. The lack of training of professionals in the sector was found exclusively from the managers' discourse, which determined, as justifications for the non-use of BI in management practices, the cost of implementation, ignoring the existence of free tools that could meet the demands of the sector. In addition, the managers reported that they do not have enough knowledge to implement such tooling in the sector and, because of that, they would have to turn to the market partners. Regarding the positioning of the IT sector as operational and not strategic, this issue was detected exclusively in the discourse of the team of analysts of the sector. IMAP IT managers do not support this statement, nor do external experts, who still emphasize that managers handle positioning the IT sector in the organization. Finally, the culture of IT managers was emphasized exclusively by external experts, who assert that managers handle positioning IT strategically and, therefore, should have a less operational and more strategic sector.

Table 4. Triangulation among survey participants

Possible Reasons	IMAP IT Managers	IMAP IT Industry Analysts	External Specialists
Insufficient time available to plan your decisions / actions	x	x	x
Lack of investments	x	x	
Deployment costs	x	x	
Team size		x	x
Lack of training of professionals in the sector	x		
IT sector is operational and not strategic		x	
Culture of IT managers			x
BI tools are for lay people			

(Source: Authors)

FINAL CONSIDERATIONS

Regarding the results of the research, data analysis showed that the level of maturity reached was level 01, termed "empirical management", meeting all the requirements of this level, reaching a total of 25% of maturity. Managers understand that the adoption of BI processes is important in their management practices, but, even so, they do not have a tool that will assist them in the decision-making process; decisions are based solely on personal experience; there are no investments to implement these solutions in the sector; their processes are being structured in an individualized way, not worrying about the integration with the other sectors; there is no data quality policy in place; and finally, there is no BI team responsible for managing a portfolio of solutions that help managers.

The research revealed that IMAP IT managers do not use BI in their management practices for the following reasons: (a) Lack of investments: the culture of the organization is to invest in solutions that can have a right financial return, such as in customer service customers, not in the internal management sectors; (b) Implementation costs: IT managers considered that the high costs of existing BI solutions are a critical barrier to negotiating with the board of directors for the definition of new investments; (c) Insufficient time available to plan their decisions / actions; (d) Lack of training of professionals in the sector.

Thus, all assumptions have been verified and proven, except for the assumption that IT managers assess which BI tools are targeted at both lay and end users. Contrary to the assumption made, IMAP IT managers advocate the use of BI tools in their practices, viewing it as a valuable strategic resource for generating insights for their management.

As a contribution, this research leaves as a legacy a BI maturity model proposed specifically for the IT area, since the existing models are geared towards the analysis of an organization, a concern not yet seen in the literature.

As limits of the study, we can highlight the use of the particular case study. This study had a deepening through the studied case, however the conclusions cannot be generalized. They serve only the context of the organization being researched.

For future work, other research could investigate the maturity of BI processes from other IT sectors, from organizations of different genres and sizes, to draw a comparative profile among managers who use BI in their management practices versus managers who make decisions based on subjective experiences.

In this sense, this work does not end, demanding more research that also aims to improve the proposed maturity model, to contribute to the improvement in the decision process, through BI tools.

REFERENCES

Barbieri, C. (2011). *BI2 – Business Intelligence: modelagem e qualidade*. Rio de Janeiro, Brazil: Elsevier.

Bardin, L. (2002). *Análise de Conteúdo*. Lisboa: Edições 70.

Becker, J., Knackstedt, R., & Pöppelbuß, J. (2009). Developing maturity models for IT management – a procedure model and its application. *Business & Information Systems Engineering*, *1*(3), 213–222. doi:10.100712599-009-0044-5

Chuah, M., & Wong, K. (2011). A review of business intelligence and its maturity models. *African Journal of Business Management*, *5*(9), 3424–3428.

Côrte-Real, N. (2010). *Avaliação da maturidade da Business Intelligence nas organizações*. Universidade Nova de Lisboa.

Eckerson, W. W. (2009). *TDWI's business intelligence maturity model*. Chatsworth, CA: The Data Warehousing Institute.

Ferraz, I. N. (2009). O Uso do Balanced Scorecard na Ótica do Business Intelligence. In *Encontro da Associação Nacional de Programas de Pós-Graduação em Administração, Anais do 37o EnANPAD*. São Paulo, Brazil: ANPAD.

Geiger, J. G. (2016). *Data warehousing supporting business intelligence*. Retrieved from http://www.cutter.com/freestuff/biareport.html

Getz, A. (2016). *Business Intelligence (BI) Maturity Model*. Retrieved from http://biinsider.com/portfolio/bi-maturity-model/

Hatcher, D., & Prentice, B. (2004). The evolution of information management. *Business Intelligence Journal*, 9(2), 49–56.

Henschen, D. (2007). HP touts neoview win, banking solutions, BI maturity model. *Intelligent Enterprise*, 10(10).

Hewlett Packard. (2016). *The HP business intelligence maturity model: Describing the BI journey*. Hewlett-Packard Development Company. Retrieved from http://h20195.www2.hp.com/V2/GetPDF.aspx/4AA1-5467ENW.pdf

Inmon, W. H. (1997). *Como construir o Data Warehouse*. Rio de Janeiro, Brazil.

Kimball, R., & Ross, M. (2002). *The data warehouse toolkit – Guia completo para modelagem dimensional*. Brasil: Campus.

Lahrman, G., ... (2011). Business intelligence maturity: Development and evaluation of a theoretical model. In *Proceedings of the Hawaii International Conference on Systems Sciences, Annals of the 44th HICSS*, Hawaii.

Larson, B. (2009). *Delivering business intelligence with Microsoft SQL server 2008*. McGraw-Hill.

Logica. (2009). The BI framework: How to turn information into competitive asset. *Logica*, 35-37, 42-44.

McGeever, C. (2000). *Business intelligence*. Arizona: Computerworld.

Moreno, A. S., & Carvalho, W. S. (2017). Business intelligence, Capacidades Dinâmicas, E Capacidades Operacionais de Marketing: Um Estudo Empírico no Setor de Telecom. In *Encontro da Associação Nacional de Programas de Pós-Graduação em Administração, Anais do 38o EnANPAD*. São Paulo, Brazil: ANPAD.

Raber, D., Winter, R., & Wortmann, F. (2016). Using qualitative analyses to construct a capability maturity model for business intelligence. In *Proceedings of the Hawaii International Conference on Systems Sciences, Annals of the 45th HICSS*, Hawaii, 4219-4228.

Rajteric, I. H. (2010). Overview of business intelligence maturity models. *Management*, 15(1), 47–67.

Rayner, N., & Schelegel, K. (2008). *Maturity model overview for business intelligence and performance management*. Stamford, CT: Gartner.

Rocha, A. & Vasconcelos, J. (2004). Os modelos de maturidade na gestão de sistemas de informação. *Revista da Faculdade de Ciência e Tecnologia da Universidade de João pessoa*, (1), 93-107.

Rosini, A. M., & Palmisiano, A. (1998). *Administração de Sistemas de Informação e a Gestão do Conhecimento*. São Paulo, Brazil: Thomson.

Santos, P. C. & Prado, M. S. (2014). *Business Intelligence: Um estudo sobre o nível de maturidade em empresas de confecções de lingerie*.

Serra, L. A. (2002). *Essência do Business Intelligence*. São Paulo, Brazil: Berkeley.

Silva, F. R., Santos, A., & Gonçalo, C. R. (2017). A Influência dos Sistemas de Business Intelligence e a Business Analytics na Medição de Desempenho e Práticas Estratégicas. In *Encontro Nacional de Administração da Informação, Anais do EnADI*. Curitiba, Brazil: ANPAD.

Tan, C., Sim, Y., & Yeoh, W. (2011). A maturity model of enterprise business intelligence. *Communications of the IBIMA*, p. 1-11.

Yin, R. K. (2001). *Estudo de caso: planejamento e métodos*. Porto Alegre, Brazil: Bookman.

KEY TERMS AND DEFINITIONS

Business Intelligence: Process of collecting, treating and using information to assist managers in the decision-making process.

Datawarehouse: A large repository that stores collected and treated data from transactional sources, which assist managers in decision making.

Decision Making Process: It is the power to choose the most appropriate way to reach a particular goal.

ETL: It is one of the stages of the Business Intelligence processes responsible for extracting, processing and loading the data from the transactional sources and later stored in the data repositories.

Information: Information is configured in a resource that assigns meaning to reality through its codes and data set. It also allows solving problems and making decisions, based on the rational use of this knowledge acquired through it.

Maturity Model: Defines a maturity structure at successive levels towards continuous improvement of business processes.

Chapter 16
From the Interview "Eye in the Eye" to the "Eye in the WhatsApp":
The Impact of Social Media on the Praxis of the Press Office in Organizational Communication Projects

Cintia Medeiros
Universidade Salvador, Brazil

Vanessa Brasil Campos Rodriguez
Universidade Salvador, Brazil

Manoel Joaquim Barros
https://orcid.org/0000-0003-0719-2802
Universidade Salvador, Brazil

Sérgio Maravilhas-Lopes
https://orcid.org/0000-0002-3824-2828
IES-ICS, Federal University of Bahia, Brazil

ABSTRACT

This study analyzes how technology and social media have transformed the praxis of press advisory activity and projects within the scope of the Communication of Organizations. To this end, it finds impacts on the functions of the activity facing this new scenario caused by the emergence of social media, updating the required profile of the new press advisor. The study adopts the conceptualization of the functions of the press officer in the organizational communication made by Duarte (2009), analyzing 17 of these functions in this new context. Authors studied the praxis of each function, before and after the advent of social media. They chose these functions because they stand for the dynamics of the Press Office, from the strategic to the operational level. The study found which social media are most used by press officers to publicize actions of their organization.

DOI: 10.4018/978-1-5225-9993-7.ch016

INTRODUCTION

Since the second half of the twentieth century, the diffusion of Information Technology has caused a profound change in daily life, in addition to society's actions and thinking; with reflections on culture, scientific research, science and all segments that guide human life. In the dawn of the recent millennium, we saw a new wave in digital innovation, with the emergence and spread of networks and social media (SM). The Organizational Communication (OC), a discipline that studies how organizations within the global society process the communication phenomenon (Kunsch, 2009), received direct reflexes from this new scenario.

The Press Office (PO) is one of the tools of organizational communication, which, according to Kunsch (2009), is divided into four principal areas: marketing, internal, administrative and institutional. Institutional Communication has PO in its compound. This study proposes to analyze the transformations that occurred in the PO activity, due to the use of SM.

The so-called digital social media (DSM), platforms that allow the creation and sharing of content between people, have changed the traditional model of PO activity. The flow of information, which was previously unidirectional, from the sender to the receiver, pulverized itself, without control, nor borders. Social media, such as Facebook[1] and Twitter[2], began to affect some functions of the activity. Currently, social media are based on the mass media, in a process of demassification of these vehicles, which, compared to the earlier universe, the offline[3] world, sounds primitive.

Thus, an important change in the way this activity runs occurred. Press releases[4], previously just printed, are now digital. The press offices currently work with an online database. Important resources, such as face-to-face meetings, are mostly held in the field of videoconferencing, needing these professionals to master this technology, from new skills, with a more strategic than operational profile.

The first question, from this study, was to understand "how did social media transform the praxis of press advisory activity in the context of the Communication of Organizations Projects"? Its overall aim was to find and analyze the impacts caused in the activity of PO, in front of this new scenario, provided by social media. In addition, the research analyzed new tools, adopted by the segment, in counterpoint with those qualified as traditional, studying their impacts on the functions of the activity, to understand the required profile of the new press officer.

This study was qualitative and quantitative. For the qualitative approach, we use the Content Analysis technique for exploratory, discovery, and verification purposes, confirming or not assumptions or presumptions. For the purposes of our analysis, we adopted the conception of the functions of the press officer in organizational communication, made by Duarte (2009), of which, we evaluated 17 functions, considering the new scenario provided by social media.

The concept of social media (SM) used the delimitation of Terra (2007), which considers texts, images, audios, blogs[5], microblogs[6], communities, message boards, online discussion forums, podcasts[7], wikis[8], vlogs[9] and the like, that allow the interaction between the users and caused a change in the praxis of the Press Office.

PRESS OFFICE ON ORGANIZATIONAL COMMUNICATION

During the period of the Industrial Revolution, the first concepts of organizational communication (OC) began to form due to profound changes in labor relations in the social, political, and economic fields,

Figure 1. Communication composite
(Source: Kunsch, 2003, Adapted)

as well as the consolidation and growth of economic organizations, advertising, business journalism and public relations (Kunsh, 2009). In Brazil, organizational communication began to be timidly delineated in the second half of the 20th century because of the country's economic, political and social development fields. According to Farias (2009), the term organizational communication can be found in the 1990s, generating the concept that began to be disseminated from a more remote concept of the 1970s. Kunsch (2003, p. 149) states that organizational communication, and corporate communication "are terminologies used interchangeably in Brazil to designate all communication work carried out by organizations in general."

In its definition, Kunsch (2009, p. 55) conceptualizes organizational communication as "the discipline that studies how the communicational phenomenon is processed in organizations within the framework of the global society. It analyzes the system, the functioning and the communication process between the organizations and their diverse publics". This concept dialogues with the definition of the theme proposed by Barichello (2009), for whom the communication process includes the strategic communication proposal of the organization and the transit of the messages through media, until the subjective interpretation of the diverse publics.

The previous flow chart reveals a convergence between all OC sub-areas, which feed and interact based on a well-defined global policy, following the proposed lines, the organization's objectives, based on strategic actions and tactics of communication. The integrated organizational communication, in this way, directs the convergence of these areas, proposing a synergistic action of the activities.

Communication is present in all sectors and flows of information and processes, considered as a strategic element in organizational competitiveness. Information became an asset of companies. One of the main tools of the Organizational Communication is the Press Office (PO), whose function is to inform the most diverse public, both inside and outside the organization, about events in the company.

Duarte (2009, p. 236) states that PO practice is "historically defined by the management of information flows and relationships between sources and journalists". The author reminds us, however, that it was after 1980 that the great organizations started to value more the professionals of the area, foreseeing an important performance in the several areas of the system, expanding their functions. Farias (2009, p. 93) goes further in conceptualizing the functions of the press office:

[...] a set of actions that seek to obtain free of charge the disclosure of news events, in a positive way, on a certain object. Meanwhile, media relations strategies are defined by: a set of perceptions, actions and negotiations that allow to fine tune the news themes (generated by the organizations and the relations they establish with their audiences), media outputs and the press and the goals of the organizations, in order to create elements guiding the perception of the organization through orchestrated actions.

Nassar and Figueiredo (2007) point out that the press officer acts in an overly broad way, not just writing and sending the press releases. It must be integrated with the organization's overall aims, especially by increasing the capacity to interact with its target audience (communicators working in print, television, radio or digital media vehicles), helping spread a positive image.

Duarte (2009) suggests that large organizations have expanded communication activities, organizing the joint action of the different areas of the system. Among the solutions to articulate this integrated system, with the participation of journalists, advertisers, public relations, and planning, among others, there is usually the creation of a more forceful structure, called communication aid, resulting, in most cases, from the expansion of the press.

Duarte (2009) typifies, in Table 1, some of the functions exercised by the press officer.

SOCIAL MEDIA AND PRESS OFFICE

Under the conditions set up by the new world environment, there is a change in the praxis of the activity of the press officer, due to the instantaneousness and unlimited scope of information dissemination, which drastically reduced the time between the fact and its dissemination. This forced the advisor to appropriate new knowledge, such as the use of digital tools, to be able to work in particular with social media.

Segura (2012) reminds us that such changes, driven by digital technologies, have altered the structure of production and consumption of journalistic information, the relationship between consumer and company, reflected in the daily PO. However, it is important to stress that the modern technologies did not extinguish the tools common to the activity. According to Mafei and Cecato (2011), Twitter did not kill the press release; nor the interactivity ended with printed newspapers and should remain in communication strategies. For the authors, what happens is a sum of new tools, added to the traditional ones.

Unlike traditional media, where a limited number of agents (advertisers, advertising agencies and vehicles) concentrates control, in digital media, this control is pulverized among hundreds of millions of people. According to Romano et al (2014), on the Internet, communication is not directional as in traditional media, making content control unfeasible. They remind us that the emergence of SM has,

From the Interview "Eye in the Eye" to the "Eye in the WhatsApp"

Table 1. Functions developed by the press office

Answer the press
Follow-up of the consultant's interviews
Administration of the press office
Support for events that may interest the press
Archive of journalistic material about the organization
Production of articles
Production of internal newspaper
Evaluation of results
Database production
Production of mailing list of journalists segmented by publishers
Training courses for journalists in a given area
Clipping and news analysis
Promotion of competitions and reports
Maintaining strategic contacts with the press
Prepare dossiers
Promotion of meetings between adviser and the press
Holding of collective interviews
Conduct survey
Reporting
Write publieditorials
Do crisis management
Make texts for website
Formation of spokespersons
Promote targeted visits

(Source: Duarte, 2009, Adapted)

in fact, favored the communication process, easing the sharing of information on various subjects and, therefore, should be included in any communication mix.

Mangold and Faulds (2009) consider SM as extensions of traditional word-of-mouth, suggesting that the significant differences between traditional media and social media focus on the speed of the communication process and the number of people affected. In fact, the virtual world offers an effective alternative to the traditional composite of communication, however, needing a new posture of PO professionals, facing these new demands.

Edelman[10] and Technorati (2006) state that in communication we are shifting from the traditional top down pyramid to a more fluid, collaborative and horizontal paradigm. The State of Blogosphere study (Terra, 2011), conducted by the Technorati Institute[11], on the state of the North American blogosphere, brings a comparative between the established model and the emergent model for the PO, showing how much the communications standard is headed towards collaboration, participation of people, interaction and relevance, common attributes to web 2.0[12].

Duarte (2002, p. 238), listing assignments of the press officer, as follow-up interviews, clippings and news analysis, suggests that, with the Internet, these attributions underwent profound changes in their models, "and the professional left to be limited to the relationship with the journalists to be an administrator of the information of interest of diverse publics of the organization ". For him, still:

the aims are no longer mere exposure in the media to incorporate the notion of strategic positioning of the organization with the internal public to society, whether for marketing purposes, information, or simply image. Tasks and challenges were broadened, needing greater capacity to create and administer different communication tools (Duarte, 2002, p. 236).

There are hundreds of social media around the world and, daily, the market releases new ones. The most used in Brazil, according to the survey conducted by Social Media (2016), are, in this order: Facebook, WhatsApp[13], Flickr[14], Messenger[15], Twitter, Web Chat[16], Youtube[17], Instagram[18], Snapchat[19] and Linkedin[20].

Of the studies that sought to build a relationship between PO activity and social media, Sant'Anna and Fernandes (2008) research was one of the first to prove that the resources made possible by digital communication were still, at that point, little used by organizations. Brown (2009) suggested that the democratization of communication was the main driver of the radical change that organizations underwent in their public relations function, notably in press relations services, due to the unique character of reach, wealth and personalization from Internet. In this context, Terra (2009) found the emergence of the media user, calling it a "fifth power", since it translated into a content producer and perception generator, which interfered in an important way with communication planning and reputation of organizations, through online social networks.

Velloso and Yanase (2014) proved one of the recurrent uses of the social media relationship application and the PO. They analyzed the path taken by the consumer to access the manifestation, negative, of the after-sales service of the organizations. In this way, they have shown that, although this channel has become a means of increasing its use by the consumer population, although not neglected, the organization's competence for this communicational management strategy has been surprisingly erratic. Traesel and Maia (2014), however, point out that the State has reacted in this perspective, pressured by the impact of social media on public opinion, with important reflexes in the electoral landscape, seeking to develop a more seductive dialogue with the population, with a view to reaching of a positive public image.

All these changes in the field of journalistic media have demanded of the PO new skills linked to media planning, the production of differentiated contents, the development of new languages and the capacity to manage telepresence teams (Mick, 2015). Duarte, Rivoire, and Ribeiro (2016) find evidence showing that social networks are now widely disseminated in news production, being used in its different stages, from the provision of agenda, through content generation, to journalistic dissemination. Ribeiro (2017) suggests that the integration of social media and PO forged a change of attitudes in the professionals, to promote the use of digital strategies and convergence of media, without renouncing traditional methods, related to press practice. Finally, Duarte and Carvalho (2017) support the emergence of online PO, strengthened by the ability of new media to expand the possibilities of storage, updating, distribution of content, access, interaction and participation of interest groups in communication processes. In this context, anyone potentially can generate and circulate easily accessible, distributed content anywhere on the planet. It is the transition from the unidirectional perspective to that of dialogue and interaction, where the former emitters and receivers become communicators.

ANALYSIS MODEL

The study was qualitative and quantitative. For the qualitative research, we use the theoretical contribution of the Content Analysis (Bardin, 2010) for exploratory, discovery and verification purposes, confirming, or not, assumptions or pre-established assumptions.

For the research, the study followed the script proposed by Vergara (2012). The method asks for key words, which, in this study, were the "impact of transformations" and "social media", which served as a unit of analysis. The analysis grid focused on 17 different items, which pertain to the roles of the Press Advisor. The relevance of the elements on the studied subject was verified and cataloged these answers for a general appreciation.

For the application of the research, we chose the use of the Likert scale, a type of psychometric response scale (quantitative measure of the strength and duration of mental phenomena) applied in questionnaires, used in opinion polls. When responding to a questionnaire based on this scale, respondents specify their level of agreement with a statement, ranging from one to five levels of responses. Since 0 (zero) was applied to less frequent and 5 (five) to more frequent. The Likert Scale is the sum of the responses given to each item, usually accompanied by an analogous visual scale.

The study was based on the production of a semi-structured questionnaire, with in-depth questions, which were sent via email to 30 communication professionals working in the Brazilian market. This universe focused on professionals who had been active for at least ten years, following the transformation that occurred in the segment, with the arrival and spread of the use of social media in the segment. To define the temporal cut, the study considered a sample of respondents from the year 2004, when two of the most popular SM examples were launched: Orkut and Facebook.

In the collection instrument, questions were presented that deal with the way in which the actors perceive the effects of social media in the transformation of the praxis of the Press Office.

The issues focused on the moments before and after the use of SM in the activity. A five-point scale was used and the figure corresponding to its proposal was placed in front of each one. To measure the change in the praxis of the activity, only the answers referring to numbers 4 and 5 were considered, which showed full agreement with the research statement. We present the proposed scale model in Table 2.

For each group of questions was added the question: What is the degree of influence of SM in this transformation? In this case, the use of the five-point scale was also requested, with the figure corresponding to its proposal being placed in front of each one. As in the earlier group, in order to measure the degree of influence of SM in this transformation, we considered only the answers referring to numbers 4 and 5, which demonstrate total agreement to the proposed item of the research.

Table 2. Scale model for praxis change assessment

Level of Change	Level
The statement does not apply to the question	1
In general the statement applies little to this question	2
The statement applies to this question	3
The statement applies a lot to this question	4
The statement applies completely to this question	5

(Source: Authors)

In this measurement of impact, in order to reach the percentage that defines the degree of influence of SM in the practice of PO, we used calculations involving the simple average criteria, where the occurrence of the values has the same importance for each of the 17 questions, which were summed and divided by the number of terms added. We present the proposed scale model in Table 3.

Finally, to complete the study procedures, respondents were asked to score from 1 to 5 the social networks that press consultants most use for dissemination and contact with journalists.

DISCUSSION OF RESULTS

The results of the study revealed that social media had a significant impact on the praxis of the press office. The overall result of the quantitative analysis shows that the measured impact was 69.58%. We saw that, of the total of 17 functions studied, nine had a more clear transformation. Another eight were partially altered with the use of SM in the activity. Next, we ordered those that had the greatest impact.

The survey shows that, before SM, the telephone was, for 80% of the aides surveyed, the main way to stay connected with journalists. After this advent, the use of the telephone became the main form of contact with journalists for only 40% of this universe. The degree of influence of social media in this transformation, according to the survey, is 87%.

The study wanted to know if, before social media, crisis management was easily conducted through media control, and obtained 83% of respondents' consent. The assertion that crisis management after SM is more difficult to conduct and have its messages controlled, was accepted for 90% of respondents, with the degree of influence of SM in this transformation of 93%.

The statement "Before the SM, the press officer's feedback for the journalists was fast" had 47% agreement of the respondents. As for the statement "after the SM, the press officer's feedback for journalists is fast," 83% of the respondents agreed, with the degree of influence also of 83% of SM, in this transformation.

The study sought to know if, before SM, the search for material for the production of the articles or data for the contextualization of materials was more restricted. The proposal received the consent of 87% of the interviewees. On the other hand, "the search for material for the production of articles or data for the contextualization of materials is easier thanks to the access to the Internet", received the agreement of 94% of the universe surveyed, of 80%.

The study evaluated whether the measurement of PO work results, before SM, was performed by centimetric (measurement by centimeters in printed journals): 90% responded positively. Already for

Table 3. Scale model for social media impact assessment

Impact	Level
None	1
Little	2
Medium	3
High	4
Complete	5

(Source: Authors)

the proposition that the evaluation of the results of the PO work, after the SM, was measured by Image Quality Indexes (IQI), was affirmative for 74% of respondents, with degree of influence of social media, in this transformation, according to the survey, also of 74%.

The research analyzed whether the production of mailing lists, before the SM, was directed to print newspapers, radio and television, receiving the agreement of 100% of the universe interviewed. Another point addressed in the study was whether the production of mailing lists, after the SM, is multiple, contemplating offline media, online. The affirmation won the unanimity, with 100% of the total answers, and with degree of influence of the social media, in this transformation, of up to 93%.

The study sought to understand if, before SM, the clipping was done manually (using paper and scissors to cut, from newspapers or magazines, all news of interest of the advised). 87% of respondents agreed with the statement. Already the proposition "after SM, clipping is done online", had 100% of agreement, with degree of influence, of social media, in this transformation, 70%.

The research sought to understand whether, prior to SM, "support for events, when performed by counsel, was done in a limited way." The affirmative received the confirmation of only 37% of the interviewees. The sentence "if, after the SM, the support for events, when carried out by my advisor, is broad, integrating several actions," received the agreement of 90% of the interviewees, with degree of influence of social media, in this transformation, of 80%.

The study revealed that actions, which need more agility in solving demands, are the most impacted. This result corroborates Duarte's (2009) statements that social networks are new tools available to the PO activity, and the channels of relationship with the public should now be based on agility and interactivity, provided by them. Social media can be important allies of the press, without restrictions of time and space. As Gabriel (2009) states, social networks have existed for at least 3,000 years BC, and the difference between our current world networks is that they have caused the collapse in time and space.

Due to information technology and the massive and uncontrolled use of SM, the time scale has changed. Speed and agility are key to preventing a crisis from spreading. It was no coincidence that 90% of respondents agreed with the statement that SM made crisis management more difficult to conduct, as this condition clearly reflects today.

The consumer has become a constant source of information. In real time, the denunciations are carried out and the information is transmitted frantically. At each sharing, new information or judgment formation is added. The strengthening of social media, such as Facebook, Instagram and Twitter, led the communication to a new paradigm. Within this virtual environment, all care is needed to try to prevent the crisis from spreading like a gunpowder fuse.

One positive issue, the use of SM, was the relationship action with journalists. This shows that the fear was not confirmed that the absence of more frequent face-to-face contact would reduce the degree of interaction with journalists. This data also reveals that the new scenario has, paradoxically, boosted the relationship between advisors and journalists. Conversation platforms were expanded. Exchanges of information are faster and the urgency of responses, in times of instantaneous news and information, makes this contact continuous and narrow.

The question becomes important in that it reveals that it is not just the face-to-face personal relationships that foster good relationships. They are conquered and potentialized also in the virtual universe. The question to achieve success, in this context, continues to be to generate true information and appropriate content.

This fact corroborates with Sullivan (2012) that even in the digital age, the old rules still apply: truthfulness, precision, openness, and verification.

The feedback function for the demands of journalists, even in situations unfavorable to the organization, was well evidenced. Research has shown that the time spent for this action, in the second decade of the twenty-first century, is more than double that needed before the advent of SM. About 15 years ago, communication was slow, often done only over the landline. Now, the possibilities of contact have increased exponentially. In the city of Salvador (Bahia), it is common for radio journalists or journalists, who present live programs with popular participation via telephone or social networks, to question media advisors, requesting an immediate response to the demand. Some advisors can give the answer within the same-day schedule.

The production of the mailing list is the most impacted function of all. This is because it has increased the universe of people with whom companies and organizations must communicate. 100% of the interviewees agreed that this function has suffered great impacts, and currently contemplates a series of new actors, who should be considered, as bloggers.

Before the advent of social media, the mailing list tried to contemplate the so-called off-line media (radio, TV, and print). At present, this universe has expanded in an overwhelming way. There are hundreds of online (digital) media and blogs, in a robust database, which maps all the actors that are in dissemination. At press conferences or social events, journalists are joining bloggers and digital influencers, people who are successful on the internet, especially through videos on the YouTube channel, unveiling their opinions on a wide range of subjects.

In the last century, the data used in the texts were searched for in encyclopedias or other printed books, available in libraries, or in magazines and newspapers. Nowadays, with search engines like Google, it is possible to get the information in a few minutes and without the need for the journalist to travel. A relevant transformation that reveals the instantaneousness assumed by communication. The contact functions, using the telephone, the way of making clipping and the accomplishment of events, contemplate the picture of actions quite impacted.

The study showed that the use of telephone to contact, for example, fell by half with the advent of SM. The degree of influence of social media in this transformation, according to the survey, is 87%. This situation, however, created another difficulty: how to reduce the use of the telephone, how could the advisors confirm that a message was received? Journalists receive dozens of messages daily through the computer. What is now perceived is that the follow up action has been used extensively by SM, such as WhatsApp and Facebook, especially in inbox messages. In case of using the inbox, this process happens only between advisors and journalists, who already have set up relationships and are part of the group of friends of another contact on Facebook pages.

The clipping, which was done offline, is now performed, most of the times, online. This is because newspapers and magazines have been publishing their editions in digital format. This fact allowed the emergence of another segment in communication, electronic clipping companies (radio and TV) and social networks. Monitoring is very comprehensive and occurs in real time. And the evaluation of results has changed from the method of measurement to the measurement of IQI: a tool that measures and analyzes how the communication policies of companies and institutions are reflected in the editorial media.

The remaining eight functions analyzed had a smaller impact with the advent of SM, however, they are not less significant actions.

E-mail has gained more momentum in recent years as a contact tool. Before SM, e-mail was considered, by 70% of respondents, as the main means of communication, used for contact with journalists. This percentage today is 80%. The degree of influence of social media in this transformation, according to the survey, is not negligible, reaching 83%.

From the Interview "Eye in the Eye" to the "Eye in the WhatsApp"

In the affirmative, "while conducting a face-to-face interview, my adviser was able to persuade people who listened to him," 67% of respondents agreed with this premise. Already in the affirmative "when doing a virtual interview, my adviser is able to persuade the people who listen to him", only 40% of the respondents agreed. Here, we can see a negative effect of the use of SM in the activity. No less than 60% of respondents consider that face-to-face interviews are better for getting the message they want. And the degree of influence of social media in this transformation, according to the survey, was 50%. This was a relevant fact that was highlighted in the research. Even with the impact of social media, the category prefers to be face to face with the interviewee, without the interference of an electronic medium of communication.

For the issue of meeting promotion, the study analyzed whether the promotion of meetings between the adviser and the press was more frequent and closer to the SM: 53% considered that the affirmative applied. On the other hand, the affirmation if "the promotion of meetings between the adviser with the press is frequent and closer with the use of the SM", 60% nodded to the affirmation. And the degree of influence of social media in this transformation, according to the survey, was 57%.

The assertion that "counseling often conducted a directed visit" was considered assertive by 57% of respondents. With the advent of SM, the percentage of respondents who agree that counseling often conducts targeted visitation has dropped to 50%. The degree of influence of social media in this transformation, according to the survey, reached 33%.

The assertion that "before SM, our advice usually carried out the formation of spokesperson", was accepted by 70% of the interviewees. Already the question "after SM, our advice usually performs the formation of spokesperson," was accepted by 77% of the interviewees, with a degree of influence of SM of 53%.

The skills of the press officer were analyzed in the survey. The goal was to understand if the requirements of good text and good relationships having were maintained, even in a scenario marked by new technologies, which requires other skills of the professional. The statement that "before SM, PO needed to have good contacts and proficient writing" was considered positive by 93% of the universe surveyed. Already the statement that "the PO needs to have good contacts and texts", even with the advent of social media, received confirmation of 100% of the interviewees, with a degree of influence of SM of 50%.

The study sought to understand whether, prior to SM, survey ratios were limited to the board level of the company. Only 40% agreed with the statement. Already the question whether "the survey of guidelines, after SM, covers all areas of the organization," received the agreement of 83% of respondents, with a degree of social media influence of only 57%.

The use of e-mail, despite having gained breath, as a contact tool, did not achieve a radical transformation. Before SM, e-mail was considered by 70% of respondents as the main means of communication used to contact journalists. This percentage is currently 80%. It is noticed that, although it is still a widely used tool, others have added to this function.

The study pointed out that SM, most commonly used by PO, to stay in touch with journalists and give information from their advisors were, firstly, Facebook, followed by WhatsApp, Twitter, virtual press rooms, web chats, Instagram, Snapchat, Periscope, Messenger, e-mail, Website and Linkedin, respectively.

The interview was considered a valuable tool for all those involved in the process (interviewee, interviewer and assessor). This is because, if the journalist attributes to the interviewee a high level of reliability and content, the journalist becomes part of the agenda of sources (denomination to designate the list of possible respondents) of this professional, being constantly demanded to express opinions about subjects of the segment that acts.

In this sense, an executive or entrepreneur of a large corporation, for example, can be considered a reliable source, both to analyze its segment, as the economic moment by which the country is going through or even about tax or labor laws, that influence the economic activity. A reputed physician, for example, can become an important source for commenting on news of his specialty or on public health policies. The research showed that the commentator-vehicle relationship, before SM, allowed a closer proximity between the parties and that, now, this monitoring is faster, but cooler. However, the important thing is that the essence of the activity has not changed, only the feeling about it.

Regarding the face-to-face interview, there was a negative impact due to the use of SM in the activity. The research showed that the advisors have a better feeling of face-to-face interviews to establish a persuasive message process. The message to respondents is most effective when hand-to-hand, eye-to-eye, handshake. Even if technology has offered the possibility of contacting more people at one time, like virtual collectives, human contact still prevails, in order to establish a relationship of more trust and credibility, that is, it is lost on the one hand, it is gained on the other.

The "meeting promotion" function did not change with SM, even though a greater frequency of use of technological tools as a contribution to the activity was clear. Also, the "directed visits" function registered little change due to SM. The interesting thing about this item was the high percentage of respondents who said that they had little relation to the advent of SM.

The drop in the percentage of "directed visits" is linked to issues such as displacement in the centers. Currently, everything needs to be scheduled and only important visits are received in newsrooms.

Regarding spokesperson training, or media training, the research revealed that, even though the activity was affected after the advent of SM, it is perceived that this training is a service that is being used by organizations. The data reveals the maturity of companies facing the need to position themselves professionally in front of the press. Another important fact of the study reveals that, despite all the technological arsenal currently available in the field of organizational communication, there are elements that stay as permanent axes of the activity. Like the skills of the press officer, who needs to be multifaceted, to develop a more strategic than operational profile. Its base, however, does not change. At any time, this professional needs to have good editorial style, content, and maintain an always profitable relationship with the press.

The "stand up" function remained unchanged, with the advent of SM. Interestingly, in this specific item, the action of the press officer to seek matters to publicize the organization's projects has changed little.

The result of the study on "event realization" showed that, even with constraints, the support of event counseling is quite present. Currently, this participation becomes even more efficient, due to the possibility of using the digital mailing list better, follow up more quickly, creating virtual invitations.

Finally, the survey asked: "What are the main social media that press officers currently use to keep in touch with journalists and disseminate information about their advisors?" Facebook was billed as the primary means of communication between advisor and journalist, followed by WhatsApp, Twitter, virtual press rooms, web chats, Instagram, Snapchat, Periscope, Messenger, e-mail, website, and, finally, Linkedin.

Juliana Marinho, manager of Communication Nordeste of the JCPM Group (João Carlos Paes Mendonça), confirmed in an interview for our research, the transformation of the speed and dynamics of the entire process of communication between press and companies in recent years. A few years ago, she says, there were several vehicles to relate to and a different workload than the current one. "We perceive the results of the research presented in everyday life. The way you communicate has changed. Deadlines today are tight. Some news has come out about the company, and you need to respond immediately.

You cannot postpone the answer any longer, "she said. Marinho believes that there is a greater distance in professional and personal relations today, due to the virtual contacts. This is because there are many vehicles for organizations to relate to, more people to communicate with, and more themes to address.

FINAL CONSIDERATIONS

The study showed ruptures in some actions of the PO activity before a new scenario, with important demands. The activity of the press officer needs fundamental requirements such as good training and field skills. However, because of the instantaneousness and wide scope of the information diffusion, the assessor was asked to appropriate new knowledge and the use of digital tools to act in his professional field.

We are facing a new world. Saad (2009) confirms this context when he says that we are seeing, at the beginning of this century, a scenario never seen before innovation and absorption of technologies to leverage human communication. In the first two decades of the twenty-first century, we have seen the overthrow of some concepts that have remained unchanged for years and have been studied worldwide. The main social communication theorists (Shannon, Weaver, Johnson, Lasswell), who based the foundations of science in the early twentieth century, focused their studies on the emitter, receiver, message, and channel elements. Currently, communication does not only present one direction, that is, it does not act through a line connecting only the source and its destination. Communication is cyclical and feedback from the receiver is critical. In this way, it becomes multidirectional, decentralized, massive and abundant. It is the information society, advocated by Castells (2006, 2011). A society with a new logic of social model, based on networks, with emphasis on the role of information in its development. In this context, the area of corporate communication had to adapt to these new challenges and the activity of PO was not left out of these changes. In this study, one of the most enriching situations was the possibility of collecting with actors who experienced this moment of transition. These actors had to adapt to the transformations caused by technological innovations, and they needed to absorb new skills. However, one of the advantages of the present study was to have access to professionals who have worked in press services for at least 10 years, and have followed the development of these actions closely. They practically saw a revolution in their area.

Due to the speed with which the transformations occur in the contemporary world, it is possible to affirm that this study can gain a new breath in the medium term. As a sign for future studies, we suggest the monitoring of existing media and the constant observation regarding the possible emergence of new social media, since its impact on the praxis of the activity may cause new and unexpected transformations.

REFERENCES

Bardin, L. (2010). *Análise de Conteúdo*. Lisboa: Edições 70.

Brown, R. (2009). *Public relations and the social web*. Kogan Page.

Castells, M. (2011). *The rise of the network society*. John Wiley & Sons.

Castells, M., & Cardoso, G. (2006). *The network society: From knowledge to policy*. Washington, DC: John Hopkins Center for Transatlantic Relations.

Duarte, J. (Org.). (2002). *Assessoria de imprensa e relacionamento com a mídia: Teoria e técnica*. São Paulo, Brazil: Atlas.

Duarte, J., & Carvalho, N. (2017). *Sala de imprensa on-line*. CFN.

Duarte, J., Rivoire, V., & Ribeiro, A. A. (2016). Mídias sociais *on-line* e prática jornalística. *Universitas. Arquitetura e Comunicação Social, 13*(1), 1–10.

Edelman & Technorati. (2006). *Public Relationships*. Winter.

Farias, L. A. (2009). Estratégias de relacionamento com a mídia. In M. M. K. Kunsch (Org.), Gestão estratégica em comunicação organizacional. São Caetano do Sul: Difusão.

Gabriel, M. (2015). *Redes Sociais Estratégicas*. Retrieved from http://pt.slideshare.net/marthagabriel/redes-sociais-estrategias-e-mensurao-por-martha-gabriel>

Kunsch, M. M. K. (2003). *Planejamento de relações públicas na comunicação integrada*. São Paulo, Brazil: Summus.

Kunsch, M. M. K. (Org.). (2009). Comunicação organizacional, histórico, fundamentos e processos. São Paulo, Brazil: Saraiva.

Mafei, M., & Cecato, V. (2011). *Comunicação Corporativa*. São Paulo, Brazil: Contexto.

Mangold, W., & Faulds, D. (2009). Social media. *Business Horizons, 52*(4), 357–365. doi:10.1016/j.bushor.2009.03.002

Mick, J. (2015). Trabalho jornalístico e convergência digital no Brasil. *Pauta Geral –. Estudos em Jornalismo, Ponta Grossa, 2*(1), 15–37. doi:10.18661/2318-857X/pauta.geral.v2n1p15-37

Nassar, P., & Figueiredo, R. (2007). *O que é Comunicação Empresarial?* São Paulo, Brazil: Brasiliense.

Ribeiro, M. E. (2017). *O papel do assessor de imprensa em um mundo movido pelas tecnologias digitais*. Retrieved from www.academia.edu

Romano, F. M., Chimenti, P., Rodrigues, M. D. S., Hupsel, L. F., & Nogueira, R. (2012). O Impacto das mídias sociais digitais na comunicação organizacional das empresas. *Future Studies Research Journal, São Paulo, 6*(1), 53–82.

Saad, E. C. (2009). Comunicação digital e novas mídias institucionais. In M. M. K. Kunsch (Ed.), *Comunicação organizacional*. São Paulo, Brazil: Saraiva.

Sant'Anna, I. B. C. & Fernandes, N. C. (2008). A comunicação institucional nos *websites* corporativos: Um estudo exploratório. *Anagrama*, 1(4).

Segura, D. P. (2014). *O Impacto das tecnologias digitais sobre o processo de assessoria de imprensa*. São Paulo, Brazil: ECAUSP.

Sullivan, M. H. (2012). *Uma assessoria de imprensa responsável na era digital*. Edição da Série Manuais. Bureau de Programas de Informações Internacionais do Departamento de Estado dos Estados Unidos.

Terra, C. F. (2007). *Blogs corporativos. São Caetano do Sul*. SP: Difusão.

Terra, C. F. (2009). Usuários-mídia. In Congresso da ABRAPCORP. *Anais do III Congresso da ABRAPCORP*, São Paulo, ABRAPCORP.

Terra, C. F. (2011). *Mídias sociais...e agora? São Caetano do Sul*. SP: Difusão.

Traesel, F. A. & Maia, N. L. (2014). As organizações nas mídias sociais. In Congresso Brasileiro de Ciências da Comunicação. *Anais do XXXVII Congresso Brasileiro de Ciências da Comunicação*, Foz do Iguaçu.

Velloso, V. F., & Yanaze, M. H. (2014). O consumidor insatisfeito em tempo de redes sociais. *Educação. Cultura e Comunicação*, 5(9), 7–20.

Vergara, S. C. (2012). *Métodos de Pesquisa em Administração*. São Paulo, Brazil: Atlas.

KEY TERMS AND DEFINITIONS

Identity and Corporate Image: Image is how the company is seen by the look of your audience. Identity does not vary from one audience to another. It consists of the attributes that define the company, such as its personnel, products and services. Identity is the visual manifestation of your reality, transmitted through the name, logo, slogan, products, services, facilities, brochures, uniforms, among others.

Information Technology: Since the second half of the twentieth century, the diffusion of Information Technology has brought about a profound change in customs, everyday life, in the way society acts and thinks, reflecting on culture, scientific research, science and all segments of society, human life, with an extraordinary advance in the field of technology.

Organizational Communication: Organizational Communication is a discipline that studies how the communicational phenomenon is processed within organizations. In Brazil, organizational communication began to be timidly delineated in the second half of the 20th century, because of the economic, political and social development fields of the country.

Press Office: The press office is responsible for informing the most diverse public, both inside and outside the organization, about the events of the company. This activity is part of the toolkit of the so-called institutional communication. Historically, the press office is defined by the management of information flows and relationships between sources and journalists.

Social Media: Texts, images, audio and video in blogs, microblogs, communities, message boards, online discussion forums, podcasts, wikis, vlogs and the like, that allow interaction among users.

ENDNOTES

[1] Facebook - One of the most popular social networks in the world. https://www.facebook.com/.
[2] Twitter - A microblogging that allows messages of up to 140 characters. Links can be posted with texts and photos, videos and other materials. https://twitter.com/.
[3] Offline: In journalistic jargon, it refers to print media, or television and radio. The term comes from English and refers to the vehicle (or user) that is disconnected or does not have access to the Internet.

4. Press Release: material with information and news about the activities of companies.
5. A blog is an internet page that hosts the record of a discussion about certain topics, through chronological posts.
6. A microblog differs from a blog by the size of the posts added.
7. A podcast is a digital media that offers content in the format of audios.
8. The wiki is a web page that allows you to update your content in a collaborative way.
9. A vlog is a blog who posts video files.
10. Edelman is a North American company, a global leader in communication and marketing.
11. Technorati is the publisher of a platform that serves as an advertising solution for thousands of web pages in your network.
12. Web 2.0 is a term used to appoint a second generation of communities and services offered on the internet, having as concept the Web through applications based on social networks and information technology.
13. WhatsApp Messenger is a free and multi-platform instant messaging encryption application for smartphones.
14. Flickr is a web hosting and video hosting platform.
15. Messenger is an instant messaging service on the Facebook platform.
16. Web-Chat is a system that allows users to communicate in real time using easily accessible interfaces.
17. YouTube is an American site for video sharing.
18. Instagram is an application and service for sharing photos for mobile, desktop and internet.
19. Snapchat is a multimedia messaging application for mobile.
20. Linkedin is a social networking service oriented to businesses and jobs, which runs through websites and mobile applications.

Chapter 17
Patent Information Project to Leverage Innovation:
The Use of Social Media for Its Selective Dissemination

Sérgio Maravilhas

https://orcid.org/0000-0002-3824-2828

IES-ICS, Federal University of Bahia, Brazil

ABSTRACT

This chapter describes the project for the development and implementation of a theoretical support model for the creation of an information system that will allow the dissemination of scientific and technical information contained in patent documents using the web sites of industrial property official entities. The support of information resources, available through libraries and information services in universities, will be crucial for the project and the success of university research centres (URC) in Science, Technology, and Medicine (STM). To achieve a coherent program of dissemination and make possible the access to patent information by the URC, social media network (SMN) tools (like RSS, Blogs, Wikis, Newsletters) will be used. The tools will also effectively achieve control to constantly improve the system implemented.

INTRODUCTION

Patent documents are a scientific and technical world-wide information resource and cover all areas of human knowledge.

Containing information that, in 80% of the cases is not published in any other source, available for free from databases and digital libraries via the Internet, we consider that it must be disseminated by the researchers of STM to support and leverage the work of these technicians and specialists, stimulating their creative vein and, thus, enable the resolution of problems and stimulate new inventions and innovations with its visualization.

DOI: 10.4018/978-1-5225-9993-7.ch017

To this end, we developed a model to build a system that enables the selective dissemination of information on request and in accordance with the areas designated by each researcher, based on the patent system own indexing codes, supported by a social media platform.

After a brief background, we describe the features that make this resource an important aid in the work of researchers and describe some application examples in different types of industries, including the pharmaceutical industry and the health sector.

Then, we describe the main points of the project to develop to allow dissemination of this information by STM researchers, as well as the advantages to obtain.

As a final goal, we seek with this tool to create value for society in general, finding partners and companies that can take advantage of the knowledge generated and developed in academia, thus, enabling increasing the human development index of the country.

BACKGROUND

Every organization needs information to innovate and gain competitive advantage in their markets. Obtaining the latest and greatest technologies has motivated the search for information that will allow maintaining productivity, competitiveness, superiority and status against competitors (Burgelman, Christensen *et al.*, 2009; Christensen, Anthony *et al.*, 2004).

There is a very important information resource available free of charge – the patent information - which is easily accessible via the Internet, allowing to stimulate creativity that can motivate new innovations (Maravilhas, 2009; Maravilhas & Borges, 2009).

Patent information repositories available through databases and digital libraries are a huge source of scientific and technical information.

There are about 90 million patent documents published worldwide, most of all containing information not available in any other source (Albrecht, Bosma *et al.*, 2010; Bregonje, 2005; Greif, 1987; Marcovitch, 1983). Even information that can be found in other documents like scientific articles, technical reports, conference proceedings, theses and dissertations, among others, are not described with the same degree of detail and, sometimes, take longer to be available to the public.

In addition, approximately 1 million (2.2 million in 2011 and 2.6 million in 2013) new documents are created every year (Mueller & Nyfeler, 2011, p. 384), and its publication allows its analysis even before the protection is granted.

More than 30% of existing patents are in public domain (having reached the expiry date of their protection, or for non-payment of annual maintenance fees), or are not being exploited due to lack of funds or technical and/or financial inability of the holder (Godinho, 2003; Idris, 2003; Maia, 1996).

Analysing the technical information contained in patent documents, researchers, scientists and entrepreneurs can exploit precious ideas about the state of the art in any area of science and technology (Maravilhas & Borges, 2011a; 2011b).

For Idris (2003), the main reason to analyze patent information relates to the current information contained in patent documents, which can help avoid erroneous investment due to research duplications previously conducted by others.

The insufficient use of patent information has caused a considerable expenditure of research funds in research and development (R&D) projects, whose results may be threatened by the return on investment made in patented technologies already existing.

According to some estimates, European industry wastes between 20 and 32 billion euros annually (Brünger-Weilandt, Geiß *et al.*, 2011; Ribeiro, 2007), mainly by not consulting patent information, leading to a duplication of efforts, reinventing existing inventions and re-developing products and processes currently available on the market that could easily be identified if patent information was consulted (Skarzynski & Gibson, 2010).

In certain conditions, you can use patent information to develop new products and processes if the resulting invention does not infringe existing patents (Trott, 2008).

This is a perfectly legitimate - and one of the most important - justifications for the existence of the patent system (Idris, 2003, p. 88).

The advantage of consulting this technical information is to encourage creativity and innovation to transform existing inventions into innovations (Maravilhas, 2009; O'Dell, 2004).

Whether using patented technologies that are not being exploited, integrating these inventions in the corporate business plan for further development, encouraging the creation of companies to exploit these inventions, identifying technologies and finding those who want to produce them, selling customized solutions or components to include in final products, all these suggestions allow the profitability of know-how, that is not being properly exploited (Maravilhas & Borges, 2009).

Patent information is not the only source of information that can stimulate innovation, but it is one of the most important.

This is due to the degree of detail, depth, geographic and time coverage, constituents (description of the invention, diagrams, drawings, graphs), analysis of the state of the art, additional information produced by experts (the patent examiner and his search report), among others.

This source of information covers all scientific and technical activities of human knowledge, and is coded to allow its easy retrieval and use (Trott, 2008).

It promotes insights, not only from the scientific area in which we are investigating, but also from additional areas that can increase the value of the solution developed.

Also, it can allow its use in initially not considered applications and resolve intermediate crossing areas between different scientific disciplines (Skarzynski & Gibson, 2010).

It allows, also, improving the time to introduce new products in the markets, lower costs associated with R&D, and monetize the installed capacity in some industries and / or scientific activities (Maravilhas & Borges, 2009).

Searching this information will allow finding certain inventions for free exploration, without the obligation to pay any license if the patent is in the public domain and free to use.

Such is the case of generic drugs - that are the result of the free use of active substances that have reached their protection limit and are free to be used -, which has been done with undeniable success by several international companies.

Most of the information contained in patent documents is not published anywhere else, being an essential source of information to meet the current technical information needs (Bregonje, 2005; Greif, 1979, 1987; Marcovitch, 1983).

Patent literature is the biggest source of technological information, available worldwide, making it the largest technical knowledge priceless repository.

The disclosure of the secrets contained in the technical documentation of a patent application reveals valuable information to the public about the state of the art in a given area thus, promoting, through this knowledge, technological development (Maravilhas & Borges, 2009).

Whenever a research program is started, it is advisable to research all the scientific and technological work already done in that area (Naetebusch, Schoeppel *et al.*, 1994; Schoeppel e Naetebusch, 1995).

In addition, it is always worth a few years back in search to get to know many of the ideas contained in patent documents whose legal validity has expired.

They may contain valuable inventions or technical information that at the time they were invented they were way ahead of the possibilities of that epoch and have not been able to be performed, or that the market was not ready to appreciate and value (Marcovitch, 1983, p. 492).

Currently, those can be seen as opportunities that are worth being explored, taking advantage of the current knowledge and recent technological developments and, perhaps, become extremely profitable.

Apple did that and launched two of the most iconic products of our era: the iPod and the iPad (Skarzynski & Gibson (2010).

Based on the knowledge of the existence of some invention, interested parties may start licensing agreements for the exploitation of these technologies in certain geographical areas or markets where they can have greater penetration capability than the holder of the patent (Maia, 1996; Maravilhas, 2009).

Regarding market analysis and competition, if we perform Technology Watch (TW) and Competitive Intelligence (CI), patent information can provide many amazing ideas that may help understand the strategies of competition, new trends, new products or technological improvements made to existing products (Ashton e Klavans, 1997; Ojala, 1989; Wilson, 1987a; 1987b).

Marcus (1995) points out many obvious parameters that can be found in patent information, such as: a) Inventor name; b) Patent owner; c) Priority dates; d) Patent family, among others.

But, if the researcher uses his imagination, he can find a huge variety of commercially valuable information such as:

1) Possibility to find or identify potential customers or business partners; 2) Provide support information to a business meeting, allowing to know the technical potential of the company with whom we will meet; 3) Identify trends in R&D, new technologies and new products; 4) Identify trends and movements between companies, which might point to mergers and acquisitions; 5) Facilitate technology transfer or technology licensing, allowing to acquire what the company needs, or sell what the company developed; 6) Prevent the duplication of R&D projects, avoiding the inadvertent copy; 7) Identify experts in a particular field of technology or scientific area; 8) Establish new applications and uses for existing products and technologies; 9) Find solutions to technical problems; 10) Support the generation of ideas for new products or processes; 11) Identify Marketing trends; 12) Establish the expiration date of a patent, which will allow the free use of that invention; 13) Identify potential competitors; 14) Monitorizing the activities of competitors; 15) Establish the technological state of the art (Marcus, 1995, p. 65, 66).

All these possibilities can trigger a lucrative business opportunity for companies who know how to explore them, as the pharmaceutical industry shows through the development of new substances, the use of generic drugs and the use of substances developed for solving a problem in the solution of other problems.

SOLVING PROBLEMS WITH PATENT INFORMATION

According to Marcus (1995), patent information, in addition to providing an excellent source of information for generating ideas, also has the advantage of being a source of inspiration when it is necessary to find a creative solution to assist in solving problems of technological nature.

Patent Information Project to Leverage Innovation

Drucker (1987) refers to several representative examples of using an invention for applications other than that which was initially intended. Drugs developed to solve human problems eventually can be adapted to solve problems for veterinary; the creation of DDT during World War II to protect American soldiers against tropical parasites was later widely used in agriculture to protect crops and cattle; 3M developed several products for industry and only much later realized that they could be used to solve domestic problems with minor changes and improvements, as the case of Scotch Tape and Post it.

The Teflon, from DuPont, invented in 1938, and marketed since 1946, is waterproof and is the material with the lowest known friction coefficient. Widely used in aircraft wings to improve aerodynamics and prevent waste to accumulate (Drucker, 1987; Lattès, 1992), its advantages can be applied in another area of activity that requires these no sticking qualities: the kitchen. DuPont started putting this coating into pans, preventing food cling during making. When the patent expired, the technology was free to be used, leveraging competition with this knowledge and this extra advantage without any costs, like companies such as Tefal did.

APPLICATION OF PATENT INFORMATION IN DIFFERENT SCIENTIFIC AND INDUSTRIAL AREAS

Most innovations result from creative processes, individual and collective, spontaneous or intentional (Baxter, 2000). Creativity is based on understanding that produces imaginative new ideas and new ways of seeing reality (Birch & Clegg, 1999).

Innovation involves change, and requires a combination of creativity, reasoning and ability to act (Tidd, Bessant *et al.*, 2003).

The assiduous consultation of information sources, which includes patent information with all that is new and relevant in the scientific areas in question, and even in complementary areas, can allow a knowledge base which triggers a response when faced with a technical difficulty to overcome.

Also, if others before us had the same problem and have been successful, we can benefit from this knowledge and find a more profitable way to overcome these difficulties (Maravilhas & Borges, 2011a).

It may even happen that from the existing knowledge someone could find a better solution.

Another advantage promoted by patent information concerns the use of technology to solve problems in different areas from which the invention has been patented.

The Cosworth factory, of high performance engines, adapted an invention of an electromagnetic pump, developed for the area of nuclear energy, to force the molten metal to enter the molds, eliminating the air that usually originates that the metal parts get porous (Tidd, Bessant *et al.*, 2003, p. 266).

Another well-known example is the case of Nylon. This product was developed in the 30s of the twentieth century by DuPont, USA. We keep using Nylon in various articles, but the technology is about 80 years old and its patent is expired (Maravilhas & Borges, 2011a).

Teflon, invented in 1938, also by DuPont and marketed from 1946, is waterproof and is the material with the lowest known friction coefficient, meaning that it is non-sticky.

Widely used in aircraft wings to increase aerodynamics and prevent debris from accumulating, which causes vibration and turbulence, it saw its advantages being applied in another area of activity that requires these non-sticky qualities: the kitchen.

DuPont began to put this coating in pans, preventing food cling during the confection. With its patent expired, this technology can be used freely, what the competition did, like Tefal of SEB Group, taking advantage of this knowledge at no cost, allowing it to be the European market leader in non-sticky pans.

Companies like Samsung, Motorola and Apple also use this feature to support their R&D projects, with profitable results from its products (Rivette & Kline, 2000).

Another example relates to generic drugs. These are no more than the active substance of a pharmaceutical product whose patent has expired, thus becoming public domain, which can be manufactured and marketed by any other company, provided they do not use the registered trademark, if still in force, from the owner of the expired patent.

This feature is of vital importance because that information can be used by those who wish to obtain information on the characteristics of the manufacture of those materials or products, and shall not be involved in any legal proceedings, avoiding large sums in R&D. That way, anyone can manufacture those materials without having incurred in the costs and the risks that a new R&D program entails.

The advantage for us - consumers - is also clear: to obtain essential products to improve the health and quality of life, with prices lower than those charged by the company that had to recover the huge investments made to obtain the same product.

Portugal has companies[1] like Hovione, Generis, Farmoz (Tecnimede), Labesfal, Ratiopharma, GP – *Genéricos de Portugal*, Bluepharma and Almus (from the National Association of Pharmacies) who search for patent information to know what patents will expire and allow the manufacture of generic drugs.

Drugs developed for a function can serendipiously[2] be used to solve different problems.

Another well-known product of the public, with increasing use, is Aspartame.

Aspartame is an artificial sweetener developed by Searle, a company that was subsequently acquired by the American giant Monsanto and marketed under the trademark Nutrasweet.

It is about 200 times sweeter than sucrose or sugar cane. It is widely used in light soft drinks (e.g. Diet Coke), desserts and as a tabletop sweetener.

Thanks to patent information, any interested company would know how to synthesize the *aspartyl-phenylalanine methyl Ester* and, more importantly, that its patent expired in 1992, which means that it can be manufactured and marketed with any other trade name (Flynn, 2002, pp. 136-138).

By being attentive to the information that the patent expired in 1992, other manufacturers are currently market leaders in sweeteners using Aspartame.

Porter, A. & Cunningham, S. (2005, 18) share some questions that can be answered by patent information, performing tech mining, such as:

- "What R&D is being done on this technology?
- Who is doing this R&D? Toward what probable market objectives? (sometimes we focus on a particular competitor to profile what it is doing in multiple areas)
- How does this technology fit organizational aims?
- What are the prospects for successful commercialization?"

Table 1 describes in more depth what can be done with patent information using tech mining.

Table 1. Various types of technology analyses that can be aided by tech mining

A	*Technology Monitoring* (also known as technology watch or environmental scanning) – cataloguing, characterizing, and interpreting technology development activities
B	*Competitive Technological Intelligence* (CTI) – finding out "Who is doing what?"
C	*Technology Forecasting* – anticipating possible future development paths for particular technologies
D	*Technology Roadmapping* – tracking evolutionary steps in related technologies and, sometimes, product families
E	*Technology Assessment* – anticipating the possible unintended, indirect, and delayed consequences of particular technological changes
F	*Technology Foresight* – strategic planning (especially national) with emphasis on technology roles and priorities
G	*Technology Process Management* – getting people involved to make decisions about technology
H	*Science and Technology Indicators* – time series that track advances in national (or other) technological capabilities

(Source: Porter & Cunningham, 2005, 18)

PROJECT DESCRIPTION

Development of a model containing a selective dissemination strategy of scientific and technical information available in patent documents, using the installed capacity of Libraries and Information Services in Universities (LISU), as well as their technical experts in research in information resources, which will be crucial to support the success of the URC, for the implementation of a system supported by social media, allowing such dissemination and the information visualization.

This project consists of two parts: the first part aims to develop a model, supported by an information strategy, to create a way to disseminate patent information by university researchers from STM, according to their research needs, sending them only the information sought (selective dissemination of information).

This first phase could be developed between one to three years, depending on the conditions that may arise in the development of the review process of the situational context.

For a coherent dissemination program to make possible the access to patent information by URC, it is necessary to develop a model that allows properly determining all the actions and resources needed to achieve the proposed objectives, as well as effectively control to allow constantly improving the system implemented.

The purpose of this TW and CI is to stimulate creativity, leading to the development of new products and processes and the improvement of the levels and ratios of innovation with cost efficiency.

To properly set the alternatives to follow, the steps for the implementation of the model are:

1. **Macro Environment Analysis**: By which we collect and organize all the necessary information for decision-making and choice of alternatives, such as: Who is our target audience, their needs and motivations? How will the desired information be delivered to our customers or public: email, social media (RSS, Wikis, Blogs, Newsletters, etc.)? Sociodemographic data (number of researchers, research areas, technological innovations developed, etc.)?
2. **Strategic Planning**: This will be the basis of its development and should be planned according to the selection of decision-making alternatives, according to the information collected in the previous paragraph.

It will be here that the alternatives to adopt, plan and execute will be developed and should specify the necessary resources for its effective realization (financial, human and material), as well as their temporal variables (annual), but taking into account the organization's strategy in the long-term.

For the model to work properly, people are required to the appropriate functions. It is also important to implement the organizational model for the system to work (customer support, administrative support, databases to query, response time to customer inquiries, technical support, etc.).

The model to be developed should have its goals quantified so that it can be controlled over its lifecycle. This control lets you know whether the objectives are being met or whether to take corrective actions to acquire it.

Since this model will be an integral part of the university's business plan, its implementation should be decided at the highest level to be accepted without reservation by all stakeholders.

The goal is to enable LISU to monetize their information resources, disseminating patent information by interested parties.

It also allows increasing competitiveness and consequent economic benefits to URC that access this information, by leveraging competitive advantages supported by innovation.

The second part, dependent on the analysis and subsequent synthesis carried out in the first part, aims at the development and subsequent implementation of a technological system based on social media (RSS, Blogs, Wikis, Newsletters, etc.), which will put in place the model in the previous point, allowing effective dissemination and visualization of patent information by researchers from URC of STM, by request, according to the desired classes and scientific fields, but also in its complementary research areas. This will allow knowing the new patented inventions, as well as inventions that are not protected in their territory, and who ended their protection period and can be used freely without any cost.

This part will last four to five years and, from the second year of the project, may be carried out in parallel with the first part, if necessary.

What Is It, For Whom, and How Will It Work

The system to be implemented, available online, using a social media solution to disseminate patent information, will be prepared so that each researcher can be recorded and choose what information wants to receive and visualize, avoiding excess and information overload, from the scientific and technical areas of his/her interest.

Classification codes CPC (Cooperative Patent Classification) and IPC (International Patent Classification), existing to index the class of patents to its technical field, will filter the information and choose which classes each researcher wants to receive according to their needs.

Information about new patent applications, new patents granted, patents in public domain (whose legal protection has expired and can be used without any constraints), and patents that are not protected in our territory (the patent has territorial and geographical scope), will be disseminated through the system, so that researchers can use it in their work, stimulating new solutions and enabling problem solving, and/or create new innovations, or improve an existing one.

LISU already have contracts with commercial suppliers such as Derwent CAS, STN, ProQuest, Questel-Orbit, among others that can be used to maximize the efficiency of researchers.

Other solutions are available for free, like: Espacenet, Google Patents, FreePatents online, Patentscope, USPTO, among others that can be used to support R&D activities at no cost.

The system may be maintained by technicians from LISU or an office to support innovation and technology transfer, functioning in the university.

It will be directed mainly to STM researchers as they will be the ones that will have a greater possibility of extracting the benefits of that information, using it in their projects and maximizing the results obtained.

FUTURE RESEARCH DIRECTIONS

Future projects consider not only the translation of the essential information of the patent documents for English, French, Spanish and German, but also making the search process more user friendly by allowing to search within human indexed keywords, by means of a thesaurus, avoiding the need to know chemical compounds or substances' name, which is not so easy for beginners.

All these measures aim to improve the number of people that effectively search patent information with benefits for the whole society.

CONCLUSION

The main features of patent information have been described, showing the value of this resource for STM research areas.

Several ways of implementing this information resource were exemplified, always using real examples of its use in various types of industry.

The project to develop and its constituents were explained, as well as the timing of work to be done.

Selective dissemination of information and its visualization at the request of researchers, to avoid excess and information overload that would make the process useless or unprofitable, is possible thanks to the indexing of patent documents in classes using the CPC and IPC codes, allowing researchers to choose the classes to receive the information via social media tools such as RSS, Blogs, Newsletters and Wikis.

The goal of this tool for selective dissemination of information is to create value for society in general, finding partners and companies that can take advantage of the knowledge generated and developed in universities allowing, thus, to increase the human development index of the country.

We conclude by pointing out that the use of patent information could result in a significant improvement of R&D results, and its profitability is ensured by the scale of use of the resources invested.

REFERENCES

Albrecht, M. A., Bosma, R., van Dinter, T., Ernst, J.-L., van Ginkel, K., & Versloot-Spoelstra, F. (2010). Quality assurance in the EPO patent information resource. *World Patent Information, 32*(4), 279–286. doi:10.1016/j.wpi.2009.09.001

Ashton, W. B., & Klavans, R. A. (1997). *Keeping abreast of science and technology: Technical intelligence for business*. Columbus, OH: Battelle Press.

Baxter, M. (2000). *Projeto de Produto: Guia prático para o design de novos produtos* (2ª ed.). S. Paulo, Brazil: Edgard Blücher.

Birch, P. & Clegg, B. (1999). *Criatividade em Negócios: Um Guia para Gestores* (1ª ed.). Viseu: Pergaminho.

Bregonje, M. (2005). Patents: A unique source for scientific technical information in chemistry related industry? *World Patent Information, 27*(4), 309–315. doi:10.1016/j.wpi.2005.05.003

Brünger-Weilandt, S., Geiß, D., Herlan, G., & Stuike-Prill, R. (2011). Quality: Key factor for high value in professional patent, technical and scientific information. *World Patent Information, 33*(3), 230–234. doi:10.1016/j.wpi.2011.04.007

Burgelman, R., Christensen, C., & Wheelwright, S. (2009). *Strategic management of technology and innovation* (5th ed.). Singapore: McGraw-Hill.

Christensen, C., Anthony, S., & Roth, E. (2004). *Seeing what's next: using the theories of innovation to predict industry change*. Harvard Business School Press.

Di Minin, A., & Faems, D. (2013). Special issue on intellectual property management. *California Management Review, 55*(4), 7–14. doi:10.1525/cmr.2013.55.4.7

Drucker, P. (1987). *Inovação e gestão* (2nd ed.). Lisboa, Portugal: Presença.

Flynn, R. (2002). NutraSweet enfrenta a concorrência: O papel crítico da inteligência competitiva. In Prescott & Miller (Eds.). *Inteligência competitiva na prática: Estudos de casos diretamente do campo de batalha*. Rio de Janeiro, Brazil: Campus, 135-149.

Godinho, M. (2003). *Estudo Sobre a Utilização da Propriedade Industrial em Portugal* (Vol. I). Lisboa, Portugal: Instituto Nacional da Propriedade Industrial.

Greif, S. (1979). The role of patent protected imports in the transfer of technology to developing countries. *International Review of Industrial Property and Copyright Law, 10*(2), 124–143.

Greif, S. (1987). Patents and economic growth. *International Review of Industrial Property and Copyright Law, 18*(2), 191–213.

Idris, K. (2003). *Intellectual property: A power tool for economic growth*. Geneva, Switzerland: World Intellectual Property Organization.

Lattès, R. (1992). *O Risco e a Fortuna: a grande aventura da Inovação*. Lisboa, Portugal: Difusão Cultural.

Maia, J. M. (1996). *Propriedade Industrial: Comunicações e Artigos do Presidente do INPI*. Lisboa, Portugal: Instituto Nacional da Propriedade Industrial (INPI).

Maravilhas, S. (2009). A Informação de Patentes: Vantagens da sua utilização como estímulo à criatividade, I&D, inovação e competitividade das empresas portuguesas. In IAPMEI (Ed.), *Parcerias Científicas para a Inovação*. Lisboa, Portugal: IAPMEI, 91-110.

Maravilhas, S., & Borges, M. (2009). O Impacto das Bibliotecas Digitais de Patentes no Processo de Inovação em Portugal. In Borges & Sanz-Casado (Eds.), A ciência da informação criadora de conhecimento. (Vol. II). Coimbra, Portugal: Actas do IV Encontro Ibérico EDIBCIC 2009, 47-63.

Maravilhas, S., & Borges, M. (2011a). Os recursos de informação usados na I&D em Portugal: Caracterização dos centros de investigação do ensino superior público das áreas de Ciência, Tecnologia e Medicina. In P. Guerrero & V. Moreno (Eds.), *Límites, fronteras y espacios comunes: Encuentros y desencuentros en las Ciencias de la Informacíon* (pp. 321–333). Badajoz, Spain: Actas do V Encontro Ibérico EDICIC.

Maravilhas, S., & Borges, M. (2011b). A utilização da informação de patentes pelos centros de investigação do ensino superior público: O seu impacto no processo de inovação em Portugal. In P. Guerrero & V. Moreno (Eds.), *Límites, fronteras y espacios comunes: Encuentros y desencuentros en las Ciencias de la Informacíon* (pp. 364–376). Badajoz, Spain: Actas do V Encontro Ibérico EDICIC.

Marcovitch, J. (1983). *Administração em ciência e tecnologia*. São Paulo, Brazil: Edgard Blücher.

Marcus, D. (1995). Benefits of Using Patent Databases as a Source of Information. In Lechter (Ed.), Successful Patents and Patenting for Engineers and Scientists. New York, NY: The Institute of Electrical and Electronics Engineers (IEEE) Press.

Mueller, H., & Nyfeler, T. (2011). Quality in patent information retrieval: Communication as the key factor. *World Patent Information*, *33*(4), 383–388. doi:10.1016/j.wpi.2011.06.012

Naetebusch, R., Schoeppel, H. R., & Fichtner, H. (1994). Patent information in a large electrical company, exemplified by the situation at Siemens. *World Patent Information*, *16*(4), 198–206. doi:10.1016/0172-2190(94)90003-5

O'Dell, D. (2004). *A Resolução Criativa do Problema: Guia para a Criatividade e Inovação na Tomada de Decisões*. Lisboa, Portugal: Instituto Piaget.

Ojala, M. (1989). A patently obvious source for competitor intelligence: The patent literature. *Database*, *12*(4), 43–49.

Porter, A., & Cunningham, S. (2005). *Tech Mining: Exploiting New Technologies for Competitive Advantage*. Hoboken, NJ: Wiley.

Ribeiro, D. (2007). Propriedade Intelectual: Mais de 30% da investigação em Portugal é redundante. *Jornal de Negócios, Quinta-feira*(24 de Maio), 34.

Rivette, K., & Kline, D. (2000). *Rembrandts in the Attic: Unlocking the Hidden Value of Patents* (1st ed.). Boston, MA: Harvard Business School Press.

Schilling, M. (2013). *Strategic management of technological innovation* (4th ed.). Singapore: McGraw-Hill.

Schoeppel, H. R., & Naetebusch, R. (1995). Patent searching in a large electrical company, as exemplified by the situation at Siemens. *World Patent Information*, *17*(3), 165–172. doi:10.1016/0172-2190(95)00016-S

Seymore, S. (2009). Serendipity. *North Carolina Law Review*, *88*, 185.

Skarzynski, P., & Gibson, R. (2010). *Inovar no Essencial: Transforme o modo como a sua empresa inova* (1st ed.). Lisboa, Portugal: Actual.

Tidd, J., Bessant, J., & Pavitt, K. (2003). *Gestão da Inovação: Integração das mudanças tecnológicas, de mercado e organizacionais*. Lisboa, Portugal: Monitor.

Trott, P. (2008). *Innovation Management and New Product Development* (4ªed.). Essex, UK: Prentice Hall | Financial Times.

Wilson, R. (1987a). Patent analysis using online databases: I. Technological trend analysis. *World Patent Information*, 9(1), 18–26. doi:10.1016/0172-2190(87)90189-X

Wilson, R. (1987b). Patent analysis using online databases: II. Competitor activity monitoring. *World Patent Information*, 9(2), 73–78. doi:10.1016/0172-2190(87)90131-1

KEY TERMS AND DEFINITIONS

Creative Imitation: The strategy followed by some companies that imitate something already existing but adding value. Often the imitator can foresee, better than the original creator, how the product or service can be best suited to meet the needs of consumers, changing it to match this observation.

Creativity: Creativity is based on reasoning that produces imaginative new ideas and new ways of looking at reality. Creativity is an individual process, arises from the idea that popped into someone's head. Relates facts or ideas without previous relationship and is discontinuous and divergent. No Creative Process exists if there is no intention or purpose. The essence of the Creative Process is to seek new combinations.

Innovation: The application of new knowledge, resulting in new products, processes or services or significant improvements in some of its attributes. When a new solution is brought to the market to solve a problem in a new or better way than the existent solutions.

Invention: The creation or discovery of a new idea, including the concept, design, model creation or improvement of a particular piece, product or system. Even though an invention may allow a patent application, in most cases it will not give rise to an innovation.

Patent Information: During the process of registration and grant of a patent, the official entities like the USPTO, EPO or WIPO, will generate one or more legal documents that are called patent literature. These documents contain information that is referred to as patent information.

Serendipity: Serendipity is the ability to make important discoveries by accident. Not all the ideas for new products or processes appear voluntarily and intentionally. Sometimes a mixture of luck and preparation provides valuable discoveries. A serendipitous discovery results from the combination of a happy coincidence with perspicacity.

ENDNOTES

[1] Vulcano, a Portuguese company, started as a licensee of Bosh's water heater systems and now they are the biggest Bosh R&D plant in Europe and the second in the world (Australia is the biggest).

[2] "Serendipity, the process of finding something of value initially unsought, has played a prominent role in modern science and technology. These "happy accidents" have spawned new fields of science, broken intellectual and technological barriers, and furnished countless products that have altered the course of human history. (…) [A]ccidental discovery is a common and widely acknowledged path to invention in unpredictable fields." (Seymore, 2009, pp. 185; 188).

Chapter 18
Internet Technology Application in Production of Internet Radio Programs in Vietnam

Kien Truong Thi
Academy of Journalism and Communication, Vietnam

ABSTRACT

This chapter introduces the concept and features of Internet radio, and expresses the status of Internet technology application in the production of Internet radio programs in Vietnam. Some solutions are proposed to help Vietnamese radio managers to improve the efficiency of Internet technology application in producing the programs.

BACKGROUND

The Trend of Internet Radio

The new 2018 Global Digital suite of reports from We Are Social and Hootsuite reveals that there are now more than 4 billion people around the world using the Internet. That means, more than 50% of the world's population has used the Internet.

With the radio application which has already integrated on smartphones, networked computers and other hand-held visual audio devices, it comes as no surprise that many people wonder whether Internet radio can replace traditional radio or not.

The answer is: "The number of Internet radio audiences is growing and the time they spend for the media is increasing rapidly. Internet radio and streaming services are not completely capable of replacing broadcast radio, but it is changing the listening time and also creating incremental audio consumption" (Xapp Media, 2015, p.13).

Thus, according to experts, although it is not able to replace completely traditional radio, Internet radio will reset the listening time, listening method and at the same time listening content.

DOI: 10.4018/978-1-5225-9993-7.ch018

The research by XAPP Media (USA) indicated: 44% of listeners supposed that the time previously reserved for AM / FM radio was replaced by time for Internet Radio (Xapp Media, 2015). In the US, in 2007, 12% of people listened to online radio at least once a month. Six years later, the number of Americans listening online was 33%, approximately 86 million. To 2016, this number was 50% and one year later - in 2017, there were more than half of the US population, which is equivalent to 53% of people listening to radio online every week. There were 57% of the US population using online radio (Statistics, 2018) (Figure 1).

Thus, the time Americans spend listening to Internet radio broadcasting increases gradually each year, specifically shown in Figure 2.

In 2017, Americans spent 879 minutes (equivalently to 14.65 hours) each month listening to on-line radio (Statistics, 2018). "The audience is expected to grow to 183 million by 2018, significantly outpacing population growth" (Xaap Media, 2015, p.5). The popular Internet radio device in the US is smartphone. In 2015, 95% of the public received Internet radio via cellphones, only 5% listened via computers (Xapp Media, 2015).

Figure 1. Share of weekly online radio listeners in the United States from 2000 to 2018

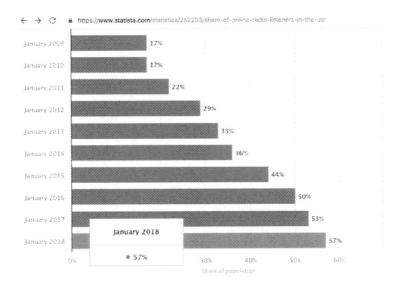

Figure 2. Share of monthly online radio listeners in the United States from 2000 to 2017

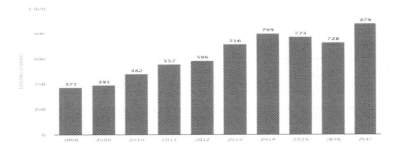

The appearance of the Internet, the popularity of smartphones, computers and the benefits of using Internet radio have explained the increase in the number of listeners of this type of communication. Internet radio has broken the inherent limitations of receiving via traditional radio like: listening only once, following time regulation (linear time); completely passive reception; storage –information inability... Moreover, Internet radio brings huge economic benefits in compare to broadcasting on traditional electromagnetic waves.

"The large and growing public will provide value to advertisers in coming years (…) Internet radio has become an important distribution channel for the music industry willing to become an important advertising channel for marketers" (Xapp, 2015, p.4) (Figure 3).

BIA/KELSEY's survey showed that Internet radio revenue has increased gradually over the years in the US. Therefore, developing Internet radio is not only the trend for radio and consumers, but also for advertisers and businesses.

The Conception of Internet Radio Broadcast

Due to the fact that in Vietnam, foreign authors find it very difficult to access to research materials on Internet radio, whereas in Vietnam, according to our understanding, there have not been any foreign documents about Internet radio which are sold or introduced in libraries, so we have to approach some Internet broadcasting (Internet radio) concepts gathered by Wikipedia. Kiraly and Jozsef state that:

Internet radio (also web radio, net radio, streaming radio, e-radio, IP radio, online radio) is a digital audio service transmitted via the Internet. Broadcasting on the Internet is usually referred to as webcasting, since it is not transmitted broadly through wireless means. It can either be used as a stand-alone device running through the Internet, or as a software running through a single computer.

Figure 3. Total revenue from Internet radio ad has gradually increased over the years (Xapp, 2015)

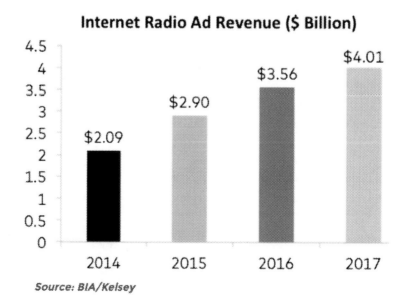

(...) Internet radio involves streaming media, presenting listeners with a continuous stream of audio that typically cannot be paused or replayed, much like traditional broadcast media; in this respect, it is distinct from on-demand file serving. Internet radio is also distinct from podcasting, which involves downloading rather than streaming. (Wikipedia, 2019 quoted Kiraly and Jozsef).

From a technical perspective, Internet radio is a bit like standard radio in terms of quality and user experience, but the similarities end there. It's based on a technical process that digitizes audio and splits into small pieces for transmission across the Internet. The audio is "streamed" through the Internet from a server and reassembled on the listener's end by a software player on an Internet-enabled device. Internet radio is not *true* radio broadcast by the traditional definition - it uses bandwidth rather than the airwaves - but the result is an incredible simulation. The public can listen by using a music player or browser that supports streaming audio to download and listen simultaneously. Several audio formats used include MP3, OGG, WMA, RA, AAC Plus...

Benjamin M. Compaine, Emma Smith supposed:

Internet radio broadcasters are defined as entities that deliver entertainment and/or news and information content as an audio stream via the Internet. These audio streams may be delivered live or archived to be accessed on demand, but in both cases the audio files were initially created as programming to be delivered to an audience of more than one. (Benjamin M. Compaine, Emma Smith, 2001)

Thus, from the above views on Internet radio, it is easy to conclude:

Internet radio is a digitized sound, transmitted on the Internet, bring information to listeners via phones, computers or network- connected devices. The interference between radio and the Internet is a form of "hybridization" between radio broadcasting via airwaves and the Internet.

In Vietnam, author Nguyen Thi Thu determined: Internet radio is a modern radio method transmitting information to the public via the Internet, in multimedia language form (in which, audio is the main language) (Nguyen Thi Thu, 2014). The author Nguyen Dinh Hau conceived: Internet radio is the process of expressing audio products in the form of multimedia (in which sound is the main data) and transmitted to the public (Nguyen Dinh Hau, 2016).

Although in their research, authors Nguyen Thi Thu and Nguyen Dinh Hau not only mention the form of multimedia radio, but also mention other forms of radio on the Internet, the concept given by them has not covered all those formats yet.

Base on the perspectives of some foreign and Vietnamese researchers, we propose the concept of Internet radio as follows:

Internet radio is the combination of analog radio and the Internet, uses digital audio being suitable for the foundation of Internet technology and terminal devices such as computers, phones, laptops..., in order to disseminate information in multimedia language, in which, audio (voice, music, noises) is the main language, to impact on many senses of the audiences.

However, this concept is only correct for the radio broadcast program format of radio stations. In fact, not only the radio stations produce Internet radio programs, but even printed and online newspapers have their own Internet radio programs, too. Therefore, through the practical survey on the production and broadcast Internet radio programs' methods of printed and online newspaper agencies in Vietnam, from January, 2016 to March, 2010, we would like to introduce the Internet radio format concept of the printed and online newspaper agencies, as follows:

Internet radio format of printed newspapers and online newspapers is audio files built under the form of a radio program, digitized to fit the Internet technology platform and terminal devices (computers, smartphones...). The information transmission language is multimedia language, in which, audio (voice, music, noises) is the main language, impacting on many senses of the audiences.

The similarities between the two types of Internet radio mentioned above are:

- They are built under the radio program format (includes: signature tune, greetings, announcements, news/articles/music, ending greetings).
- Be digitized to broadcast on the Internet
- Can be used multimedia language
- Have the ability to create hyperlinks...

The differences are:

- The radio broadcasts of Internet radio stations are often retrieved from AM and FM radio broadcasts, therefore, in comparison to the printed and online newspaper agencies' radio programs, the number of the radio station agencies' programs is more diverse and the quality is much better.
- The Internet radio programs' duration of the radio stations is longer than the Internet radio programs' duration of the printed and online newspapers.
- The radio production team of the radio stations is usually radio journalists, while the radio production team of the print and online newspapers is usually print and online newspaper journalist.

APPLICATION OF INTERNET TECHNOLOGY IN THE PRODUCTION OF RADIO PROGRAMS: ANALYSED FROM THE THEORETICAL PERSPECTIVE

Application of Internet technology, from the technical perspective, Vietnamese radio technicians have applied it in producing 3 Internet radio formats: live streaming, replayed broadcast and audio on-demand. In terms of creative mode, there are single-media radio and multimedia radio. Live radio programs are usually single-media radio programs. Multimedia radio programs are replayed programs. Single media programs can be live broadcasted. Theoretically, it can be understood as follows:

Live Streaming (Live Audio Broadcast, Live Radio Broadcast)

Live radio broadcast on the Internet, actually are traditional radio programs (which are produced to broadcast via analog radio), digitized to suit the Internet technology platform to broadcast and listen online.

Live broadcast almost preserves the integrity of traditional analog radio' characteristics, strengths and limitations, such as:

- Using synthesized sound (including speech, sound, music) is the unique language to convey information. These are single-media radio works.
- The only way to receive information is via ears
- Provides information in linear chronological order
- The listener can listen only one time and passively, so their information memory capacity is very low
- The storage capacity of information is almost zero.

However, much better than live programs on AM, FM radio, live programs on the Internet have many advantages such as: No bandwidth registration is required; no airwave receiving stations and signal transmitter means are needed; transferring information not limited by territorial factor...

Replayed Radio, On -Demand Radio

Replayed radio, on - demand radio can be understood similarly in terms of technical aspects but differentiate in terms of ability to meet the user requirements.

Replayed Radio

This is the term used to indicate the programs which are broadcasted on the analog radio and stored on the Internet for listening again. Just click, select the program, the program will be replayed.

For replay programs, listeners have the right to choose to listen to programs/works according to their preferences and needs. At this point, the replay radio asymptotic the audio on-demand format, because listeners have right to "select to listen again". However, it is asymptotic but not completely on-demand radio, because there are only a few programs that are replayed (podcasts) online, and those programs are not selected "according to the request by the audience", but by the radio stations themselves.

Replayed radio programs on the Internet have some differences in comparison to live radio programs:

- Can use multimedia language to convey information, in which, synthetic sound is the main language
- Audiences can listen by ears and read/watch by eyes (with multimedia programs/podcasts)
- Information is not necessarily in a linear order, because listeners can rewind to listen again
- Listeners can listen many times depending on the needs; can be able to remember information very well because they are proactive in receiving information
- Ability storing information for a long time on the Internet

On – Demand Radio

Like replayed radio, on-demand radio are broadcasted programs, stored online so that listeners can choose to listen on demand. It has similar characteristics and strengths to replay radio (as mentioned above), but there are still some different characteristics:

- Listeners have to subscribe to buy on- demand service with the radio stations or radio service provider agencies and pay a certain fee for listening to the programs that they choose ("required subscriber audio")
- Radio stations or service provider agencies must be licensed to provide services.

When buyers subscribe the on-demand service with the radio stations, the programs registered by buyers will be updated on their terminal devices (computers, phones, tablet computer (iPads, etc.), by either automatically or downloading. Listeners not only can listen again at any time, but also can store on mobile devices in DRM format (Digital Rights Management) to listen offline. Users can also use the Radio Maximus Portable device - one of the devices for listening and recording- to automatically record according to the schedule they have set, marked and displayed the icons of the channels.

On-demand radio "allows users to decide convenient time and place, (...) that can have a strong influence on receivers' listening habits" (Franc Kozamernik and Michael Mullance, 2016).

Multimedia Radio

Appearing on the basis of Internet technology, terms such as *multimedia radio works, multimedia radio programs* are inevitable. These are works / programs that use a variety of languages to convey information: audio, images, video, text, diagrams, tables, etc., in which, synthetic sound must still be the mainstream language form. Other forms of languages are only complementary. This helps distinguish from other online multimedia press works (In Vietnam, multimedia works in online newspapers often use text as the main language; other languages are complementary languages. However, many foreign newspapers as CNN, BBC online, video language and image language can be equivalent to writing language).

Like the Franc Kozamernik and Michael Mullane's opinion: "Internet radio redefines radio content. It not only introduces new music and information formats, but can also decorate them with text, graphics and videos" (Franc Kozamernik and Michael Mullane, 2005). And Ben Williams - director of Media company Beyon Broadcasting (UK) said:

Through radio sites, listeners can also watch additional videos and images for radio broadcasts, which plays an important role, not less than the sound. And this is the new requirement for radio program producers (Minh Thuy, 2016 quoted Ben Williams).

Multimedia works/programs are always based on replayed technology. Currently, production techniques in Vietnam are not capable of producing multimedia programs when the programs are being broadcasted lively.

Internet radio has both the characteristics of traditional analog radio and new features because of specific types of multimedia radio:

- Is a digitized information service for transmission to listeners via the Internet in the form of images, audio, video, text... with terminal devices such as computers and phones connected to the network.
- Using multimedia to transmit information, in which, audio is the main language

Multimedia (multilingual) in Internet radio programs includes: videos, pictures, graphics, text. Currently, the Voice of Vietnam's journalists have begun to focus on producing multimedia radio programs. For example, in *the News and Current Affairs* program (broadcast on February 24th, 2019) of the Voice of Vietnam, in addition to sound - the main language component, there are also an illustrated photo (an illustration for the most important news in the program), text (to brief introduces main content that will be broadcast in the program) (Figures 4 and 5).

On VOV2 channel of the Voice of Vietnam, there is Media page, is a video collection providing more information for VOV2' radio programs.

With Internet radio, because of the ability to transmit multimedia information, so, obviously, Internet radio gives people the right to receive information by multiple senses: can listen by ears and read / see by eyes – the things that traditional radio cannot have.

Figure 4.

Figure 5.

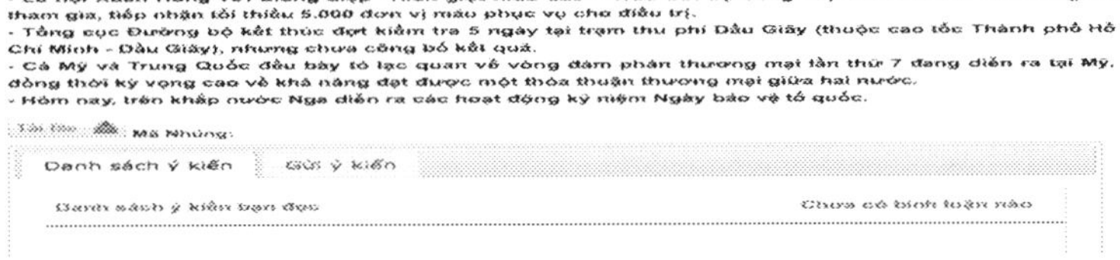

Instant and Non-Periodic, Non-Linear Properties

Unlike traditional analog radio, listeners are depending on the scheduled time frame and broadcasting duration, receive information according to the rules of linear time, now with Internet radio, especially with replay radio programs and radio on demand, periodic and linear factors can almost be removed. Thanks to the ability to update information instantly, no longer limited by the broadcast time, the producing process is simple and easy of the Internet, journalists can update information right at the time happen of the stories/facts/events, with any amount of information without having to wait for the program's broadcast time frame (if the journalists/radio stations want). Additional information can be video, visual, audio, or even written language.

Linearity, the dependence on the rule of time in broadcasting and receiving information is also removed, because with Internet radio, listeners can listen to the program at any segments they want, can rewind the content to receive information without depending on the rule of time.

Of course, in Vietnam, because 100% of Internet radio programs are mainly live broadcasted and replayed from the station's analog programs, therefore the instant and non-periodic, non-linear natures are also only at certain limits.

Has a Global Nature

Nearly one third of the world's population is using the Internet. Therefore, any radio station is broadcasted on the Internet, which is obviously "wear a coats of global nature". Any people around the world, if they have a networked computer and phone and radio address, they can receive the Internet radio programs. Internet radio overcomes all barriers of territorial, tariff, cultural and political restrictions to connect the public on a global scale. Especially, an Internet radio station will be trully an international radio station if it is able to meet some additional criteria, such as:

- Official prestige and reliability information (voices of the State, Government, large organizations)
- Using international language (English), or using multilingual - in addition to the national language - to convey information
- Being well known and accepted by the international public (such as BBC, CNN, RFI...).

High Interaction Ability

Previously, when Internet radio had not been born yet, the interaction between listeners and radio stations in general, with programs and journalists in particular was very difficult to implement. Firstly, listening via the "loudspeaker" will make the public catch information implicitly, lack of evidence when they want to criticize, comment and evaluate information, which lead to the psychology of afraid to interact; the second reason is due to the slow way of interaction (done by calling, sending messages or sending email or hand-writing letter...). However, when the Internet radio appeared, it created sites with clear addresses, the interaction was also convenient, easier and faster than ever. The public can interact by sending feedback in the comment boxes... attached to each program and each article at every sites; the public can even interact deeper with the program by sending their news and articles.

The Ability to "Hyperlink" and Store Information

Like other online newspapers, the Internet radio program also takes advantages of the hyperlinked ability by creating "links" or "keywords" to lead the audience to related content with the programs/works.

Internet radio is able to save information stably as any online newspapers. Listeners can easily access the broadcasted information/programs by typing "keywords" in the "Search" section. All content in text, video, audio, pictures and graphics... will appear in a few seconds.

Having Significant Identification

With traditional analog radio, due to the characteristic "millions of auditors have to listen to the programs simultaneously", so the identification of auditors' private reactions is a little limited. However, with Internet radio, because of the replay function and on request, the public does not need to listen to the program simultaneously with the broadcast time.

The proactive in receiving information of the public is enhanced. Each person has their own choice of the listening time, speed and way that suitable with the personal needs and preferences. The emotions of the public are evoked completely privately, lead to private emotional reactions: love or hate, like or dislike, believe or not believe, feedback or no feedback...

It's foretold that many nations around the world will practices step by step and forward to abandon FM waves to transfer completely to digital broadcasting. The main reasons given for the transition to digital technology are: (1) Better sound quality; (2) Integrating and synchronizing radios with other types of communication such as text, images, web; (3) Ability to play podcasts; (4) Operating costs are many times cheaper (in the case of Norway, it is 8 times cheaper) than traditional broadcasting (Nguyen The Ky - Member of the Party Central Committee, General Director of the Voice of Vietnam VOV, 2018).

APPLICATION OF INTERNET TECHNOLOGY IN RADIO PROGRAM PRODUCTION FROM A PRACTICAL PERSPECTIVE IN VIETNAM

Internet technology helps the Vietnamese radio develop with many remarkable achievements, at many angles. Vietnam now has 1 national radio station - *the Voice of Vietnam*, and a network of 63 local radio stations (each province in Vietnam has one radio station). *The Voice of Vietnam* is under the administration of the Government of the Socialist Republic of Vietnam. Local radio stations in Vietnam have typical characteristic: the governing organ is the People's Committees of provinces and cities, which are combined with local television stations to become local radio and television stations (except Ho Chi Minh City has its own radio station and private television station). All Vietnamese radio stations are funded by the State. The Ministry of Information and Communications is the State management agency in the field of press, which has responsibility to general manage all activities of radio stations from the Central to local levels. The Central Propaganda and training Commission is the agency on behalf of the Communist Party of Vietnam to lead and direct the direction of the press agencies activities in general and of Vietnamese radio stations in particular.

The Voice of Vietnam was established on July 7, 1945. This is considered a multimedia agency, with many types of press: radio (a major type), television (The Vietnamese Journey, Vietnam television channels VTC), print newspaper (the Voice of Vietnam newspaper), online newspaper (vov.vn). Particularly in the field of radio, *the Voice of Vietnam* currently has 5 channels: broadcast on AM and FM channels, including: VOV1 (The News and Current Affairs), VOV2 (The Cultural and Social Affairs), VOV3 (The Music. Information and Entertainment), VOV4 (Ethnic Languages Channel), VOV5 (World Services), VOV Traffic Hanoi, VOV traffic Ho Chi Minh city...

To overcome the limitations of traditional radio, at the same time, take full advantages of Internet communication technology on the Internet in the age of 4.0 technology, catch up with the current world trend, on January 3, 1999, the Voice of Vietnam has officially been on the Internet. In 2005, the Voice of Vietnam started new technology: digital technology for the production of radio programs by specialized software of computers. Presently, all channels of the Voice of Vietnam are put on the Internet. Radio programs are also broadcast on the national Internet site and also transmitted through domestic and international satellites. In addition to bringing the program to the Internet, the Voice of Vietnam also invest a system of equipment for production, editorial and broadcast, invest modern and digitized receiving facilities such as tablets, smartphones... Thanks to it, the Voice of Vietnam' programs have been grow so far both in quality and quantity, the number of listeners has increased, too.

Beside of the Voice of Vietnam, many local radio stations also broadcast its programs on the Internet. By January 1, 2019, there are 35/63 local Vietnamese radio stations joined the Internet, in other words, there are 35 local radio stations broadcasting radio programs on the Internet. In which, the Radio Stations of: Ho Chi Minh City, Hanoi, Ba Ria-Vung Tau, Soc Trang, Ninh Binh, Quang Ninh, Thua Thien Hue, Phu Yen... are the typical stations. These radio stations often transfer their radio programs on Internet at the stations' websites. For example, the Internet radio program of Hanoi Radio and Television Station is located at: hanoitv.vn (click "Radio section" on the website); Internet radio program of Hai Phong Radio and Television Station at: thhp.vn (click "Radio section")...

On February 13, 2013, the Radio online website (Radioonline.vn) of Vietnam was officially launched, integrating the program of 19 Vietnamese radio stations (until December 1, 2019). Radioonline.vn includes radio programs of local and national radio stations in the replay radio form, or live broadcast... Viewers simply click on the icon of the radio to be able to listen directly or listen again to the radio programs. In addition to the radio stations broadcast on the Internet of the radio stations, online press agencies and many print press agencies in Vietnam have also applied Internet technology to post on their online newspapers/websites the radio programs/categories, such as: Vietnamnet.vn, Tuoitre.vn, Hanoimoi.com.vn, nhandan.org.vn... (In Vietnam, in addition to press agencies, there are many companies and enterprises applying mobile radio and Internet radio services such as VDC Media, VASC Orien, Viettel, Mobifone... However, in this article, we do not mention these communication methods). The development of information technology and the Internet has created great opportunities for Vietnamese radio to take advantages and approach new listeners - listeners on the computer or mobile around the world.

Our survey shows that popular Internet radio formats used in Vietnamese press agencies are as follows (we divide it into two cases):

Case 1: Internet Radio Format of the Vietnamese Radio Stations

Live Broadcast (Live Streaming)

This is the popular Internet radio format at the Voice of Vietnam. Until January 1, 2019, five channels of *the Voice of Vietnam* have live broadcast programs, which coincide with the broadcast time of analog radio stations on AM and FM channel.

The following image illustrates for the News and Current Affair program of the News and Current Affair channel VOV1 (the Voice of Vietnam) is broadcast live at the address Vov1.vov.vn (Figure 6).

To listen directly programs which are being simultaneously broadcast in real time on the Voice of Vietnam, listeners just need to click on the arrow icon on the horizontal bar, which usually located at the top of the page.

The local radio stations such as *Radio the Voice of Ho Chi Minh City People station* and *Hanoi Radio and Television Station* also apply the live streaming formats on the Internet. *Radio the Voice of Ho Chi Minh City People station* brings all three channels on FM 99.9Mhz, FM 95.6Mhz, AM 610Khz to stream live on the network; *the Hanoi Radio and Television* brings the FM channel 90 Mhz, FM 96 Mhz, JOYFM to broadcast live on radioonline.vn. In addition, *the Hai Duong Radio and Television, the Ba Ria - Vung Tau Radio and Television, Da Nang Radio and Television, Bac Giang Radio and Television, Phu Yen Radio and Television, Ninh Binh Radio and Television, Kien Giang Radio and Television, Tien Giang Radio and Television...* also have the live radio format.

Some radio stations are posted on radioonline.vn, but there is no live broadcast format as those of Ninh Thuan province, Ca Mau province, Thua Thien Hue province, Binh Phuoc province... According to the author's survey in January 1, 2019, among 35 local radio stations are posted on their own website in Vietnam, only 11/35 radios can be lived as an Internet program format, occupy 31.4% rate. If compared to the total of 63 radio stations in Vietnam today, the radio stations use live-broadcast format is 11/63 stations, accounting a very low rate: 17.5%.

Replayed Radio

In Vietnam, not all analog radio programs are broadcasted online to replay; at the same time, the storage time of program is not infinite because the capacity of the storage device is limited. Radio the Voice of Vietnam, the Radio the Voice of Ho Chi Minh City People, Hanoi Radio and Television Station (of the Hanoi Capital) put most of their programs online to replay - the highest percentage of the Internet replay programs in compared to other radio stations.

Figure 6. Live streaming program

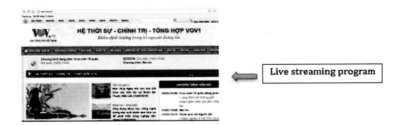

According to Nguyen The Ky, General Director of *the Voice of Vietnam*, until January 1, 2019, *the Voice of Vietnam* has about 60% of the programs posted online to replay. *The Voice of Ho Chi Minh City People* posted all program broadcast schedule on its website; and about 54% of its programs are replayed on the Internet. Listeners can listen to the broadcast again easily. Similarly with *the Voice of Ho Chi Minh City People station*, the number of *the Hanoi Radio and Television Station*' replay radio programs accounts for about 70% of the total programs.

Our survey showed that from June 2013 to January 1, 2019, the replayed radio programs of local radio often is the News and Current Affairs program. 100% of the News and Current Affairs program of 35 local radio stations are broadcasted on the Internet, accounting for 100%. After that, the local stations choose the programs that to be considered as the strength of stations...

On - Demand Radio

Our survey indicates that, in Vietnam, there is no type of on demand radio programs. All radio programs on the Internet are selected intentionally to the radio stations' will but not according to the audience's buying and listening needs. In Vietnam, there is no paid radio program yet. All programs are free. The budget for radio stations - including national radio stations and 63 local radio stations - is taken from the State budget. The source of advertising for Vietnamese radio is not plentiful, because the public is not yet eager to listen.

Multimedia Radio

As we have mentioned in the theory section, the multimedia radio program is the program format that fully demonstrates the characteristics and advantages of Internet applications in radio program production. See Figure 7 for further detail.

Figure 7.

Internet Technology Application in Production of Internet Radio Programs in Vietnam

Presenting on the Internet includes the elements: sound (the main language), text (to display the information fields: headline, the program topic introduction, the "hyperlink" keyword, help search related information), photos, interactive audience mailbox. Listeners can also download audio files to the computer to listen again.

By that way, the receivers can listen again to the program at the appropriate time, clearly grasp the topic of news and articles in the program to decide whether to listen or not; can listen to it again on demand; can comment on the works, provide additional information to programmers (if they want); can find more information related to the topic from many different sources, etc. In some Internet radio interviewing and talk show programs, instead of using documentary images, producers use photos about studio with the appearance of the host and guests. In some other radio reportages, producers use the journalists' photo of the scene.

Currently, *the Voice of Vietnam* applies multimedia programs just at the simplest level. That means the Station has not focused on producing the "true" multimedia programs. There are text, photos, audio files in the multimedia Internet radio programs of *the Voice of Vietnam*, but the text is only used to brief introduction the main content of programs. Text attached to programs is usually less than 10 lines, about 40-200 words, including program's titles (program topics) and brief introduction. Illustrated photo usually consists of only one photo (can be a photo of the scene or photo of the studio, or a character image appears in the program...). The key factor of the program is still the audio file.

In addition to national radio station of Vietnam known as *the Voice of Vietnam*, the multimedia radio program format has not been produced at any local radio station.

Case 2: Internet Radio Format of Online Newspapers and Printed Newspapers

In fact in Vietnam, the Internet radio programs of press agencies (not belong to radio stations) have also appeared. These programs are simply produced as follows:

News -> editor -> broadcaster (journalist) read -> record files -> audio editor -> upload online.

According to our survey, before January 1, 2019, many online newspapers in Vietnam updated their online radio programs online by day, such as *hanoimoi.com.vn* (the Party newspaper of Hanoi Capital), *tuoitre.vn* (the newspaper of Ho Chi Minh City Union), *laodong.vn* (the newspaper of the Vietnam General Confederation of Labor), *nhandan.org.vn* (the biggest Party newspaper in Vietnam today).

For example: the image of the Tuoitre audio/ radio (the Youth radio) page of tuoitre.vn that was released in July 2016 as shown in Figure 8.

The tuoitre audio (the Youth audio) page belongs to the Entertainment Culture page, has 60 minutes duration. Tuoitre audio was built in the form of multimedia, including photos, text, audio. In which, the task of the text is presents the content abstract to be broadcasted in the program.

However, since January 1, 2019, the above newspapers have removed the radio form on the network, focusing on developing online television, video clips... A few online newspapers presently upload their online audio/radio content. Take *dantri.vn* as an example and see Figure 9.

Figure 8.

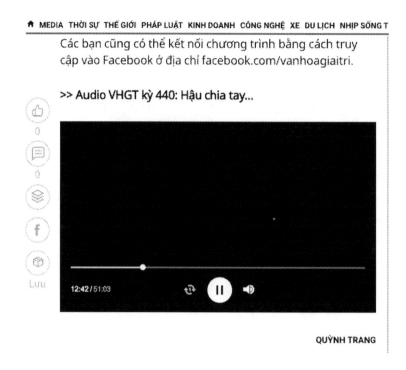

Figure 9.

The reason why many online newspapers in Vietnam stopped using the online audio program is because these programs have low follower rate. For other newspapers, although there has not been an official investigation, but perhaps, the demand for receiving online audio from the public is not high, which is the reason for Vietnamese online newspapers to have "temporary divorce" with this form of communication.

Internet Technology Application in Production of Internet Radio Programs in Vietnam

RECEPTION OF VIETNAMESE LISTENERS

The educational level of the Vietnamese public is increasingly enhanced, both the motivation and basis for radio stations to constantly innovate, improve the quality of radio programs in general and Internet radio in particular. By the end of 2018, Vietnam has 235 universities and institutes (including 170 public schools, 60 private and people-founded schools, 5 schools with 100% foreign capital), 37 scientific research institutes assigned the mission of doctoral training, 35 pedagogical colleges. The number of people has qualifications from college to master is increasing. It is proportional to the level and the need to receive information of themselves. The Vietnamese public today not only needs to receive information, but also needs to choose information, higher is needs to participate in creating works and programs. The term "citizen journalism" has become popular worldwide, including in Vietnam. And therefore, radio must to adapt to the "citizen radio" trend. The Internet is an "extended arm" of "citizen broadcasting".

Our survey also shows that Vietnamese listeners are now beginning to receive Internet radio programs, expressed through the increasing number of people to access programs.

Vietnam's government-run VTV television network continues to dominate the country's media landscape. However, the current survey also points to a powerful generational shift in media use toward online news sources and less dependence on state TV.

About one in four Vietnamese (24.8%) now say they listen to the radio weekly or more, down marginally from 27.6% in the 2012-2013 study. The FM band is by far the most commonly used, with 3.8% of Vietnamese saying they listen to AM radio and 1.0% listening to shortwave radio on a weekly basis. Notably, Vietnamese are now about as likely to have listened to the radio on mobile phones (10.3%) as on conventional radio sets (11.1%) in the past week. (The Broadcasting Board of Governor, (n.d.).

For journalists, the follower number is an indicator to help them finding methods to change and improve the program quality. For radio managers in particular, and the state administration of newspapers agencies in Vietnam in general, Internet radio is a new form of communication that needs to be studied and applied strongly at all radio stations nationwide.

In this article, we conducted a small survey: surveying 600 Hanoi public - who have jobs already (random survey by distributing questionnaires at some Hanoi offices, December, 2018) about whether or not they listen to radio in Hanoi, the results as shown in Table 1.

Surveying 238 audiences listening to radio in Hanoi, the rate of audiences listening to Internet radio as shown in Table 2.

Among 84 people who listen to Internet radio, the access frequency is as shown in Table 3.

Table 1. The rate of listening to radio of Hanoi citizens

	Number of People	**Percentage**
No listen	362 people	60%
Listen	238 people	40%
Total	*600 people*	*100%*

Source: Authors' self-investigation

Table 2. The rate of audiences listening to Internet radio in Hanoi

Access Frequency	Number of People	Ratio Over Radio Listeners in Particular (238 people)	Ratio Over the Total Number of Surveyed People in General (600 people)
Listen to Internet radio	84 people	35.3%	14%
Do not listen to Internet radio	154 people	64.7%	25.7%

Source: Authors' self-investigation

Table 3. Access frequency of Internet radio listeners

Access Frequency	Number of People	Percentage
Regular access (4-7 times / a week)	57 people	67.9%
Occasionally access (1-3 times/a week)	18 people	21.4%
Rarely access (several times / month)	9 people	10.7%
Total	**84 people**	**100%**

Source: Authors' self-investigation

The survey results show that Internet listeners often listen to news (48/84 people, 57.1%), then listen to music (21 people, 25%), and unintentional listen (15 people, 17.9%).

The above survey results shown that the rate of radio listeners is high (40%), but the public's ratio receiving Internet radio in Vietnam is still low (84/600 people, accounting for 14%). The reason for the high number of Hanoi citizen radio listeners is because many people use cars to move and listen to radio on the way to work (on buses, taxi, personal cars). However, the small survey mentioned above also shows that most radio listeners on cars still do not have the habit of listening to radio on the Internet.

According to our survey from December 2018 to February 2019, on the "Feedback"/"Comment" box attached to the Internet radio programs at vov1.vov.vn (the Voice of Vietnam), the response rate and interaction of the audience is not very low: *100% the Internet radio programs of the Voice of Vietnam have no responses/comments yet.* It proves that, the Vietnamese public is not yet interested in interaction. This is a major challenge for broadcasters to promote and attract listeners to Internet radio programs.

SUGGESTIONS TO IMPROVE THE EFFICIENCY OF INTERNET APPLICATION FOR MANAGERS IN PRODUCTION OF INTERNET RADIO PROGRAM IN VIETNAM

In Vietnam, the General Director of the Voice of Vietnam, As. Prof. Dr. Nguyen The Ky is a pioneering leader in innovating production of Internet radio programs in the digital era. He is the first leader in the field of radio decided to digitize all programs and upload all radio programs of the Station on the Internet. In National radio station known as *the Voice of Vietnam*, journalists also master the program digitization processes and Internet technical to present radio programs on the Internet. However, the reality shows that Internet radio is not only, and should not, be the only type of AM and FM radio digitized and replayed or live streamed on the Internet. The true Internet radio must be multimedia broadcast programs, produced specifically for broadcasting on the Internet, which we tentatively name it is "specific Internet radio programs". And therefore, the leaders and managers of radio agencies need to see the needed qualities for specific Internet radio. Radio agencies need to take full advantages of the Internet to organize and produce the radio program formats as follows:

Increasing the Production of Multimedia Radio Programs

Raw materials for building specific Internet radio program include:

- Information via audio (including speech, sounds, music)
- Information via video clip
- Information via pictures, tables, charts
- Information via texts.

In which, sound must still be the main language. However, audio flow on the Internet needs to be different from the audio flow of analog radio. If analog audio is a continuous time flow and linear streaming, then the sound on the Internet should be cut into small pieces, just about 3 to 5-minute duration. In line with arrays and sound pieces are titles and content in written language. This work ensures the public can be quickly accessed to the pieces of information that they are interested in, and does not waste time listening to all programs.

The attached image is also the image of reporters at studio or the field photos where the news happened... as *the Voice of Vietnam* has performed recently, however, the number of pictures should be used more in order to make the Internet radio programs more lively. Viewers can also receive information via photo channel.

Video clips should also be attached to ensure the diverse information. However, unlike normal video clips, the video clips included in the Internet multimedia radio programs need to be short. The capacity should be from less than 1 minute/clip. Its duty is illustration for audio content in the program. It is unnecessary to attach video clips with all news, articles in the radio program.

For information text, the content should not just the brief introduction about the program' main contents like it is currently, but it should be more detailed and specific. Furthermore, producers do attach special articles in multimedia radio program through reading channel to meet the public's demand to receive information.

And therefore, although audio information is the main means, but the Internet radio public still have opportunity to receive information by multiple senses. The key factor of the *"specific multimedia Internet radio"* is the ability to increase the information attractiveness and effectiveness of the program.

In order to implement multimedia radio programs, local radio leaders need:

Building a Multimedia Radio Reporter Staff

Currently, at the Voice of Vietnam, the capacity of journalists is quite high. 80% of reporters have college, university or higher levels degrees. Most reporters of the Voice of Vietnam are good at writing skill, can produce by themselves a simple analog radio program with just a smartphone and a networked computer contains audio editing software. They can also upload audio files/ radio program on the Internet by themselves, build images on laptops, build digital audio... However, not all journalists at the Voice of Vietnam can produce the *"specific multimedia Internet radio programs"*. So, building and shaping the radio journalist staff with standard skills needs to get the right and timely orientation of radio leaders. The synchronization of qualifications and capabilities of a staff includes journalists, editors and technicians at Vietnamese radio stations should be regarded as essential requirements. The radio station' leaders in Vietnam need to pay special attention to human resource investments for two blocks: content and technical blocks.

Technically, the radio and Television stations need to establish and build the digital technology pre-production and post-production system, transmission and broadcasting follows the HD standard; the network administration system takes the Internet as a center and establishes processes from editing and managing documents, sharing data through the Internet to all press types at the same time. Broadcast transmission technology also needs to be more diversified when exploiting on many means: satellite, cable, Internet, ground digital... Two samples are Bac Kan and Tuyen Quang radio and television have officially standardized HD input data and start broadcasting standard frames 16: 9.

Four elements that local radio stations need to pay attention to are: Exploiting scientific and technological achievements; taking the Internet as a center; training multimedia skills for journalists; exchanging information - from competition to cooperation. In order to create a foundation for this trend, first of all, the local radio and television stations need to focus on improving 2 main issues: Content and technology. (Le Duy Hoa, 2016).

Facilities Investment

Currently, the average Internet connection speed in Vietnam is 3.3 Mbps. Accordingly, during the period from May 30th, 2017 - May 29th, 2018, the global Internet speed measured from 163 million independent tests showed a result increased to 23%. The average broadband connection speed is 9.1 Mbps. Vietnam ranked 75th with an average download speed of 6.72 Mpbs (measured from June 2016 to July 2017) (Tan Phong, 2018).

The fact, Vietnam's technical infrastructure and Internet technology platform completely allows radio stations to invest in facilities to develop the quality of Internet radio programs. Radio stations should start producing the Internet radio format for mobile phones for mobile divides, because this is the most popular media means in Vietnam today.

A key factor in rising Internet use in Vietnam is the growing availability of web-enabled phones and other mobile devices. Over the past decade, the government has supported an aggressive expansion of the country's mobile data infrastructure to spur economic growth. Rising living standards have also fueled the expansion of Vietnam's smartphone market in recent years.

One result is that mobile phones have become the primary means of accessing the Internet in Vietnam, with eight in 10 weekly web users saying they used their mobiles to go online in the past seven days. By contrast, less than half (45.5%) say they have used a desktop computer to do so, and just over one-fourth (26.5%) have used a laptop (Broadcasting Board of Governor. n.d).

Internet Technology Application in Production of Internet Radio Programs in Vietnam

The problem depends only on the innovative thinking of radio leaders and the budget investment of the governing organ or the broadcasting agency themselves.

Update Radio Programs on the Network More Frequency

Internet radio programs are "extending" arms of radio stations. Therefore, the leaders and managers of radio agencies need to build a policy to allow update of radio programs online more frequency. As stated above, in Vietnam, there are only 35 local radio stations put the program on the Internet platform. Such numbers are not commensurate with the potential of a large online community with the number of Internet subscribers reaching more than 60 million people. While Vietnamese people don't have the habit of receiving information via radio, it is necessary for radio stations to create a habit of receiving Internet radio for the public. Stations need to enhance to advertise on analog radio about on online radio programs and their benefits. Simultaneously, increasing upload the live radio and re-played programs; focusing on the ways to increase the application of on demand radio programs.

However, to develop the format of on demand programs base on the basis of Internet technology is not simple, because the Vietnamese public currently has too many free communication channels to choose, so it is not necessary for them to buy on demand radio programs. In Vietnam today, according to the report of the Ministry of Information and Communications at *the Conference on summarizing the work of the first 6 months of 2018* (held in Ha Noi, July, 2018), until July, 2018, the country has 875 printed newspapers and magazines; 159 online newspapers and journals; 67 central and local radio and television stations. The number of radio and television programs is up to 182 channels for free, including: 103 television channels, 79 radio channels; 73 television channels and 9 radio channels pay fee. 54 foreign television channels are licensed to broadcast on pay TV systems... In this context, to have the properly on demand radio programs, it is necessary to improve the attractiveness of the Internet radio programs. As the statement by the General Director of *the Voice of Vietnam*:

The information market in Vietnam currently is developing fast and highly competitive. Radio information is in the fast segment and has fierce competition. Radio reporters are not far away from the core values of information - fast, accurate, so, knowing how to promote specific competitive advantages then multi-platform technology is really an opportunity for radio to develop (Nguyen The Ky, 2018).

Radio stations need to bring programs according to public demand, especialy international audience not what they already have – analog radio programs. The Vietnam radio's face needs to be presented to the world with a sharper to the world friends about a dynamic and beautiful Vietnamese image, with an advanced culture imbued with national identity. These must be quality Internet radio programs, delivered in English, or even better, in Chinese, French, Russian... Currently, in Vietnam, only the National radio station - *the Voice of Vietnam*, broadcasts English radio program on the Internet at the address: VOV5.vov.vn (VOV World). Listeners can click the address: vov.vn -> click vov5 -> click icon "radio 5 listen live" to listen lively the programs for oversea Vietnamese or for foreigners. This channel is broadcasted in 5 languages: English, French, Chinese, Russian, especially for ASEAN - Africa - Middle East communities. Or listeners can also listen to the English program just by clicking the "24/7 English" icon.

Changing the Management Methods (At the National Level as Well as Ministries, Branches and Localities Levels)

Vietnamese press managers need to realize the importance of grasping the Internet and telecommunications application opportunities in managing the field of Radio. In Vietnam, the latest 2016 Press Law, with its six chapters and 61 articles – 25 more than the old law - goes into effect on January 1st, 2017. However, the Law does not mention a series of emerging press formats, such as: Internet radio, multimedia radio, replayed radio, on -demand radio... Moreover, the Article 5 *"The State Policy on Journalism development"* and the Article 6 *"The State Management content on Journalism"* also are not mentioned about the State management on Internet radio. The issues about application of emerging Internet technologies in journalism producing and emerging technology management have not been mentioned in the Law, too. In other words, so far, Vietnam has not had a legal framework on Internet radio in order to help the State press agencies oversee the radio program production activities effectively.

In this situation, the Ministry of Information and Communications - the State management agency on press in Vietnam, needs to have orientation to be able to create a legal corridor for the type of Internet communication to develop, simultaneously, minimize the risk of information insecurity in the network environment. First of all, it is necessary to follow the *Network Security Law 2019* to adjust the scope and content of information on the Internet environment. Article 8 of Vietnam's *Law on Network Security 2019* prohibits acts of Abusing or abusing network security protection to infringe upon sovereignty, interests, national security, social order and safety, legal rights and interests of organizations or individuals or to profiteers. When radioonline.vn was born, besides the official radio channels of national and local radio stations, there will be foreign radio channels / programs broadcast in Vietnam, such as the *NHK channel* of Japan, the *VABC channel* - the overseas Vietnamese Radio channel, which are not affected by any party or organization; *My country voice* Radio channel... Because of the complexity and variety of radio channels, which leads to the complexity of the content, it is necessary to manage the content tightly and reasonably, to avoid exiting errors, especially the political mistakes. Secondly, there must be a communication manager staff who understand about information technology and radio journalism skills. Because the new Internet radio format appears, the production method is also completely new, so the management of production methods should also be set. Meanwhile, the management team at the Ministry of Information and Communications has been lacking in quantity and not yet strong in modern broadcasting skills.

The Ministry of Information and Communications should build solutions to encourage technology companies to participate in the communication market, especially providing solutions to ensure network information security. And because the Internet radio programs on a network-based environment will definitely bear the same status of online newspapers: being hacked and crippled at any time, the Radio agencies need to have a solid technology infrastructure system and technology experts to promptly handle the emergency situations on network security issues.

Applying Technology in Managing the Radio Stations; Investing New Technology Platforms at the Radio Stations; Investing in Improving the Network Technology Level for Radio Journalists

Radio stations need to invest in technological innovation - not only to buy robot journalism software, but also to buy infrastructure software to support for radio operation.

For example, radio stations should focus on investment in developing the terminal devices in producing, publishing and disseminating works and programs: producing Internet radio products suitable for smartphones, iPads and tablets.

Focusing on Communication Culture and Ethics Management for the Public

The Ministry of Information and Communication of Vietnam needs to pay to issues relating to communication ethics and conscious activities on the Internet environment, especially social network and social media environments. In the context of Internet radio, the public has the opportunity to participate in the debate, discussion and even participate in the production process. Public self-determination is very high. Therefore, the authorities need to direct the public so that all comments and discussions on the public's network are free within the framework, in line with recognized ethical and social laws.

CONCLUSION

In the digital age, all types of journalism must move to develop. If the press agencies know how to capture new technologies, there is no type of journalism goes backwards, including print or radio. Incorporating the emerging technologies flow, Internet radio is a trend of modern radio in Vietnam and the world. In Vietnam, Internet radio applications based on Internet technology are just the first steps, simple and primitive. Therefore, Vietnam needs a revolution to develop new types of communication with many advantages. *The Voice of Vietnam*, as a national radio agency, needs to be a pioneering and powerful unit in the application and production of Internet radio programs. Specific multimedia radio programs need to be invested and built by multimedia radio journalists. Furthermore, bring radio on Facebook, Youtube, Fanpage… is also an inevitable choice that Vietnamese radio stations need to apply to strongly attract the Vietnam and international audience.

REFERENCES

Broadcasting Board of Governor. (n.d.). *Young Vietnamese increasingly turning to online news sources over state TV*. Retrieved from https://www.bbg.gov/wp-content/media/.../Vietnam-research-brief.docx

Cerf, V. & Huddle, S. (2002). *Internet radio communication system*. Retrieved from https://en.wikipedia.org/wiki/Internet_radio

Compaine, B. M., & Smith, E. (2001). *Internet radio: A new engine for content diversity?* (p. 2). Retrieved from http://ebusiness.mit.edu/research/papers/131%20Compaine,%20Internet%20Radio.pdf

Deitz, C. (2018). *Internet radio - What is internet radio and how to listen to it*. Retrieved from http://radio.about.com/od/createyourownradioshow/a/blyourshowhub.htm

Le Huy Hoa. (2016). *Develop communication convergence model for radio stations*. Journalist Magazine No. 387 – May, 2016.

Kozamernik, F. & Mullane, M. (2015). *An introduction to internet radio,* EBU technical review. Retrieved from https://tech.ebu.ch/docs/techreview/trev_304-webcasting.pdf

XAPP Media. (2015). *Internet Radio Trends Report 2015.* pp.12, 11, 9, 13, 4, 5. Retrieved from https://xappmedia.com/wp-content/uploads/2015/01/Internet-Radio-Trends-Report-2015_january.pdf

Nguyen, K. (2018). *Do not take advantage, overcome challenges, radio will lose position.* Retrieved from vov.vn.

Phong, T. (2018). *Vietnam Internet speed is nearly 3 times faster than China.* Retrieved from http://chungta.vn/tin-tuc/cong-nghe/toc-do-internet-viet-nam-nhanh-gap-gan-3-lan-trung-quoc-66055.html)

Statistics & Facts. (n.d.). *U.S. Online Radio.* Available at https://www.statista.com/topics/1348/online-radio/

Statistics. (n.d.). Available at https://www.statista.com/statistics/252203/share-of-online-radio-listeners-in-the-us/; www.statista.com/statistics/253329/weekly-time-spent-with-online-radio-in-the-us/

Thu, N. T. (2014). *The problem of using multimedia materials for radio on the internet in Vietnam today.* (Master thesis), Academy of Journalism and Communication, p. 27.

Thuy, M. (2016). *Internet: Great opportunity of radio,* Retrieved from https://vov.vn/xa-hoi/dau-an-vov/Internet-co-hoi-lon-cua-phat-thanh-502781.vov

Van Kien, P., Hai, P. Q., Thang, P. C., & Hau, N. D. (2016). Some trends of modern communication press. Publisher of Information and Communication, p.151.

Wikipedia. (n.d.). Internet radio. Available at https://en.wikipedia.org/wiki/Internet_radio

Chapter 19
Blockchain Technology Applied to the Cocoa Export Supply Chain:
A Latin America Case

Mario Chong
Universidad del Pacífico, Peru

Eduardo Perez
Universidad del Pacífico, Peru

Jet Castilla
Universidad del Pacífico, Peru

Hernan Rosario
Universidad del Pacífico, Peru

ABSTRACT

This chapter recommends applying block chain technology to the cocoa supply chain. Using this technology, it will be possible to show and guarantee the traceability of the final product. Traceability in the cocoa chain begins in the production stages (harvest and post-harvest) to obtain relevant data related to cocoa beans and their producers, promptly, until finding the raw material origin and inputs used during the process. The material provider's name must be considered, as well as the manufacturer's expiration date, the batch number, and the production area's reception date. This is why authors recommend using Block chain, which is a data structure that stores information chronologically in interlinked blocks. It works as a digital master book and the participants reach an agreement to register any information in the blocks. Throughout the chapter, authors show how to apply this technology.

DOI: 10.4018/978-1-5225-9993-7.ch019

INTRODUCTION

Cocoa is a food rich in minerals, vitamins and fiber that has numerous benefits. Additionally, it has nutritional and therapeutic properties that are used for various products processing.

Presently, cocoa's supply tends to be less than its demand. The estimated projections made by international experts' point to a stock reduction in the main global production centers (African and Asian countries), thereby a price increase is predicted in the coming years (Romero 2015).

Cocoa is mainly known as input to produce chocolate (Morales *et al.* 2015). It is valued globally for its benefits. Its commercialization is influenced by European chocolate producers' demand.

In Peru, cocoa is produced since the XI century (Barrientos Felipa 2015). It has grown in an orderly and competitive manner, entering in an increasingly demanding global market that is willing to pay better prices even from New York and London Stock Exchange.

Due to the growth in cocoa production in some towns, many producers have financially benefited and, at the same time, the chain has contributed to improve the gross domestic product (GDP) (Swisscontact 2016). Most of the producers are small and integrated, directly or indirectly, to the international market with the challenge of achieving economic sustainability. The area from San Martin until Puno is tropical, ideal for cocoa cultivation. In 2015, Peru produced more than 85 thousand tons of cocoa. However, it is still predicted that there will be production potential and logistic competence.

Any strategy design within the framework of international supply must consider that local production is gathered in a small number of producers, exporters, local and international transportation services providers as well as public institutions (Barrientos Felipa 2015). In that sense, it is necessary to develop a stronger traceability mechanism that has the ability to be unalterable, generating trust in international buyers. For that purpose, this research will attempt to apply the block chain technology to the cocoa supply chain. This technology, invented in 2010, allows to show the traceability of any final product without centralized servers (Ge, Brewster, Spek, Smeenk y Top 2017).

This paper is divided into three sections: the performance of the cocoa value chain and the issue to be addressed, the fieldwork and proposals to improve the value chain, and, finally, the research findings.

For this reason, the objective of the project is focused in the proposal of applying block chain to the cocoa export supply chain, in order to identify and attack the factors that limit the international markets' access to information in real time.

As a consequence of the foregoing, it is necessary to answer the question, what are the factors that affect the cocoa logistics chain? The specific questions answered throughout the project are the following:

How important is it to count on a product and/or input traceability in real time? How could information uncertainty be reduced?

It is important to indicate that the hypothesis which this project maintains is that the logistics chain's participants don't have an adequate tool capable of providing digital traceability of information in real time.

Cocoa in Peru

Different cocoa varieties exist in Peru. Around 60% of cocoa varieties in the world. This is because they were introduced from the Caribbean, Central America and Ecuador, also because there were crossings with native varieties. The main cultivation areas in Peru are the valleys of La Convención (Cuzco), Huallaga (Huanuco and San Martin), Marañon (Cajamarca and Amazonas), Tambo (Junin) and the Apurimac-Ene River or VRAE (Ayacucho, Cuzco and Junin).

Figure 1. Agricultural exports 2013 -2017(value in millions of dollars)
Source: Sunat, 2018. Elaboration Minagri - DGESEP-DEA, 2018.

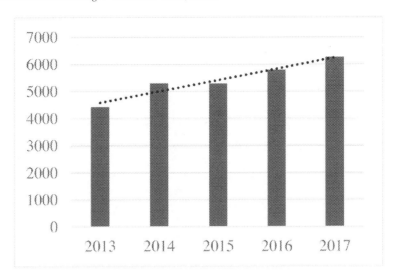

The Peruvian crops are focused in the departments of San Martin and Huanuco. In San Martin, with 28.984 hectares, there are 8% of 'creole' and 'native', 90% of 'CCN51' and 2% of 'trinitary' plus 'foreign'. In Huanuco, for a total amount of 4.201 hectares, there are 45% of 'creole' and 'native', 50% of 'CCN51' and 5% of 'trinitary' plus 'foreign'.

According to the general manager of La Cooperativa Agraria Industrial Naranjillo (The Industrial Agricultural Cooperative Naranjillo), one of the largest national cocoa producers, Huanuco has more advantages than San Martin. The cultivators from Huanuco have many cocoa varieties mixed in their plots. 50% 'trinitary', 40% 'CCN51' and 10% 'native'.

Other types of creole cocoa are produced in Peru such as white bean, also known as 'porcelain', grown in the provinces of Morropon and Huancabamba, department of Piura, as well as the 'chuncho' cocoa bean, produced in the province of La Convencion, in the department of Cuzco.

Agricultural Exports Market in Peru

At the end of the year 2017, the agro-exports reached US$ 6.255 million and had an increase of 8% in comparison to the year 2016 and the agricultural trade balance registered a surplus of US$ 1.645 million (Dirección General de Seguimiento y Evaluación de Políticas del Ministerio de Agricultura y Riego 2017). Below we have a comparative table about the agricultural exportations' development since 2013.

Agricultural Trade Balance

In 2017, there was a positive agricultural trade balance of US$ 1.645 million (Dirección General de Seguimiento y Evaluación de Políticas del Ministerio de Agricultura y Riego 2017).

The products that stood out for their higher FOB export values in comparison to 2016 were the following:

Figure 2. Agricultural trade balance 2013 - 2017. FOB value in millions of dollars
Source: Sunat, 2018. Elaboration Minagri - DGESEP-DEA, 2018.

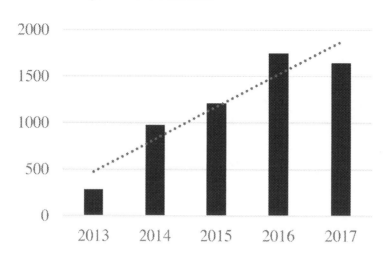

Export to International Markets

Among the remarkable ones, we have the Netherlands, Belgium, United States and Italy with 37%, 13%, 8% and 7% of participation, respectively.

Traceability

The European parliament, in the regulation (EC) 178/2002, defines traceability as the possibility to trace and follow a food, feed, food-producing animal or substance intended to be, or expected to be incorporated into a food or feed, through all stages of production, processing and distribution (Europeas 2002).

According to the European parliament regulation, food traceability is important to be able to detect if there is any product contamination in the production chain and be able to take immediate action such as removing the product specifically and precisely. Another topic about traceability is that it can help with historic information for some specific foods research.

There are concepts associated to traceability as tracking and tracing. Enríquez (2010) states that tracking is the capability to follow a product's route through the supply chain as its movements through

Table 1. Products that stood out for their higher FOB values

Agricultural Product	FOB Value Variation
Fresh avocado	+46%
Fresh blueberries and cranberries	+52%
Animal feeding formula	+40%
Quinoa	+18%
Fresh artichokes	+14%

Source: Sunat, 2018. Elaboration Minagri - DGESEP-DEA, 2018.

Figure 3. International markets receiving exports from Peru
Source: Ministerio de Transportes y Comunicaciones, 2015.

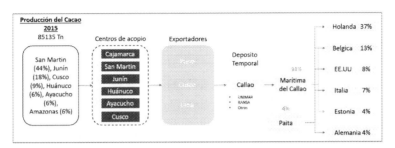

organizations. On the other hand, tracing is the capability to identify a particular unit and/or batch's origin inside a productive chain. For that reason, the route is made from back to front.

Traceability is an important aspect in the current global trends mainly in public health policies. Quispe (2009) points that traceability grew in importance when developed countries started to show concern towards food safety and the quality that needed to be offered from a public health point of view. This is due to the occurrence of pollution cases that put said countries' population at risk. Quispe (2009) mentions other traceability benefits: impact reduction of a deficient shipment through allowing a quick location of the product to be removed, stock control improvement through amount reduction of the stored inventory and consumer trust increase in the products' safety and quality. Also, due to traceability, higher standards would be reached, standards demanded by companies that request products properly identified and labeled. Finally, it also becomes a tool for business decision making due to the access to permanent product information.

Table 2. Cocoa value chain

Input and Research	Good practices with the seedling to guarantee agricultural performance.
Crop	Cocoa plantations grow at less than 23° of the Equator. Cocoa is sensitive to diseases and climate, therefore, it needs correct fertilization and crop protection.
Harvesting	On average, two harvest a year. The pods are picked and cut manually with scissors or machete. After the harvest, the pods are broken and the beans are extracted.
Fermenting and drying	The beans are fermented to develop the cocoa flavor and aroma. They are covered with banana leaves or placed in wooden boxes. The drying process is as important as the fermentation, for this reason it is necessary for the beans to dry slowly for them to lose humidity and acidity.
Trading and exporting	The beans are packed, bagged and stored. The buyer does a quality check.
Roasting and Grinding	The beans are cleaned, shelled and roasted. The inner part of the cocoa bean is called nib and in the heating process, the "cocoa liquor" is formed. It is important to highlight that this cocoa liquor does not have alcohol and solidifies at room temperature. This paste (solidified liquor) can be sold as unsweetened baking chocolate or be used for chocolate manufacturing.
Pressing	The cocoa liquor is poured in presses. With this process, we can separate the cocoa butter and cocoa cake.
Chocolate Manufacturing	Cocoa liquor is mixed with cocoa butter, sugar and, if desired, with powdered milk. It is poured in molds and tempered. Among other ingredients, we have nuts and dried fruits to then sold to bakeries and confectioners.
Consumption	Chocolate is enjoyed in hundred different forms.

Fuente: Swisscontact, 2016.

EAN-UCC Standards for Traceability

- The multindustrial system standard, EAN-UCC, offers commercial solutions for all industries across the world. When companies adopt and integrate the EAN-UCC identification and commercial communication standards, they possess a total visibility of merchandise and services in the administrative, supply and logistics processes. This includes all parts of the chain: manufacturing, distribution, transportation and final consumers. Currently, companies that come from a wide range of industries benefit from the EAN-UCC system, including the retail, food, healthcare, transportation, public purchases and defense sector, services and computer and equipment manufacturer (Argentina 2003).
- The Global Trade Item Number, GTIN, is a code to identify commercial products. Its most common representation is the bar code that allows us to read electronically in sale point, which is a reception point at the warehouse or at any other point in the business processes where it is required. GTIN is used to identify uniquely any product or item whose specific preset information is needed and can be quoted, required or billed at any point in any supply chain. This definition includes raw materials, finish products, inputs and services (GS1-Peru 2018).

Bar Code

Bar code is an automatic information technology that allows to identify items and services, regardless of its origin or destination. It permits an automatic information reading through a numeric and/or alphabetic code, which is represented graphically with a rectangular symbol composed by bars and parallel spaces. A fix or handheld scanner performs the reading that identifies the item with no error.

The GS1 bar code is a standard coding system that identifies each reference in a unique an unambiguous way. No two codes are ever the same all over the world, such as no two fingerprints are ever the same. More than 25 years ago, GS1 Peru worked in our country with the purpose of enhancing the standard identification quality in consumption units, groups and locations. This is done to facilitate the operations between producers, wholesalers, distributors, retailers and final consumers. It is based on the tools that GS1 has developed to optimize and make more efficient the supply chain (GS1 2018). Given the companies diversity, types of products and variety of distribution forms of the market, below, it is present some basic rules that must be respected when assigning codes. Product coding impact is especially relevant in the physical and database management.

The application of coding is considered under the following general concept: "every product modification perceived by the final consumer has a different GS1 code".

Block Chain Technology Concept

Block chain is a data structure that stores information chronologically in blocks that are linked to each other. It operates like a digital master book where the participants reach an agreement to record any information in the blocks (MIT Center for Transportation & Logistics, 2018). Block chain was created as a support platform for the first digital currency that is Bitcoin. The timeline is presented below.

Block Chain Elements

- **Network**: Companies and organizations that share information and transactions.
- **Assets**: It is not just money but also physical items that need to be digitized, for instance, bill of lading, notice of receipt, among others.
- **Ledger**: Support where all relevant data is recorded.

The benefits that block chain has are the following:

1. Processing capacity (MIT Center for Transportation & Logistics 2018)
   - Participants can have access in real time
   - Anti-handling test. If any actor of the chain writes some mistaken information, this cannot be deleted. An amendment must be drafted.
   - The access cost is decreasing ever more.
2. Scalable with other technologies (MIT Center for Transportation & Logistics 2018).
   - Smart contracts, with data coding, the obligations from a contract may be written in the block chain and, if they are met, the terms are executed. For example: payment, resupply terms, package, weight, volume accordance terms, etc.
   - IoT, Internet of Things, meaning, a smart contract that could be issued by a machine that the final user can utilize, such as the soda machine. This vending machine, as it runs out of supply, orders a replenishment and, at the same time pays the provider.
3. Underlying radical philosophy (MIT Center for Transportation & Logistics 2018)
   - Disseminated Consensus: instead of tracing with independent or centralized servers, in the block chain, data is distributed and decentralized. Open source.
   - Transparency: each participant can know what is in every link of the chain.

Application 1: Traceability

The block chain technology helps consumers and buyers in the supply chain see the property history and relevant information about the asset in the transaction (MIT Center for Transportation & Logistics 2018). It combats falsification and fraud and prevents health and security dangers.

In Diamonds

In the precious stones industry, there is a risk of not knowing the stone authentically, or if the investments made for exploration are financed by money laundering funds. For this reason, a very important company in this sector is implementing the block chain technology to give access to every piece of information and relevant history about the diamond's origin to the gem buyers, ring producers and sellers. Security controls are performed since the initial extraction to the final point of sale.

In Tuna

Block chain is used not only for luxury products but also for edibles. The complexity of modern food chains has distanced consumers and producers. There are a lot of lawsuits about foods with false certificates, risk of fraud (selling low quality products labeled as high quality products) and adulterations (Ge, Brewster, Spek, Smeenk & Top 2017).

In the case of tuna, every fish is registered in the block chain and the company responsible for cutting is recorded as well as the packing company so that, at the end, the final consumer at the supermarket can see the traceability (who fished it, who filleted it and what other relevant processes the fish had) by scanning the QR code in the fillet can.

Application 2: International Trade

Block chain application to international trade will allow documentation automatization. The actors will have safer faster transactions; the logistics cost would be potentially reduced.

According to CNBC (2018), HSBC made its first commercial financial transaction using block chain. According to information from CNN, published on May 15th 2018, FedEx is testing the block chain technology for critical load shipment. According to information from Reuters, published on February 16th 2018, Maersk and IBM are preparing to launch a block chain platform for global trade.

The Soy Exportation Case From Argentina-Cargill

Argentinian soy is shipped to Malaysia. The banks in this operation are HSBC and ING. Platform Corda was used, a platform developed by a consortium of 100 banks. The time needed for the credit documents approval was 24 hours, whereas, traditionally, it would take five to ten days. In this transaction all parts were linked in the block chain, which is why no document reconciliation was needed.

Block Chain Technology Operating

How does block chain technology works? Block chain means a chain of blocks. Every transaction is recorded in the blocks. Every block has a hash, meaning a block identifier code. Besides, every block has the previous hash so that it's linked and can continue making more blocks and order transactions. Thus, the next block makes the previous block stronger and thereby, the entire chain (Gupta 2017).

It is useful to be able to model businesses on a flow of activity level. That is how it is possible to identify the participants inside the supply chain (Gil, Aspegren y Gómez Lopez 2017). Below, we show the flows for each participant in the chain:

Participant 1: The Farmer

The farmer, being the first link in the chain, plays a very important role which creates the first block in the system and who guarantees that the mineral cadmium level is less than 0.1%. The farmer records in the block chain the amount in weight of what he is going to give the exporter, the origin and certificate of analysis (COA) results. After doing this, he goes ahead and distributes the merchandise to the exporters' warehouses.

Figure 4. Participant 1, Farmer
Source: Prepared by the authors

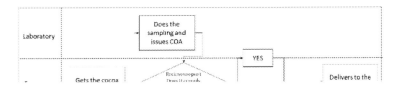

Participant 2: The Exporting Company

The exporter is in Lima city. It may have several warehouses where he keeps the cocoa seeds stock and where it may produce other by-products. This exporter enters the block chain twice: the first time to record what it has received, and the second time to initiate the export dispatch with the merchandise shipment to the temporary storage in the port. In this last step, he must record the commercial invoice number and the shipping document number with the customs temporary storage address. It ends the flow by passing the mantle to the customs agency.

Participant 3 and 4: Temporary Storage and Customs Agency

The supply chain flow goes on and by receiving the merchandise, the temporary storage comes into place. This company, a privately managed company in Peru, will have to record in the block chain key information such as the received weight and merchandise status. Meanwhile, the customs agency starts the export dispatch before Aduana Marítima del Callao (Maritime Customs of Callao) and is granted the release to export.

The temporary storage drives the merchandise into the container ship. Once embarked, the customs agency settles the dispatch.

It must record in the block chain the bill of lading (B/L) issue date as well as the shipment date. It will take 20 to 30 days for the merchandise to arrive to its destination. For this research purposes, we assume the destination country to be Italy.

Figure 5. Participant 2, the exporter
Source: Prepared by the authors

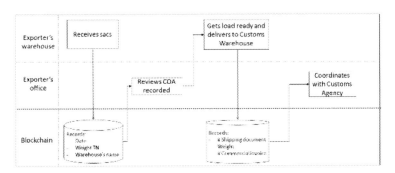

Figure 6. Participants 3 and 4
Source: Prepared by the authors

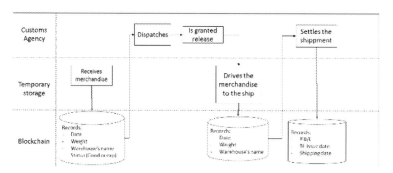

Participants 5 and 6: Customs Warehouse and Importer

Finally, the destination's customs warehouse is who receives the merchandise. In this point must record weight, name, port of discharge and the load status when arrived. It is important to state that blockchain automatically registers the transaction's record date and the notifications recorded. Once the importer clears the shipment the temporary storage must register the departure with weight, name and port. The importer is the last one that records in the block chain its merchandise reception with weight and name in his warehouse. This way, everyone should see cocoa's traceability in a distributed and unalterable way.

Analysis of the Results

In this section, we analyze the results that were obtained from the block chain implementing in the supply chain.

Each participant can register their activities in the block chain without entering a new system. For example, the customs agency in Peru will be able to use its platform 'Sistema de Aduanas' (Sintad) in an ordinary and automatic way.

According to certain clients, exporter and consigner, Sintad can record key information in the block chain. However, during the research, we found that not all farmers and cooperative count on a stock, warehouses or distribution management software. Because of this a website could be enabled where any participant could register movements in their block chain and see a product traceability in a practical way. Below a view of this platform.

In Figure 7, we can see the complete traceability from the farmer until the merchandise arrival to the importer's warehouses. This is the interface where the major book can be seen. The first column is entitled 'code' and shows the amount of blocks. In the column 'type of document', it could be an analysis certificate, an international transportation document or any other key document all participants have the right to know. It can also include notifications, entries, and exits from the warehouses involved.

The issue date refers to any specific document whereas the transaction date shows the day (it could also be set to show the time) the participant creates the block. The description column shows the key information that the participants need to know in the supply chain flow. For this research purposes, we have prioritized the elements evaluation result in cocoa like cadmium, the weight in metric tons, the warehouses names and the merchandise status upon reception. Every node in the supply chain has

Figure 7. Block chain website platform view
Source: Prepared by authors, 2018.

freedom to decide and agree on what information and documents are relevant and important to keep in the traceability. This could be subject to the commercial and legislative juncture.

If we go back to the start point, the farmer that didn't have access to a management software, would see the following table to register his transaction:

Research Description

This chapter will show the research made in order to know how block chain would be implemented operationally in the supply chain.

Hyperledger Fabric, from now on "HF", is one of the platforms used to build a block chain according to the supply chain participants needs (Medina 2018). HF was donated by IBM. In this platform, the smartcontracts are programmed in language Go from Google, which is not so popular yet. An extensive knowledge in programming is required to design the model in HF, which is why the access bar is higher for the interested public as well as for los propios. Nevertheless, in the Hyperledger project we found the 'composer' tool that gives us the APIS necessary to have a faster programming. Smartcontracts are not programmed in Go language but in Java Script, which is the most common and used language.

Some characteristics about Hyperledger Fabric are the following:

- **Permissible**: The nodes do not necessarily have Access to the complete block chain, in other words, the participants themselves can come to an agreement to grant Access to external users.
- **Multi-Channel Capability**: There can be several private networks inside a block chain.
- There is no mining as for bitcoin. There is no reward for making a transaction.
- It handles cryptography and digital certificates.

On the other hand, Hyperledger Composer, from now on HC, permits the blockchain to have a fast and simple development. The value is to accelerate the development through APIS not to access HF directly. For this research, we will apply HC to work on the demo.

It is in Hyperledger Fabric where all blockchain operations take place. Hyperledger Composer is alike the development framework.

It has an advance complexity to create a Project BC. This is why HC allows to make the BC in an easier way. The composer's most important function is modelling businesses.

Table 3. Business actors and stages' description

Actors/Stages	Description
Participants	In our chain, there are three: the provider, the exporter and the importer.
Roles	Provider: to deliver the raw material, cadmium free, to the exporter, to register the bags' delivery and the COA in the BC. Exporter: to gather the entry loads, to group COAs, to ship with final export documentation. Importer: to start the import dispatch. To receive merchandise.
Actors/stages	**Description**
Transactions	Documentation numbers and dates.
Assets	COA.
Interaction with the BC	It will be six accesses. One for each.

Source: Prepared by the authors, 2018.

Five different files can be created:

- ".cto" participants, assets and transactions using composer language.
- ".js" where the smartcontracts' logic is defined in Java.
- ".acl" participants' access is defined.
- ".qry" consults are defined.
- ".bna" encapsules the foregoing.

Blockchain Technological System Design

Inside the composer tab in Playground, we designed the supply chain model mentioned in the previous chapter. This file has an extension ".cto". It is here where the following must be indicated:

- Type of document for B/L or COA.
- The participants as the farmer, the exporter, the temporary storage, the customs agency, the customs warehouse at the destination and the importer at the destination.
- The product, which is cocoa seed for this research.
- The transaction or participants' communication about which key activity was just performed.

Here below, we present the system's view:

Once we have the BNA, we test it by creating the participants and transactions in javascript. Then, we present the view where we can see the model working correctly.

It can be noticed that the transactions created are traceable in the system. It is worth noting that this will not be the view that the blockchain's participants will have, since an advanced proficiency in programming is required. What you are seeing is the technical origin of blockchain programming.

Finally, if we see each transaction's record in the system, we will have a block with programming language as showed in the next image.

Figure 8. Hyperledger composer application system view

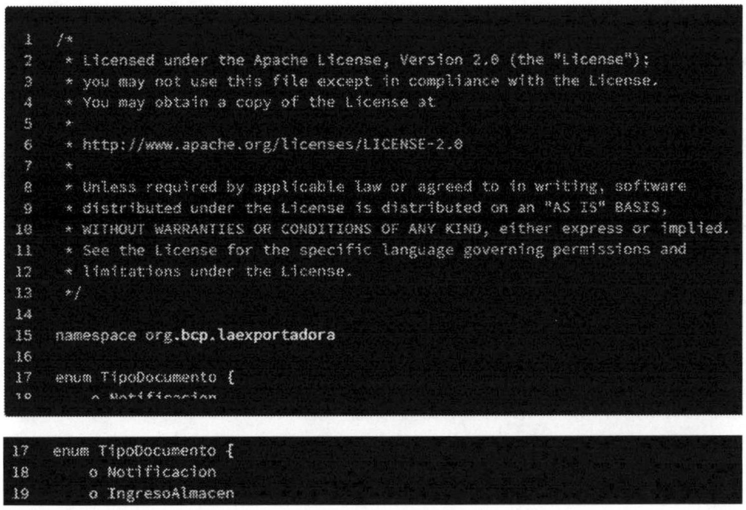

```
1  /*
2   * Licensed under the Apache License, Version 2.0 (the "License");
3   * you may not use this file except in compliance with the License.
4   * You may obtain a copy of the License at
5   *
6   * http://www.apache.org/licenses/LICENSE-2.0
7   *
8   * Unless required by applicable law or agreed to in writing, software
9   * distributed under the License is distributed on an "AS IS" BASIS,
10  * WITHOUT WARRANTIES OR CONDITIONS OF ANY KIND, either express or implied.
11  * See the License for the specific language governing permissions and
12  * limitations under the License.
13  */
14
15  namespace org.bcp.laexportadora
16
17  enum TipoDocumento {
```

```
17  enum TipoDocumento {
18      o Notificacion
19      o IngresoAlmacen
```

CONCLUSION

In Peru, blockchain technology is in its beginnings (Asparza 2018). There is still lack of knowledge and usage standards. For this reason, many entrepreneurs consider the tool to be too complex. It is necessary to point that, during the research, no technical blockchain course was found in Peru. This forces the current programmers to look for external sources and to adapt their realities to overseas company models.

Blockchain building can not only be made by the staff through ERP, as supported in this research, but also through other devices such as drones (Jimenez 2018). SAI Technologies Company has started offering, since 2019, plague diagnosis and identification in farmland through a spectrometer. The drone flies through the farmer's cocoa hectares taking radiographies and key information. This information

Figure 9. Hyperledger composer view playground
Source: Prepared by the authors, 2018.

Figure 10. Model functioning view
Source: Prepared by the authors, 2018.

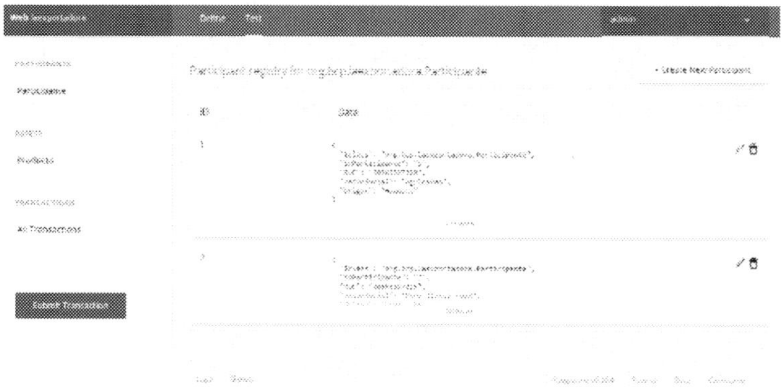

Figure 11. Technical origin of blockchain programming
Source: Prepared by the authors, 2018.

is analyzed and recorded in the supply chain's blockchain. This initiative would be a help in improving cocoa traceability because it would begin from the sowing to the yield development.

Society's trust in institutionalism has reduced. Truth is ever more valuable (Esparza and Nicastro 2018). Blockchain will be in charge of defending the truth. No more medicine theft, expired products delivery or donation misappropriation. This technology will force companies to be transparent before their clients, especially if it belongs to the food industry. Finally, the hypothesis, which states that the

Figure 12. Block display with programming language
Source: Prepared by the authors, 2018

cocoa chain supply members do not have a digital traceability management tool, is sustained. Currently, every node handles its own form of traceability through the EAN code managed only by the exporter. The order to get to a healthy traceability system would be to standardize the EAN codes with the client, add GTS from GS1 and, finally, record in the blockchain the key data improve the visibility supply chain.

We are in an economy where every company is seen as an isolated entity and not as part of a supply chain. It is necessary to create ecosystems that, in their own chain, promote cooperation instead of competitiveness. In Peru there is a certain degree of uncertainty, given that there is no legislation about this topic. It is still unknown if there will be legal obstacles or incentives in the future for the use of blockchain.

REFERENCES

Asparza, M. (2018). *Dialogo sobre Blockchain en Perú*. [Interview]

Barrientos Felipa, P. (2015). *La cadena de valor del cacao en Perú y su oportunidad en el mercado mundial*. Medellín, Colombia: Universidad de Medellín. doi:10.22395eec.v18n37a5

Catálogo Flora Valle Aburra. (n.d.). Theobroma cacao. In *Portal institucional Catálogo Flora Valle Aburra*. Retrieved from https://catalogofloravalleaburra.eia.edu.co/species/63

Chopra, S., & Meindl, P. (2013). *Administración de la Cadena de Sumnistro*. D.F, México: Pearson.

Comercio, E. (2017). La increíble y terrible historia tras el origen del chocolate. In *El Comercio*. March 19, 2017. Retrieved from https://elcomercio.pe/redes-sociales/youtube/increible-terrible-historia-origen-chocolate-145532

Diario Oficial de las Comunidades Empresa. (2002). Reglamento (Ce) No 178/2002 del Parlamento Europeo y del Consejo. In *Diario Oficial de las Comunidades Empresa*. Retrieved from https://eur-lex.europa.eu/legal-content/ES/TXT/PDF/?uri=CELEX:32002R0178&from=IT>

Dirección General de Seguimiento y Evaluación de Políticas del Ministerio de Agricultura y Riego. (2017). *Comercio Exterior Agrario*. Lima, Peru: Ministerio de Agricultura y Riego.

Dirección General de Seguimiento y Evaluación de Políticas del Ministerio de Agricultura y Riego. (2017). Enero-diciembre 2017. In *Portal institucional Ministerio de Cultura.* Retrieved from http://siea.minag.gob.pe/siea/sites/default/files/nota-comercio-exterior-diciembre17_3.pdf>

Enríquez, G. (2010). *Trazabilidad: cuando proteger la vida se configura.* Lima, Peru.

Escudero, J. (2018). *Aseguramiento de Calidad en implementación de tecnologías.* [Interview].

Esparza, M., & Nicastro, M. (2018). *Blockchain is life.* Lima, Peru: Saxo.

GS1. (2003). Tranzabilidad. *Portal institucional GS1.* Retrieved from https://www.gs1.org.ar/documentos/TRAZABILIDAD.pdf>

GS1. (2016). *Global traceability compliance criteria for food application standard.*

GS1. (2018). Visibilidad y tranzabilidad. *Portal GS1.* Retrieved from http://www.gs1pe.org/content/visibilidad-y-trazabilidad>

GS1. (2018). Bienvenido a GS1 Perú. *Portal GS1.* Retrieved from http://www.gs1pe.org/content/codificaci%C3%B3n-gln-c%C3%B3digos-de-localizaci%C3%B3n

GS1. (2018). Código de barras. *Portal GS1.* Retrieved from http://www.gs1pe.org/content/gs1-c%C3%B3digo-de-barras

Ge, L., Brewster, C., Spek, J., Smeenk, A., & Top, J. (2017). *Blockchain for agriculture and Food; Findings from a pilot study.* Wageningen, The Netherlands: Wageningen Economic Research. doi:10.18174/426747

Genesis, G. (2011). La historia del cacao en la época prehispánica. *Portal La Historia del Cacao.* Retrieved from http://lahistoriadelcacao.blogspot.com/

Gil, N., Aspegren, H., & Gomez Lopez, A. M. (2017). *Report on blockchain asset registries.* Cambridge, MA: MIT.

Gupta, M. (2017). *Blockchain for dummies.* Hoboken, NJ: John Wiley & Sons.

Internacional, C. C.-N.-S. (2017). Catalogo flora Valle Aburra. [Online]. Retrieved from https://catalogofloravalleaburra.eia.edu.co/

International Trade Centre. (2018). Trade Map. *Portal institucional International Trade Center.* Retrieved from https://www.trademap.org/Country_SelProduct_TS.aspx?nvpm=3|||||180|||4|1|1|2|2|1|2|2|1>

Jiménez, M. (2018). *Operatividad de la Blockchain.* [Interview].

Keck, A. R. (2018). *Experiencias sobre AgUnity - Blockchain en agricultores del Cacao.*

Magazine (2014). *Pastry revolution.* [Online]. Retrieved from http://www.pastryrevolution.es/pasteleria/productores/

Manos, B., & Manikas, I. (2010). Traceability in the Greek fresh produce sector: Drivers and constraints. *British Food Journal, 112*(6), 640–652. doi:10.1108/00070701011052727

Map, T. (2017). *Trade map.* [En línea]. Available at https://www.trademap.org/Index.aspx

Medina, Y. (2018). *Programación Blockchain*. Lima, Peru: Hyperledger Composer.

Ministerio de Agricultura y Riego. (2018). Comercio exterior agrario. *Portal institucional Ministerio de Agricultura y Riego*. Retrieved from http://siea.minag.gob.pe/siea/sites/default/files/nota-comercio-exterior-diciembre17_3.pdf

MIT Center for Transportation & Logistics. (2018). *Blockchain in Supply Chains*. Boston: MA.

Moe T. (1998). Perspectives on traceability in food manufacture. *Trends in Food Science and Technology*. doi:10.1016/S0924-2244(98)00037-5

Morales, O., Borda, A., Argandoña, A., Farach, R., Garcia Naranjo, L., & Lazo, K. (2015). *La Alianza Cacao Perú y la cadena productiva del cacao fino de aroma*. Lima, Peru: Esan.

Morales, O. & Borda, A. (2015). La alianza cacao Perú y la cadena productiva del cacao fino de aroma. *Gerencia para el desarrollo, 49,* p. 19-30.

PMI. (2013). *Guía de los fundamentos para la dirección de proyectos. Pensilvania*. Project Management Institute.

Quispe, S. (2009). *Trazabilidad y gestión agroalimentaria*. Lima, Peru: Agroenfoque.

Ràbale, A. (2006). *Buyer – supplier relationship's influence on traceability implementation in the vegetable industry*. [Online] doi:10.1016/j.pursup.2006.02.003

Pastry Revolution. (2014). 10 grandes productores de cacao. In *Pastry Revolution*. Retrieved from http://www.pastryrevolution.es/pasteleria/productores/Z

Rivas, T. (2011). *Trazabilidad en la Industria Alimentaria*. Salamanca, Spain: Universidad de Salamanca.

Romero, C. A. (2015). *Estudio del Cacao en el Perú y el mundo*. Lima, Peru: Ministerio de Agricultura y Riego.

Schwagele, F. (2005). *Traceability from a European perspective*. [Online]. doi:10.1016/j.meatsci.2005.03.002

Swisscontact. (2016). *Desarrollo de la cadena de valor de cacao*. Zúrich, Switzerland: Swiss Foundation for Technical Cooperation.

Vela, E. V. (2017). *Competitividad del Comercio Exterior*. Lima, Peru.

Vera, G. (2018). Tipos de cacao: forastero, cacao y trinitario. In *Cocina y Vino*. Retrieved from http://www.cocinayvino.com/mundo-gourmet/tipos-cacao-forastero-criollo-trinitario/

World Bank. (2018). Datos Banco Mundial. *Portal institucional Banco Mundial*. Retrieved from https://datos.bancomundial.org/indicador/LP.EXP.DURS.MD?end=2016&locations=PE-BR-CL-CO-EC&start=2016&view=bar>

Chapter 20
Agile Teams in Digital Media:
A 13 Week Retrospective

Rachel Ralph
https://orcid.org/0000-0001-8661-8343
Centre for Digital Media, Canada

Patrick Pennefather
https://orcid.org/0000-0002-1936-1872
University of British Columbia, Canada

ABSTRACT

As we move towards the third decade of the 21st century, the development of emerging technologies continues to grow alongside innovative practices in digital media environments. This chapter presents a comparative case study of two teams (Team A and Team B) in a professional master's program during a 13-week, project-based course. Based on the role of documentation and the reflective practitioner, team blogs representing learner experiences of Agile practices were analyzed. This case study chapter focused on one blog post of a mid-term release retrospective. The results of this case study are framed around Derby and Larson's (2006) Agile retrospectives framework, including: set the stage, gather data, generating insights, deciding what to do, and closing the retrospective. The case study results suggest the need for public documentation of retrospectives and how this can be challenging with non-disclosure agreements. Also, the authors identify the importance of being a reflective practitioner. Future research on educational and professional practices needs to be explored.

INTRODUCTION

As we move into the third decade of the 21st century, the development of emerging technologies continues to exponentially grow as with the demand for innovative practices in digital media environments. A variety of Agile approaches that champion iterative product development more commonly used in software companies, challenge previously established hierarchical waterfall techniques (Appelo, 2011, 2016). As we progress, a better understanding of best practices within Agile project management need

DOI: 10.4018/978-1-5225-9993-7.ch020

to be articulated. One such practice is reflection through the application of retrospective tools. The field would benefit from understanding not only what retrospective tools teams are successfully (or not successfully) using, but also the impact of these reflective tools on iterative versions of digital prototypes. As researchers, we can better understand Agile practices by investigating special cases. Part of that understanding requires teams to document their reflective processes so that these can be examined, interpreted, and shared publicly.

Chapter Overview

This chapter will present a case study of two teams (Team A and Team B) in a professional master's program. Both teams applied Agile practices during a 13-week project-based course. Team blogs representing learner experiences of Agile practices will be analyzed, highlighting reflections that occurred during a mid-term release retrospective. To remain integral to the structure of an Agile retrospective, the chapter will be presented using Derby and Larsen (2006) framework. In particular, the chapter will detail the following sections considering the reflective nature of an Agile retrospective: *setting the stage, gathering data, gathering insights, deciding what to do, and closing the retrospective.*

To *set the stage* (commonly referred to as the Background section), we offer a definition of an Agile retrospective, drawn from the theoretical literature of the reflective practitioner and discuss the role of documentation through retrospective activities. In the *gathering data* (also known as Results) section, we will detail the methods used to capture mid-term team retrospectives through a comparative case study of two team blogs. The *gathering insights* section most resembles a typical data discussion in which team-collected data will be discussed based on themes that emerged from the data. Learner actions will be examined and explained in the larger context of what actions they proposed to improve throughout subsequent project development sprints. In the *deciding what to do* section, limitations of this case study, key values of reflective practices, and recommendations for future research will be identified. In the *closing the retrospective* section we will offer readers some concluding remarks on the role of retrospectives in Agile project development cycles.

SET THE STAGE

To set the stage for an Agile software development process it is important for teams to understand the definition, purpose, and features of a retrospective. For Derby and Larsen (2006), a retrospective allows teams to reflect on work previously completed, so as to best prepare them for the work that will be done in the future. We will set the stage by discussing: how teams conduct retrospectives, the role of reflection, and the importance of documentation.

Understanding How Teams Conduct Retrospectives

A retrospective is an opportunity for team members to reflect on any aspect of the project at the end of a small-time cycle. These shorter time cycles are often referred to as *sprints* in Agile methodological processes, defined by the activities that occur within a set period of time in order to allow for frequent review and refinement. For example, understanding how teams complete tasks that together contribute to a prototype's feature, and how a particular feature is reflective of the core experience that the team is designing for users. The retrospective provides the team with the ability to adapt their approach and activities in subsequent sprints. During the retrospective team members reflect on the process and decide what actions they will need to take to continue a project's development moving forward. While Agile teams propose their own structures for conducting a retrospective, typically guided by a ScrumMaster®, most do not define their approach to reflection. In other words, how-to-reflect during a retrospective, is generally not discussed. To better understand why this is the case we draw from Polanyi (1966) notion that "all knowledge has tacit dimensions" .

At one extreme it is almost completely tacit, that is, semiconscious and unconscious knowledge held in peoples' heads and bodies. At the other end of the spectrum, knowledge is almost completely explicit, or codified, structured, and accessible to people other than the individuals originating it. (Leonard & Sensiper, 1998, p. 113)

Individual tacit knowledge is the type of know-how that comes through experience (Cook & Brown, 1999).Team members with more experience taking part in retrospectives develop a codified way in which it is conducted, often making it difficult for those new to a team to engage fully or to understand the value of reflection in a time-pressured work environment. For this reason, we find it may be helpful for teams to more fully understand the role of group reflection in Agile and the types of reflection that occur.

The Role of Reflection in a Retrospective

Retrospective tools provide individuals working in Agile processes an opportunity to reflect on their team's recently completed tasks that led to their current prototype: a representation of their project that demonstrates the core user experience. Inherent in Agile is an emphasis on "regular reflection as a means to sustainable development pace and continuous learning" (Babb, Hoda, & Nørbjerg, 2014, p. 51). However, the reflective activities surrounding a retrospective are group-based; they are more "challenging", "demanding" and collaborative than individual reflections that are usually characterized as "solitary" experiences (Osterman & Kottkamp, 1993, p. 24). Through persistent group engagement of reflective practice that occurs during cyclic retrospectives "practitioners can develop a greater level of self-awareness about the nature and impact of their performance, an awareness that creates opportunities for professional growth and development" (Osterman & Kottkamp, 1993, p. 24). Further, an opportunity to act on insights gathered during a retrospective and apply them immediately to subsequent project sprints.

Retrospectives often afford teams two types of reflections common to reflective practitioners. The first is a reflection based on individual and group activities that have already occurred. These types of "on-action" (Schön, 1987) reflections tend to focus on how the team managed aspects of the project co-construction, their relationships with each other and their interactions with client partners. The second type of reflection is more difficult to capture and measure. This type of reflection occurs "in-action" and is characterized by individuals and teams on projects acting in-the-moment. In particular, on their processes and habits during that time. Schön's depiction of the reflective practitioner is closely related to Cook and Brown (1999) "knowing-in action" (p. 384) and "that takes fuller account of the competence practitioners sometimes display in situations of uncertainty, complexity, uniqueness, and conflict" (Boyles, 2006, p. 29). Both depictions on reflection can be traced to Dewey's epistomological ways of knowing or "inquiry in a world that is not static" (Boyles, 2006, p. 61). A type of knowing defined as "inquiry into things 'lived' by people…experimenting with solving problems such that the action entailed in the solving of the problem is [in itself] inquiry" (p.61).

For the current case study, both "on-action" and "in-action" reflection occurred. While the first is evidenced in the capture of data using a visual retrospective tool, the second type of reflection was captured within weekly blog posts that individual team members wrote. Before exploring the teams' procedures, we need to understand the importance of documenting retrospectives.

Reinventing the Documentation Process in Agile Retrospectives

"Documents are part of political and social life, but also part of material culture" (Berenbeim, 2012, p. 2). Documents and documentation have been created and collected in various forms, since the beginning of time with primitive historical drawings to illuminated texts of the middle ages to the 10 second snapchat photos of today. Historically, documents were produced and presented, such as the Magna Carta, with ceremony and celebration and given honour and homage, such as the Declaration of Independence and the Charter of Rights and Freedoms (Berenbeim, 2012). Historical documents used images to enhance their pieces, like Richard II decorating documents with large Rs and various other illuminated texts (Berenbeim, 2012).

The current state of documentation is lost, especially in the Agile development world of the software industry, and is contradictory to historical artefacts that demonstrate long-form writing. One of the four pillars of the Agile manifesto is "working software over comprehensive documentation" (Beck et al., 2001). This promotes the need for teams to spend most of their time co-constructing prototypes and working technology and a value of spending less time documenting. Whereas we have more space and ability to share documentation online and otherwise, our need to be concise seems to be a growing trend within product development environments. While documentation is encouraged in development processes team members breathe a collective sigh of relief if they don't have to. This may be influenced by society's attention spans. Consider the 280-character limit of a tweet. Yet, even 280 characters seems to be too much in some observed team correspondence. This can be observed in professional messaging applications on Slack or on apps like Snapchat, Instagram, or TikTok that promote the use of images or very short videos with limited or no text.

The initial demand for less documentation is the dissatisfaction with traditional methods that were very bureaucratic and exhaustive in the nature of their documentation, which, left little time for actual product development (Rubin & Rubin, 2011; Selic, 2009; Shafiq & Waheed, 2018). Selic (2009) rightfully questions, whether or not working software and good documentation need to be mutually exclusive. As products are being developed presently, and in the near future, we need to consider that:

- *Modern software is increasing in complexity;*
- *Software systems are typically revisited and revised more often than other types of systems;*
- *Code maintenance is usually delegated to less experienced junior staff who are unfamiliar with the code.* (Selic, 2009, p. 11)

If our attention spans are short, our need to produce more is high, and the technical limitations continue to grow, what type of documentation should be used? Some researchers suggest that extensive face to face collaboration reduces the need for extensive documentation (Hess, Diebold, & Seyff, 2017; Rubin & Rubin, 2011; Shafiq & Waheed, 2018). However, researchers and *agilists* also suggest the need for precise documentation that is understandable, more focused on domain knowledge, and includes design rationales alongside of Agile iterative practices (Rubin & Rubin, 2011; Selic, 2009; Shafiq & Waheed, 2018). In other words, documentation needs to be succinct but does not exclude design features. Ideally documentation does not include lengthy technical pieces. Some *agilists* believe that true design documents are source codes only (Reeves, 1992). However, source code alone is not a replacement of design documents (Rubin & Rubin, 2011). As with other Agile practices, the documentation should be iterative and collaborative. As the time allowed for individual and team documentation is reduced, its need within Agile practices is not a prioritized task.

Even though Agile retrospectives, regular reviews, and other final project reflections have occurred in many companies, these written documents seem to be limited to internal circulation only. External or public presentations of retrospectives, reviews, or any type of reflection are nearly impossible to locate in scholarly literature.

In the context of this case study, students were encouraged to extend their retrospective reflections beyond the internal team through public blog posts. By posting weekly, the teams developed a practice of public reflection which could be emulated by professional industries and other institutions. Through the public documentation of retrospectives, researchers and professionals alike may have the ability to understand the purpose of documentation and their role within distinct Agile practices. Beyond articulating "what happened" during a retrospective, team members could articulate how retrospective data was received by the team or by individuals. This could give outside viewers of online posts a sense of the experience and an informed choice as to what retrospective practices they might bring to their own teams.

This section identified the background to this chapter by briefly discussing the documentation, role of the reflective practitioner, and retrospectives. The following section will present the results of the two blogs in this case study.

GATHER DATA

Gathering data helps teams have an aligned understanding of what happened during a sprint or project (Derby & Larsen, 2006). Using this technique, this section will present the blog data on a comparative timeline for two MDM program teams based on a 13-week model. Team A Blog *Wet Toast Saga* and Team B Blog *Berry Good*. The self-reported blogs capture moments throughout the 13 weeks as they iterate through real-world client facing projects. In particular, this section will focus on their mid-term release retrospective. The two teams both participated in the same steps during their retrospective. Based on the qualitative data collected, we will present this comparative data side by side.

Typical Agile retrospective structures can be a simple informal conversation about what went well and what a team might need to improve upon, or they can consist of a more complex formalized approach, as suggested by Derby and Larsen (2006). The formalized version below (*Figure 1*) includes several

Figure 1. Agile retrospective activities

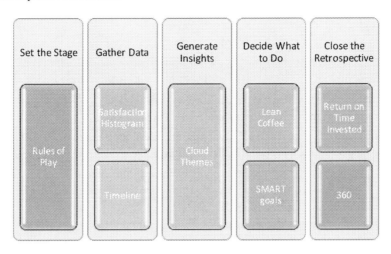

stages. Ideally, when all stages are complete, the team as a whole should have a better understanding of past practices of the project and improvements for subsequential sprints. Additionally, individuals will have identified areas of success and areas for self-improvement. We will draw from Derby and Larsen (2006), but note that at time the sequence of events that occur in a retrospective are not always linear.

In the context of the professional master's program's 13-week project-based course, teams set the stage by first establishing rules of play for all of the retrospectives. These can include rules that guide individuals to what aspects of a project process they can critique (process vs individuals) and how that critique is articulated (through a language that is respectful). Rules establishing the moderation of a retrospective are useful and teams can choose a team member who is diplomatic. Rules can also implicate the media used to capture the reflections. Rules of play tend to center around how team members conduct themselves. Critiques are "live organic events…[that] are 'out-loud' rather than silent internal discussions" that provide opportunities to learn about the team and yourself (Barrett, 2019, p. 8). For example, some teams prefer to write short key messages on stickies that can be more fully articulated when talked through. Once team members are clear what they are able to reflect on, the quality of their critique, and how they will communicate their reflections, the retrospective can move forward. In this case study, once the rules of play were established, students were guided into two activities: satisfaction histogram and timeline.

Satisfaction Histogram

A satisfaction histogram is designed to quickly rate how individuals in the room relate to their current team dynamic, at the moment of their retrospective. Each person anonymously writes a number on a piece of paper and folds it in half to give to the facilitator. They choose from the following (Derby & Larsen, 2006):

How satisfied are we?

5 = I think we are the best team on the planet! We work great together.
4 = I am glad I'm a part of the team and satisfied with how our team works together.

3 = I'm fairly satisfied. We work well together most of the time.
2 = I have some moments of satisfaction, but not enough.
1 = I'm unhappy and dissatisfied with our level of teamwork.

Results from histograms are collected into a graph and trends are briefly discussed amongst team members and moderated by a ScrumMaster®, in this case, an instructor. For the first activity, the teams completed a satisfaction histogram. Each team presented this data in two ways. Team A presented their satisfaction using only an image (*Figure 2*). Team B not only included the image of their histogram (*Figure 3*), but also a detailed description of what each number meant as well as the process to collect this information "taking the temperature of the room. We individually wrote a number from 1 to 5 on

Figure 2. Team A's satisfaction histogram

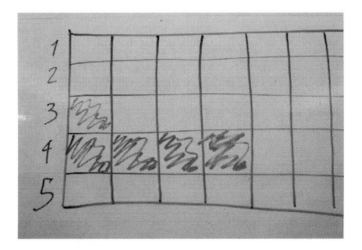

Figure 3. Team B's satisfaction histogram

a post-it and put it in a box". Team B also stated "As we can see, 3 of our team members voted for "3", while the 3 others voted for "4"." Team B indicated the significance of these numbers and how "taking the temperature" was important to set the stage for the retrospective as well as gathering data about individuals on the team. Team B described how their histogram represented satisfaction regarding the current project and the collaboration within the team. Team A did not provide any of these insights.

Timeline

The next activity used by Agile coaches is a timeline (Adkins, 2010; Derby & Larsen, 2006). This activity gathers data in a longer project to describe the work from many perspectives and see if there any discernible patterns (Derby & Larsen, 2006). During the timeline, the team members write down key moments during the project and put it on the timeline. For example, first client meeting, first user test, or pivot. These sticky notes are put along an x axis. Once all of the key moments are placed, it is time to understand the feelings of the team members. Along the y axis, each team member is given a different colour sticky note and writes how they were feeling (extremely happy to neutral to sad/frustrated) during that key moment and why and places it along the y axis of timeline. After each member writes their feelings, they read other team members posts.

Team A and B switched roles in providing data to the blog reader. This time Team A provided slightly more detailed data about "putting major events (both positive and negative) on a timeline" (*Figure 4*). They did not describe anymore details of specifically how this activity took place or what were some overall feelings of the team other than the image provided. However, this was more detail than Team B. Team B only provided the heading of Gathering Data followed by a gif (an animated photo that continuously repeats through several photos). Their gif included about 9 images of the team writing sticky notes and putting it on a whiteboard (*Figure 5*). This is the only piece of data for this section that Team B provides. There is no description of the activity or the gif.

Figure 4. Team A's timeline with positive and negative feelings related to events

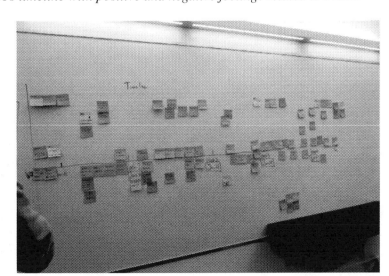

Figure 5. Selected Team B images from gif of their timeline

Cloud Themes to Generate Insights

The next part of Derby and Larsen (2006) approach involves generating insights from the gathered data by grouping the "feelings" uncovered in the histogram and timeline, into "cloud" themes. Each team generates different themes, for example: client expectations, team dynamics, process challenges, and so on. This helps the team decide what were the most important parts during product development. Once the themes are generated, the team needs to decide on what to focus on for improvement in subsequent sprints. This is done through dot voting. With dot voting, each member gets 3 dots to vote inside the cloud they think is the most important.

As posted on the blogs, both teams generated several cloud themes. Team A explained that they grouped post-its based on themes and provided the read with an image (*Figure 6*). They did not provide any other detail.

Team B only provided an image (*Figure 7*). They did not provide any written details about what this activity included. They did not explain what they did or how they did it. However, compared to Team A, they provided a large high definition (HD) image that was easily viewed on their blog and able to zoom into each area. Their insight themes included: directionless and frustration, skeptical of product and client, skeptical of process, hopeful, energy, good ideas, learning, and team building. There were

Figure 6. Team A's gathered insights

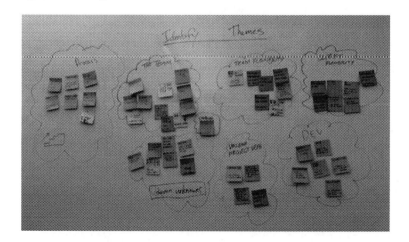

Figure 7. Team B's insights in cloud themes

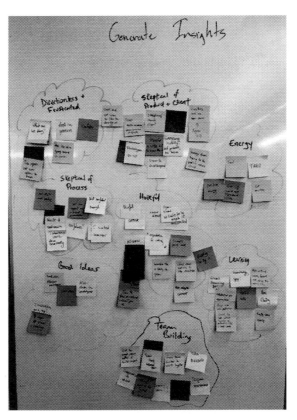

about 5-10 notes per theme that were also readable. For example, within the skeptical of product and client one of the notes said "vision is over scoped". Even though there was a lack of description of this activity, a reader could interpret the ideas based on the details provided in each sticky note, providing the reader knows how to zoom in on a digital device.

Lean Coffee and SMART Goals

Once the dot votes are tallied, the highest votes help the team decide what to do. Using a table, like a scrum or kanban board (*Figure 8*). The themes with the most votes become new notes and are placed into the *to talk* column. As the team discusses the theme, it is moved to the *talking* column. With a very strict timebox, the team makes some fast decisions about the topics. From each theme the team should come up with specific actionables to take into the next sprint. They are encouraged to create SMART goals based on each theme (Specific, Measurable, Attainable, Relevant, Timely). Goals that are "SMART" "are more likely to reach fruition" (Derby & Larsen, 2006, p. 108). If the team runs out of time, this becomes a *parking lot* item (a place on the team's larger kanban board) to be discussed in the next sprint. If several actionables arise, then team members sign up for only one theme that they have the energy to work on.

Figure 8. Lean coffee board

To talk	Talking	Done	Actions

Each team posted about taking action through lean coffee and SMART goal. Team A described sitting through an unblocking exercise by listing pain-points and coming up with an actionable solution for each of them. They also described the timeboxed restrictions. In particular, each solution in 10, 5, and 2 minute time slots. The team then described how they had to come up with full detail and deadline dates for each solution and sign up for a task to be responsible for. They also provided two images (sideways on the screen) of their actionable with team signatures at the bottom (*Figure 9*).

Team B also described this activity by addressing major teams in a timeboxed discussion. They also described writing actionable for each team and to follow up in the next sprint. They did not provide any images for this section. They did, however, tell readers to "see lean coffee" but did not provide a link or indication of where to find this information. After exploring the rest of the blog, there were no other blog posts dedicated to learn coffee. Again, both teams only provided a minimal description of this activity, which could be fairly unclear to readers who may be interested in trying out this method of reflection.

Figure 9. Two actionables created by Team A

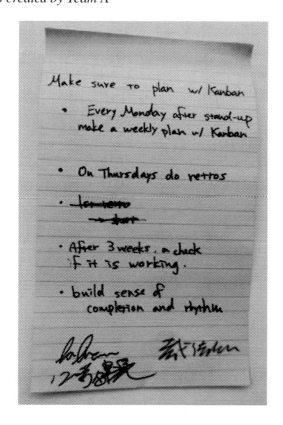

This section described detailed experiences of teach team for the first parts of the retrospective. The next section will present the final parts, including something called a 360 review.

360s

Once actionables are created, teams can participate in a 360 review. During this time, each person is encouraged to reflect on themselves, but also on their teammates. Each team member will have an opportunity to be the centre of the 360 *(Figure 10)*. The goal is to reflect on the past iteration(s) and how each member worked during the sprint. The *plusses* (represented by a plus sign) are things that you do well, *deltas* (represented by a triangle) are things you can improve on, ideas (represented by a light bulb) are how you can improve on the deltas, and the thank you (represented by a bouquet of flowers) are for thanking the team or thanking a person for something. This activity can be difficult at the beginning of a project as team members may not feel comfortable to share deltas, but rules of play can protect any fears for rudeness or lack of honesty. Understanding appropriate skills of effective critiques is critical to the success or failure of a 360 review. In particular, critiques are more effective when:

- Feedback is provided by peers rather than authority figures (i.e., instructors or managers)
- Members respond and take part in discussion
- Conversation is thoughtful, engaged, and focused
- All members can speak freely
- All members receive honest and constructive, but respectful feedback
- Members have an attitude of cooperation (Barrett, 2019; Rabinowitz, 2018)

Figure 10. 360 feedback cycle

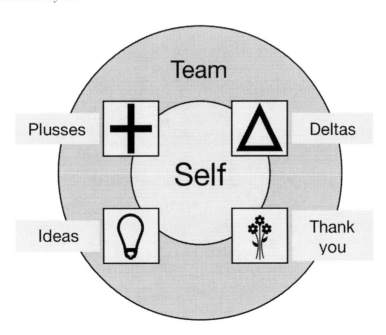

At this point, each team ended their retrospective with a 360. Team A provided a more detailed description of this activity:

Our final exercise for the day was the 360 itself — what we had all been waiting for. We each had different coloured post-its assigned to us, and we spent about 2–5 minutes writing out as may post-its as we felt were necessary in the four categories for each of our teammates and ourselves. After we had all written our post-its, the person who was being peer reviewed would go first to explain what they had written about themselves. Once that was done, each team member would come up and talk about what they had written for that person and if any other members of the team had a similar comment, they would shout "Bingo". We didn't feel the "Bingo" element added any benefit in particular as it removed the chance for the second person to explain why they had written that comment, so we might skip that for the next 360.

This was one of the most detailed descriptions provided by the team. Even though their image is blurry and difficult to read (*Figure 11*), the details provided in the written description explained how the team participated in the activity. They explained how they spent time writing out ideas for the four categories, but did not explain what the four categories were. This was also the first post that provided some insights into how the team was feeling. In particular, their excitement for the 360 "what we had all been waiting for" and how they did not like "bingo" as "it removed the chance for the second person to explain why they had written that comment".

Team B, again, did not provided any written description of their 360 experience. They did not describe what the 360 was or their feelings about the process. Even though the provided a HD quality photo, unlike Team A, there were many small details that were too difficult to zoom into (*Figure 12*).

At the end of this activity, sometimes doing a quick return on time invested (ROTI) is important for facilitators to check in to see how team members felt about all of the activities by asking if this time would get a 5: this was the best, high return, received greater benefit than time invested to 1: lost time,

Figure 11. Team A's 360

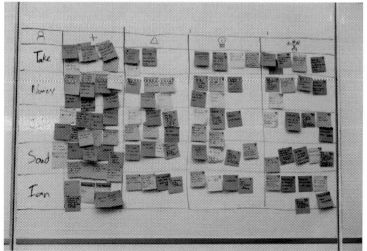

Figure 12. Team B's 360

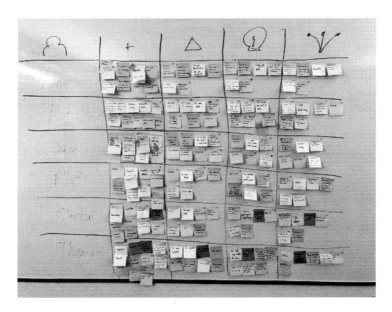

no benefit received from time investment; or somewhere in between. Neither team described this activity, which could suggest that this activity did not occur.

On the blogs, after the 360, each team provided a very brief reflection on the whole process. Team A described the process as providing the team with a "renewed sense of vigor" and "confirmation" of alignment. They also indicated "no big surprises". This was a very vague and very brief reflection of the process. Almost the entire blog post was dedicated to a very brief, semi-descriptive ideas around the process rather than their attitudes towards the process.

Team B did provide a few more "takeaways" through a bullet point list, including the opportunity to think about the process when working as a team and ability to identify pain points. They did also indicate they had difficulties keeping track of the actionables. It would be difficult to know how they created these actionables (on a whiteboard, on sticky notes like Team A, something else) as they did not provide details in their blog post.

Overall, each team presented the process of an Agile retrospective using images and/or text descriptions. However, most of these descriptions lacked detail and lacked insights into the process as a whole. There could be several reasons to the lack of detail and insights, perhaps the assumption of images conveying messages, external pressures, or the inability to reflect on practices.

This section provided a comprehensive comparative case study analysis of blog posts from two teams. The next section will discuss insights through a thematic analysis.

GENERATING INSIGHTS

Gathering insights is about evaluating the data, reflecting on what was captured in the data collection process to make it meaningful (Adkins, 2010; Derby & Larsen, 2006) In particular, we will provide details based on the following themes from the two blog posts: images versus text, non-disclosure agreements, and reflective practice.

Images vs Text: Is a Picture Worth 1000 Words?

The idiom "A picture is worth a 1000 words" from the early 20th century was used to encourage the use of photographs in publications (The idioms, 2019). In the 21st century, we have found images and videos have conveyed thousands of words, all that can be interpreted in several ways. With popular sites like Instagram and Tumblr only using images or the use of emojis to convey feelings, we can ask, can pictures alone describe meaning? Do we need text to explain the full idea? Or can words alone describe a situation? Based on the experience of reading this blog, we identify the lack of understanding or the easy ability to misunderstand ideas due to lack of description. And if a picture is a 1000 words, are my words different from yours?

To illustrate the challenge, Team A described their satisfaction histogram using one image alone. They did not provide the title or a description of the image. All you can see is a series of boxes with some numbers and scribbles inside the box. This can be interpreted as a score of some sort. Perhaps it was the grade their received, a number representing their participation level, or a figure that represented how they were feeling that day? Team B also relied on images alone to convey meaning in their post. In particular, in their "gather data" section, those were the only words used. There were no other headings or subheadings and no description at all. Instead, the reader has to interpret their actions in this activity based on a gif. This may be a little more insightful than a single image, but perhaps we have to contend with 9000 words through 9 pictures. With each image flashing we can interpret activity and movement, but also more confusion. The reader needs to figure out what the participants are doing and why based on less than a second clip of several images. Also, due to the nature of these images it is quite difficult and almost impossible to interpret what the teams are writing or doing. Team B also only used an image was in "generate insights" where they identified themes. This image, however, may be considered more successful and provide more insight, as it was a large HD image that was easily read. Despite the HD attempt, some nuances of the experience are still lost. Team A did not provide HD images, in particular when they posted their actionables, both images are blurry and flipped horizontally.

Quality of images can be problematic especially when messages are being conveyed through images alone. In earlier days of online images, we had technological limitations of standard definition and low resolutions. However, with speed, HD options, and high resolutions, we have been spoiled in quality and have very high expectations. If an image is not in HD, we are not satisfied. In the case of Team B, the use of the gif was innovative, but lead to poor photos that are quite difficult to interpret. In the case of Team A, several of their images were low resolution, blurry, and sometimes sideways. There could be several reasons for this, such as inability to take quality photographs, unclear purpose of photos (do we want to see the team? Or do we want to see the results of the activity?), and other privacy concerns.

This brings back the question of what type of documentation should be used. Images or text? As stated earlier with working software and documentation, do images and text need to be mutually exclusive? As mentioned by several researchers, precise documentation will be the key to software development success (Braaf, Manias, & Riley, 2011; Rubin & Rubin, 2011; Selic, 2009; Shafiq & Waheed, 2018). In the case of blog documentation of a mid-term retrospective, precise documentation is essential. It is the key to understand not only what happened, but how the team members reflective on their practice and what lessons were learned. This is where images and text in conjunction are important as one without the other leaves the reader with significant gaps on how they can use these elements in their own practice. If the purpose of these blogs is to share these experiences to other potential *agilists* or future students, too much is left to interpretation. Perhaps, the lack of details provided was purposeful, or perhaps it was lack of attention to detail, unfortunately both teams do not provide blog readers with either reason.

This section discussed the role of images and/or text in posts and the need for more detail. The next section will provide a potential reason for lack of detail.

Non-Disclosure Agreements (NDAs) Silencing Teams

Images only, blurry images, vague descriptions, no description at all. Were these decisions intentional? It could be considered following agile practice and limit documentation, but this leads to a lack of understanding and the ability to develop best practices (Braaf et al., 2011; Rubin & Rubin, 2011; Selic, 2009; Shafiq & Waheed, 2018). Did both teams intend to be vague or were they functioning as cliché lazy students doing the minimal amount of work necessary to get the grade? If it is the former, we can ask why? And one answer is related to privacy concerns and intellectual property through an NDA. A NDA is a "legal contract between two or more parties that signifies a confidential relationships" that does not allow the parties to discuss or present anything to an outside parties such as the general public and especially not the competitors (Kenton, 2018). These agreements are quite common in the digital media industry as many companies are trying to create the "next big thing". The next billion dollar idea, the next Facebook, the next Google, the next Instagram. This can be quite problematic when trying to create and share best practices, as possibly seen in the two blogs.

Both teams were quite vague and we never really had an idea of what the project was or what actionables were created to move forward. As the blogs were available publicly and not password protected, the teams working on the projects were possibly protecting themselves and their clients from having information public. But to what extent do NDAs limit our understanding and ability to share knowledge? If we have to sign NDAs for everything we do, then how can we learn from others? How can we learn to avoid certain mistakes rather than wasting time (and money) going down a bug path? Perhaps instead of *agilists* focusing on "customer collaboration over contract negotiation" (Beck et al., 2001), we should have "shared practices over non-disclosure agreements"? No longer should we restrict knowledge from the privileged few, but give access to all. But the entrepreneur might ask, "how do I make money if I'm giving it away for free?" We are not suggesting giving away your product or idea, but there must be a way to share this information and practices and learning that will mobilize knowledge that would benefit the entire digital media industry.

This section briefly discussed a possible reason for lack of descriptive pieces in each blog. The next section will then question if the teams were reflective practitioners.

Did the Two Teams Embody the Role of a Reflective Practitioner?

As mentioned, "regular reflection as a means to sustainable development pace and continuous learning" (Babb et al., 2014, p. 51) and yet more "challenging", "demanding" (Osterman & Kottkamp, 1993, p. 24) when working in groups. In this case study, did the two teams embody (or not) the role of a reflective practitioner?

Looking at the blogs, on the surface it seems that both teams were fairly reflective. We can presume the group activities are a first past at reflection. Each team focused on relationships with each other and their client partners and their co-construction. The satisfaction histogram, timeline, and cloud themes were "on-action" reflection activities related to the retrospective (Schön, 1987). They are participating in the activity and documenting it (as seen either in some sort of text or image). However, as seen in the detailed analysis, the teams were very light on deep reflection. Perhaps they were able to gather insights during the process and move forward into their subsequent sprints, but their action plans presented on the blog were vague and up for interpretation. Neither team explicitly stated their goals or plans of taking the information learned and moving forward. This is the challenge, as suggested earlier, with "in-action" reflection of the weekly blog. They made attempts to capture, but struggled with measuring. There were minimal reflections by Team B in their final bullet points, and even less in Team A with the doing the 360 and "what we had all been waiting for". Why do we need to become reflective practitioners? "The purpose of reflective practice is to enhance awareness of our own thoughts and action, as a means of professional growth" (Osterman & Kottkamp, 1993, p. 2). One of the presumed goals of a professional masters would hopefully be professional growth. And part of that isn't just making things or doing assignments, rather embodying learning and the idea of becoming a professional are essential components to overall success. Unfortunately, not all professionals are "reflective" of their own practice.

Part of the challenge is that young professionals not used to capturing reflections may require more guidance and structure; an understanding of what to reflect on. According to Glaze (2002), while "there is an extensive body of literature on reflection, there is little written about students' experiences of developing reflective skills" (p. 267). Some students feel "inhibited" writing about practice (p. 268) and typical of many students may feel "concerned about disclosing difficult situations in case of censure" (p. 268). The NDA students signed may also have contributed to a sense of censorship. Once students become used to the initial challenge of reflection, there is a particular acceptance they come to terms with (p. 268). Over the course of a 13-week project, students also deepen the quality of their reflections and are able to contribute more insights to the project development process.

This section suggested the challenges of becoming a young professional and some factors inhibiting growth. The next section will discuss limitations of this case study and the need for future research.

DECIDING WHAT TO DO

Deciding what to do allows teams to make concrete goals for the next iteration (Derby & Larsen, 2006). Some limitations of this study and the need for future research. Additionally, we will discuss how this chapter can inform industry discourse.

As with many qualitative studies, this case study had limited participants, only 2. Even though this can be limiting, rich data was provided. More qualitative studies exploring the perspectives of instructors and students on Agile practices are needed to get a better understanding of general practices across the field. Future research could explore more blogs (or other public forums) and more teams (including varying team sizes) across various programs (undergraduate and graduate, traditional and professional programs). Another limitation of this study was only exploring the public presentations of the blogs and not speaking to individual team members through interviews. However, the purpose of this case study was to explore only what the teams presented publicly, including the limitations this set on full understanding of the process. Future research could do follow-up interviews to explore why limited evidence was provided. Furthermore, a longitudinal research study could be established to identify whether the practices taught in school is used in the workplace.

Not only can this chapter inform future research, this current study can also inform professional practice. Agile practices are presented in the "what to do" or the "how to do" but quite often remise "lessons learned". This case study can suggest to industry professionals on how to reflect on their own practices. By exploring what has been presented here, industry professionals can promote discussions about the importance of appropriate documentation of retrospectives. They could also identify the potential benefits of public discourse around these practices.

CLOSING THE RETROSPECTIVE

To close the retrospective on this chapter, we will provide some overall conclusions on using Agile practices. The value of a formalized retrospective process allows individuals to reflect on their approach to project development and an opportunity to improve the process in subsequent sprints. While many Agile processes shy away from too much written documentation we propose a balance that can be achieved when retrospectives are timeboxed and the data collected through them support teams in making incremental versus monumental changes.

Along with the practice of participating in retrospectives, attention also needs to be focused on providing team members with sufficient guidance and encouragement to transform into reflective practitioners. While many types of reflection are possible within a retrospective, approaches like (Derby & Larsen, 2006) offer teams a structured approach that can help them increase their contributions. This may lead to teams generating new insights into the project development process that might have previously remained uncommunicated. Further research that attempts to operationalize the reflective process that occurs during a retrospective may generate new approaches that can in turn, inform individual performance during project development.

ACKNOWLEDGMENT

We would like to acknowledge the two teams and their public blogs. This research received no specific grant from any funding agency in the public, commercial, or not-for-profit sectors.

REFERENCES

Adkins, L. (2010). *Coaching agile teams: A companion for ScrumMasters, agile coaches, and project managers in transition*. Boston, MA: Addison-Wesley Professional.

Appelo, J. (2011). *Management 3.0: Leading agile developers, developing agile leaders*. Boston, MA: Addison-Wesley Professional.

Appelo, J. (2016). *Managing for happiness: Games, tools, and practices to motivate any team*. Hoboken, NJ: John Wiley & Sons.

Babb, J., Hoda, R., & Nørbjerg, J. (2014). Embedding reflection and learning into agile software development. *IEEE Software*, *31*(4), 51–57. doi:10.1109/MS.2014.54

Barrett, T. (2019). *Crits: A student manual*. London, UK: Bloomsbury.

Beck, K., Beedle, M., van Bennekum, A., Cockburn, A., Cunningham, W., Fowler, M., . . . Sutherland, J. (2001). The agile manifesto. Retrieved from http://agilemanifesto.org/

Berenbeim, J. L. (2012). *Art of documentation: The sherborne missal and the role of documents in english medieval art*, (Doctoral Dissertation), Harvard University. (3513897)

Boyles, D. R. (2006). Dewey's epistemology: An argument for warranted assertions, knowing, and meaningful classroom practice. *Educational Theory*, *56*(1), 57–68. doi:10.1111/j.1741-5446.2006.00003.x

Braaf, S., Manias, E., & Riley, R. (2011). The role of documents and documentation in communication failure across the perioperative pathway. A literature review. *International Journal of Nursing Studies*, *48*(8), 1024–1038. doi:10.1016/j.ijnurstu.2011.05.009 PMID:21669433

Cook, S. D., & Brown, J. S. (1999). Bridging epistemologies: The generative dance between organizational knowledge and organizational knowing. *Organization Science*, *10*(4), 381–400. doi:10.1287/orsc.10.4.381

Derby, E., & Larsen, D. (2006). *Agile retrospectives: Making good teams great*. Dallas, TX: Pragmatic Bookshelf.

Glaze, J. E. (2002). Stages in coming to terms with reflection: Student advanced nurse practitioners' perceptions of their reflective journeys. *Journal of Advanced Nursing*, *37*(3), 265–272. doi:10.1046/j.1365-2648.2002.02093.x PMID:11851797

Hess, A., Diebold, P., & Seyff, N. (2017, Sept. 4-8). Towards requirements communication and documentation guidelines for agile teams. In *Proceedings 2017 IEEE 25th International Requirements Engineering Conference Workshops (REW)*. 10.1109/REW.2017.64

Kenton, W. (2018). Non-disclosure agreement - NDA. Retrieved from https://www.investopedia.com/terms/n/nda.asp

Leonard, D., & Sensiper, S. (1998). The role of tacit knowledge in group innovation. *California Management Review*, *40*(3), 112–132. doi:10.2307/41165946

Osterman, K. F., & Kottkamp, R. B. (1993). *Reflective practice for educators: Improving schooling through professional development*. Newbury Park, CA: Corwin Press.

Polanyi, M. (1966). The logic of tacit inference. *Philosophy (London, England)*, *41*(155), 1–18. doi:10.1017/S0031819100066110

Rabinowitz, P. (2018). Techniques for leading group discussions. *Community Tool Box*. Retrieved from https://ctb.ku.edu/en/table-of-contents/leadership/group-facilitation/group-discussions/main

Reeves, W. J. (1992). What is software design? *C++ Journal, 2*(2).

Rubin, E., & Rubin, H. (2011). Supporting agile software development through active documentation. *Requirements Engineering*, *16*(2), 117–132. doi:10.100700766-010-0113-9

Schön, D. A. (1987). *Educating the reflective practitioner: Toward a new design for teaching and learning in the professions*. San Francisco, CA: Jossey-Bass.

Selic, B. (2009). Agile documentation, anyone? *IEEE Software*, *26*(6), 11–12. doi:10.1109/MS.2009.167

Shafiq, M., & Waheed, U. S. (2018). Documentation in agile development a comparative analysis. In *Proceedings 2018 IEEE 21st International Multi-Topic Conference (INMIC)*. 10.1109/INMIC.2018.8595625

The idioms. (2019). The meaning and origins of the expression: A picture is worth a thousand words. Retrieved from https://www.theidioms.com/a-picture-is-worth-a-thousand-words/

Compilation of References

Abanda, F. H., Vidalakis, C., Oti, A. H., & Tah, J. H. M. (2015). A critical analysis of Building Information Modelling systems used in construction projects. *Advances in Engineering Software*, *90*, 183–201. doi:10.1016/j.advengsoft.2015.08.009

Abernathy, W., & Clark, K. B. (1985). Innovation: Mapping the winds of creative destruction. *Research Policy*, *14*(1), 3–22. doi:10.1016/0048-7333(85)90021-6

ABNT. ABNT - Associação Brasileira de Normas Técnicas. Disponível em <http://www.abnt.org.br/

ABPMP (Association of Business Process Management Professionals). (2009). *Guide to business process management: Common knowledge body*, Version 2.0, 2009.

ABPMP. (2014). *BPM CBOKTM V3.0. guide to the business process management common body of knowledge*. 2. Brasil: Association of Businees Process Management Professsionals Brasil.

Achmad, T., Rusmin, R., Neilson, J., & Tower, G. (2009). The iniquitous influence of family ownership structures on corporate performance. *Journal of Global Business Issues*, *3*, 41–49.

Adeoti-Adekeye, W. (1997). The importance of management information systems. *Library Review*, *46*(5), 318–327. doi:10.1108/00242539710178452

Adkins, L. (2010). *Coaching agile teams: A companion for ScrumMasters, agile coaches, and project managers in transition*. Boston, MA: Addison-Wesley Professional.

Adolpho, C. (2011). *Os 8Ps do marketing digital*. Novatec Editora.

Agarwal, R., & Hoetker, G. (2007). A faustian bargain? The growth of management and its relationship with related disciplines. *Academy of Management Journal*, *50*(6), 1304–1322. doi:10.5465/amj.2007.28165901

Agrawal, A. (2016). The dumbest legal mistakes early startups make. Retrieved from https://www.inc.com/aj-agrawal/the-dumbest-legal-mistakes-early-startups-make.html

Aidin, S. (2015). New venture creation: controversial perspectives and theories. *Economic Analysis*, *48*(1984), 101–109. Retrieved from http://ssrn.com/abstract=2711731

Albagli, S. (2007). Tecnologias da Informação, Inovação e Desenvolvimento. In: VII Cinform Encontro Nacional de Ciência da Informação, Salvador, 16.

Albert, S., & Whetten, D. (1985). Organizational identity. In L. L. Cummings & B. M. Staw (Eds.), *Research in organizational behavior*, 7, pp. 263–295. Greenwich, CT: JAI Press.

Albrecht, M. A., Bosma, R., van Dinter, T., Ernst, J.-L., van Ginkel, K., & Versloot-Spoelstra, F. (2010). Quality assurance in the EPO patent information resource. *World Patent Information*, *32*(4), 279–286. doi:10.1016/j.wpi.2009.09.001

Compilation of References

Albuquerque, J. D. (2012). Flexibility and business process modeling: A multidimensional and national relationship. *RAE: Journal of Business Administration, 52*(3), 313–329.

Aldrich, H., & Fiol, C. M. (1994). Fools rush in. The institutional context of industry creation. *Academy of Management Review, 19*(4), 645–670. doi:10.5465/amr.1994.9412190214

Alem, A. C., & Cavalcanti, C. E. (2005). O BNDES e o apoio à internacionalização das empresas brasileiras: Algumas reflexões. *Revista do BNDES, 12*(24), 46–76.

Allan, B. (2004). Project management: tools and techniques for today's ILS professional. London: Facet Publishing.

Alotaibi, A. B., & Al Nufei, E. A. F. (2014). Critical success factors (CSFS) in project management: critical review of secondary data. *International Journal of Scientific & Engineering Research, 5*(6), 325–331. Retrieved from http://www.ijser.org/researchpaper%5CCRITICAL-SUCCESS-FACTORS-CSFs-IN-PROJECT-MANAGEMENT-CRITICAL-REVIEW-OF-SECONDARY-DATA.pdf

Alturas, B. (2013). *Introdução aos sistemas de informação organizacionais* (1st ed.). Lisboa: Sílabo.

Alvarenga, N., & Rivadávia, C. D. (2005). *Gestão do conhecimento em organizações: proposta de mapeamento conceitual integrativo*. São Paulo, Brazil.

Amaral, L., & Varajão, J. (2000). *Planeamento de sistemas de informação* (2nd ed.). Lisboa: FCA.

American Marketing Association. (2016). Definition of Marketing. Retrieved from https://www.ama.org/AboutAMA/Pages/Definition-of-Marketing.aspx

Amram, M., & Kulatilaka, N. (2000). Strategy and Shareholder Value Creation: The Real Options Frontier. *The Bank of America Journal of Applied Corporate Finance, 13*(2), 8–21. doi:10.1111/j.1745-6622.2000.tb00051.x

Anantatmula, V. S. (2008). The role of technology in the project manager performance model. *Project Management Journal, 39*(1), 34–48. doi:10.1002/pmj.20038

Anderson, E. (2015). Do the top U.S. corporations often use the same words in their vision, mission and value statements? *Journal of Marketing Management, 6*(1), 1–15.

Anderson, R. C., & Reeb, D. M. (2003a). Founding-family ownership and firm performance: Evidence from the S&P 500. *The Journal of Finance, 58*(3), 1301–1328. doi:10.1111/1540-6261.00567

Andres, C. (2008). Large shareholders and firm performance: An empirical examination of founding-family ownership. *Journal of Corporate Finance, 14*(4), 431–445. doi:10.1016/j.jcorpfin.2008.05.003

Andrew, J., & Sirkin, H. (2008). *Payback: Como conquistar o retorno financeiro da inovação* (1st ed.). Lisboa, Portugal: Actual.

Antonucci, Y. L., & … . (2009). *Business process management common body of knowledge*. Terre Haute, IN: ABPMP.

Antunes, A., & Mercado, A. (2000). *A aprendizagem tecnológica no Brasil: a experiência da indústria química e petroquímica* [s.l.]. Editora E-papers.

Antunes, V. M., & Pugas, P. G. O. (2018, April). A Logística De Distribuição No Setor Moveleiro: Um Estudo De Caso Em Uma Empresa De Grande Porte. *Ciências Gerenciais em Foco, 5*(2), 1.

Anunciação, P. F. (2011). Results of the 1st study of information systems governance in the large companies in Portugal. In *3rd International Conference of ceGSI – Governance of Information Systems: Models, Ethics & Performances*, INA – National Institute of Administration, Oeiras, Oct. 11 *(In Portuguese)*

Anunciação, P. F. (2012 a)). *Results of the 1st study of information systems governance in the large companies in Portugal*, Calouste Gulbenkian Foundation, Feb. 14 *(In Portuguese)*

Anunciação, P. F. (2013). *Results of the 2nd study of Information Systems Governance in the large companies in Portugal*, National Parliament – New Auditorium, March 19 *(In Portuguese)*

Anunciação, P. F. (2012 b)). Results of the 2nd study of information systems governance in the large companies in Portugal, In *4th International Conference of ceGSI – Information Systems Governance: Economy & Security*, National Security Office, Lisbon, Portugal, Oct. 9 *(In Portuguese)*

Anunciação, P. F., & Zorrinho, C. (2006). *Organizational Urbanism – How to manage technological shock*. Lisboa, Portugal: Sílabo Publishing. (In Portuguese)

ANVISA. Agência Nacional de Vigilância Sanitária. Disponível em http://portal.anvisa.gov.br/

Appelo, J. (2011). *Management 3.0: Leading agile developers, developing agile leaders*. Boston, MA: Addison-Wesley Professional.

Appelo, J. (2016). *Managing for happiness: Games, tools, and practices to motivate any team*. Hoboken, NJ: John Wiley & Sons.

Ariely, D., Huber, J., & Wertenbroch, K. (2005). When do losses loom larger than gains? *JMR, Journal of Marketing Research*, *42*(2), 134–138. doi:10.1509/jmkr.42.2.134.62283

Arndt, J. (1978). How broad should the marketing concept be? *Journal of Marketing*, *42*(1), 101–103.

Arregle, J. L., Hitt, M. A., Sirmon, D. G., & Very, P. (2007). The development of organizational social capital: Attributes of family firms. *Journal of Management Studies*, *44*(1), 73–95. doi:10.1111/j.1467-6486.2007.00665.x

Ashton, W. B., & Klavans, R. A. (1997). *Keeping abreast of science and technology: technical intelligence for business*. Columbus, OH: Battelle Press.

Ashton, W. B., & Klavans, R. A. (1997). *Keeping abreast of science and technology: Technical intelligence for business*. Columbus, OH: Battelle Press.

Asparza, M. (2018). *Dialogo sobre Blockchain en Perú*. [Interview]

Associação Brasileira de Normas Técnicas. (2008). *NBR ISO 9001:2008. 2*. ABNT - Associação Brasileira de Normas Técnicas. [s.l.]

Astrachan, J. H., & Carey Shanker, M. (2006). Family businesses' contribution to the US economy: a closer look. In P. Poutziouris, K. X. Smyrnios, & S. B. Klein (Eds.), *Handbook of research in family business*. Cheltenham, UK: Edward Elgar. doi:10.4337/9781847204394.00011

Astrachan, J. H., & Jaskiewicz, P. (2008). Emotional returns and emotional costs in privately held family businesses: Advancing traditional business valuation. *Family Business Review*, *21*(2), 139–149. doi:10.1111/j.1741-6248.2008.00115.x

Babb, J., Hoda, R., & Nørbjerg, J. (2014). Embedding reflection and learning into agile software development. *IEEE Software*, *31*(4), 51–57. doi:10.1109/MS.2014.54

Back, W., & Moreau, K. (2001). Information management strategies for project management. *Project Management Journal*, *32*(March), 10–20. doi:10.1177/875697280103200103

Bagozzi, R. (1975). Marketing as Exchange. *Journal of Marketing*, *39*(4), 32–39. doi:10.1177/002224297503900405

Compilation of References

Bahrin, M.; Othman, F.; Azli, N.; Talib, M. (2016). Industry 4.0: A review on industrial automation and robotic. *Journal Teknologi, 78*(6-13),137–143.

Baker, W. E., & Sinkula, J. M. (2002). Market orientation, learning orientation and product innovation: Delving into the organization's black box. *Journal of Market Focused Management, 5*(1), 5–23. doi:10.1023/A:1012543911149

Balachandra, R., & Raelin, J. A. (1984). When to kill that R & D project. *Research Management, 27*(4), 30–34. doi:10.1080/00345334.1984.11756846

Barach, J. A., & Gantisky, J. B. (1995). Successful succession in family business. *Family Business Review, 8*(2), 131–155. doi:10.1111/j.1741-6248.1995.00131.x

Barbieri, C. (2011). *BI2 – Business Intelligence: modelagem e qualidade*. Rio de Janeiro, Brazil: Elsevier.

Bardin, L. (2002). *Análise de Conteúdo*. Lisboa: Edições 70.

Bardin, L. (2010). *Análise de Conteúdo*. Lisboa: Edições 70.

Barnes, L. B., & Hershon, S. A. (1976). Transferring power in family business. *Harvard Business Review*, (July-August), 105–114.

Barrett, T. (2019). *Crits: A student manual*. London, UK: Bloomsbury.

Barrientos Felipa, P. (2015). *La cadena de valor del cacao en Perú y su oportunidad en el mercado mundial*. Medellín, Colombia: Universidad de Medellín. doi:10.22395eec.v18n37a5

Basil Achilladelis, N. A. (2001a). The dynamics of technological innovation: the case of the pharmaceutical industry. *Research Policy, 30*(4), 535–588. doi:10.1016/S0048-7333(00)00093-7

Baxter, M. (2000). Projeto de Produto: Guia prático para o design de novos produtos (2ª ed.). S. Paulo, Brazil: Edgard Blücher.

Beal, A. (2012). *Gestão estratégica da informação: como transformar a informação e a tecnologia da informação em fatores de crescimento e de alto desempenho nas organizações*. São Paulo, Brazil: Atlas.

Beavers, W. R., & Hampson, R. B. (1995). Misurare la competenza famigliare: il modello sistemico di Beavers, tr. it. In *Ciclo vitale e dinamiche familiari, a cura di F*. Milano, Italy: Walsh, Franco Angeli.

Beck, K., Beedle, M., van Bennekum, A., Cockburn, A., Cunningham, W., Fowler, M., . . . Sutherland, J. (2001). The agile manifesto. Retrieved from http://agilemanifesto.org/

Becker, J., Knackstedt, R., & Pöppelbuß, J. (2009). Developing maturity models for IT management – a procedure model and its application. *Business & Information Systems Engineering, 1*(3), 213–222. doi:10.100712599-009-0044-5

Beckhard, R., & Dyer, W. G. Jr. (1983). Managing continuity in the family-owned business. *Organizational Dynamics, 12*(1), 5–12. doi:10.1016/0090-2616(83)90022-0

Beck, T., & Demirguc-Kunt, A. (2006). Small and medium-size enterprises: Access to finance as a growth constraint. *Journal of Banking & Finance, 30*(11), 2931–2943. doi:10.1016/j.jbankfin.2006.05.009

Beck, T., Demirguc-Kunt, A., & Martinez Peria, M. S. (2007). Reaching out: Access to and use of banking services across countries. *Journal of Financial Economics, 85*(1), 234–266. doi:10.1016/j.jfineco.2006.07.002

Bedell, R. I. (1983). Terminating R & D projects prematurely. *Research Management, 26*(4), 32–35. doi:10.1080/00345334.1983.11756785

Beehr, T. A., Drexler, J. A. Jr, & Faulkner, S. (1997). Working in small family businesses: Empirical comparisons to non-family businesses. *Journal of Organizational Behavior*, *18*(3), 297–312. doi:10.1002/(SICI)1099-1379(199705)18:3<297::AID-JOB805>3.0.CO;2-D

Benedict, B. (1968). Family firms and economic development. *Southwestern Journal of Anthropology*, *24*(1), 1–19. doi:10.1086outjanth.24.1.3629299

Berenbeim, J. L. (2012). *Art of documentation: The sherborne missal and the role of documents in english medieval art*, (Doctoral Dissertation), Harvard University. (3513897)

Berg, B. L. (2001). Focus group interviewing. In B. L. Berg (Ed.), *Qualitative research methods for the Social Sciences*, 4, pp. 111–132. Needham Heights, MA: Pearson.

Berger, A. N., & Udell, G. F. (2006). A more complete conceptual framework for SME finance. *Journal of Banking & Finance*, *30*(11), 2945–2966. doi:10.1016/j.jbankfin.2006.05.008

Berle, G. (1992). *O Empreendedor do Verde: Oportunidade de negócios em que você pode ajudar a salvar a Terra e ainda ganhar dinheiro*. San Paulo, Brazil: Makron Books.

Berrone, P., Cruz, C. C., Gomez-Mejia, L. R., & Larraza Kintana, M. (2010). Socioemotional wealth and corporate response to institutional pressures: Do family-controlled firms pollute less? *Administrative Science Quarterly*, *55*(1), 82–113. doi:10.2189/asqu.2010.55.1.82

Berrone, P., Cruz, C., & Gomez-Mejia, L. R. (2012). Socioemotional wealth in family firms: Theoretical dimensions, assessment approaches, and agenda for future research. *Family Business Review*, *25*(3), 258–279. doi:10.1177/0894486511435355

Berry, D. (1988). The marketing concept revisited: It's setting goals, not making a mad dash for profits. *Marketing News*, *22*(15), 26–28.

Bessa, M. J. C., & Carvalho, T. M. X. B. (2005). *Tecnologia da informação aplicada à logística*, 11, pp. 120–127.

Best, D. (1996a). Business process and information management. In The fourth resource: Information and its management. Hampshire: Aslib/Gower.

Best, D. (1996b). *The fourth resource: information and its management*. Hampshire: Aslib/Gower.

Beuren, I. (1998). *Gerenciamento da informação: um recurso estratégico no processo de gestão empresarial*. São Paulo: Atlas.

Bhide, A. (1999). *The origin and evolution of new businesses*. The Oxford University Press.

Birch, P. & Clegg, B. (1999). Criatividade em Negócios: Um Guia para Gestores (1ª ed.). Viseu: Pergaminho.

Birk, J. (1990, October). A corporate project management council. *Proceedings of The IEEE International Engineering Management Conference, Gaining the Competitive Advantage* (pp. 180-181). IEEE.

Black, F., & Scholes, M. (1973). The Pricing of Options and Corporate Liabilities. *Journal of Political Economy*, *81*(3), 637–654. doi:10.1086/260062

Boer, F. P. (2000). Valuation of Technology Using "Real Options". *Research Technology Management*, *43*(July/August), 26–30. doi:10.1080/08956308.2000.11671365

Boonstra, A. (2013). How do top managers support strategic information system projects and why do they sometimes withhold this support? *International Journal of Project Management*, *31*(4), 498–512. doi:10.1016/j.ijproman.2012.09.013

Borges, M. A. G. (2008). A informação e o conhecimento como insumo ao processo de desenvolvimento. [RICI]. *Revista Ibero-americana de Ciência da Informação*, *1*(1), 175–196.

Borko, H. (1968). Information Science - what is it? *American Documentation*, *19*(1), 3-5.

Boyles, D. R. (2006). Dewey's epistemology: An argument for warranted assertions, knowing, and meaningful classroom practice. *Educational Theory*, *56*(1), 57–68. doi:10.1111/j.1741-5446.2006.00003.x

Braaf, S., Manias, E., & Riley, R. (2011). The role of documents and documentation in communication failure across the perioperative pathway. A literature review. *International Journal of Nursing Studies*, *48*(8), 1024–1038. doi:10.1016/j.ijnurstu.2011.05.009 PMID:21669433

Brandenburger, A., & Nalebuff, B. (1996). *Co-opetition: A revolution mindset that combines competition and cooperation: The game theory strategy that's changing the game of business*. New York: Bantam Doubleday.

Bregonje, M. (2005). Patents: A unique source for scientific technical information in chemistry related industry? *World Patent Information*, *27*(4), 309–315. doi:10.1016/j.wpi.2005.05.003

Broadcasting Board of Governor. (n.d.). *Young Vietnamese increasingly turning to online news sources over state TV*. Retrieved from https://www.bbg.gov/wp-content/media/.../Vietnam-research-brief.docx

Brocke, J. V., & Sinnl, T. (2011). Culture in business process management: A literature review. *Business Process Management Journal*, *17*(2), 357–377. doi:10.1108/14637151111122383

Brown, J., & Duguid, P. (2000). *The social life of information* (1st ed.). Boston: Harvard Business School Press.

Brown, R. (2009). *Public relations and the social web*. Kogan Page.

Brünger-Weilandt, S., Geiß, D., Herlan, G., & Stuike-Prill, R. (2011). Quality: Key factor for high value in professional patent, technical and scientific information. *World Patent Information*, *33*(3), 230–234. doi:10.1016/j.wpi.2011.04.007

Brush, C. G., Carter, N. M., Gatewood, E. J., Greene, P. G., & Hart, M. M. (2006). The use of bootstrapping by women entrepreneurs in positioning for growth. *Venture Capital*, *8*(1), 15–31. doi:10.1080/13691060500433975

Bryan, D. (2013). The Great (Farm) Depression of the 1920s. *American History USA*. Retrieved from https://www.americanhistoryusa.com/great-farm-depression-1920s/

Bryant, A., & Charmaz, K. (2007). *The Sage handbook of grounded theory*. London, UK: Sage. doi:10.4135/9781848607941

Burgelman, R., Christensen, C., & Wheelwright, S. (2009). *Strategic management of technology and innovation* (5th ed.). Singapore: McGraw-Hill.

Burlton, R. (2010). Delivering *business strategy through process management*. In J. Vom Brocke & M. Rosemann (Eds.), *Handbook on business process management: strategic alignment, governance, people and culture*, 2, 1, (pp. 5–37). Berlin, Germany: Springer. doi:10.1007/978-3-642-01982-1_1

Butler, J. T. (1995). Patent searching using commercial databases. In Lechter (Ed.), Successful Patents and Patenting for Engineers and Scientists. New York, NY: The Institute of Electrical and Electronics Engineers (IEEE) Press.

Cachon, P. G. (2012). What Is Interesting in Operations Management? *Manufacturing & Service Operations Management: M & SOM*, *14*(2), 166–169. doi:10.1287/msom.1110.0375

Calarge, F. A., Satolo, E. G., & Satolo, L. F. (2007). Aplicação do sistema de gestão da qualidade BPF (boas práticas de fabricação) na indústria de produtos farmacêuticos veterinários. Gestão &amp. *Produção*, *14*(2), 379–392.

Calof, J. (2006). The SCIP06 academic program - Reporting on the state of the art. *Journal of Competitive Intelligence and Management, 3*(4).

Calof, J. L., & Beamish, P. W. (1999). The right attitude for international success. *Business Quarterly, 59*(1), 105–110.

Campos, A. (2013). *Modeling of processes with BPMN*. Rio de Janeiro, Brazil: Brasport Books and Multimedia.

Caroli, E., & Van Reenen, J. (2001). Skill-biased organisational change. Evidence from a panel of British and French establishments. *The Quarterly Journal of Economics, CXVI*(4), 1449–1492. doi:10.1162/003355301753265624

Cartlidge, A.; Hanna, A.; Rudd, C.; Macfarlane, I.; Windebank, J., & Rance, S. (2007). *An introductory overview of ITIL,* The UK Chapter of the ITSMF.

Cartwright, D. & Zander, A. (Eds.). (1984). Group dynamics: Research and theory (2nd ed., pp. 414-448). Evanston, IL: Row, Peterson, & Company.

carvalho, g. M.; vianna, n. M. C. Sistema de gerenciamento de armazéns: um estudo de caso sobre sua implementação no setor aeronáutico. Politécnico: ufrj, 2018.

Cascino, S., Pugliese, A., Mussolino, D., & Sansone, C. (2010). The influence of family ownership on the quality of accounting information. *Family Business Review, 23*(3), 246–265. doi:10.1177/0894486510374302

Castells, M. (2001). *A sociedade em rede: A era da informação: Economia, sociedade e cultura* (5th ed.). São Paulo: Paz e Terra.

Castells, M. (2004). *A galáxia internet: Reflexões sobre internet, negócios e sociedade*. Lisboa: Fundação Calouste Gulbenkian.

Castells, M. (2011). *The rise of the network society*. John Wiley & Sons.

Castells, M., & Cardoso, G. (2006). *The network society: From knowledge to policy*. Washington, DC: John Hopkins Center for Transatlantic Relations.

Catálogo Flora Valle Aburra. (n.d.). Theobroma cacao. In *Portal institucional Catálogo Flora Valle Aburra*. Retrieved from https://catalogofloravalleaburra.eia.edu.co/species/63

Cater, J. J. III, & Justis, R. T. (2009). The development of successors from followers to leaders in small family firms: An exploratory study. *Family Business Review, 22*(2), 109–124. doi:10.1177/0894486508327822

Cattela, R. (1981). Information as a corporate asset. *Information & Management, 4*(1), 29–37. doi:10.1016/0378-7206(81)90023-9

Cavalcanti, A. M., Oliveira, M. R. G., Gracas Vieira, M., & Cavalcanti Filho, A. (2012). O característico de inovação setorial: uma métrica para avaliar potencial crescimento de inovação nas micro e pequenas empresas. In: Encontro Nacional de Engenharia de Produção, 32., Bento Gonçalves, 2012. Anais... ENEGEP.

Cerf, V. & Huddle, S. (2002). *Internet radio communication system*. Retrieved from https://en.wikipedia.org/wiki/Internet_radio

Chan, D. (1998). Functional relations among constructs in the same content domain at different levels of analysis: A typology of composition models. *The Journal of Applied Psychology, 83*(2), 234–246. doi:10.1037/0021-9010.83.2.234

Chang, J.-R., & Lin, H.-S. (2011). Underground Pipeline Management Based on Road Information Modeling to Assist in Road Management. *Journal of Performance of Constructed Facilities, 25*(August), 326–335.

Compilation of References

Charef, R., Alaka, H., & Emmitt, S. (2018). Beyond the third dimension of BIM: A systematic review of literature and assessment of professional views. *Journal of Building Engineering, 19*(April), 242–257. doi:10.1016/j.jobe.2018.04.028

Charmaz, K. (1995). *Grounded theory. a practical guide through. qualitative analysis*. London, UK: Sage.

Chaves, G. C., & .... (2007a, February). Evolution of the international intellectual property rights system: Patent protection for the pharmaceutical industry and access to medicines. *Cadernos de Saude Publica, 23*(2), 257–267. doi:10.1590/S0102-311X2007000200002 PMID:17221075

Chenail, R. (2009). Qualitative research like politics can also be local: A review of interdisciplinary standards for systematic qualitative research. *The Weekly Qualitative Report, 2*(11), 61–65.

Chirico, F. (2008). Knowledge accumulation in family firms: evidence from four case studies. *International Small Business Journal, 26*(4), 433–462. doi:10.1177/0266242608091173

Chirico, F., & Salvato, C. (2008). Knowledge integration and dynamic organizational adaptation in family firms. *Family Business Review, 21*(1), 169–181. doi:10.1111/j.1741-6248.2008.00117.x

Cho, H., Lee, K. H., Lee, S. H., Lee, T., Cho, H. J., Kim, S. H., & Nam, S. H. (2011). Introduction of construction management integrated system using BIM in the Honam high-speed railway lot no. 4-2. *Proceedings of the 28th International Symposium on Automation and Robotics in Construction, ISARC 2011*, (4), 1300–1305.

Chong, H. Y., Lopez, R., Wang, J., Wang, X., & Zhao, Z. (2016). Comparative Analysis on the Adoption and Use of BIM in Road Infrastructure Projects. *Journal of Management Engineering, 32*(6), 0501602. doi:10.1061/(ASCE)ME.1943-5479.0000460

Choo, C. (2003). *Gestão de informação para a organização inteligente: A arte de explorar o meio ambiente*. Lisboa: Editorial Caminho.

Chopra, S., & Meindl, P. (2013). *Administración de la Cadena de Sumnistro*. D.F, México: Pearson.

Chrisman, J. J., Chua, J. H., & Litz, R. (2003). A unified systems perspective of family firm performance: An extension and integration. *Journal of Business Venturing, 18*(4), 467–472. doi:10.1016/S0883-9026(03)00055-7

Chrisman, J. J., Chua, J. H., & Litz, R. (2004). Comparing the agency costs of family and nonfamily firms: Conceptual issues and exploratory evidence. *Entrepreneurship Theory and Practice, 28*(4), 335–354. doi:10.1111/j.1540-6520.2004.00049.x

Chrisman, J. J., Kellermanns, F. W., Chan, K. C., & Liano, K. (2010). Intellectual foundations of current research in family business: An identification and review of 25 influential articles. *Family Business Review, 23*(1), 9–26. doi:10.1177/0894486509357920

Chrisman, J. J., Sharma, P., & Taggar, S. (2007). Family influences on firms: An introduction. *Journal of Business Research, 60*(10), 1005–1011. doi:10.1016/j.jbusres.2007.02.016

Christensen, C., Anthony, S., & Roth, E. (2004). *Seeing what's next: using the theories of innovation to predict industry change*. Harvard Business School Press.

Chuah, M., & Wong, K. (2011). A review of business intelligence and its maturity models. *African Journal of Business Management, 5*(9), 3424–3428.

Chua, J. H., Chrisman, J. J., & Bergiel, E. B. (2009). An agency theoretic analysis of the professionalized family firm. *Entrepreneurship Theory and Practice, 33*(2), 355–372. doi:10.1111/j.1540-6520.2009.00294.x

Chua, J. H., Chrisman, J. J., & Sharma, P. (1999). Defining the family business by behavior. *Entrepreneurship Theory and Practice, 23*(4), 19–39. doi:10.1177/104225879902300402

Cleveland, H. (1985). *The knowledge executive: Leadership in an information society* (1st ed.). New York: Dutton.

Clewlow & Strickland. (1998). *Implementing Derivatives Models*. John Wiley & Sons.

Coelho, P. M. N. (2016). *Rumo à Indústria 4.0. Tese de Mestrado*. Faculdade de Ciências e Tecnologia. Departamento de Engenharia Mecânica. Universidade de Coimbra.

Comercio, E. (2017). La increíble y terrible historia tras el origen del chocolate. In *El Comercio*. March 19, 2017. Retrieved from https://elcomercio.pe/redes-sociales/youtube/increible-terrible-historia-origen-chocolate-145532

Compaine, B. M., & Smith, E. (2001). *Internet radio: A new engine for content diversity?* (p. 2). Retrieved from http://ebusiness.mit.edu/research/papers/131%20Compaine,%20Internet%20Radio.pdf

Cooke-Davies, T. (2002). The "real" success factors on projects. *International Journal of Project Management*, *20*(3), 185–190. doi:10.1016/S0263-7863(01)00067-9

Cook, S. D., & Brown, J. S. (1999). Bridging epistemologies: The generative dance between organizational knowledge and organizational knowing. *Organization Science*, *10*(4), 381–400. doi:10.1287/orsc.10.4.381

Corbetta, G., & Montemerlo, D. (1999). Ownership, governance and management issues in small and medium sized family businesses: a comparison of Italy and the United States. *Family Business Review*, *12*(4), 361–374. doi:10.1111/j.1741-6248.1999.00361.x

Corbin, J., & Strauss, A. (2008). *Basics of qualitative research* (3rd ed.). Thousand Oaks, CA: Sage.

correia, p.; marcelino, s.; pizolato, c. Análise das ferramentas de armazenagem e estocagem. **Anais da SEMCITEC - Semana de CiÃªncia, Tecnologia, InovaÃ§Ã£o e Desenvolvimento de Guarulhos**, v. 1, n. 1, 6 ago. 2018.

Côrte-Real, N. (2010). *Avaliação da maturidade da Business Intelligence nas organizações*. Universidade Nova de Lisboa.

Cousins, L. (1991). Marketing plans or marketing planning? *Business Strategy Review*, *2*(2), 35–54. doi:10.1111/j.1467-8616.1991.tb00151.x

Coutinho, L. (1992). A terceira revolução industrial e tecnológica: As grandes tendências de mudança. *Economia e sociedade. Revista do Instituto de Economia da UNICAMP.*, *1*(1), 69–87.

Cox, J., Ross, S., & Rubinstein, M. (1979). Option Pricing: A Simplified Approach. *Journal of Financial Economics*, *7*(2), 229–264. doi:10.1016/0304-405X(79)90015-1

Crotty, R. A. Y. (2012). *The Impact of Building Information Modelling Transforming Construction*. New York: SPON Press.

Cruz, E. (1990). *Planeamento estratégico: Um guia para a PME* (3rd ed.). Lisboa: Texto Editora.

D'Acenção, L. C. M. (2001). *Organização, sistema e método: análise, desenho e informatização de processos administrativo. 1*. São Paulo, Brazil: Editora Atlas.

da Silva, A. M. (2005). A Gestão da informação abordada no campo da Ciência da Informação. Páginas a&b: arquivos e bibliotecas, 16, 89-113.

da Silva, A. M. (2006). A Informação: da compreensão do fenômeno e construção do objecto científico. Porto: Edições Afrontamento; CETAC.COM.

da Silva, A. M. (2009). A Gestão da Informação na perspectiva da pesquisa em Ciência da Informação: retorno a um tema estratégico. In Coletânea Luso Brasileira: governança estratégica, redes de negócios e meio ambiente: fundamentos e aplicações. Anápolis: Universidade Estadual de Goiás.

Compilation of References

da Silva, A. M., & Ribeiro, F. (2002). Das "Ciências" Documentais à Ciência da Informação: ensaio epistemológico para um novo modelo curricular. Porto: Edições Afrontamento.

Damian, I. P. M., Borges, L. S., & Pádua, S. I. D. (2015). The importance of tasks and critical success factors for business process management. *Journal of Management of UNIMEP 13*(2), p. 162. Retrieved from http://www.raunimep.com.br/ojs/index.php/regen/editor/submissionediting/899#scheduling

Dantas, J. & Carrizo Moreira, A. (2011). O Processo de Inovação: Como Potenciar a Criatividade Organizacional Visando uma Competitividade Sustentável (1ª ed.). Lousã: Lidel.

Dantas, J. (2001). *Gestão da Inovação*. Porto, Portugal: Vida Económica.

Daspit, J. J., Madison, K., Barnett, T., & Long, R. G. (2017, in press). The emergence of bifurcation bias from unbalanced families: Examining HR practices in the family firm using Circumplex Theory, in Human Resource Management Review, Special Issue on "The Role of Family Science Theories for Human Resource Management in Family Firms".

Davenport, T. (1997). *Information ecology: Mastering the information and knowledge environment*. New York: Oxford University Press.

Davenport, T. H., & Prusak, L. (1998). *Conhecimento Empresarial*. Rio de Janeiro, Brazil: Campus.

Davenport, T., Marchand, D., & Dickson, T. (2004). *Dominando a gestão da informação*. Porto Alegre: Bookmann.

Davis, J. H., Allen, M. R., & Hayes, H. D. (2010). Is blood thicker than water? A study of stewardship perceptions in family business. *Entrepreneurship Theory and Practice*, *34*(6), 1093–1116. doi:10.1111/j.1540-6520.2010.00415.x

Davis, P. S., & Harveston, P. D. (1998). The influence of family on the family business succession process: A multi-generational perspective. *Entrepreneurship Theory and Practice*, *22*(3), 31–53. doi:10.1177/104225879802200302

De Bes, F., & Kotler, P. (2011). *The innovation bible*. São Paulo, Brazil: Leya.

de Carvalho, H. G. (2000). Inteligência Competitiva Tecnológica para PMEs Através da Cooperação Escola-Empresa. Tese de Doutorado apresentada ao Programa de Pós-Graduação em Engenharia de Produção da Universidade Federal de Santa Catarina - UFSC. Florianópolis.

De Massis, A., Chua, J., & Chrisman, J. J. (2008). Factors preventing intra-family succession. *Family Business Review*, *21*(2), 183–199. doi:10.1111/j.1741-6248.2008.00118.x

de Ven, A. V., & Hargrave, T. (2000). *Social, technical, and institutional change*. Oxford, UK: Oxford University Pr.

Dearlove, D. (1998). *Key management decisions: Tools and techniques of the executive decision-maker*. Wiltshire: Financial Times/Pitman.

Debackere, K., Luwel, M., & Veugelers, R. (1999). Can technology lead to a competitive advantage? A case study of Flanders using European patent data. *Scientometrics, 44*(3), 379-400.

Debelak, D. (2006). *Business models made easy*. Entrepreneur Press.

Deitz, C. (2018). *Internet radio - What is internet radio and how to listen to it*. Retrieved from http://radio.about.com/od/createyourownradioshow/a/blyourshowhub.htm

den Hertog, J. F., & Huizenga, E. (2000). *The knowledge enterprise. Implementation of intelligent business strategies*. London, UK: Imperial College Press. doi:10.1142/p117

Departamento Intersindical de Estatística e Estudos Socioeconômicos (DIEESE). (2008). Política de Desenvolvimento Produtivo: nova política industrial do governo de 2008. n. 67. São Paulo, Brazil.

Derby, E., & Larsen, D. (2006). *Agile retrospectives: Making good teams great*. Dallas, TX: Pragmatic Bookshelf.

DGP/CNPq - Diretório dos Grupos de Pesquisa no Brasil, do CNPq (2014). *Censo dos Grupos de pesquisa no Brasil, do CNPq*. Retrieved from http://dgp.CNPq.br/planotabular/

Di Minin, A., & Faems, D. (2013). Special issue on intellectual property management. *California Management Review*, *55*(4), 7–14. doi:10.1525/cmr.2013.55.4.7

Diario Oficial de las Comunidades Empresa. (2002). Reglamento (Ce) No 178/2002 del Parlamento Europeo y del Consejo. In *Diario Oficial de las Comunidades Empresa*. Retrieved from https://eur-lex.europa.eu/legal-content/ES/TXT/PDF/?uri=CELEX:32002R0178&from=IT>

Dibrell, C., & Moeller, M. (2011). The impact of a service-dominant focus strategy and stewardship culture on organizational innovativeness in family-owned businesses. *Journal of Family Business Strategy*, *2*(1), 43–51. doi:10.1016/j.jfbs.2011.01.004

Dicionário eletrônico de terminologia em Ciência da Informação. (n.d.). Retrieved from http://www.ccje.ufes.br/arquivologia/deltci/

Dirección General de Seguimiento y Evaluación de Políticas del Ministerio de Agricultura y Riego. (2017). *Comercio Exterior Agrario*. Lima, Peru: Ministerio de Agricultura y Riego.

Dirección General de Seguimiento y Evaluación de Políticas del Ministerio de Agricultura y Riego. (2017). Enero-diciembre 2017. In *Portal institucional Ministerio de Cultura*. Retrieved from http://siea.minag.gob.pe/siea/sites/default/files/nota-comercio-exterior-diciembre17_3.pdf>

Diretrizes de Política Industrial, Tecnológica e de Comércio Exterior de 2003. (2003). UNICAMP. Retrieved from http://www.inovacao.unicamp.br/politicact/diretrizes-pi-031212.pdf

Dixit, A. K., & Pindyck, R. S. (1994). *Investment Under uncertainty*. Princeton, NJ: Princeton University Press.

Doloi, H. (2013). Cost overruns and failure in project management: Understanding the roles of key stakeholders in construction projects. *Journal of Construction Engineering and Management*, *139*(3), 267–279. doi:10.1061/(ASCE)CO.1943-7862.0000621

Donckels, R., & Fröhlich, E. (1991). Are family businesses really different? European experiences from Stratos. *Family Business Review*, *4*(2), 149–160. doi:10.1111/j.1741-6248.1991.00149.x

Donnelly, J. H., Gibson, J. L., & Ivancevich, J. M. (2000). *Administração: Princípios de gestão empresarial* (10th ed.). Lisboa: McGraw-Hill.

dos Santos, S. A., Leite, N. P., & Ferraresi, A. A. (2007). Gestão do conhecimento: institucionalização e práticas nas empresas e instituições: pesquisas e estudos. Maringá, PR: Unicorpore.

Dou, H. (2004). Benchmarking R&D and companies through patent analysis using free databases and special software: A tool to improve innovative thinking. *World Patent Information*, *26*(4), 297–309. doi:10.1016/j.wpi.2004.03.001

Doyle, P. & Stern, P. (2006). *Marketing Management and Strategy*, 4/E, 2006, Financial Times Press.

Dresner, S. (2008). *The principles of sustainability* (2nd ed.). Chippenham, UK: Earthscan.

Drucker, P. (1987). *Inovação e gestão* (2nd ed.). Lisboa, Portugal: Presença.

Drucker, P. F. (1974). *Management: tasks, responsibilities, practices* (p. 65). Oxford, UK: Butterworth.

Drucker, P. F. (2007). *Management challenges for the 21st century*. New York, NY: Routledge.

Compilation of References

Duarte, J. (Org.). (2002). Assessoria de imprensa e relacionamento com a mídia: Teoria e técnica. São Paulo, Brazil: Atlas.

Duarte, J., & Carvalho, N. (2017). *Sala de imprensa on-line*. CFN.

Duarte, J., Rivoire, V., & Ribeiro, A. A. (2016). Mídias sociais *on-line* e prática jornalística. *Universitas. Arquitetura e Comunicação Social*, *13*(1), 1–10.

Dukerich, J. M., Golden, B., & Shortell, S. M. (2002). Beauty is in the eye of the beholder: The impact of organizational identification, identity, and image on the cooperative behaviors of physicians. *Administrative Science Quarterly*, *47*(3), 507–533. doi:10.2307/3094849

Dunn, B. (1999). The family factor: The impact of family relationship dynamics on business-owning families during transitions. *Family Business Review*, *12*(1), 41–60. doi:10.1111/j.1741-6248.1999.00041.x

Dunn, R., & Harwood, K. (2015). Bridge Asset Management in Hertfordshire - now and in the future. In *Proceedings of Asset Management Conference 2015*. London, UK: IET. 10.1049/cp.2015.1722

Dutton, J. E. & Dukerich, J. M. (1991). Keeping an eye on the mirror: image and identity in organizational adaptation, *Academy Of Management Journal, 34,* 3, 517-554.

Dyck, B., Mauws, M., Starke, F. A., & Mischke, G. A. (2002). Passing the baton: The importance of sequence, timing, technique and communication in executive succession. *Journal of Business Venturing*, *17*(2), 143–162. doi:10.1016/S0883-9026(00)00056-2

Earl, M. (1998). *Information management: The organizational dimension*. New York: Oxford University Press.

Ebben, J., & Johnson, A. (2006). Bootstrapping in small firms: An empirical analysis of change over time. *Journal of Business Venturing*, *21*(6), 851–865. doi:10.1016/j.jbusvent.2005.06.007

Eckerson, W. W. (2009). *TDWI's business intelligence maturity model*. Chatsworth, CA: The Data Warehousing Institute.

Eddleston, K. A., & Kellermanns, F. W. (2007). Destructive and productive family relationships: A stewardship theory perspective. *Journal of Business Venturing*, *22*(4), 545–565. doi:10.1016/j.jbusvent.2006.06.004

Eddleston, K. A., Kellermanns, F. W., & Sarathy, R. (2008). Resource configuration in family firms: Linking resources, strategic planning and technological opportunities to performance. *Journal of Management Studies*, *45*(1), 26–50.

Eddleston, K. A., Kellermanns, F. W., & Zellweger, T. M. (2012). Exploring the entrepreneurial behavior of family firms: Does the stewardship perspective explain differences? *Entrepreneurship Theory and Practice*, *36*(2), 347–367. doi:10.1111/j.1540-6520.2010.00402.x

Eddleston, K. A., & Kidwell, R. E. (2012). Parent–child relationships: Planting the seeds of deviant behavior in the family firm. *Entrepreneurship Theory and Practice*, *36*(2), 369–386. doi:10.1111/j.1540-6520.2010.00403.x

Eddleston, K. A., & Morgan, R. M. (2014). Trust, commitment and relationships in family business: Challenging conventional wisdom. *Journal of Family Business Strategy*, *5*(3), 213–216. doi:10.1016/j.jfbs.2014.08.003

Edelman & Technorati. (2006). *Public Relationships*. Winter.

Edmunds, A., & Morris, A. (2000). The problem of information overload in business organizations: A review of the literature. *International Journal of Information Management*, *20*(1), 17–28. doi:10.1016/S0268-4012(99)00051-1

Edquist, C. (Ed.). (1997). Systems of innovation; technologies, institutions, and organisations. London, UK: Pinter European Commission. 1996. The Green Book on Innovation. Brussels, Belgium: EU.

Enríquez, G. (2010). *Trazabilidad: cuando proteger la vida se configura*. Lima, Peru.

Eriksson, H.-E., & Penker, M. (2000). *Business modeling with uml: business patterns at Work. 1*. USA: OMG PRESS.

Escudero, J. (2018). *Aseguramiento de Calidad en implementación de tecnologías*. [Interview].

Esparza, M., & Nicastro, M. (2018). *Blockchain is life*. Lima, Peru: Saxo.

Esty, D. & Winston, A. (2008). Do Verde ao Ouro: Como Empresas Inteligentes usam a Estratégia Ambiental para Inovar, Criar Valor e Construir uma Vantagem Competitiva (1ª ed.). Cruz Quebrada: Casa das Letras.

Eteokleous, P., Leonidou, C., & Katsikeas, C. (2016). *Corporate social responsibility in international marketing: Review, assessment, and future research. International Marketing Review*, *33*(4), 580–624. doi:10.1108/IMR-04-2014-0120

Etzkowitz, H. (2009). *Hélice Tríplice: Universidade-Empresa-Governo, Inovação em movimento*. Porto Alegre, Brazil: EDIPUCRS.

Evers, N. (2003). The process and problems of business start-ups. *The ITB Journal*, *4*(1). doi:10.21427/D7WT8K

Fahey, L., & Prusak, L. (1988). The eleven deadliest sins of knowledge management. *California Management Review*, *40*(3), 265–276. doi:10.2307/41165954

Fanning, B., Clevenger, C. M., Ozbek, M. E., & Mahmoud, H. (2014). Implementing BIM on Infrastructure : Comparison of Two Bridge Construction Projects. *Practice Periodical on Structural Design and Construction*, *20*(4), 04014044. doi:10.1061/(ASCE)SC.1943-5576.0000239

Farias, L. A. (2009). Estratégias de relacionamento com a mídia. In M. M. K. Kunsch (Org.), Gestão estratégica em comunicação organizacional. São Caetano do Sul: Difusão.

Faulkner, D., & Bowman, C. (1995). *The essence of competitive strategy* (1st ed.). Exeter: Prentice Hall.

Faulkner, T. W. (1996). Applying Options Thinking To R&D Valuation. *Research Technology Management*, *39*(3), 50–56. doi:10.1080/08956308.1996.11671064

FDA guidance for industry: process validation: general principles and practices - ECA Academy. Disponível em https://www.gmp-compliance.org/guidelines/gmp-guideline/fda-guidance-for-industry-process-validation-general-principles-and-practices

Feltham, T. S., Feltham, G., & Barnett, J. J. (2005). The dependence of family businesses on a single decision-maker. *Journal of Small Business Management*, *43*(1), 1–15. doi:10.1111/j.1540-627X.2004.00122.x

Fendt, J., & Sachs, W. (2008). Grounded theory method in management research. Users' perspectives. *Organizational Research Methods*, *11*(3), 430–455. doi:10.1177/1094428106297812

Ferina, N., Khurotul, A. E., & Sukowidyanti, A. P. (2018). Does family social support affect startup business activities? 2, 41–54. DOES

Ferraz, I. N. (2009). O Uso do Balanced Scorecard na Ótica do Business Intelligence. In *Encontro da Associação Nacional de Programas de Pós-Graduação em Administração, Anais do 37o EnANPAD*. São Paulo, Brazil: ANPAD.

Ferreira, H. P. (2004). Sistema de gestão da qualidade - estudo de caso: Far-Manguinhos.

Fiol, C. M., Pratt, G. M., & O'Connor, E. J. (2009). Managing intractable identity conflicts. *Academy of Management Review*, *34*(1), 32–55. doi:10.5465/amr.2009.35713276

Fletcher, D., De Massis, A., & Nordqvist, M. (2016). Qualitative research practices and family business scholarship: A review and future research agenda. *Journal of Family Business Strategy*, *7*(1), 8–25. doi:10.1016/j.jfbs.2015.08.001

Compilation of References

Fleury, A. C. C., & Fleury, M. T. L. (2004). Entrepreneurial strategies and skills training: A kaleidoscopic puzzle of Brazilian industry (3rd ed.). São Paulo, Brazil: Atlas.

Flynn, R. (2002). NutraSweet enfrenta a concorrência: O papel crítico da inteligência competitiva. In Prescott & Miller (Eds.). Inteligência competitiva na prática: Estudos de casos diretamente do campo de batalha. Rio de Janeiro, Brazil: Campus, 135-149.

Foo, M.-D. (2011). Emotions and entrepreneurial opportunity evaluation. *Entrepreneurship Theory and Practice*, *35*(2), 375–393. doi:10.1111/j.1540-6520.2009.00357.x

Fox, P. & Hundley, S. (2011). The importance of globalization in higher education. In New Knowledge in a New Era of Globalization. doi:10.5772/17972

Frank, H., Lueger, M., Nose´, L., & Suchy, D. (2010). The concept of "Familiness": Literature review and systems theory-based reflections. *Journal of Family Business Strategy*, *1*(3), 119–130. doi:10.1016/j.jfbs.2010.08.001

Freear, J., Sohl, J. E., & Wetzel, W. E. J. (n.d.-a). *Early stage software ventures: what is working and what is not*. Report for the Massachusetts Software Council.

Freear, J., Sohl, J. E., & Wetzel, W. E. J. (n.d.-b). Who bankrolls software entrepreneurs. In Frontiers of Entrepreneurship Research.

Freear, J., Sohl, J. E., & Wetzel, W. (2002). Angles on angels: Financing technology-based ventures-a historical perspective. *Venture Capital*, *4*(4), 275–287. doi:10.1080/1369106022000024923

Freeman, C. (1987). *Technology policy and economic performance: lessons from Japan*. London, UK: Pinter.

Freudenberger, H. J., Freedheim, D. K., & Kurtz, T. S. (1989). Treatment of individuals in family business. *Psychotherapy (Chicago, Ill.)*, *26*(1), 47–53. doi:10.1037/h0085404

Fuld, L. M. (2006). The secret language of competitive intelligence. New York, NY: Crown Publishing, Random House.

Fuld, L. M. (1995). *The new competitor intelligence: the complete resource for finding, analysing, and using information about your competitors*. New York, NY: Wiley & Sons.

Fuld, L. M. (1995). *What competitive intelligence is and is not. Your competitors*. New York, NY: Wiley & Sons.

Gabriel, M. (2015). *Redes Sociais Estratégicas*. Retrieved from http://pt.slideshare.net/marthagabriel/redes-sociais-estrategias-e-mensurao-por-martha-gabriel>

Galambos, L. (2010). The role of professionals in the Chandler paradigm. *Industrial and Corporate Change*, *19*(2), 377–398. doi:10.1093/icc/dtq009

Galpin, T. J. (1996). The human side of change: a practical guide to organization redesign. San Francisco, CA: Jossey-Bass.

Gatlin, L. (1972). *Information theory and the living system*. Columbia University Press.

Gedajlovic, E., Lubatkin, M. H., & Schulze, W. S. (2004). Crossing the threshold from founder management to professional management: A governance perspective. *Journal of Management Studies*, *41*(5), 899–912. doi:10.1111/j.1467-6486.2004.00459.x

Geiger, J. G. (2016). *Data warehousing supporting business intelligence*. Retrieved from http://www.cutter.com/freestuff/biareport.html

Ge, L., Brewster, C., Spek, J., Smeenk, A., & Top, J. (2017). *Blockchain for agriculture and Food; Findings from a pilot study*. Wageningen, The Netherlands: Wageningen Economic Research. doi:10.18174/426747

Genesis, G. (2011). La historia del cacao en la época prehispánica. *Portal La Historia del Cacao*. Retrieved from http://lahistoriadelcacao.blogspot.com/

Geske, R. (1979). The valuation of Compound Option. *Journal of Financial Economics*, 7(1), 63–81. doi:10.1016/0304-405X(79)90022-9

Getz, A. (2016). *Business Intelligence (BI) Maturity Model*. Retrieved from http://biinsider.com/portfolio/bi-maturity-model/

G, Hamel,. & Prahalad, C. K. (1990). The core competence of the corporation. *Harvard Business Review*, 68(3), 79–91.

Gil, N., Aspegren, H., & Gomez Lopez, A. M. (2017). *Report on blockchain asset registries*. Cambridge, MA: MIT.

Glaze, J. E. (2002). Stages in coming to terms with reflection: Student advanced nurse practitioners' perceptions of their reflective journeys. *Journal of Advanced Nursing*, 37(3), 265–272. doi:10.1046/j.1365-2648.2002.02093.x PMID:11851797

Gleick, J. (2011). The information: A history, a theory, a flood. St. Ives: 4th Estate.

Godinho, M. (2003). *Estudo Sobre a Utilização da Propriedade Industrial em Portugal* (Vol. I). Lisboa, Portugal: Instituto Nacional da Propriedade Industrial.

Goh, J. (2018). Internet of Things and the pump market: How the industry will evolve. *Market Insight. IHS Markit*. Retrieved from https://technology.ihs.com/599560/internet-of-things-and-the-pump-market-how-the-industry-will-evolve

Gomes, E., & Braga, F. (2001). *Inteligência competitiva: Como transformar informação em um negócio lucrativo*. Rio de Janeiro: Campus.

Gómez-Mejía, L. R., Cruz, C., Berrone, P., & De Castro, J. (2011). The bind that ties: socioemotional wealth preservation in family firms. *The Academy of Management Annals*, 5(1), 653–707. doi:10.1080/19416520.2011.593320

Gomez-Mejia, L. R., Haynes, K. T., Nunez-Nickel, M., Jacobson, K. J. L., & Moyano-Fuentes, J. (2007). Socioemotional wealth and business risks in family-controlled firms: Evidence from Spanish olive oil mills. *Administrative Science Quarterly*, 52(1), 106–137. doi:10.2189/asqu.52.1.106

Gomez-Mejia, L. R., Larraza-Kintana, M., & Makri, M. (2003). The determinants of executive compensation in family-controlled public corporations. *Academy of Management Journal*, 46, 226–237.

Gomez-Mejia, L. R., Makri, M., & Larraza-Kintana, M. (2010). Diversification decisions in family-controlled firms. *Journal of Management Studies*, 47(2), 223–252. doi:10.1111/j.1467-6486.2009.00889.x

Gonçalves, J. E. L. (2000). *Companies are large collections of processes*. RAE: Revista dein Business Administration 40(1), p. 1.

Gorecki, J. (2015). Information technology in project management, 77.

Grant, D. B. Gestão de Logística e Cadeia de Suprimentos. [s.l: s.n.].

Grant, R. (2002). *Contemporary strategy analysis: concepts, techniques, applications*. Wiley-Blackwell.

Greenhalgh, S. (1994). Deorientalizing the Chinese family firm. *American Ethnologist*, 21(4), 746–775. doi:10.1525/ae.1994.21.4.02a00050

Greif, S. (1979). The role of patent protected imports in the transfer of technology to developing countries. *International Review of Industrial Property and Copyright Law*, 10(2), 124–143.

Greif, S. (1987). Patents and economic growth. *International Review of Industrial Property and Copyright Law*, 18(2), 191–213.

Greitzke, S. (2007). Intelligent pumps for building automation systems. *World Pumps*, *490*, 26–32.

GS1. (2003). Tranzabilidad. *Portal institucional GS1*. Retrieved from https://www.gs1.org.ar/documentos/TRAZABILIDAD.pdf>

GS1. (2016). *Global traceability compliance criteria for food application standard*.

GS1. (2018). Bienvenido a GS1 Perú. *Portal GS1*. Retrieved from http://www.gs1pe.org/content/codificaci%C3%B3n-gln-c%C3%B3digos-de-localizaci%C3%B3n

GS1. (2018). Código de barras. *Portal GS1*. Retrieved from http://www.gs1pe.org/content/gs1-c%C3%B3digo-de-barras

GS1. (2018). Visibilidad y tranzabilidad. *Portal GS1*. Retrieved from http://www.gs1pe.org/content/visibilidad-y-trazabilidad>

Guarnieri, P., & ... . (2006b, April). WMS -Warehouse management system: Adaptation proposed for the management of the reverse logistics. *Production*, *16*(1), 126–139. doi:10.1590/S0103-65132006000100011

Guidolin, S. M.; Monteiro Filha, D. C. (2010). Cadeia de suprimentos: o papel dos provedores de serviços logísticos. set.

Gündüz, M., Nielsen, Y., & Özdemir, M. (2013). Quantification of delay factors using the relative importance index method for construction projects in Turkey. *Journal of Management in Engineering*, *29*(2), 133–139. doi:10.1061/(ASCE)ME.1943-5479.0000129

Gupta, M. (2017). *Blockchain for dummies*. Hoboken, NJ: John Wiley & Sons.

Habbershon, T. G., Williams, M., & MacMillan, I. C. (2003). A unified systems perspective of family firm performance. *Journal of Business Venturing*, *18*(4), 451–465. doi:10.1016/S0883-9026(03)00053-3

Haberman, M. (2001). The role of intellectual property and patent information in successful innovation, production and marketing. Case study 1: The non-spill drinking vessel. *World Patent Information*, *23*(1), 71–73. doi:10.1016/S0172-2190(00)00105-8

Hallberg, D., & Tarandi, V. (2011). on the Use of Open Bim and 4D Visualisation in a Predictive Life Cycle Management System for Construction Works. *Journal of Information Technology in Construction*, *16*, 445–466.

Hall, P. (1980). *Great planning disasters*. London, UK: Weidenfeld and Nicolson. doi:10.1016/S0016-3287(80)80006-1

Hamel, G. & Prahalad, C. K. (1989). Strategic intent (3rd ed.). Harvard Business Review.

Hammer, M., & Champy, J. (1994). *Reengineering the Corporation: A manifesto for business revolution*. Londres, UK: Nicholas Brealy.

Hansen, B. (1980). Economic aspects of technology transfer to developing countries. *International Review of Industrial Property and Copyright Law*, *11*, 430–440.

Hardcastle, J. L. (2015). IoT Shakes Up Global Pumps Industry. *Environmental Leader*. Retrieved from https://www.environmentalleader.com/2015/07/iot-shakes-up-global-pumps-industry/

Harrison, R. T., Mason, C. M., & Girling, P. (2004). Financial bootstrapping and venture development the software industry. *Entrepreneurship and Regional Development*, *16*(4), 307–333. doi:10.1080/0898562042000263276

Hatcher, D., & Prentice, B. (2004). The evolution of information management. *Business Intelligence Journal*, *9*(2), 49–56.

Haveman, H. A., & Khaire, M. V. (2004). Survival beyond succession? The contingent impact of founder succession on organizational failure. *Journal of Business Venturing*, *19*(3), 437–463. doi:10.1016/S0883-9026(03)00039-9

Haynes, G. W., Walker, R., Rowe, B. R., & Hong, G. S. (1999). The intermingling of business and family finances in family-owned businesses. *Family Business Review, 12*(3), 225–239. doi:10.1111/j.1741-6248.1999.00225.x

HBR. (2018). *Marketing estratégico. 10 artigos essenciais*. Actual Editora.

Heikkila, R. (2013). Development of BIM based rehabilitation and maintenance process for roads. In *Proceedigs fo 30th International Symposium on Automation and Robotics in Construction and Mining and Petroleum Industries* (pp. 1216–1222). Montreal, Canada: ISARC. 10.22260/ISARC2013/0136

Hékis, H. R., Medeiros Araújo de Moura, L. C., Pires de Souza, R., & De Medeiros Valentim, R. A. (2013, Oct. 1). Sistema de informação: Benefícios auferidos com a implantação de um sistema WMS em um centro de distribuição do setor têxtil em Natal/RN. *RAI Revista de Administração e Inovação, 10*(4), 85–109. doi:10.5773/rai.v10i4.920

Heldman, K. (2005). Gerência de projetos: guia para o exame oficial do PMI. Rio de Janeiro: Elsevier.

Hendieh, J., & Osta, A. (2019). Students' attitudes toward entrepreneurship at the Arab Open University-Lebanon. *Journal of Entrepreneurship Education, 22*(2).

Henschen, D. (2007). HP touts neoview win, banking solutions, BI maturity model. *Intelligent Enterprise, 10*(10).

Herath, H. S. B., & Park, C. S. (1999). Economic Analysis of R&D Projects: An Options Approach. *The Engineering Economist, 44*(1), 1–35. doi:10.1080/00137919908967506

Hermann, M., Pentek, T., & Otto, B. (2015). Design principles for Industrie 4.0 scenarios: a literature review. *Working Paper nº 1. Technische Universität Dortmund*. Retrieved from https://www.researchgate.net/publication/307864150_Design_Principles_for_Industrie_40_Scenarios_A_Literature_Review

Hernaus, T., Bach, M. P., & Vuksic, V. B. (2012). *Influence of strategic approach to BPM on financial and non-financial performance, Baltic Journal of Management, 7*(4), 376–396.

Herrando, M. Cicle optimització processos SAP amb MM, WM i RF. (2019). Disponível em http://www.marcherrando.com/2010/03/cicle-optimizacio-processos-sap-amb-mm.html

Hess, A., Diebold, P., & Seyff, N. (2017, Sept. 4-8). Towards requirements communication and documentation guidelines for agile teams. In *Proceedings 2017 IEEE 25th International Requirements Engineering Conference Workshops (REW)*. 10.1109/REW.2017.64

Hewlett Packard. (2016). *The HP business intelligence maturity model: Describing the BI journey*. Hewlett-Packard Development Company. Retrieved from http://h20195.www2.hp.com/V2/GetPDF.aspx/4AA1-5467ENW.pdf

Hildebrandt, D. (1997). *Internet tools of the profession: a guide for information professionals*. Chicago: Special Libraries Association.

Hinton, M. (2006). *Introducing information management: The business approach* (1st ed.). Burlington: Elsevier Butterworth-Heinemann.

Hitt, M. A., Beamish, P. W., Jackson, S. E., & Mathieu, J. E. (2007). Building theoretical and empirical bridges across levels: Multilevel research in management. *Academy of Management Journal, 50*(6), 1385–1399. doi:10.5465/amj.2007.28166219

Hoffman, S. G. (2011, Nov. 1). The new tools of the science trade: Contested knowledge production and the conceptual vocabularies of academic capitalism. *Social Anthropology, 19*(4), 439–462. doi:10.1111/j.1469-8676.2011.00180.x

Holyoak, J., & Torremans, P. (1995). *Intellectual property law*. London, UK: Butterworths.

Compilation of References

House, R., Rousseau, D. M., & Thomas-Hunt, M. (1995). The meso paradigm: A framework for the integration of micro and macro organizational behavior. In L. L. Cummings & B. M. Staw (Eds.), *Research in organizational behavior* , 17, pp. 71–114. Greenwich, CT: JAI Press.

Huber, L. (2005). Risk-based validation of commercial off-the-shelf. *Computer Systems*, *2005*(6).

Hull, J. C. (1997). *Options, Futures and Other Derivatives Securities* (3rd ed.). New York: Prentice Hall.

Hunsicker, J. (1989). Strategies for European survival. *The McKinsey Quarterly*, (Summer): 37–47.

Idris, K. (2003). *Intellectual property: a power tool for economic growth*. Geneva, Switzerland: World Intellectual Property Organization.

Idris, K. (2003). *Intellectual property: A power tool for economic growth*. Geneva, Switzerland: World Intellectual Property Organization.

IEC. (2006). Guidance on failure modes and effects analyses (FMEAs). Disponível em https://webstore.iec.ch/publication/3571

Ika, L. A., Diallo, A., & Thuillier, D. (2012). Critical success factors for World Bank projects: An empirical investigation. *International Journal of Project Management*, *30*(1), 105–116. doi:10.1016/j.ijproman.2011.03.005

Imai, M. (2006). *Gemba Kaizen*. Warszawa, Poland: MT Biznes.

Information Technology. (2013). Retrieved from http://www.globalization101.org/information-technology/

Ingram, P., & Lifschitz, A. (2006). Kinship in the shadow of the corporation: The interbuilder network in Clyde River shipbuilding, 1711-1990. *American Sociological Review*, *71*(2), 334–352. doi:10.1177/000312240607100208

Inmon, W. H. (1997). *Como construir o Data Warehouse*. Rio de Janeiro, Brazil.

Internacional, C. C.-N.-S. (2017). Catalogo flora Valle Aburra. [Online]. Retrieved from https://catalogofloravalleaburra.eia.edu.co/

International council for harmonisation of technical requirements for pharmaceuticals for human use. Disponível em http://www.ich.org/home.html

International Trade Centre. (2018). Trade Map. *Portal institucional International Trade Center.* Retrieved from https://www.trademap.org/Country_SelProduct_TS.aspx?nvpm=3|||||1801|||4|1|1|2|2|1|2|2|1>

Irava, W. J., & Moores, K. (2010). Clarifying the strategic advantage of familiness: Unbundling its dimensions and highlighting its paradoxes. *Journal of Family Business Strategy*, *1*(3), 131–144. doi:10.1016/j.jfbs.2010.08.002

ISGec – Information System Governance European Club. (2009). Information Systems Governance Manifest. Retrieved from http://www.cegsi.org/index.php/qui-sommes-nous/o-que-e-o-cegsi

ISGec – Information System Governance European Club. (2010). Why the corporations are asking for an information systems governance? Retrieved from http://www.cegsi.org/index.php/documents/telechargement-du-document-la-gouvernance-des-systemes-d-information-pourquoi/la-gouvernance-des-systemes-d-information-pourquoi

ISGec – Information System Governance European Club. (2011). The importance of the information systems approach in the governance of organizations. Retrieved from http://www.cegsi.org/index.php/documents/l-importance-de-l-approche-par-les-systemes-d-information/the-importance-of-the-information-systems-approach-for-governance-organizations (In French)

ISPE. Welcome to the Brazil Affiliate I ISPE. Disponível em http://www.ispe.org.br/ispe_about

Ivanoff, S. D. & Hultberg, J. (2006). Understanding the multiple realities of everyday life: basic assumptions in focus group methodology, *Scandanavian Journal of Occupational Therapy*, 13.

Jacobs, R. L., & Washington, C. (2003). Employee development and job performance: A review of literature and directions for future research. *Human Resource Development International*, *6*(3), 343–355. doi:10.1080/13678860110096211

Jacobzone, S. (2016a). Pharmaceutical policies in OECD countries. Disponível em http://www.oecd-ilibrary.org/social-issues-migration-health/pharmaceutical-policies-in-oecd-countries_323807375536

Jacobzone, S. 2016b. Pharmaceutical policies in OECD countries. Disponível em http://www.oecd-ilibrary.org/social-issues-migration-health/pharmaceutical-policies-in-oecd-countries_323807375536

Jamil, G. L., & Carvalho, L. F. M. (2019). Improving Project management decisions with Big data Analytics. In *Handbook of Research on Expanding Business opportunities with information systems and analytics*. Hershey, PA: IGI Global Publishers. doi:10.4018/978-1-5225-6225-2.ch003

Jashapara, A. (2007). Moving beyond tacit and explicit distinctions: A realist theory of organisational knowledge. *Journal of Information Science*, *33*(6), 752–766. doi:10.1177/0165551506078404

Jegorov, A., Husak, M., Kratochvil, B., & Cisarova, I. (2003). How many "new" entities can be created from one active substance? the case of Cyclosporin A. *Crystal Growth & Design*, *3*(4), 441–444. doi:10.1021/cg0300127

Jeston, J., & Nelis, J. (2006). *Business process management: practical guidelines to successful implementations*. Oxford, UK: Butterworth-Heinemann.

Jiménez, M. (2018). *Operatividad de la Blockchain*. [Interview].

Johansson, F. (2007). O Efeito Medici: O que nos podem ensinar os Elefantes e as Epidemias acerca da Inovação (1ª ed.). Cruz Quebrada: Casa das Letras.

Jolly, A., & Philpott, J. (2009). *The handbook of European intellectual property management: developing, managing & protecting your company's intellectual property* (2nd ed.). Glasgow, Scotland: Kogan Page.

Jordan, M. P. (1998). How can problem-solution structures help writers plan and write technical documents? In L. Beene & P. White (Eds.), Solving Problems in Technical Writing. Academic Press.

Jorge, M. (1995). *Biologia, informação e conhecimento*. Lisboa: F. C. Gulbenkian.

Jorissen, A., Laveren, E., Martens, R., & Reheul, A.-M. (2005). Real versus sample-based differences in comparative family business research. *Family Business Review*, *18*(3), 229–246. doi:10.1111/j.1741-6248.2005.00044.x

Juran, J. M. Juran Plancjando para a qualidade. 2. ed. São Paulo - SP: Editora e livraria Pioneira, 1992.

Kahaner, L. (1997). *Competitive intelligence: How to gather, analyze, and use information to move your business to the top* (1st ed.). New York: Simon & Schuster.

Kalakota, R., & Robinson, M. (2001). *M-Business: the race to mobility*. New York, NY: McGraw-Hill.

Kalseth, K. (1991). Business information strategy: The strategic use of information and knowledge. *Information Services & Use*, *11*(3), 155–164. doi:10.3233/ISU-1991-11307

Kamardeen, I. (2010). 8D BIM Modelling Tool for Accident Prevention Through Design. In *Proceedings of the 26th Annual ARCOM Conference*, (vol. 2, pp. 281–289). Leeds, UK: ARCOM.

Kang, J. H., Anderson, S. D., & Clayton, M. J. (2007). Empirical Study on the Merit of Web-Based 4D Visualization in Collaborative Construction Planning and Scheduling. *Journal of Construction Engineering and Management, 133*(6), 447–461. doi:10.1061/(ASCE)0733-9364(2007)133:6(447)

Kang, L. S., Kim, H. S., Moon, H. S., & Kim, S. K. (2016). Managing construction schedule by telepresence: Integration of site video feed with an active nD CAD simulation. *Automation in Construction, 68*, 32–43. doi:10.1016/j.autcon.2016.04.003

Kang, L., Pyeon, J., Moon, H., Kim, C., & Kang, M. (2013). Development of Improved 4D CAD System for Horizontal Works in Civil Engineering Projects. *Journal of Computing in Civil Engineering, 27*(3), 212–230. doi:10.1061/(ASCE)CP.1943-5487.0000216

Kannan, P. K., & Li, H. A. (2017). Digital marketing: A framework, review and research agenda. *International Journal of Research in Marketing, 34*(1), 22–45. doi:10.1016/j.ijresmar.2016.11.006

Kaplan, R., & Norton, D. (2007). *The execution premium: linking strategies to operation for competitive advantage.* Harvard Business School Press.

Katz, G., & Gartner, W. B. (1988). Properties of emerging organisation. *Academy of Management Review, 13*(3), 429–441. doi:10.5465/amr.1988.4306967

Kaye K. (1991). Penetrating the cycle of sustained conflict, *Family Business Review, 4*(1), 2, 1-44.

Kaye, K. (1992). The kid brother. *Family Business Review, 5*(3), 237–256. doi:10.1111/j.1741-6248.1992.00237.x

Kaye, K. (1996). When the family business is a sickness. *Family Business Review, 9*(4), 347–368. doi:10.1111/j.1741-6248.1996.00347.x

Kearns, G. S., & Lederer, A. L. (2003). A resource-based view of strategic IT alignment: How knowledge sharing creates a competitive advantage. *Decision Sciences, 34*(1), 1–29. doi:10.1111/1540-5915.02289

Keck, A. R. (2018). *Experiencias sobre AgUnity - Blockchain en agricultores del Cacao.*

Kellermanns, F. W., Eddleston, K. A., & Zellweger, T. M. (2012). Extending the socioemotional wealth perspective: a look at the dark side, *Entrepreneurship Theory and Practice, 36*, Special Issue on Family Business, 1175–1182.

Kemma, A. G. Z. (1993). Case Studies on Real Options, Financial Management: 259-270. Lint, O., E. Pennings. R&D as An Option on Market Introduction. *R & D Management, 28*(4), 279–287.

Kenton, W. (2018). Non-disclosure agreement - NDA. Retrieved from https://www.investopedia.com/terms/n/nda.asp

Kets de Vries, M. F. (1977). The entrepreneurial personality: A person at the crossroads. *Journal of Management Studies, 14*(1), 34–57. doi:10.1111/j.1467-6486.1977.tb00616.x

Kets de Vries, M. F. (1993). The dynamics of family-controlled firms: The good and the bad news. *Organizational Dynamics, 21*(3), 59–71. doi:10.1016/0090-2616(93)90071-8

Khan, I., & Rahman, Z. (2015). A review and future directions of brand experience research. *International Strategic Management Review, 3*(1), 1–14. doi:10.1016/j.ism.2015.09.003

Kidwell, R. E., Eddleston, K. A., Cater, J. J. III, & Kellermanns, F. W. (2013). How one bad family member can undermine a family firm: Preventing the Fredo effect. *Business Horizons, 56*(1), 5–12. doi:10.1016/j.bushor.2012.08.004

Kidwell, R. E., Kellermanns, F. W., & Eddleston, K. A. (2012). Harmony, justice, confusion, and conflict in the family firm: Implications for ethical climate and the "Fredo effect.". *Journal of Business Ethics*, *106*(4), 503–517. doi:10.100710551-011-1014-7

Kienzler, M., & Kowalkowski, C. (2017). Pricing strategy: A review of 22 years of marketing research. *Journal of Business Research*, *78*, 101–110. doi:10.1016/j.jbusres.2017.05.005

Kimball, R., & Ross, M. (2002). *The data warehouse toolkit – Guia completo para modelagem dimensional*. Brasil: Campus.

Kim, C., Kim, H., Park, T., & Kim, M. K. (2011). Applicability of 4D CAD in Civil Engineering Construction: Case Study of a Cable-Stayed Bridge Project. *Journal of Computing in Civil Engineering*, *25*(1), 98–107. doi:10.1061/(ASCE)CP.1943-5487.0000074

Kim, H., Orr, K., Shen, Z., Moon, H., Ju, K., & Choi, W. (2014). Highway Alignment Construction Comparison Using Object-Oriented 3D Visualization Modeling. *Journal of Construction Engineering and Management*, *140*(10), 05014008. doi:10.1061/(ASCE)CO.1943-7862.0000898

Kim, J. I., Kim, J., Fischer, M., & Orr, R. (2015). BIM-based decision-support method for master planning of sustainable large-scale developments. *Automation in Construction*, *58*, 95–108. doi:10.1016/j.autcon.2015.07.003

King, W., Grover, V., & Hufnagel, E. (1989). Using information and information technology for sustainable competitive advantage: Some empirical evidence. *Information & Management*, *17*(2), 87–93. doi:10.1016/0378-7206(89)90010-4

Kluska, R. A., da Lima, E. P., & da Costa, S. E. G. (2015). *A proposal for structure and use of business process management (BPM)*. Revista Produção Online 15(3), 886–913. Retrieved from http://web.a.ebscohost.com/ehost/pdfviewer/pdfviewer?Vid=5&sid=e40d1be6-302b-4759-885b-ac4e11392bcc%40sdc-v-sessmgr05

Korosteleva, J., & Mickiewicz, T. (2011). Start-up financing in the age of globalization. *Emerging Markets Finance & Trade*, *47*(3), 23–49. doi:10.2753/REE1540-496X470302

Kotler P., Keller L., Tuck A., Ang H., Tan T., & Leong S. (2017. *Marketing Management*, An Asian Perspective, 7/E, Pearson.

Kotler, P., Keller, L., & Lane, K. (2015). Marketing management, Global Edition, 15/E, 2016, Pearson.

Kotler, P., Hermawan, K., & Setiawan, I. (2017). *Marketing 4.0: Moving from tradition to digital*. Hoboken, NJ: John Wiley & Sons.

Kozamernik, F. & Mullane, M. (2015). *An introduction to internet radio,* EBU technical review. Retrieved from https://tech.ebu.ch/docs/techreview/trev_304-webcasting.pdf

Kozlowski, S. W. J., & Klein, K. J. (2000). A multilevel approach to theory and research in organizations contextual, temporal, and emergent processes. In K. Klein & S. Kozlowski (Eds.), Multilevel theory, research, and methods in organizations: Foundations, extensions, and new directions, San Francisco, CA: Jossey-Bass.

Kraatz, M. & Block, E. (2008). Organizational implications of institutional pluralism. In R. Greenwood, C. Oliver, R. Suddaby, & K. Sahlin-Andersson (Eds.), Handbook of organizational institutionalism, 243-275. London, UK: Sage. doi:10.4135/9781849200387.n10

Krupp, F. & Horn, M. (2009). Reinventar a Energia: Estratégias para o Futuro Energético do Planeta (1ª ed.). Alfragide: Estrela Polar.

Kubota, S. (2011). Utilization of 3D Information on Road Construction Projects in Japan. In *Proceeding of 2nd International Conference on Construction and Project Management*, (vol. 15, pp. 21–25). Singapore: IACSIT Press.

Compilation of References

Kucharska, W. & Kowalczyk, R. (2016). Trust, collaborative culture and tacit knowledge sharing in project management – a relationship model, In *Proceedings of the 13th International Conference on Intellectual Capital, Knowledge Management & Organisational Learning, ICICKM.* (pp. 159-166). doi:10.13140/RG.2.2.25908.04486

Kuhn, T., & Jackson, M. H. (2008). Accomplishing knowledge: A framework for investigating knowing in organisations. *Management Communication Quarterly, 21*(4), 454–485. doi:10.1177/0893318907313710

Kulasekara, G., Jayasena, H. S., & Ranadewa, K. A. T. O. (2013). Comparative effectiveness of quantity surveying in a building information modelling implementation. In Proceedings of Socio-Economic Sustainability in Construction, (pp. 101–107). Colombo, Sri Lanka: Academic Press.

Kumar, B., Cai, H., & Hastak, M. (2017). An Assessment of Benefits of Using BIM on an Infrastructure Project. In *Proceedings of International Conference on Sustainable Infrastructure 2017* (pp. 88–95). New York: ASCE. 10.1061/9780784481219.008

Kumar, N. (2004). *Marketing as strategy: Understanding the CEO's agenda for driving growth and innovation.* Boston: Harvard Business School Press.

Kunsch, M. M. K. (Org.). (2009). Comunicação organizacional, histórico, fundamentos e processos. São Paulo, Brazil: Saraiva.

Kunsch, M. M. K. (2003). *Planejamento de relações públicas na comunicação integrada.* São Paulo, Brazil: Summus.

Kymmell, W. (2008). *BIM - planning and manging construction projects with 4D CAD and simulations. Construction.* New York: The McGraw-Hill Companies, Inc.

Lahrman, G., ... (2011). Business intelligence maturity: Development and evaluation of a theoretical model. In *Proceedings of the Hawaii International Conference on Systems Sciences, Annals of the 44th HICSS*, Hawaii.

Lambrecht, J. (2005). Multigenerational transition in family businesses: A new explanatory model. *Family Business Review, 28*(4), 267–282. doi:10.1111/j.1741-6248.2005.00048.x

Lambrecht, J., & Lievens, J. (2008). Pruning the family tree: An unexplored path to family business continuity and family harmony. *Family Business Review, 21*(4), 295–313. doi:10.1177/08944865080210040103

Lansberg, I. (1988). The succession conspiracy. In C. E. Aronoff, J. H. Astrachan, & J. L. Ward (Eds.), *Family business sourcebook II* (pp. 70–86). Marietta, GA: Business Owner Resources.

Lansberg, I., & Astrachan, J. H. (1994). Influence of family relationships on succession planning and training: The importance of mediating factors. *Family Business Review, 7*(1), 39–59. doi:10.1111/j.1741-6248.1994.00039.x

Larson, B. (2009). *Delivering business intelligence with Microsoft SQL server 2008.* McGraw-Hill.

Larson, E. W., & Gray, C. F. (2010). *Project management the managerial process* (5th ed.).

Lattès, R. (1992). *O Risco e a Fortuna: a grande aventura da Inovação.* Lisboa, Portugal: Difusão Cultural.

Laudon, K. C., & Laudon, J. P. (2014). *Management information systems - managing the digital firm* (13th ed.). Harlow, Essex: Pearson.

Le Breton-Miller, I., Miller, D., & Steier, L. P. (2004). Toward an integrative model of effective FOB succession. *Entrepreneurship Theory and Practice, 28*(4), 305–328. doi:10.1111/j.1540-6520.2004.00047.x

Le Coadic, Y-F. (1994). La Science de l'Information. Paris: PUF.

Le Huy Hoa. (2016). *Develop communication convergence model for radio stations.* Journalist Magazine No. 387 – May, 2016.

Leadbeater, C. 2003. Open innovation in public services. The Adaptive State - Strategies for personalizing the public realm. In T. Bentley & J. Wilsdon (Eds.), *Demos Collection,* 37-49. Retrieved from http://www.demos.co.uk/files/HPAPft.pdf

Lee, K. M., Lee, Y. B., Shim, C. S., & Park, K. L. (2012). Bridge information models for construction of a concrete box-girder bridge. *Structure and Infrastructure Engineering, 8*(7), 687–703. doi:10.1080/15732471003727977

Leibowitz, B. (1986). Resolving conflict in the family owned business. *Consultation, 5*(3), 191–205.

Leonard-Barton, D. (1998). *Nascentes do saber: criando e sustentando as fontes de inovação.* Rio de Janeiro, Brazil: Fundação Getúlio Vargas.

Leonard, D., & Sensiper, S. (1998). The role of tacit knowledge in group innovation. *California Management Review, 40*(3), 112–132. doi:10.2307/41165946

Lewis, J. P. (2001). Fundamentals of project management: developing core competencies to help outperform the competition (2nd ed.). New York: AMACOM - American Management Association.

Lewis, A. M., Valdes-Vasquez, R., Clevenger, C., & Shealy, T. (2014). BIM Energy Modeling: Case Study of a Teaching Module for Sustainable Design and Construction Courses. *Journal of Professional Issues in Engineering Education and Practice, 141*(2), C5014005. doi:10.1061/(ASCE)EI.1943-5541.0000230

Liautaud, B., & Hammond, M. (2001). e-Business intelligence: Turning information into knowledge into profit. New York: McGraw-Hill.

Liau, Y. H., & Lin, Y. C. (2017). Application of Civil Information Modeling for Constructability Review for Highway Projects. In *Proceedings of International Symposium on Automation and Robotics in Construction (ISARC)* (vol. 1, pp. 410-415). Taipei, Taiwan: IAARC. 10.22260/ISARC2017/0057

Lins, S. (2003). *Transferring tacit knowledge: A constructivist approach.* Rio de Janeiro, Brazil: E-papers.

Linstead, S., Marechal, G., & Griffin, R. W. (2014). Theorizing and researching the dark side of organization. *Organization Studies, 35*(2), 165–188. doi:10.1177/0170840613515402

Liu, W., Guo, H., Li, H., & Li, Y. (2014). Using BIM to improve the design and construction of bridge projects: A case study of a long-span steel-box arch bridge project. *International Journal of Advanced Robotic Systems, 11*(8), 1–11. doi:10.5772/58442

Logica. (2009). The BI framework: How to turn information into competitive asset. *Logica,* 35-37, 42-44.

Lomnitz, L. A., & Pérez-Lizaur, M. (1987). *A Mexican elite family, 1820-1980: Kinship, class and culture.* Princeton, NJ: Princeton University Press.

Lubatkin, M. H., Durand, R., & Ling, Y. (2007). The missing lens in family firm governance theory: A self-other typology of parental altruism. *Journal of Business Research, 60*(10), 1022–1029. doi:10.1016/j.jbusres.2006.12.019

Lück, H. (2013). Metodologia de projetos: uma ferramenta de planejamento e gestão (9th ed.). Petrópolis: Editora Vozes.

Luehraman, T. A. (1998). Strategy as Portfolio of Real Options. *Harvard Business Review,* 89–99. PMID:10185434

Lundvall, B. A. (Ed.). (1992). *National systems of innovation: towards a theory of innovation and interactive learning.* London, UK: Pinter.

Lundvall, B. A., & Johnson, B. (1994). The learning economy. *Journal of Industry Studies*, *1*(2), 23–42. doi:10.1080/13662719400000002

Macedo, M., & Barbosa, A. (2000). *Patentes, pesquisa & desenvolvimento: um manual de propriedade intelectual*. Rio de Janeiro, Brazil: Fiocruz. doi:10.7476/9788575412725

Mafei, M., & Cecato, V. (2011). *Comunicação Corporativa*. São Paulo, Brazil: Contexto.

Magalhães, J. L.; Quoniam, L.; Boechat, N. (2013a). Pharmaceutical market and opportunity in the 21st for generic drugs: a Brazilian case study of Olanzapine. *Problems of Management in the 21st Century, 6*.

Magalhães, J. L.; Quoniam, L.; Boechat, N. (2013b). Pharmaceutical market and opportunity in the 21st for generic drugs: a Brazilian case study of Olanzapine. *Problems of Management in the 21st Century, 6*.

Magazine (2014). *Pastry revolution*. [Online]. Retrieved from http://www.pastryrevolution.es/pasteleria/productores/

Maia, J. M. (1996). Propriedade Industrial: Comunicações e Artigos do Presidente do INPI. Lisboa, Portugal: Instituto Nacional da Propriedade Industrial (INPI).

Maia, J. M. (1996). Propriedade Industrial: Comunicações e Artigos do Presidente do INPI. Lisboa: Instituto Nacional da Propriedade Industrial (INPI).

Manchanda, K. & Muralidharan, P. (2014, January). Crowdfunding: a new paradigm in start-up financing. In *Global Conference on Business & Finance Proceedings, 9*(1), p. 369. Institute for Business & Finance Research.

Mangold, W., & Faulds, D. (2009). Social media. *Business Horizons*, *52*(4), 357–365. doi:10.1016/j.bushor.2009.03.002

Manos, B., & Manikas, I. (2010). Traceability in the Greek fresh produce sector: Drivers and constraints. *British Food Journal*, *112*(6), 640–652. doi:10.1108/00070701011052727

Mantel, S. J. Jr. (2008). *Project management in practice* (3rd ed.). Danvers: John Wiley & Sons.

Map, T. (2017). *Trade map*. [En línea]. Available at https://www.trademap.org/Index.aspx

Maravilhas, S. (2009). A Informação de Patentes: Vantagens da sua utilização como estímulo à criatividade, I&D, inovação e competitividade das empresas portuguesas. In IAPMEI (Ed.), Parcerias Científicas para a Inovação. Lisboa, Portugal: IAPMEI, 91-110.

Maravilhas, S. (2013b). Social media tools for quality business information. In Information quality and governance for business intelligence. Hershey, PA: IGI Global.

Maravilhas, S., & Borges, M. (2009). O Impacto das Bibliotecas Digitais de Patentes no Processo de Inovação em Portugal. In Borges & Sanz-Casado (Eds.), A ciência da informação criadora de conhecimento. (Vol. II). Coimbra, Portugal: Actas do IV Encontro Ibérico EDIBCIC 2009, 47-63.

Maravilhas, S. (2013a). A web 2.0 como ferramenta de análise de tendências e monitorização do ambiente externo e sua relação com a cultura de convergência dos media. *Perspectivas em Ciência da Informação*, *18*(1), 126–137. doi:10.1590/S1413-99362013000100009

Maravilhas, S., & Borges, M. (2011a). Os recursos de informação usados na I&D em Portugal: Caracterização dos centros de investigação do ensino superior público das áreas de Ciência, Tecnologia e Medicina. In P. Guerrero & V. Moreno (Eds.), *Límites, fronteras y espacios comunes: Encuentros y desencuentros en las Ciencias de la Informacíon* (pp. 321–333). Badajoz, Spain: Actas do V Encontro Ibérico EDICIC.

Maravilhas, S., & Borges, M. (2011b). A utilização da informação de patentes pelos centros de investigação do ensino superior público: O seu impacto no processo de inovação em Portugal. In P. Guerrero & V. Moreno (Eds.), *Límites, fronteras y espacios comunes: Encuentros y desencuentros en las Ciencias de la Informacíon* (pp. 364–376). Badajoz, Spain: Actas do V Encontro Ibérico EDICIC.

Marchand, D. (1990). Infotrends: A 1990s outlook on strategic information management. *Industrial Management Review*, 5(4), 23–32.

Marchand, D., & Horton, F. Jr. (1986). *Infotrends: Profiting from your information resources*. Wiley.

Marchewka, J. T. (2002). *Information technology project management: providing measurable organizational value*. Retrieved from /citations?view_op=view_citation&continue=/scholar%3Fhl%3Den%26start%3D10%26as_sdt%3D0,5%26scilib%3D1%26scioq%3Dentrepreneurship%2Bin%2BEgypt&citilm=1&citation_for_view=EJ3GnDgAAAAJ:SP6oXDckpogC&hl=en&oi=p

Marchiori, P. Z. (2002). A Ciência e a Gestão da Informação: compatibilidades no espaço profissional. *Ciência da Informação. Brasília, 31*(2), maio/ago., 72-79.

Marcovitch, J. (1983). *Administração em ciência e tecnologia*. São Paulo, Brazil: Edgard Blücher.

Marcus, D. (1995). Benefits of using patent databases as a source of information. In Lechter (Ed.), Successful Patents and Patenting for Engineers and Scientists. New York, NY: The Institute of Electrical and Electronics Engineers (IEEE) Press.

Marcus, D. (1995). Benefits of Using Patent Databases as a Source of Information. In Lechter (Ed.), Successful Patents and Patenting for Engineers and Scientists. New York, NY: The Institute of Electrical and Electronics Engineers (IEEE) Press.

Markides, C. (2000). *All the right moves: A guide to craft a breakthrough strategy*. Boston: Harvard Business School Press.

Markides, C. (2008). *Game-changing strategies: How to create new market space in established industries by breaking the rules* (1st ed.). San Francisco: Wiley.

Marques, V. (2015). *Vídeo marketing 360*. Actual Editora.

Marques, V. (2016). *Redes sociais 360. Como comunicar online*. Actual Editora.

Marques, V. (2018). *Marketing digital 360, 2.ª Edição*. Actual Editora.

Martínez, J. I., Stöhr, B. S., & Quiroga, B. F. (2007). Family ownership and firm performance: Evidence from public companies in Chile. *Family Business Review, 20*(2), 83–94. doi:10.1111/j.1741-6248.2007.00087.x

Marzouk, M., Hisham, M., Ismail, S., Youssef, M., & Seif, O. (2010). On the use of Building Information Modelling in infrastructure bridges. In Proceedings of 27th International Conference-Applications of IT in the AEC Industry (CIB W78) (vol. 136, pp. 1-10). Cairo, Egypt: CIBW78.

Marzouk, M., El-zayat, M., & Aboushady, A. (2017). Assessing Environmental Impact Indicators in Road Construction Projects in Developing Countries. *Sustainability, 9*(5), 843. doi:10.3390u9050843

Marzouk, M., & Hisham, M. (2014). Implementing earned value management using bridge information modeling. *KSCE Journal of Civil Engineering, 18*(5), 1302–1313. doi:10.100712205-014-0455-9

Masson, S. M. (2005). Projeto SIMAI-SIMAP: proposição de adoção da metodologia PMBOK para o gerenciamento de Sistema de Informação na Administração Municipal. In CONGRESSO DE ARQUIVOLOGIA DO MERCOSUL, 6º, Campos do Jordão, 2005 - Anais do VI CAM. São Paulo: CEDIC/PUC-SP.

Masson, S. M., & da Silva, A. M. (2001). Uma abordagem sistêmica da informação municipal: o projecto SIMAP e um caso de aplicação ainda incipiente: o SIMAI. *Cadernos de Estudos Municipais*, 33-62.

Mazhar, M. (2016). *Challenges faced by startup companies in software project management.*

McCollom, M. E. (1990). Problems and prospects in clinical research on family firms. *Family Business Review*, *3*(3), 245–262. doi:10.1111/j.1741-6248.1990.00245.x

McComack, K., Willems, J., Bergh, J., Deschoolmeester, D., Willaert, P., Stemberger, M. I., ... Vlahovic, N. (2009). A global investigation of key turning points in business process maturity. *Business Process Management Journal*, *15*(5), 792–815. doi:10.1108/14637150910987946

McGee, J., & Prusak, L. (1995). *Gerenciamento estratégico da informação: Aumente a competitividade e a eficiência de sua empresa utilizando a informação como uma ferramenta estratégica*. Rio de Janeiro: Campus.

McGeever, C. (2000). *Business intelligence*. Arizona: Computerworld.

McGuire, B., Atadero, R., Clevenger, C., & Ozbek, M. (2016). Bridge Information Modeling for Inspection and Evaluation. *Journal of Bridge Engineering*, *21*(4), 04015076. doi:10.1061/(ASCE)BE.1943-5592.0000850

McKenny, A. F., Payne, G. T., Zachary, M. A., & Short, J. C. (2014). Multilevel analysis in family business studies. In L. Melin, M. Nordqvist, & P. Sharma (Eds.), *The SAGE handbook of family business* (pp. 594–608). Thousand Oaks, CA: Sage. doi:10.4135/9781446247556.n30

McMullan, W. E. & Long, W. A. (1990). *Developing new ventures: the entrepreneurial option.*

Medina, Y. (2018). *Programación Blockchain*. Lima, Peru: Hyperledger Composer.

Meffert J., Mendonça, P., McKinsey & Co. (2017). Eins oder Null, Planeta.

Melin, L., & Nordqvist, M. (2007). The reflexive dynamics of institutionalization: The case of the family business. *Strategic Organization*, *5*(3), 321–333. doi:10.1177/1476127007079959

Mendrot, R. A., Oliveira, E. A. D. A., de Moraes, M. B., & Monteiro, R. (2017). The use of technical tools of project management and knowledge management to stimulate success in innovation projects. *Magazine Alcance—Eletrônica 24*(4).

Meredith, J. R.., & Mantel, S. J., Jr. (2006). Project management: a managerial approach (6th ed.). Danvers: John Wiley & Sons.

Merton, R. C. (1973). Theory of Rational Option Pricing. *The Bell Journal of Economics and Management Science*, *4*(1), 141–183. doi:10.2307/3003143

Michael, A., & Salter, B. (2006). *Mobile marketing: Achieving competitive advantage through wireless technology* (1st ed.). Oxford: Butterworth-Heinemann. doi:10.4324/9780080459653

Mick, J. (2015). Trabalho jornalístico e convergência digital no Brasil. *Pauta Geral –. Estudos em Jornalismo, Ponta Grossa*, *2*(1), 15–37. doi:10.18661/2318-857X/pauta.geral.v2n1p15-37

Microsoft Excel. (n.d.). Retrieved from https://www.techopedia.com/definition/5430/microsoft-excel

Miguel, A. (2006). Gestão moderna de projectos: melhores técnicas e práticas (2nd ed.). Lisboa: FCA – Editora de Informática.

Miller, D., Steier, L., & Le Breton-Miller, I. (2003). Lost in time: Intergenerational succession, change, and failure in family business. *Journal of Business Venturing*, *18*(4), 513–531. doi:10.1016/S0883-9026(03)00058-2

Ministerio de Agricultura y Riego. (2018). Comercio exterior agrario. *Portal institucional Ministerio de Agricultura y Riego*. Retrieved from http://siea.minag.gob.pe/siea/sites/default/files/nota-comercio-exterior-diciembre17_3.pdf

Minonne, C., & Turner, G. (2012). Business process management – Are you ready for the future? *Knowledge and Process Management*, *19*(3), 111–120. doi:10.1002/kpm.1388

Mintzberg, H. (1987, July). Crafting strategy. *Harvard Business Review*, 66–75.

MIT Center for Transportation & Logistics. (2018). *Blockchain in Supply Chains*. Boston: MA.

Mitchell, G. R., & Hamilton, W. F. (1988). Managing R&D as a Strategic Option. *Research Technology Management*, *31*(3), 15–22. doi:10.1080/08956308.1988.11670521

Moe T. (1998). Perspectives on traceability in food manufacture. *Trends in Food Science and Technology*. doi:10.1016/S0924-2244(98)00037-5

Mooij, M. (2015). Cross-cultural research in international marketing: Clearing up some of the confusion. *International Marketing Review*, *32*(6), 646–662. doi:10.1108/IMR-12-2014-0376

Moon, H., Dawood, N., & Kang, L. (2014). Development of workspace conflict visualization system using 4D object of work schedule. *Advanced Engineering Informatics*, *28*(1), 50–65. doi:10.1016/j.aei.2013.12.001

Morales, O. & Borda, A. (2015). La alianza cacao Perú y la cadena productiva del cacao fino de aroma. *Gerencia para el desarrollo, 49*, p. 19-30.

Morales, O., Borda, A., Argandoña, A., Farach, R., Garcia Naranjo, L., & Lazo, K. (2015). *La Alianza Cacao Perú y la cadena productiva del cacao fino de aroma*. Lima, Peru: Esan.

Morck, R., Shleifer, A., & Vishny, R. W. (1988). Management ownership and market valuation: An empirical analysis. *Journal of Financial Economics*, *20*, 293–315. doi:10.1016/0304-405X(88)90048-7Morck, R., Wolfenzon, D., & Yeung, B. (2005). Corporate governance, economic entrenchment, and growth. *Journal of Economic Literature*, *43*(3), 655–720. doi:10.1257/002205105774431252

Morck, R., & Yeung, B. (2003). Agency problems in large family business groups. *Entrepreneurship Theory and Practice*, *27*(4), 367–382. doi:10.1111/1540-8520.t01-1-00015

Moreira, L. B.; Santana, A. A.; Miranda, A. R. A. (2014, Dec. 10). Os impactos da implementação do SAP R/3 em uma empresa do setor de laticínios. Revista Ciências Administrativas ou Journal of Administrative Sciences, 18(1).

Moreira, D. A., & Queiroz, A. C. S. (2007). *Inovação Organizacional e Tecnológica*. São Paulo, Brazil: Thomson Learning.

Moreno, A. S., & Carvalho, W. S. (2017). Business intelligence, Capacidades Dinâmicas, E Capacidades Operacionais de Marketing: Um Estudo Empírico no Setor de Telecom. In *Encontro da Associação Nacional de Programas de Pós-Graduação em Administração, Anais do 38o EnANPAD*. São Paulo, Brazil: ANPAD.

Morgan, D. L. (1996). Focus groups. *Annual Review of Sociology*, 22. Retrieved from http://www.jstor.org/stable/2083427

Morgan, G. (2002). *Images of the organization*. São Paulo, Brazil: Atlas.

Morgan, H., & Soden, J. (1979). Understanding MIS failures. *Database*, (5): 157–171.

Morin, E. (1996). *O Problema epistemológico da complexidade*. Publicações Europa- América.

Morris, G., & Beckett, D. (2004). Performing identities: the new focus on embodied adults learning. In P. Kell, S. Shore, & M. Singh (Eds.), *Adult education @ 21st century. Studies in the postmodern theory of education*. New York, NY: Peter Lang.

Compilation of References

Moura, L. R. (1999). Gestão e Tecnologia da Informação como instrumento de interação Universidade-Empresa. In *IBICT. Interação Universidade-Empresa II*. Brasília: Instituto Brasileiro de Informação em Ciência e Tecnologia.

Mueller, H., & Nyfeler, T. (2011). Quality in patent information retrieval: Communication as the key factor. *World Patent Information*, *33*(4), 383–388. doi:10.1016/j.wpi.2011.06.012

Mulgan, G., & Albury, D. (2003). *Innovation in the public sector. Strategy Unit, Cabinet Office, 1*, 40.

Müller, R., Glücker, J., Aubry, M., & Shaun, J. (2013). Project management knowledge flows in network of project managers and project management offices: A case study in the pharmaceutical Industry. *Project Management Journal*, *44*(2), 4–19. doi:10.1002/pmj.21326

Muntean, S. C. (2009). *A political theory of the firm: Why ownership matters*, (Unpublished doctoral dissertation). University of California, San Diego, CA.

Murphy, M. (2002). Organisational Change and Firm Performance, DSTI/DOC (2002)14. Paris, France: OECD.

Murphy, L., & Lambrechts, F. (2015). Investigating the actual career decisions of the next generation: The impact of family business involvement. *Journal of Family Business Strategy*, *6*(1), 33–44. doi:10.1016/j.jfbs.2014.10.003

Murray, B. (2003). The succession transition process: A longitudinal perspective. *Family Business Review*, *16*(1), 17–33. doi:10.1111/j.1741-6248.2003.00017.x

Myers, S. C. (1977). Determinants of Corporate Borrowing. *Journal of Financial Economics*, *5*(2), 147–175. doi:10.1016/0304-405X(77)90015-0

Naetebusch, R., Schoeppel, H. R., & Fichtner, H. (1994). Patent information in a large electrical company, exemplified by the situation at Siemens. *World Patent Information*, *16*(4), 198–206. doi:10.1016/0172-2190(94)90003-5

Nassar, P., & Figueiredo, R. (2007). *O que é Comunicação Empresarial?* São Paulo, Brazil: Brasiliense.

Nelson, R. R. (2005). Project retrospectives: Evaluating project success, failure, and everything in between. *MIS Quarterly Executive*, *4*(3), 361–372.

Nelson, R. R. (Ed.). (1993). *National innovation systems: a comparative analysis*. Oxford, UK: Oxford University Press.

Neubauer, T. (2009). An empirical study on the state of business process management. *Business Process Management Journal 15*(2), p. 2.

Nguyen, K. (2018). *Do not take advantage, overcome challenges, radio will lose position*. Retrieved from vov.vn.

Nonaka, I., & Takeuchi, H. (1997). *Creation of knowledge in the company: How Japanese companies generate the dynamics of innovation* (4th ed.). Rio de Janeiro, Brazil: Campus.

Nonaka, I., Von Krogh, G., & Voelpel, S. (2006). Organizational knowledge creation theory: Evolutionary paths and future advances. *Organization Studies*, *27*(8), 1179–1208. doi:10.1177/0170840606066312

Nordqvist, M., Hall, A., & Melin, L. (2009). Qualitative research on family businesses: The relevance and usefulness of the interpretive approach. *Journal of Management & Organization*, *15*(3), 294–308. doi:10.1017/S1833367200002637

Nusim, S. (2016). Active pharmaceutical ingredients: development, manufacturing, and regulation (2nd ed.). Boca Raton, FL: CRC Press.

O'Brien, J., & Marakas, G. M. (2011). *Management information systems*. Mcgraw-Hill.

O'Connor, A. (1993). Successful strategic information systems planning. *Journal of Information Systems*, *3*(2), 71–83. doi:10.1111/j.1365-2575.1993.tb00116.x

O'Dell, D. (2004). *A Resolução Criativa do Problema: Guia para a Criatividade e Inovação na Tomada de Decisões*. Lisboa, Portugal: Instituto Piaget.

Ogorelc, A. (1999). Higher education in tourism: An entrepreneurial approach. *Tourism Review*, *54*(1), 51–60.

Ohtonen, J. & Lainema, T. (2011). Critical success factors in business process management – a literature review. In Proceedings of IRIS (pp. 572-585). London, UK.

Ojala, M. (1989). A patently obvious source for competitor intelligence: The patent literature. *Database*, *12*(4), 43–49.

Oliveira, A. (2010). Governance of information systems – Why, ISGec Conference, Madrid, Spain.

Oliveira, A. (2004). *Análise do investimento em sistemas e tecnologias da informação e da comunicação*. Lisboa: Sílabo.

Oliveira, A. (2004). *Analysis of investments in information and communication technologies and systems*. Sílabo Publishing. (In Portuguese)

Oliveira, A. (2009). *Information & information systemas – Facts, myths, mystifications, dangerous half-truths & lack of common sense*. Refertelecom Publishing. (In Portuguese)

Oliveira, D. (2006). *Administração de processos: conceitos, metodologia e práticas*. São Paulo, Brazil: Editora Atlas.

Olson, D. H. (2000). Circumplex model of marital and family systems. *Journal of Family Therapy*, *22*(2), 144–167. doi:10.1111/1467-6427.00144 PMID:6840263

OMG. (2011). *O. M. G. Business process model and notation (BPMN)*. Needham, MA: OMG.

Omoregie, A., & Turnbull, D. E. (2016). Highway infrastructure and Building Information Modelling in UK. *Proceedings of the Institution of Civil Engineers. Municipal Engineer*, *169*(4), 220–232. doi:10.1680/jmuen.15.00020

Oomen, O., & Ooztaso, A. (2008). Construction project network evaluation with correlated schedule risk analysis model. *Journal of Construction Engineering and Management*, (1), 49–63.

Organização para Cooperação e Desenvolvimento Econômico (OCDE). (2005). Manual de Oslo: diretrizes para coleta e interpretação de dados sobre inovação de 2005. (Traduzido pela FINEP - Financiadora de Estudos e Projetos, 3ed.)

Orna, E. (1999). *Practical information policies* (2nd ed.). Cambridge: Gower.

Osterman, K. F., & Kottkamp, R. B. (1993). *Reflective practice for educators: Improving schooling through professional development*. Newbury Park, CA. Corwin Press.

Osterwalder, A., Bernarda, G., Pigneur, Y., & Smith, A. (2015). *Criar Propostas de Valor*. Dom Quixote.

Osterwalder, A., & Pigneur, Y. (2010). *Business model generation: A handbook for visionaries, game changers, and challengers*. Wiley.

Oswald, S. L., Muse, L. A., & Rutherford, M. W. (2009). The influence of large stake family control on performance: Is it agency or entrenchment? *Journal of Small Business Management*, *47*(1), 116–135. doi:10.1111/j.1540-627X.2008.00264.x

Oti, A. H., Tizani, W., Abanda, F. H., Jaly-Zada, A., & Tah, J. H. M. (2016). Structural sustainability appraisal in BIM. *Automation in Construction*, *69*, 44–58. doi:10.1016/j.autcon.2016.05.019

Pache, A.-C. & Santos, F. M. (2010). Inside the hybrid organization: an organizational level view of responses to conflicting institutional demands (August 2010). *Research Center ESSEC Working Paper 1101*. doi:10.2139srn

Paim, I. (2003). A Gestão de informação e do conhecimento. Belo Horizonte: Escola de Ciência da Informação-UFMG.

Paoleschi, B. (2018). *Estoques e armazenagem*. Editora Saraiva.

Parada, M. J., Nordqvist, M., & Gimeno, A. (2010). Institutionalizing the family business: The role of professional associations in fostering a change of values. *Family Business Review*, *23*(4), 355–372. doi:10.1177/0894486510381756

Paramkusham, R. B., & Gordon, J. (2013, Fall). Inhibiting factors for knowledge transfer in information technology projects. *Journal of Global Business and Technology*, *9*, 2.

Park, T., Kang, T., Lee, Y., & Seo, K. (2014). Project Cost Estimation of National Road in Preliminary Feasibility Stage Using BIM/GIS Platform. *Computing in Civil and Building Engineering*, 423–430.

Parker, D. P. & Zilberman, D. (1993). University technology transfers: impacts on local and U.S. economies. *Contemporary Policy Issues. XI*, 87-99.

Pask, G. (1969). *Learning strategies and teaching strategies*. Surrey: System Research Ltd.

Pastry Revolution. (2014). 10 grandes productores de cacao. In *Pastry Revolution*. Retrieved from http://www.pastryrevolution.es/pasteleria/productores/Z

Paviani, J. (2009). Epistemologia prática: ensino e conhecimento científico. Caxias do Sul: EDUCS.

Pearson, A. W., Carr, J. C., & Shaw, J. C. (2008). Toward a theory of familiness: A social capital perspective. *Entrepreneurship Theory and Practice*, *32*(6), 949–969. doi:10.1111/j.1540-6520.2008.00265.x

Peay, R. T., & Dyer, G. (1989). Power orientations of entrepreneurs and succession planning. *Journal of Small Business Management*, *27*(1), 47–52.

Pellegrino, B. & Zingales, L. (2014). Diagnosing the Italian disease. *Chicago Booth Working Paper*. Retrieved from http://faculty.chicagobooth.edu/luigi.zingales/papers/research/Diagnosing.pdf

Penttilä, H. (2006). Describing the changes in architectural information technology to understand design complexity and free-form architectural expression. *ITcon.*, *11*, 395–408.

Perez, C. (2010). Technological revolutions and techno-economic paradigms. *Cambridge Journal of Economics*, *34*(1), 185–202. doi:10.1093/cje/bep051

Pérez-González, F. (2006). Inherited control and firm performance. *The American Economic Review*, *96*(5), 1559–1588. doi:10.1257/aer.96.5.1559

Perlitz, M., Peske, T., & Schrank, R. (1999). Real Option Valuation: The New Frontier in R&D Project Evaluation? *R & D Management*, *29*(3), 255–269. doi:10.1111/1467-9310.00135

Petroski, H. (2008). *Inovação: da Idéia ao Produto*. São Paulo, Brazil: Edgard Blücher.

Philipp, N. H. (2013). Building information modeling and the consultant: Managing roles and risk in an evolving design and construction process. *Proceedings of Meetings on Acoustics Acoustical Society of America*, *133*(5), 3441–3441. doi:10.1121/1.4806085

Phong, T. (2018). *Vietnam Internet speed is nearly 3 times faster than China*. Retrieved from http://chungta.vn/tin-tuc/cong-nghe/toc-do-internet-viet-nam-nhanh-gap-gan-3-lan-trung-quoc-66055.html)

Picot, A. (1989). Information management: The science of solving problems. *International Journal of Information Management*, *9*(4), 237–243. doi:10.1016/0268-4012(89)90047-9

Pigneur, Y., & Osterwalder, A. (2015). *Criar Modelos de Negócio*. Dom Quixote.

Pinheiro, M. G., Donaires, O. S., & Figueiredo, L. R. (2011). Aplicação da Visão Sistêmica na implantação de Sistemas Integrados de Gestão ERP. Anais do 7o Congresso Brasileiro de Sistemas, 7, p. 409–421.

Platts, K. W. (1993). A process approach to researching manufacturing strategy. *International Journal of Operations & Production Management*, *13*(8), 4–17. doi:10.1108/01443579310039533

Plonski, G. A. (2005). Bases para um movimento pela inovação tecnológica no Brasil. *Revista São Paulo em Perspectiva*, *19*(1), 5–33.

PMBoK. (2018). *Project Management Body of Knowledge* (6th ed.). Project Management Institute.

PMI – Project Management Institute. (n.d.). *What is project management*. Available at https://www.pmi.org/about/learn-about-pmi/what-is-project-management

PMI. (2013). *Guía de los fundamentos para la dirección de proyectos. Pensilvania*. Project Management Institute.

Polanyi, M. (1966). The logic of tacit inference. *Philosophy (London, England)*, *41*(155), 1–18. doi:10.1017/S0031819100066110

Porter, A., & Cunningham, S. (2005). *Tech Mining: Exploiting New Technologies for Competitive Advantage*. Hoboken, NJ: Wiley.

Porter, M. (1980). *Competitive strategy: Techniques for analyzing industries and competitors*. New York: Free Press.

Porter, M. (1985). *Competitive advantage: creating and sustaining superior performance*. New York: Free Press.

Porter, M. E. (1992). Capital Disadvantage: America's failing capital investment system. *Harvard Business Review*, 65–82. PMID:10121317

Porter, M., & Millar, V. (1985). How information gives you competitive advantage. *Harvard Business Review*, (July/August): 75–98.

Pratt, M. G., & Foreman, P. O. (2000). Classifying managerial responses to multiple organizational identities. *Academy of Management Review*, *25*(1), 18–42. doi:10.5465/amr.2000.2791601

Premkumar, G., & King, W. (1994). The evaluation of strategic information systems planning. *Information & Management*, *26*(6), 327–340. doi:10.1016/0378-7206(94)90030-2

Project Management Institute. (2004). *A Guide to the Project Management Body of Knowledge: PMBOK® Guide* (3rd ed.). PMI.

Quandt, C. (2004). Inovação em clusters emergentes. *Revista Com. Ciência,* 57, ago, 1-5.

Queirós, P., & Lacerda, T. (2013). The importance of interview in qualitative research. In I. Mesquita & A. Graça (Eds.), *Qualitative research in sport, 2*. Porto, Portugal: Center for Research, Training, Innovation, and Intervention in Sport, Faculty of Sport, Porto University. (In Portuguese)

Quevedo, L. A. (2007). Conhecer para participar da sociedade do conhecimento. In Maria Lucia Maciel, Sarita Albagli (Org.). Informação e Desenvolvimento: conhecimento, inovação e apropriação social. Brasília: IBICT, UNESCO.

Quispe, S. (2009). *Trazabilidad y gestión agroalimentaria*. Lima, Peru: Agroenfoque.

Compilation of References

Ràbale, A. (2006). *Buyer – supplier relationship's influence on traceability implementation in the vegetable industry.* [Online] doi:10.1016/j.pursup.2006.02.003

Raber, D., Winter, R., & Wortmann, F. (2016). Using qualitative analyses to construct a capability maturity model for business intelligence. In *Proceedings of the Hawaii International Conference on Systems Sciences, Annals of the 45th HICSS*, Hawaii, 4219-4228.

Rabinowitz, P. (2018). Techniques for leading group discussions. *Community Tool Box*. Retrieved from https://ctb.ku.edu/en/table-of-contents/leadership/group-facilitation/group-discussions/main

Rafferty, J. (2018). Industrial Revolution. *Encyclopaedia Britannica*. Retrieved from https://www.britannica.com/event/Industrial-Revolution

Rajteric, I. H. (2010). Overview of business intelligence maturity models. *Management*, *15*(1), 47–67.

Ram, M. (1994). *Managing to survive: Working lives in small firms.* Oxford, UK: Blackwell.

Rapini, M. S., & Righi, H. M. (2007). Interação Universidade-Empresa no Brasil em 2002 e 2004: Uma aproximação a partir dos grupos de pesquisa do CNPq. *Revista Economia (Brasília)*, *8*, 263–284.

Rappa, M. (2010). Business models on the web. *Managing the Digital Enterprise*. Retrieved 25-05, 2012, from http://digitalenterprise.org/models/models.html

Rascão, J. (2008). Novos desafios da gestão da informação. Lisboa: Edições Sílabo.

Rascão, J. (2012). Novas realidades na gestão e na gestão da informação. Lisboa: Edições Sílabo.

Rascão, J. (2001). *A análise estratégica e o sistema de informação para a tomada de decisão estratégica* (2nd ed.). Lisboa: Sílabo.

Rascão, J. (2004). *Sistemas de informação para as organizações: A informação chave para a tomada de decisão* (2nd ed.). Lisboa: Sílabo.

Rascão, J. (2008). *Novos desafios da gestão da informação* (1st ed.). Lisboa: Sílabo.

Rayner, N., & Schelegel, K. (2008). *Maturity model overview for business intelligence and performance management.* Stamford, CT: Gartner.

Raza, H., Arshad, W., Soo, S., & Won, J. (2017). Flexible Earthwork BIM Module Framework for Road Project. In *Proceedings 34th International Symposium on Automation and Robotics in Construction (ISARC)* (vol. 1, pp. 410-415). Taipei, Taiwan: IAARC. 10.22260/ISARC2017/0056

Reeves, W. J. (1992). What is software design? *C++ Journal*, *2*(2).

Reich, B., Gemino, A., & Sauer, C. (2012). Knowledge management and Project-based knowledge in its projects: A model and preliminary results. *International Journal of Project Management*, *30*(6), 663–674. doi:10.1016/j.ijproman.2011.12.003

Reich, B., & Wee, S. W. (2006). Searching for knowledge in the PMBoK Guide. *Project Management Journal*, *37*(2), 11–27. doi:10.1177/875697280603700203

Reis, A. M. M., & Perini, E. (2008, April). Desabastecimento de medicamentos: Determinantes, conseqüências e gerenciamento. Ciência &amp. *Saúde Coletiva*, *13*(suppl), 603–610. doi:10.1590/S1413-81232008000700009

Reis, C. (1993). *Planejamento Estratégico de Sistemas de informação.* Lisboa, Portugal: Presença.

Requeijo, J. F., & Pereira, Z. L. (2008). *QUALITY: Planning and statistical process control*. Prefácio Publishing. (In Portuguese)

Reynolds, P., & Miller, B. (1992). New firm gestation: conception, birth and implications for research. *Journal of Business Venturing*, *7*(5), 405–417. doi:10.1016/0883-9026(92)90016-K

Ribeiro, D. (2007). Propriedade Intelectual: Mais de 30% da investigação em Portugal é redundante. *Jornal de Negócios, Quinta-feira*(24 de Maio), 34.

Ribeiro, M. E. (2017). *O papel do assessor de imprensa em um mundo movido pelas tecnologias digitais*. Retrieved from www.academia.edu

Richmond, L., Stevenson, J., & Turton, A. (2003). Essay review the pharmaceutical industry: a guide to historical records. Aldershot.

Rivas, T. (2011). *Trazabilidad en la Industria Alimentaria*. Salamanca, Spain: Universidad de Salamanca.

Rivette, K., & Kline, D. (2000). *Rembrandts in the attic: unlocking the hidden value of patents* (1st ed.). Boston, MA: Harvard Business School Press.

Rivette, K., & Kline, D. (2000). *Rembrandts in the Attic: Unlocking the Hidden Value of Patents* (1st ed.). Boston, MA: Harvard Business School Press.

Rocha, A. & Vasconcelos, J. (2004). Os modelos de maturidade na gestão de sistemas de informação. *Revista da Faculdade de Ciência e Tecnologia da Universidade de João pessoa*, (1), 93-107.

Rodrigues, J. (1999). *A Conspiração Solar do Padre Himalaya*. Porto, Portugal: Árvore - Cooperativa de Actividades Artísticas.

Rogers, D. (2017). Transformação Digital: Repensando o seu Negócio para a era Digital. S. Paulo: Autêntica Business.

Romagni, P. (1999). *10 instrumentos-chave da gestão* (1st ed.). Lisboa: D. Quixote.

Romano, F. M., Chimenti, P., Rodrigues, M. D. S., Hupsel, L. F., & Nogueira, R. (2012). O Impacto das mídias sociais digitais na comunicação organizacional das empresas. *Future Studies Research Journal, São Paulo*, *6*(1), 53–82.

Romero, C. A. (2015). *Estudio del Cacao en el Perú y el mundo*. Lima, Peru: Ministerio de Agricultura y Riego.

Rosini, A. M., & Palmisiano, A. (1998). *Administração de Sistemas de Informação e a Gestão do Conhecimento*. São Paulo, Brazil: Thomson.

Ross, J. W., Weill, P., & Robertson, D. (2006). *Enterprise architecture as strategy: creating a foundation for business execution*. Boston, MA: Harvard Business Review Press.

Rousseau, D. M. (1985). Issues of level in organizational research: Multi-level and cross-level perspectives. In L. L. Cummings & B. M. Staw (Eds.), *Research in organizational behavior* (Vol. 7, pp. 1–37). Greenwich, CT: JAI Press.

Rubenson, G. C., & Gupta, A. K. (1996). The initial succession: A contingency model of founder tenure. *Entrepreneurship Theory and Practice*, *21*(2), 21–32. doi:10.1177/104225879602100202

Rubin, E., & Rubin, H. (2011). Supporting agile software development through active documentation. *Requirements Engineering*, *16*(2), 117–132. doi:10.100700766-010-0113-9

Russell, A., Staub-French, S., Tran, N., & Wong, W. (2009). Visualizing high-rise building construction strategies using linear scheduling and 4D CAD. *Automation in Construction*, *18*(2), 219–236. doi:10.1016/j.autcon.2008.08.001

Compilation of References

Ryan, D., & Jones, C. (2012). *The best digital marketing campaigns in the world: Mastering the art of customer engagement* (1st ed.). Croydon: Kogan Page.

Ryan, D., & Jones, C. (2013). *Understanding digital marketing: Marketing strategies for engaging the digital generation* (2nd ed.). Croydon: Kogan Page.

Saad, E. C. (2009). Comunicação digital e novas mídias institucionais. In M. M. K. Kunsch (Ed.*), Comunicação organizacional*. São Paulo, Brazil: Saraiva.

Sadkowska J. (2017). The impact of the project environment uncertainty on project management practices in family firms. *PM World Journal, 6,* 7, July 2017, 1-12.

Salamzadeh, A. (2015, June). Innovation accelerators: emergence of startup companies in Iran. In *Proceedings 60th Annual ICSB World Conference* (pp. 6–9).

Salamzadeh, A., & Kawamorita Kesim, H. (2015). *Startup companies: life cycle and challenges*. doi:10.2139srn.2628861

Salovey, P., O'Leary, A., Stretton, M. S., Fishkin, S. A., & Drake, C. A. (1991). Influence of mood on judgments about health and illness. In J. P. Forgas (Ed.), *Emotion and social judgments* (pp. 241–262). Elmsford, NY: Pergamon Press.

Sandle, T. (2012, December). Application of quality risk management to set viable environmental monitoring frequencies in biotechnology processing and support areas. *PDA Journal of Pharmaceutical Science and Technology, 66*(6), 560–579. doi:10.5731/pdajpst.2012.00891 PMID:23183652

Sant'Anna, I. B. C. & Fernandes, N. C. (2008). A comunicação institucional nos *websites* corporativos: Um estudo exploratório. *Anagrama*, 1(4).

Santiago, A. L. (2000). Succession experiences in Philippine family businesses. *Family Business Review, 13*(1), 15–35. doi:10.1111/j.1741-6248.2000.00015.x

Santos, P. C. & Prado, M. S. (2014). *Business Intelligence: Um estudo sobre o nível de maturidade em empresas de confecções de lingerie.*

Santos, B. P., Lima, T.D.F.M., & Charrua-Santos, F.M.B. (2018). Indústria 4.0: Desafios e oportunidades. *Revista Produção e Desenvolvimento., 4*(1), 111–124.

SAP ERP. Seidor BRASIL - SAP Business Suite (ERP). Disponível em http://www.seidorbrasil.com.br/solucoes/saperp

Sarkar, S. (2010). *Entrepreneurship and innovation. Forte da Casa*. School Publishing.

Scabini, E. (1995). *Psicologia sociale della famiglia*. Torino, Italy: Bollati Boringhieri.

Scabini, E., & Iafrate, R. (2003). *Psicologia dei legami familiari*. Roma, Italy: Carocci.

Schilling, M. (2013). *Strategic management of technological innovation* (4th ed.). Singapore: McGraw-Hill.

Schoeppel, H. R., & Naetebusch, R. (1995). Patent searching in a large electrical company, as exemplified by the situation at Siemens. *World Patent Information, 17*(3), 165–172. doi:10.1016/0172-2190(95)00016-S

Schön, D. A. (1987). *Educating the reflective practitioner: Toward a new design for teaching and learning in the professions*. San Francisco, CA: Jossey-Bass.

Schulze, W. S., Lubatkin, M. H., & Dino, R. N. (2003). Toward a theory of agency and altruism in family firms. *Journal of Business Venturing, 18*(4), 473–490. doi:10.1016/S0883-9026(03)00054-5

Schulze, W. S., Lubatkin, M. H., Dino, R. N., & Buchholtz, A. K. (2001). Agency relationships in family firms: Theory and evidence. *Organization Science, 12*(2), 99–116. doi:10.1287/orsc.12.2.99.10114

Schwab, K. (2015). *The fouth industrial revolution*. New York: Penguin Random House.

Schwagele, F. (2005). *Traceability from a European perspective*. [Online]. doi:10.1016/j.meatsci.2005.03.002

Schwalbe, K. (2016). Information technology - project management.

Seethamraju, R. & Marjanovic, O. (2009). Role of process knowledge in business process improvement methodology: A case study, *Business Process Management Journal 15*(6), 920–936.

Segura, D. P. (2014). *O Impacto das tecnologias digitais sobre o processo de assessoria de imprensa*. São Paulo, Brazil: ECAUSP.

Selic, B. (2009). Agile documentation, anyone? *IEEE Software, 26*(6), 11–12. doi:10.1109/MS.2009.167

Sentanin, O. F., Santos, F. C. A., & Jabbour, C. J. C. (2008). Business process management in a Brazilian public research centre. *Business Process Management Journal, 14*(4), 483–496. doi:10.1108/14637150810888037

Serra, L. A. (2002). Essência do Business Intelligence. São Paulo, Brazil: Berkeley.

Seymore, S. (2009). Serendipity. *North Carolina Law Review, 88*, 185.

Shafiq, M., & Waheed, U. S. (2018). Documentation in agile development a comparative analysis. In *Proceedings 2018 IEEE 21st International Multi-Topic Conference (INMIC)*. 10.1109/INMIC.2018.8595625

Shanker, M. C., & Astrachan, J. H. (1996). Myths and realities: family businesses' contribution to the us economy—a framework for assessing family business statistics. *Family Business Review, 9*(2), 107–119. doi:10.1111/j.1741-6248.1996.00107.x

Shapero, A. (1982). *Inventors and entrepreneurs: Their roles in innovation*. College of Administrative Science, Ohio State University.

Shapiro, C., & Varian, H. (1999). *Information rules: A strategic guide to the network economy* (1st ed.). Boston: Harvard Business School Press.

Shead, S. (2013). Industry 4.0: the next industrial revolution. *Engineer (online edition)*. Retrieved from http://www.theengineer.co.uk

Shepherd, D., & Haynie, J. M. (2009). Family business, identità conflict, and an expedited entrepreneurial process: A process of resolving identity conflict. *Entrepreneurship Theory and Practice, 33*(6), 1245–1264. doi:10.1111/j.1540-6520.2009.00344.x

Shera, J. H., & Cleveland, D. B. (1977). History and foundations of Information Science. *Annual Review of Information Science and Technology, 12*, 249-275.

Sherwood, R. E. (1992). *Propriedade intelectual e desenvolvimento econômico*. S. Paulo, Brazil: EdUSP.

Shim, C. S., Lee, K. M., Kang, L. S., Hwang, J., & Kim, Y. (2012). Three-dimensional information model-based bridge engineering in Korea. *Structural Engineering International: Journal of the International Association for Bridge and Structural Engineering, 22*(1), 8–13. doi:10.2749/101686612X13216060212834

Shim, C. S., Yun, N. R., & Song, H. H. (2011). Application of 3D bridge information modeling to design and construction of bridges. *Procedia Engineering, 14*, 95–99. doi:10.1016/j.proeng.2011.07.010

Compilation of References

Shingo, S. (1996). *The Toyota production system from the point of view of production* (2nd ed.). Porto Alegre, Brazil: Bookman.

Silva, F. R., Santos, A., & Gonçalo, C. R. (2017). A Influência dos Sistemas de Business Intelligence e a Business Analytics na Medição de Desempenho e Práticas Estratégicas. In *Encontro Nacional de Administração da Informação, Anais do EnADI*. Curitiba, Brazil: ANPAD.

Silva, & Teixeira, & De Paula. (2012). Analysis of the Acquisition Process of a Autoparts Company Using Discounted Cash Flows and Real Options Models. *Perspectivas Contemporâneas, 7*, 11–43.

Sirmon, D. G., Arrègle, J.-L., Hitt, M. A., & Webb, J. W. (2008). The role of family influence in firms' strategic responses to threat of imitation. *Entrepreneurship Theory and Practice, 32*(6), 979–998. doi:10.1111/j.1540-6520.2008.00267.x

Sirmon, D. G., & Hitt, M. A. (2003). Managing resource: Linking unique resource, management and wealth creation in family firms. *Entrepreneurship Theory and Practice, 27*(4), 339–358. doi:10.1111/1540-8520.t01-1-00013

Sismeiro, L. F. L. (2014, Dec. 18). Projectos de consultoria em SAP e tecnologias microsoft: análise e desenvolvimento de soluções de software à medida.

Skarzynski, P., & Gibson, R. (2010). *Inovar no Essencial: Transforme o modo como a sua empresa inova* (1st ed.). Lisboa, Portugal: Actual.

Smil, V. (2010). *Energy Trasitions: history, requirements, prospects*. London: Praeger Publishers.

Smit, J., Kreutzer, S., Moeller, C., & Carlberg, M. (2016). Industry 4.0. *Study issued by Policy Department A: Economic and Scientific Policy. European Parliament*. Retrieved from http://www.europarl.europa.eu/studies

Smith, P. (2014). BIM & the 5D Project Cost Manager. *Proceedia Social and Behavioral Sciences, 119*, 475–484.

Sorenson, R. (1999). Conflict management strategies used by successful family businesses. *Family Business Review, 12*(4), 325–340. doi:10.1111/j.1741-6248.1999.00325.x

Sorenson, R. L., Goodpaster, K. E., Hedberg, P. R., & Yu, A. (2009). The family point of view, family social capital, and firm performance: An exploratory test. *Family Business Review, 22*(3), 239–253. doi:10.1177/0894486509332456

Sousa, M. J. (2009). *Knowledge dilemmas: the case of two Portuguese organizations*. (Doctoral thesis, University of Aveiro).

Sousa, M. J. (2010). Dynamic knowledge: An action research project. *The International Journal of Knowledge, Culture and Change Management, 10*(1), 317–331.

Sousa, M. J. (2013). Knowledge profiles boosting innovation. *Knowledge Management, 12*(4), 35–46.

Sousa, M. J., & González-Loureiro, M. (2016). Employee knowledge profiles – a mixed-research methods approach. *Information Systems Frontiers, 18*(6), 1103–1117. doi:10.100710796-016-9626-1

Souza, C., Oliveira, J., & Kligerman, D. (2014, September). Advances and challenges in standardization of free samples of drugs in Brazil. *Physis (Rio de Janeiro, Brazil), 24*(3), 871–883. doi:10.1590/S0103-73312014000300011

Spaltro, E., & de Vito Piscicelli, P. (1990). *Psicologia per le organizzazioni*. Roma, Italy: Carocci.

Spender, J. (2014). *Business strategy: managing uncertainty, opportunity, and enterprise*. Oxford University Press. doi:10.1093/acprof:oso/9780199686544.001.0001

Stair, R., & Reynolds, G. (2009). *Principles of information systems*. Course Technology.

Statistics & Facts. (n.d.). *U.S. Online Radio*. Available at https://www.statista.com/topics/1348/online-radio/

Statistics. (n.d.). Available at https://www.statista.com/statistics/252203/share-of-online-radio-listeners-in-the-us/; www.statista.com/statistics/253329/weekly-time-spent-with-online-radio-in-the-us/

Steier, L. (2001). Family firms, plural forms of governance, and the evolving role of trust. *Family Business Review*, *14*(4), 353–367. doi:10.1111/j.1741-6248.2001.00353.x

Steier, L., & Miller, D. (2010). Pre- and post-succession governance philosophies in entrepreneurial family firms. *Journal of Family Business Strategy*, *1*(3), 145–154. doi:10.1016/j.jfbs.2010.07.001

Stewart, A., & Hitt, M. A. (2012). Why can't a family business be more like a nonfamily business? modes of professionalization in family firms. *Family Business Review*, *25*(1), 58–86. doi:10.1177/0894486511421665

Stewart, T. A. (1998). *Intellectual Capital* (3rd ed.). Rio de Janeiro, Brazil: Campus.

Strauss, A., & Corbin, J. (1998). *Basics of qualitative research: grounded theory procedures and techniques* (2nd ed.). Newbury Park, CA: Sage.

Stryker, S. (1968). Identity salience and role performance. *Journal of Marriage and the Family*, *4*(4), 558–564. doi:10.2307/349494

Sullivan, M. H. (2012). *Uma assessoria de imprensa responsável na era digital*. Edição da Série Manuais. Bureau de Programas de Informações Internacionais do Departamento de Estado dos Estados Unidos.

Sundaramurthy, C. (2008). Sustaining trust within family businesses. *Family Business Review*, *21*(1), 89–102. doi:10.1111/j.1741-6248.2007.00110.x

Swisscontact. (2016). *Desarrollo de la cadena de valor de cacao*. Zúrich, Switzerland: Swiss Foundation for Technical Cooperation.

Świtalski, P. (2014). Intelligent pumps: Do they exist? *World Pumps*, *2014*(3), 34–36. doi:10.1016/S0262-1762(14)70052-5

Swogger, G. (1991). Assessing the successor generation in family businesses. *Family Business Review*, *4*(4), 397–411. doi:10.1111/j.1741-6248.1991.00397.x

Tachinardi, M. H. (1993). *A Guerra das Patentes: O conflito Brasil x EUA sobre propriedade intelectual*. Rio de Janeiro, Brazil: Paz e Terra.

Tajfel, H., & Turner, J. C. (1986). The social identity theory of intergroup behaviour. In S. Worchel & W. G. Austin (Eds.), *Psychology of intergroup relations* (pp. 7–24). Chicago, IL: Nelson-Hall Publishers.

Takeuchi, H., & Nonaka, I. (2004). *Hitotsubashi on knowledge management*. Singapore: John Wiley & Sons.

Takeuchi, H., & Nonaka, I. (2008). *Gestão do Conhecimento, tradução Ana Thorell*. Porto Alegre, Brazil: Bookman.

Takeuchi, H., & Nonaka, I. (2008). *Knowledge management*. Porto Alegre, Brazil. *The Bookman*.

Tan, C., Sim, Y., & Yeoh, W. (2011). A maturity model of enterprise business intelligence. *Communications of the IBIMA*, p. 1-11.

Tapping, D., & Shuker, T. (2010). *Lean Office: gerenciamento do fluxo de valor para áreas administrativas - 8 passos para planejar, mapear e sustentar melhorias lean nas áreas administrativas*. São Paulo, Brazil: Leopardo Editora.

Tapscott, D. (1999). *Creating value in the network economy*. Boston: Harvard Business School Press.

Tarapanoff, K. (2006). Inteligência, informação e conhecimento em corporações. Brasília: IBICT; UNESCO.

Te Chiu, C., Hsu, T. H., Wang, M. T., & Chiu, H. Y. (2011). Simulation for steel bridge erection by using BIM tools. In *Proceedings of the 28th International Symposium on Automation and Robotics in Construction, ISARC 2011* (pp. 560–563). Seoul: I.A.A.R.C.

Teall, O. (2014). Building information modelling in the highways sector: major projects of the future. *Proceedings of the Institution of Civil Engineers - Management, Procurement and Law, 167*(3), 127–133. 10.1680/mpal.13.00018

Tegner, M. G., de Lima, P. N., Veit, R. R., & Neto, S. L. H. C. (2016). Lean office and BPM: Proposition and application method for the reduction of waste in administrative areas. *Revista Produção Online, 16*(3), 1007–1032.

Terra, C. F. (2009). Usuários-mídia. In Congresso da ABRAPCORP. *Anais do III Congresso da ABRAPCORP*, São Paulo, ABRAPCORP.

Terra, C. F. (2007). *Blogs corporativos. São Caetano do Sul*. SP: Difusão.

Terra, C. F. (2011). *Mídias sociais...e agora? São Caetano do Sul*. SP: Difusão.

The 4 phases of the project management life cycle. (2017). Retrieved from https://www.lucidchart.com/blog/the-4-phases-of-the-project-management-life cycle

The idioms. (2019). The meaning and origins of the expression: A picture is worth a thousand words. Retrieved from https://www.theidioms.com/a-picture-is-worth-a-thousand-words/

Thu, N. T. (2014). *The problem of using multimedia materials for radio on the internet in Vietnam today.* (Master thesis), Academy of Journalism and Communication, p. 27.

Thuy, M. (2016). *Internet: Great opportunity of radio*, Retrieved from https://vov.vn/xa-hoi/dau-an-vov/Internet-co-hoi-lon-cua-phat-thanh-502781.vov

Tidd, J., Bessant, J., & Pavitt, K. (2003). *Gestão da Inovação: Integração das mudanças tecnológicas, de mercado e organizacionais*. Lisboa, Portugal: Monitor.

Traesel, F. A. & Maia, N. L. (2014). As organizações nas mídias sociais. In Congresso Brasileiro de Ciências da Comunicação. *Anais do XXXVII Congresso Brasileiro de Ciências da Comunicação*, Foz do Iguaçu.

Trigeorgis, L. (1996). *Real Options: Managerial Flexibility and Strategy in Resource Allocation*. Cambridge, MA: The MIT Press.

Trott, P. (2008). *Innovation Management and New Product Development* (4ªed.). Essex, UK: Prentice Hall | Financial Times.

Trout, J. (1969). Industrial Marketing Magazine, June, and then popularized by Ries, A. & Trout J. (1981). Positioning - The Battle for Your Mind. McGraw-Hill.

Tsang, E. W. K. (2002). Learning from overseas venturing experience: The case of Chinese family business. *Journal of Business Venturing, 17*(1), 21–40. doi:10.1016/S0883-9026(00)00052-5

Tumi, S. A. H., Omran, A., & Pakir, A. H. K. (2009). Causes of delay in construction industry in Libya. In *The International Conference on Economics and Administration* (pp. 265–272).

Turban, E., Rainer, R. K. Jr, & Potter, R. E. (2007). *Introduction to information systems*. Hoboken, NJ: John Wiley and Sons.

Uhlaner, L. M. (2006). Business family as a team: Underlying force for sustained competitive advantage. In P. Z. Poutziouris, K. X. Smyrnios, & S. B. Klein (Eds.), *Handbook of research on family business* (pp. 125–144). Cheltenham, UK: Edward Elgar. doi:10.4337/9781847204394.00016

Ullrich, H. (1989). The importance of industrial property law and other legal measures in the promotion of technological innovation. *Industrial Property, 28*, 102–112.

Vaidya, S., Ambad, P., & Bhosle, S. (2018). Industry 4.0 – A Glimpse. *Procedia Manufacturing 20. 2nd International Conference on Materials Manufacturing and Design Engineering, 20*, 233-238.

Van de Ven, A. H. (1986). Central problems in the management of innovation. *Management Science, 32*(5), 590–607. doi:10.1287/mnsc.32.5.590

Van de Ven, A. H., Policy, D. E., Garud, R., & Venkataraman, S. (1999). *The innovation journey*. Oxford, UK: Oxford University Press.

Van der Grinten, J., & Riezebos, R. (2011). *Positioning the brand*. Routledge.

Van Haren - Publishing. (2011). *ITIL Foundations - Best Practice*.

Van Kien, P., Hai, P. Q., Thang, P. C., & Hau, N. D. (2016). Some trends of modern communication press. Publisher of Information and Communication, p.151.

Varajão, J. (1998). *A arquitectura da gestão de sistemas de informação* (2nd ed.). Lisboa: FCA.

Vecchia, A. F. D. (2011, March 23). Sistemas Erp: A Gestão Do Processo De Implantação Em Universidade Pública.

Vela, E. V. (2017). *Competitividad del Comercio Exterior*. Lima, Peru.

Velho, S. (1996). *Relações Universidade-Empresa: desvelando mitos*. Campinas, SP: Autores Associados.

Velloso, V. F., & Yanaze, M. H. (2014). O consumidor insatisfeito em tempo de redes sociais. *Educação. Cultura e Comunicação, 5*(9), 7–20.

Vera, G. (2018). Tipos de cacao: forastero, cacao y trinitario. In *Cocina y Vino*. Retrieved from http://www.cocinayvino.com/mundo-gourmet/tipos-cacao-forastero-criollo-trinitario/

Vergara, S. C. (2012). *Métodos de Pesquisa em Administração*. São Paulo, Brazil: Atlas.

Vesper, K. H. (1990). *New venture strategies*. Prentice-Hall.

Vieira, A. da S. (1998). *Monitoração da competitividade científica e tecnológica dos estados brasileiros a partir do SEICT*. Brasília: Ibict.

Villanueva, J., & Sapienza, H. J. (2009). Goal tolerance, outside investors, and family firm governance. *Entrepreneurship Theory and Practice, 33*(6), 1193–1199. doi:10.1111/j.1540-6520.2009.00340.x

Vogler, M., Gratieri, T., Gelfuso, G. M., & Cunha Filho, M. S. S. (2017). As boas práticas de fabricação de medicamentos e suas determinantes. Vigilância Sanitária em Debate: Sociedade, Ciência & Tecnologia, 5(2), 34-41.

Von Krogh, G., Ichijō, K., & Nonaka, I. (2000). *Enabling knowledge creation: how to unlock the mystery of tacit knowledge and release the power of innovation*. New York, NY: Oxford University Press. doi:10.1007/978-1-349-62753-0

Vossebeld, N., & Hartmann, T. (2016). Modeling Information for Maintenance and Safety along the Lifecycle of Road Tunnels. *Journal of Computing in Civil Engineering, 30*(5), C4016003. doi:10.1061/(ASCE)CP.1943-5487.0000593

Ward, J., & Griffiths, P. (1996). *Strategic planning for information systems* (2nd ed.). Wiley.

Watson, D. (2005). *Business models: Investing in companies and sectors with strong competitive advantage*. Harriman House Publishing.

Compilation of References

Watt, A. (2014). *Project management*.

Westhead, P., & Howorth, C. (2007). "Types" of private family firms: An exploratory conceptual and empirical analysis. *Entrepreneurship and Regional Development*, *19*(5), 405–431. doi:10.1080/08985620701552405

Whetten, D. A., Foreman, P. & Dyer, W. G. (2014), Organizational identity and family business, In L. Melin, M. Nordqvist, & P. Sharma (Eds.), The SAGE handbook of family business (1st ed.), Thousand Oaks, CA: Sage. pp. 480–497. doi:10.4135/9781446247556.n24

Whetten, D. A. (1987). Organisational growth and decline processes. *Annual Review of Sociology*, *13*(1), 335–358. doi:10.1146/annurev.so.13.080187.002003

Whetten, D., & Mackey, A. (2002). A social actor conception of organizational identity and its implications for the study of organizational reputation. *Business & Society*, *41*(4), 393–415. doi:10.1177/0007650302238775

Wikipedia. (n.d.). Internet radio. Available at https://en.wikipedia.org/wiki/Internet_radio

Willem, A., Scarborough, H., & Buelens, M. (2008). Impact of coherent versus multiple identities on knowledge integration. *Journal of Information Science*, *34*(3), 370–386. doi:10.1177/0165551507086259

Wilson, T. (1994b). The nature of strategic information and its implications for information management. In New Worlds in Information and Documentation. Elsevier Science B. V.

Wilson, T. (2001). Information overload: Myth, reality and implications for health care. *ISHIMR - International Symposium on Health Information Management Research*.

Wilson, D. (2002). *Managing information: IT for business processes* (3rd ed.). Woburn: Butterworth-Heinemann.

Wilson, R. (1987a). Patent analysis using online databases: I. Technological trend analysis. *World Patent Information*, *9*(1), 18–26. doi:10.1016/0172-2190(87)90189-X

Wilson, R. (1987b). Patent analysis using online databases: II. Competitor activity monitoring. *World Patent Information*, *9*(2), 73–78. doi:10.1016/0172-2190(87)90131-1

Wilson, T. (1985). Information management. *The Electronic Library*, *3*(1), 62–66. doi:10.1108/eb044644 PMID:2498741

Wilson, T. (1987). Information for business: The business of information. *Aslib Proceedings*, *39*(10), 275–279. doi:10.1108/eb051066

Wilson, T. (1989a). Towards an information management curriculum. *Journal of Information Science*, *15*(4/5), 203–209. doi:10.1177/016555158901500403

Wilson, T. (1989b). The implementation of information system strategies in UK companies: Aims and barriers to success. *International Journal of Information Management*, *9*(4), 245–258. doi:10.1016/0268-4012(89)90048-0

Wilson, T. (1994a). Tools for the analysis of business information needs. *Aslib Proceedings*, *46*(1), 19–23. doi:10.1108/eb051339

World Bank. (2018). Datos Banco Mundial. *Portal institucional Banco Mundial*. Retrieved from https://datos.bancomundial.org/indicador/LP.EXP.DURS.MD?end=2016&locations=PE-BR-CL-CO-EC&start=2016&view=bar>

Wulong, G. & Gera, S. (2004). *The effect of organizational innovation and information technology on firm performance*. Statistics Canada, Catalogue No. 11-622-MIE No. 007.

XAPP Media. (2015). *Internet Radio Trends Report 2015*. pp.12, 11, 9, 13, 4, 5. Retrieved from https://xappmedia.com/wp-content/uploads/2015/01/Internet-Radio-Trends-Report-2015_january.pdf

Yeomans, M. (2006). *Oil - Petróleo: Guia Conciso para o Produto mais Importante do Mundo* (1st ed.). Lisboa, Portugal: D. Quixote.

Yin, R. K. (2001). *Estudo de caso: planejamento e métodos*. Porto Alegre, Brazil: Bookman.

Yin, R. K. (2003). *Case study research: Design and methods* (3rd ed., Vol. 5). Thousand Oaks, CA: Sage.

Zahra, S. A., Hayton, J. C., Neubaum, D. O., Dibrell, C., & Craig, J. (2008). Culture of family commitment and strategic flexibility: The moderating effect of stewardship. *Entrepreneurship Theory and Practice*, *32*(6), 1035–1054. doi:10.1111/j.1540-6520.2008.00271.x

Zahra, S. A., Hayton, J. C., & Salvato, C. (2004). Entrepreneurship in family vs. non-family firms: A resource-based analysis of the effect of organizational culture. *Entrepreneurship Theory and Practice*, *28*(4), 363–381. doi:10.1111/j.1540-6520.2004.00051.x

Zanen, P. P. A., Hartmann, T., Al-Jibouri, S. H. S., & Heijmans, H. W. N. (2013). Using 4D CAD to visualize the impacts of highway construction on the public. *Automation in Construction*, *32*, 136–144. doi:10.1016/j.autcon.2013.01.016

Zellweger, T. M., & Astrachan, J. H. (2008). On the emotional value of owning a firm. *Family Business Review*, *21*(4), 347–363. doi:10.1177/08944865080210040106

Zellweger, T. M., & Dehlen, T. (2012). Value is in the eye of the owner: affect infusion and socioemotional wealth among family firm owners. *Family Business Review*, *25*(3), 280–297. doi:10.1177/0894486511416648

Zellweger, T. M., Eddleston, K. A., & Kellermanns, F. W. (2010). Exploring the concept of familiness: Introducing family firm identity. *Journal of Family Business Strategy*, *1*(1), 54–63. doi:10.1016/j.jfbs.2009.12.003

Zellweger, T. M., Kellermanns, F. W., Chrisman, J. J., & Chua, J. H. (2011). Family control and family firm valuations by family CEOs: The importance of intentions for transgenerational control. *Organization Science*, *23*(3), 851–868. doi:10.1287/orsc.1110.0665

Zellweger, T. M., Kellermanns, F. W., Eddleston, K. A., & Memili, E. (2012). Building a family firm image: How family firms capitalize on their family ties. *Journal of Family Business Strategy*, *3*(4), 239–250. doi:10.1016/j.jfbs.2012.10.001

Zellweger, T. M., Nason, R. S., Nordqvist, M., & Brush, V. (2013). Why do family firms strive for nonfinancial goals? an organizational identity perspective. *Entrepreneurship Theory and Practice*, *37*(2), 229–248. doi:10.1111/j.1540-6520.2011.00466.x

Zhang, J. & Ma, H. (2009). Adoption of professional management in Chinese family business: A multilevel analysis of impetuses and impediments. *Asia Pacific Journal of Management*, 26, 119-139.

Zhang, S., Teizer, J., Lee, J. K., Eastman, C. M., & Venugopal, M. (2013). Building Information Modeling (BIM) and Safety: Automatic Safety Checking of Construction Models and Schedules. *Automation in Construction*, *29*, 183–195. doi:10.1016/j.autcon.2012.05.006

Zhou, Y., Ding, L., Luo, H., & Chen, L. (2010). Research and Application on 6D Integrated System in Metro Construction Based on BIM. *Applied Mechanics and Materials*, *26*(28), 241–245. doi:10.4028/www.scientific.net/AMM.26-28.241

Zorrinho, C. (1991). *Gestão da informação*. Lisboa: Editorial Presença.

Zorrinho, C. (1995). *Gestão da Informação*. Condição para Vencer. Iapmei.

Zou, Y., Kiviniemi, A., & Jones, S. W. (2016). Developing a tailored RBS linking to BIM for risk management of bridge projects. *Engineering, Construction, and Architectural Management*, *23*(6), 727–750. doi:10.1108/ECAM-01-2016-0009

About the Contributors

George Leal Jamil is a professor of several post-graduation courses from Minas Gerais, Brazil. He has a PhD in Information Science from the Federal University of Minas Gerais (UFMG), master degree in Computer Science (UFMG) and his graduate area was Electric Engineering (UFMG) and a pos-doctorate certificate at FLUP – Communication School at University of Porto, Portugal. He wrote more than thirty books in the information technology and strategic management areas, with more than ten works in books as co-author and Editor. His main research interests are information systems management, strategy, knowledge management, software engineering, marketing and IT adoption in business contexts. Contact:gljamil@gmail.com

Fernanda Ribeiro is since 1989 Professor at the Faculty of Arts of University of Oporto (Portugal). She got a graduation in History from the University of Porto in 1980 and a post-graduation in Library and Archival Studies from the University of Coimbra in 1982. In 1999, she got the PhD at the University of Porto with a dissertation entitled The Access to Information in Archives, under the supervision of Michael Cook. She has been the Head of the Department of Journalism and Communication Sciences (2010-2014) and director of the bachelor in Information Science (2003-2014). She's also scientific coordinator of the research centre CETAC.MEDIA (Centre of Studies in Communication Sciences and Techonologies). During the last 20 years she devoted her academic research to access and retrieval of information in archives, subject indexing and theory and methodology of Information Science as well as professional training in the same area.

Armando Barreiros Malheiro da Silva is Associated Professor of the Faculty of Arts of the University of Porto and member of the coordinating committee of the Information Science degree taught by the Arts and Engineering Faculties of the University of Porto. Born in Braga, PdD in Contemporary History in the University of Minho, graduated in Philosophy by Philosophy Faculty of the Catholic University of Braga and in History by the Faculty of Arts of the University of Porto. He obtained the diploma of the course of Librarian-Archivist of the Faculty of Arts of the University of Coimbra. Is member of the Center for Studies in Technology, Arts and Communication Sciences (CETAC.Media) and shares his researches in areas such as the archivist and information science; the metanalysis; political and ideological History in Portugal in the XIX-XX century; Family history and local studies.

Sérgio Maravilhas is a Lecturer at SENAI/CIMATEC and UNIFACS, Brazil. Having two Post-Docs, one from UFBA in Patents and innovation and another one from UNIFACS in FABLABS, both with PNPD/CAPES Grants. He has a PhD in Information and Communication in Digital Platforms (UA+UP), a Master in Information Management (FEUP and Sheffield University, UK), a Postgraduate Course in ICT (FEUP), a Specialization in Innovation and Technological Entrepreneurship (FEUP and North Carolina State University, USA) and a 5 years Degree in Philosophy, Educational Branch (FLUP). A Teacher and Trainer since 1998, worked at ESE-IPP as a Supervisor at the Internet@School project between 2002/2005, and a university teacher since 2005 at Aveiro University (DEGEI), and since 2010 at ULP and IESFF in Masters and MBA levels. Teaches Marketing, Research Methods, Creative Processes in Innovation, Intellectual and Industrial Property, Technology Watch, Information Management and Organizational Behaviour in universities and he's a trainer in ICT, Sales, Negotiation and Neuro-Linguistic Programming (NLP). Publishes and attends conferences mainly in the subjects of patent information, innovation, marketing, web 2.0, webradio and sustainability.

Shabir Ahmad - Post Graduate Student of Project Management in Institute of Management Sciences, Imsciences, Hayat Abad Peshawar, Pakistan

Adelaide Antunes - Professor and Senior Specialist at National Institute for Industrial Property/INPI & Chemical School of the University of Rio de Janeiro - UFRJ, Brazil.

Luiz Eduardo Bastos has a bachelor degree in Mechanical and Automotive Engineering from the Military Engineering Institute – Instituto Militar de Engenharia (1983), a Master's degree in Business Administration from the Federal University of Bahia – Universidade Federal da Bahia (2003) and a PhD in Regional and Urban Development from the Unifacs University Salvador (2013). Is currently a Consultant in Oil, Gas and Management at Union Energy and Braslift (Brasil Eletromecânica). Also has experience in the educational area as a teacher and manager, as well as in the Oil and Natural Gas industry, as engineer and consultant.

Emelia Biney was a post graduate student of the Pentecost University Graduate school, Accra, Ghana. She is currently holding MSc in International Finance and Accounting from Buckinghamshire University, UK. Her research interest is Accounting Information systems and systems integration.

Alvaro Cairrão - Phd Ciências da Comunicação

António Cardoso - Phd Ciências Empresariais

Jet Castilla – has a Master's in Supply Chain Managment at Universidad del Pacífico, Peru.

Mario Chong - Professor and Researcher at Universidad del Pacífico. He holds a PhD in Business Management from Universidad Nacional Mayor de San Marcos; a Master's in Industrial Engineering, a Master's in Systems Engineering and a degree in Industrial Engineer from Universidad de Lima. He has completed a certification in Advanced Supply Chain Management at the Massachusetts Institute of Technology (MIT). He has experience developing research projects related to the field of business such as business strategy, supply chain, operations, global business, agribusiness, and rural associativity. He is Director of the Peruvian Association of Professionals in Logistics (APPROLOG). He is an international member of the System Dynamics Society (SD) and the Association of Supply Chain Professionals and Operations Management (APICS). He has been associate dean of Business Engineering, coordinator

About the Contributors

of special projects, corporate and international program development director, academic director of the Master's programs in Business Administration (MBA), Global Business, Agribusiness and Supply Chain Management at Universidad del Pacífico.

Valéria Costa - Lawyer, graduated in law from the Federal University of Minas Gerais in 2000, master's degree in tax law from PUC Minas in 2011. MBA in tax law from FGV in 2004. Professor of tax law. Executive MBA by the DOm Cabral Foundation in 2018.

Frederico D´Orey - Phd Ciências Empresariais

José Diniz Filho - Lawyer, graduated from Milton Campos Law School in 2003, studying law in tax by UCB, Executive MBA by Fundação Dom Cabral in 2018.

Zainab Durrani is a Ms Project Management final year student at the Institute of Mangement Sciences Peshawar. She is currently an O level teacher of Business Studies and Economics. After her graduation, she has started her professional career as a Program Coordinator at the university level and started working with Google in her 6th semester of BBA as a Co-Manager at Google Business Group Peshawar. She's a Teacher, Marketer, Project manager, Event Manager and Growth hacking trainer. She has been a contributor to the startup ecosystem by uplifting it via arranging different need-based workshops and sessions for budding entrepreneurs on Google tools.

Rabia Imtiaz is a Ms Project Management Final Semester student. presently working on her thesis. She graduated from the Institute of management sciences Peshawar with a degree in MBA finance in 2015. After graduation, she worked as an assistant lecturer with an Asst. Professor at Institute of management sciences. she is a hardworking and dedicated girl. She has excellent management skills and arranged a social seminar, business festivals and conducted sessions etc. she is honest and has a professional attitude towards work.

Filippo Ferrari holds a Bachelor degree in Work and Organizational Psychology and a Masters degree in Adult Training and Education. He currently serves as a Lecturer at Bologna University, School of Economics, Management and Statistics, where he teaches Organizational Behaviour and Design and Human Resources Management. His fields of interests are International Human Resources Management, Organizational Reliability, Change Management, Adult Training.

Jorge Figueiredo - Phd Ciências da Comunicação

Zulmira Hartz - Professor and Researcher of the Institute of Hygiene and Tropical Medicine – IHMT. Global Health and Tropical Medicine (GHTM) in NOVA University of Lisbon, Portugal.

Juliana Igarashi - Master's in science by Professional Postgraduate Program in Management, Researcher and Developing in Pharmaceutical Industry, Analyst in Information Technology, Daudt Oliveira Pharma Laboratory, Brazil

Stefan Kögl has a Master's degree in Computer Science from the Vienna University of Technology as well as an MBA with specialization on Project Management from the Vienna University of Economics and Business. Throughout his professional career he has contributed to and managed numerous distributed multidisciplinary project teams in various industries.

Elizabeth Macedo - Full Professor and Researcher at Federal Fluminense University - UFF

Jorge Magalhães - Postgraduated in Competitive Intelligence for Public Health. Doctor and Master in Sciences in Management and Technological Innovation. Has over 20 years of experience in strategic management in the pharmaceutical Industrial Operations and in the last 14 years to act in R, D & I for Public Health area at FIOCRUZ. Published 03 books, 9 book chapters and several articles in journals indexed. Actually work in Technology Innovation Center (NIT-Far) at FIOCRUZ. It is leader of the CNPq Research Group Knowledge Management and Prospecting Health. The emphasis of their research permeate the identification, extraction and analysis of essential information within the "Big data" for Health, regarding the Management and Technological Innovation. The topics covered are inherent in Global Health, in which involves the pharmaceutical industry, pharmochemical and public health. Investigations are carried out through the development of prospective and technological scenarios of information science tools, Competitive Intelligence and Knowledge Management. Included in this context the analysis of BIG DATA, Web 2.0, Health 2.0 Technological Trends, market, Patents and Knowledge Translation.

Miguel Magalhães - Phd Gestão e Administração de Empresas

Patrick Pennefather - As an emerging researcher and scholar, Patrick's research efforts can be categorized into several persistent areas of inquiry: Research that extends and experiments with sound in the design of interactive experiences for live and hybrid stages; Investigating self-reflective teaching practices particularly in project-based learning environments; Research into user-centred design practices that can support digital media innovation including user-testing prototypes; Case study research within digital media production pipelines and live production pipelines focused on intersecting areas of sound, user-testing, improvised behavior and collaboration.

Manuel Pereira - Degree in Public Relations Master in Marketing and Strategic Communication Doctor of Communication Sciences

Eduardo Perez has a Master's in Supply Chain Management at Universidad del Pacífico, Peru.

Rajadurai R has obtained his Masters in Construction management from SRM institute of Technology and management and currently working as Junior Research Fellow in the DST-SERB project.

Rachel Ralph is a Faculty member at the Centre for Digital Media and a Postdoctoral Fellow at the University of British Columbia (UBC). Her research focuses on edutainment, learning and teaching, curriculum, agile practices, and virtual reality with a focus on media and technology.

Hernan Rosario has a Master's in Supply Chain Management at Universidad del Pacífco, Peru.

About the Contributors

Gilbert Silvius (1963) is professor of project and programme management at LOI University of Applied Sciences in the Netherlands, visiting professor at the University of Johannesburg in South Africa and fellow at Turku University of Applied Sciences in Finland. He initiated and developed the first MSc in Project Management program in the Netherlands and is considered a leading expert in the field of project management and information management. Gilbert has published over a 100 academic papers and several books. His areas of specialization are: Sustainability in project management, Standards and methodologies of project management, Project management maturity, Business and IT alignment and Business case management. Gilbert holds a PhD degree in information sciences from Utrecht University and masters' degrees in economics and business administration. As a practitioner, Gilbert has over 20 years' experience in organizational change and IT projects and is a member of the international enable2change network of project management experts.

Aneetha Viventhan has obtained her PhD in Building Technology and Construction Management from Indian Institute of Technology Madras, Chennai, India. She has few years of industrial experience in Qatar as a Planning Engineer and currently working as an Assistant Professor at National Institute of Technology Warangal.

Winfred Yaokumah is a senior lecturer at the University of Ghana, Accra, Ghana. He obtained his PhD in Information Technology with specialization in Information Assurance and Security. He has published extensively in several international journals, including the Information Management & Computer Security, Information Resource Management Journal, International Journal of Technology Diffusion, Journal of Information Technology Research, International Journal of Information Systems and Social Change, International Journal of IT/Business Alignment and Governance, and the International Journal of Information Systems in the Service Sector. His research interest includes information security, cyber security, e-services, information security governance, IT project management, and IT leadership.

Index

A

App 90, 92, 95

B

Blockchain 84-85, 323, 332-337, 332-339
Blogs 272, 280, 284-285, 285, 287, 293-295, 294-295, 340-341, 344, 348, 353, 355-357
Business Intelligence xxiii, 55, 57, 249-250, 255-257, 262, 266, 268-270, 270
Business Management 59, 101, 106, 116, 120, 162, 268
Business Transmission 97, 111

C

Circumplex Model 111, 120, 123
Cocoa xxiv, 323-325, 327, 331-332, 334-335, 335-337, 337
Competitive Intelligence xxiii, 22, 55, 201, 237-238, 246-250, 252, 290
Competitiveness xxiv, 7, 41, 59, 86-87, 124-125, 133, 147, 169, 185-188, 190, 193, 201-202, 239, 243, 246-247, 257, 274, 288, 294, 337
Competitors 9, 40, 42-44, 44, 50, 52, 56, 59-60, 158-159, 162, 162-163, 175, 186, 238, 246, 250, 252, 257, 288, 290, 355
Creative Imitation 298
Creativity 132, 170, 184, 288-289, 291, 293, 298

D

Data xxii, xxiii, ii, 12, 14, 18, 26, 49, 59-60, 59-61, 64-65, 71, 73-74, 76, 78, 83-84, 87, 90-92, 91-92, 103, 106, 132, 134, 136, 141-146, 151, 162, 182, 187, 194, 200, 202-208, 211, 216, 218, 223, 225, 227, 231-232, 231-233, 242, 246, 248-250, 248-251, 255-266, 268, 268-270, 270, 278-280, 282, 293, 303, 318, 323, 328, 328-329, 337, 340-341, 343-344, 346-348, 354, 356-357
Datawarehouse 270
Decision Making Process 38, 270
Directory Of Research Groups In Brazil Of Cnpq (DGP 213
Documentation 21, 30, 30-31, 38, 46-47, 57, 75-77, 103, 126, 145, 227, 231, 245, 260, 289, 330, 340-341, 343-344, 354-355, 357, 357-359
Dysfunctional Behaviour 123

E

ETL 257, 270

F

Family Business 97-108, 110-112, 110-123, 123
Family Conflicts 97
Family Control 97, 100, 120, 123
Flow Rate 90, 95
Framework xxii, 3, 22, 81, 108, 111, 117, 121, 143, 158-159, 162, 165, 167, 187, 201, 221, 227, 233, 251, 253, 260, 269, 273, 320-321, 324, 333, 340-341
Frequency Inverter 92-93, 95

H

Head 35, 49, 49-50, 90, 95, 112, 179, 184, 298

I

Identity And Corporate Image 285
Industrial Pump 83, 93, 95
Industry 4.0 xxii, 83-87, 83-95, 89-95
Information xxi, xxii, xxiii, xxiv, iii, 2, 4, 7, 2, 9, 4, 11, 6-7, 14, 18-36, 20-30, 32, 34-36, 38-53, 38-61, 59-60, 63-65, 67-69, 71, 73-76, 79-82, 84, 86, 88, 90-92, 95, 103, 114, 125-126, 132-136, 141-142, 145-148, 153, 161-162, 161-163, 169-171, 173-177, 179, 181-182, 181-187, 184-187, 194, 197, 197-206, 200-206, 209-211, 213, 215-219, 222-224, 227-230, 232, 234-235, 242, 244, 246-250, 252-262, 255-262, 264-265, 264-270, 267, 270, 272, 274-276, 279-283, 285, 285-298, 287-295, 298, 302-310, 313, 315, 317-321, 317-324, 323-324, 326-333, 335, 346, 350, 355-356
Information And Communication Technologies 22, 40, 60, 199, 201
Information And Knowledge Society 201, 213
Information Management xxi, xxiii, xxiv, iii, 20-22, 26, 32, 32-35, 34-35, 39-41, 45, 45-47, 47, 51, 53-61, 60, 95, 141, 145, 200-204, 209-211, 213, 258-260, 269
Information Science xxi, 20-22, 31-33, 32, 57, 249, 253-254
Information Technology xxiii, 4, 7, 18, 22, 24, 28, 35, 38, 40, 46, 51-52, 55, 79, 81, 95, 125-126, 132, 134-136, 145, 148, 185, 187, 197, 215-218, 229, 232, 234-235, 250, 253, 255-256, 262, 272, 279, 285, 285-286, 310, 320, 328
Informational Behaviour 21, 29, 32, 32-33
Innovation xxi, xxii, xxiii, 1-2, 1-2, 11, 17, 47, 49, 55, 57, 86, 104, 110, 112, 124, 131-133, 135-136, 135-139, 154, 160, 169-171, 174-175, 178, 183-184, 184, 186-188, 197, 199, 201-203, 205, 208, 211, 213, 216, 229, 231, 235, 237-245, 247-249, 251, 251-254, 272, 283, 287, 289, 291, 293-295, 293-298, 298, 320, 358
Invention 174-175, 177-179, 181-182, 184, 289-291, 298, 298-299
Investment Analyses 1, 11

K

Knowledge xxi, xxiii, 3-4, 18-19, 22, 24-28, 31, 33, 35-36, 41, 41-42, 47-49, 47-50, 52, 54-55, 57-60, 59-60, 90, 97-105, 107-112, 114, 118, 124, 126, 131-136, 131, 133-139, 150, 155, 157, 160-161, 163, 171, 174, 179, 181-182, 184, 193-197, 194-197, 200-206, 208-211, 213, 216, 220, 232, 234, 237, 240-248, 250-251, 250-254, 257, 267, 270, 274, 283, 287-292, 295, 298, 333, 335, 342, 344, 355, 358
Knowledge Profiles 237, 240, 253

M

Market Intelligence 40, 60
Marketing xxii, iii, 2, 9, 11, 14, 25, 38, 41, 49, 55-56, 59, 61, 113, 131, 134, 138, 158-162, 158-168, 165-167, 175, 183, 202, 230, 241, 246-247, 249-250, 249-251, 269, 272, 276, 286, 290
Marketing Plan (JEL L30) 158
Maturity Model 258-259, 262-263, 263, 265, 268, 268-270, 270
Multilevel Analysis 119, 123

N

Net Present Value 1-3, 7-8, 11, 14, 16

O

Organization xxiii, 9, 32, 34-36, 38-48, 51-53, 58-60, 59-60, 85-86, 95, 98, 103, 110-112, 119-121, 123-127, 129, 131-136, 138, 140-143, 141-143, 146, 148-149, 156-157, 159-163, 165, 167, 175, 183-184, 184, 187, 189, 193-194, 194, 196-198, 202, 213, 216-220, 222-223, 227, 237, 239, 241-246, 241, 243-246, 248-250, 252-253, 256, 259-263, 265-268, 266-268, 271, 273-274, 276, 280-282, 285, 288, 294, 296, 320, 358
Organization And Representation Of Information 32
Organizational Communication xxiii, 271-274, 282, 285
Organizational Identity 97, 113, 122-123, 123

P

Patent Information xxiii, 169-170, 173-177, 179, 181-182, 181-184, 184, 287-295, 287-298, 298
Pharma Industry 237-238, 238

Plan 4, 24, 28, 30, 41, 47-48, 60, 65-66, 68, 75, 94, 100, 102, 112, 133, 147, 153, 158-159, 161-162, 161-167, 165-167, 190, 192, 204, 221-222, 225, 232, 246, 256, 266-268, 289, 294

Press Office 271-272, 274, 274-275, 277-278, 285

Product Project 1

Production Of Information Flow 32

Project Management xxi, xxii, xxiii, xxiv, 1-4, 1-4, 8, 18-20, 20, 23-27, 31, 53, 80, 97-99, 101, 103, 105, 107-108, 111-112, 118, 120, 131, 137, 139-141, 141, 143, 151, 151-152, 191, 193, 215-218, 221-228, 231-232, 231-235, 339-340, 340

R

Real Options Theory xxi, 1-3, 5-6, 13, 17

Reflection xxi, xxiii, 240, 340-344, 341-344, 350, 353, 356-357, 356-358

Reflective Practitioner 340-342, 344, 356, 359

Research And Development xxi, 1-3, 9, 13, 17, 41, 86-87, 169, 204, 219, 229, 241, 288

Research Groups 200-201, 203-211, 213

S

Serendipity 184, 297-299, 298

Social Media xxiii, xxiv, 34, 44-45, 48, 48-51, 51, 53, 55, 271-272, 274-283, 274-285, 285, 287-288, 293-295, 321

Socio-Emotional Wealth 97, 107, 123

Strategy 3-4, 7, 17-19, 24, 30, 34-35, 38-42, 44-49, 44-56, 51-53, 58-61, 60, 97-98, 101-102, 115-117, 119, 121-123, 125, 127-129, 128, 133, 136-140, 150, 156, 159-161, 163, 167-168, 189-191, 193, 196, 198, 202, 205, 229, 231, 235, 239, 243, 247-248, 250, 253, 259-260, 263, 276, 293-294, 298, 324

Supply Chain 93, 147, 157, 323-324, 326, 328-334, 336-337

Sustainability 66, 80-81, 108, 170, 172, 175, 179, 181, 183-184, 184, 187, 229, 324

Systems Validations. Computerized systems. Risk Analysis. SAP. ERP. Warehouse management. Project Management. 141

T

Team xxi, 25-26, 49, 63-67, 69, 71-74, 76-77, 91, 98, 102, 106, 108, 110, 122, 128, 128-129, 134, 143, 149, 163, 220-222, 224, 227, 230-231, 245, 259, 262, 264, 266-268, 304, 320, 340-357, 340-358

Technology xxi, xxiii, xxiv, 4, 7, 17-19, 21-24, 28, 32, 35, 38-41, 46, 49, 51-52, 55, 58, 63, 78, 78-79, 81, 85-87, 85-88, 92, 94-95, 95, 124-126, 132, 134-136, 145, 148, 156, 161-162, 169, 171-172, 175-176, 178, 182, 182-183, 185-187, 197, 201-203, 208, 212-213, 213, 215-218, 224, 229, 232, 232-235, 238-239, 241, 243, 247, 250, 252-253, 255-256, 258-263, 265, 265-266, 271-272, 279, 282, 285, 285-288, 287-288, 290-292, 290-293, 295, 295-296, 299-300, 300, 303-304, 306, 309-310, 318-321, 323-324, 328-330, 335-336, 339, 343

U

University-Enterprise-Government Interaction (Triple Helix) 213

Purchase Print, E-Book, or Print + E-Book

IGI Global's reference books are available in three unique pricing formats:
Print Only, E-Book Only, or Print + E-Book.
Shipping fees may apply.
www.igi-global.com

Recommended Reference Books

ISBN: 978-1-5225-6201-6
© 2019; 341 pp.
List Price: $345

ISBN: 978-1-5225-7262-6
© 2019; 360 pp.
List Price: $215

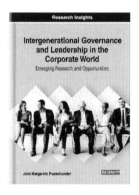

ISBN: 978-1-5225-8003-4
© 2019; 216 pp.
List Price: $205

ISBN: 978-1-5225-5763-0
© 2019; 304 pp.
List Price: $205

ISBN: 978-1-5225-6980-0
© 2019; 325 pp.
List Price: $235

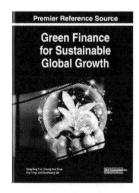

ISBN: 978-1-5225-7808-6
© 2019; 397 pp.
List Price: $215

Do you want to stay current on the latest research trends, product announcements, news and special offers?
Join IGI Global's mailing list today and start enjoying exclusive perks sent only to IGI Global members.
Add your name to the list at **www.igi-global.com/newsletters**.

Publisher of Peer-Reviewed, Timely, and Innovative Academic Research

www.igi-global.com | Sign up at www.igi-global.com/newsletters | facebook.com/igiglobal | twitter.com/igiglobal | linkedin.com/igiglobal

Ensure Quality Research is Introduced to the Academic Community

Become an IGI Global Reviewer for Authored Book Projects

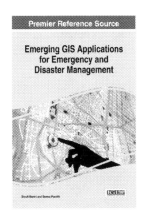 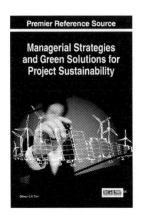  

The overall success of an authored book project is dependent on quality and timely reviews.

In this competitive age of scholarly publishing, constructive and timely feedback significantly expedites the turnaround time of manuscripts from submission to acceptance, allowing the publication and discovery of forward-thinking research at a much more expeditious rate. Several IGI Global authored book projects are currently seeking highly-qualified experts in the field to fill vacancies on their respective editorial review boards:

Applications and Inquiries may be sent to:
development@igi-global.com

Applicants must have a doctorate (or an equivalent degree) as well as publishing and reviewing experience. Reviewers are asked to complete the open-ended evaluation questions with as much detail as possible in a timely, collegial, and constructive manner. All reviewers' tenures run for one-year terms on the editorial review boards and are expected to complete at least three reviews per term. Upon successful completion of this term, reviewers can be considered for an additional term.

If you have a colleague that may be interested in this opportunity,
we encourage you to share this information with them.

IGI Global Proudly Partners With eContent Pro International

Receive a 25% Discount on all Editorial Services

Editorial Services

IGI Global expects all final manuscripts submitted for publication to be in their final form. This means they must be reviewed, revised, and professionally copy edited prior to their final submission. Not only does this support with accelerating the publication process, but it also ensures that the highest quality scholarly work can be disseminated.

English Language Copy Editing

Let eContent Pro International's expert copy editors perform edits on your manuscript to resolve spelling, punctuaion, grammar, syntax, flow, formatting issues and more.

Scientific and Scholarly Editing

Allow colleagues in your research area to examine the content of your manuscript and provide you with valuable feedback and suggestions before submission.

Figure, Table, Chart & Equation Conversions

Do you have poor quality figures? Do you need visual elements in your manuscript created or converted? A design expert can help!

Translation

Need your documjent translated into English? eContent Pro International's expert translators are fluent in English and more than 40 different languages.

Hear What Your Colleagues are Saying About Editorial Services Supported by IGI Global

"The service was very fast, very thorough, and very helpful in ensuring our chapter meets the criteria and requirements of the book's editors. I was quite impressed and happy with your service."

– Prof. Tom Brinthaupt,
Middle Tennessee State University, USA

"I found the work actually spectacular. The editing, formatting, and other checks were very thorough. The turnaround time was great as well. I will definitely use eContent Pro in the future."

– Nickanor Amwata, Lecturer,
University of Kurdistan Hawler, Iraq

"I was impressed that it was done timely, and wherever the content was not clear for the reader, the paper was improved with better readability for the audience."

– Prof. James Chilembwe,
Mzuzu University, Malawi

Email: customerservice@econtentpro.com   **www.igi-global.com/editorial-service-partners**

www.igi-global.com

Celebrating Over 30 Years of Scholarly Knowledge Creation & Dissemination

InfoSci®-Books

A Database of Over 5,300+ Reference Books Containing Over 100,000+ Chapters Focusing on Emerging Research

GAIN ACCESS TO **THOUSANDS** OF REFERENCE BOOKS AT **A FRACTION** OF THEIR INDIVIDUAL LIST **PRICE**.

InfoSci®-Books Database

The **InfoSci®-Books** database is a collection of over 5,300+ IGI Global single and multi-volume reference books, handbooks of research, and encyclopedias, encompassing groundbreaking research from prominent experts worldwide that span over 350+ topics in 11 core subject areas including business, computer science, education, science and engineering, social sciences and more.

Open Access Fee Waiver (Offset Model) Initiative

For any library that invests in IGI Global's InfoSci-Journals and/or InfoSci-Books databases, IGI Global will match the library's investment with a fund of equal value to go toward **subsidizing the OA article processing charges (APCs) for their students, faculty, and staff** at that institution when their work is submitted and accepted under OA into an IGI Global journal.*

INFOSCI® PLATFORM FEATURES

- No DRM
- No Set-Up or Maintenance Fees
- A Guarantee of No More Than a 5% Annual Increase
- Full-Text HTML and PDF Viewing Options
- Downloadable MARC Records
- Unlimited Simultaneous Access
- COUNTER 5 Compliant Reports
- Formatted Citations With Ability to Export to RefWorks and EasyBib
- No Embargo of Content (Research is Available Months in Advance of the Print Release)

*The fund will be offered on an annual basis and expire at the end of the subscription period. The fund would renew as the subscription is renewed for each year thereafter. The open access fees will be waived after the student, faculty, or staff's paper has been vetted and accepted into an IGI Global journal and the fund can only be used toward publishing OA in an IGI Global journal. Libraries in developing countries will have the match on their investment doubled.

To Learn More or To Purchase This Database:
www.igi-global.com/infosci-books

eresources@igi-global.com • Toll Free: 1-866-342-6657 ext. 100 • Phone: 717-533-8845 x100

www.igi-global.com

www.igi-global.com

Publisher of Peer-Reviewed, Timely, and Innovative Academic Research Since 1988

IGI Global's Transformative Open Access (OA) Model:
How to Turn Your University Library's Database Acquisitions Into a Source of OA Funding

In response to the OA movement and well in advance of Plan S, IGI Global, early last year, unveiled their OA Fee Waiver (Offset Model) Initiative.

Under this initiative, librarians who invest in IGI Global's InfoSci-Books (5,300+ reference books) and/or InfoSci-Journals (185+ scholarly journals) databases will be able to subsidize their patron's OA article processing charges (APC) when their work is submitted and accepted (after the peer review process) into an IGI Global journal.*

How Does it Work?

1. When a library subscribes or perpetually purchases IGI Global's InfoSci-Databases including InfoSci-Books (5,300+ e-books), InfoSci-Journals (185+ e-journals), and/or their discipline/subject-focused subsets, IGI Global will match the library's investment with a fund of equal value to go toward subsidizing the OA article processing charges (APCs) for their patrons.

    Researchers: Be sure to recommend the InfoSci-Books and InfoSci-Journals to take advantage of this initiative.

2. When a student, faculty, or staff member submits a paper and it is accepted (following the peer review) into one of IGI Global's 185+ scholarly journals, the author will have the option to have their paper published under a traditional publishing model or as OA.

3. When the author chooses to have their paper published under OA, IGI Global will notify them of the OA Fee Waiver (Offset Model) Initiative. If the author decides they would like to take advantage of this initiative, IGI Global will deduct the US$ 1,500 APC from the created fund.

4. This fund will be offered on an annual basis and will renew as the subscription is renewed for each year thereafter. IGI Global will manage the fund and award the APC waivers unless the librarian has a preference as to how the funds should be managed.

Hear From the Experts on This Initiative:

"I'm very happy to have been able to make one of my recent research contributions, 'Visualizing the Social Media Conversations of a National Information Technology Professional Association' featured in the *International Journal of Human Capital and Information Technology Professionals*, freely available along with having access to the valuable resources found within IGI Global's InfoSci-Journals database."

– **Prof. Stuart Palmer**, Deakin University, Australia

For More Information, Visit: www.igi-global.com/publish/contributor-resources/open-access or contact IGI Global's Database Team at eresources@igi-global.com.

Printed in the United States
By Bookmasters